STUDY GUIDE

WARREN BURGGREN, EDITOR
University of North Texas

CONTRIBUTORS:

BRIAN BAGATTO
University of Akron

JAY BREWSTER
Pepperdine University

LAUREL HESTER
University of South Carolina

BIOLOGICAL SCIENCE

SECOND EDITION

SCOTT FREEMAN

PEARSON
Prentice Hall

Upper Saddle River, NJ 07458

Editor-in-Chief: Sheri L. Snavely
Project Manager: Karen Horton
Executive Managing Editor: Kathleen Schiaparelli
Assistant Managing Editor: Becca Richter
Production Editor: Gina M. Cheselka
Supplement Cover Manager: Paul Gourhan
Supplement Cover Designer: Joanne Alexandris
Manufacturing Buyer: Ilene Kahn
About the Cover Photo: Great tiger moth, *Arctia caja*, as photographed by Joseph Scheer. Scheer, of New York's Alfred University, is the author of *Night Visions: The Secret Designs of Moths* (Prestel, 2003)

© 2005 Pearson Education, Inc.
Pearson Prentice Hall
Pearson Education, Inc.
Upper Saddle River, NJ 07458

Printed in the United States of America

10 9 8 7 6 5 4 3 2 1

ISBN 0-13-109028-3

Pearson Education Ltd., *London*
Pearson Education Australia Pty. Ltd., *Sydney*
Pearson Education Singapore, Pte. Ltd.
Pearson Education North Asia Ltd., *Hong Kong*
Pearson Education Canada, Inc., *Toronto*
Pearson Educación de Mexico, S.A. de C.V.
Pearson Education—Japan, *Tokyo*
Pearson Education Malaysia, Pte. Ltd.

Preface

Introductory textbooks are written to provide a foundation for successful learning. These days, in addition to the textbook, students also make use of additional learning materials associated with their text, including CDs, websites, and study guides. In our experience, a study guide can be a wonderful asset to students, helping them assess their progress and move to the all-important stage of thinking critically about the material they are learning. Unfortunately, it is also our experience that publishers often produce study guides as a hurried afterthought—simply a repackaging of the text material along with a series of multiple-choice questions to determine if biological facts and principles have been learned.

We wanted to do something different—and something far better. So, while Scott Freeman was writing *Biological Science*, Brian Bagatto, Jay Brewster, Laurel Hester, and I were discussing with Karen Horton, Project Manager, how we could produce a more interactive, effective, and enjoyable study guide that would be an important component of the *Biological Science* support package. As this study guide emerged, we realized that it would be helpful to do more than just review the material found in the textbook. We also wanted to put together a guideline for how to study biology, how to write in biology, and how to understand why biological experiments underpin everything the student learns. Consequently, Part I of the book, unique among study guides, is intended to be a "survival guide" for your biology course. Part II subsequently delves into the actual material presented in the second edition of *Biological Science*, but in a way that challenges students to think critically and analytically about the material they are reading in the text and hearing in their lectures.

Each chapter has a similar format to aid your learning, beginning with an outline of "Key Biological Concepts" that summarizes the concepts discussed in the chapter. A "Cross-Cutting Themes" section ties together concepts throughout the book, followed by a "Difficult Concepts" section that walks you through the most challenging topics and concepts. Finally, a comprehensive section called "Assessing What You've Learned" tests your understanding of the material and helps you prepare for exams. Thus, each chapter provides a formula for successful learning, knowledge, and retention.

These efforts have resulted in what we perhaps immodestly regard as a state-of-the-art study guide. Together with the second edition of Scott Freeman's *Biological Science* and the other support material available to the student and instructor, we hope to help instill our own love of biology, and biological experimentation, into future generations of biology students.

Numerous individuals have made important contributions to this Study Guide. Many of the features we included originated in a focus group meeting organized by Prentice Hall in October of 2000 at Sundance, Utah. At the meeting, introductory biology instructors from across the country had a chance to express their opinions about the kinds of material that would be most useful for students. At a subsequent meeting in Austin, Texas, a rough plan for a study guide was outlined. While developing the guide, we referred to that original plan many times; it bears the fingerprints of some very innovative instructors, including Julie Palmer, University of Texas; Ruth Buskirk, University of Texas; Carole Kelley, Cabrillo College; and Judy Heady, University of Michigan, Dearborn.

Our special thanks go also to the team of editors and other specialists at Prentice Hall. Chris Thillen copyedited this Study Guide, bringing to bear not only her language skills and attention to detail but also her insight from working on the *Biological Science* text, instructional material, and all of the media. Chris also expertly formatted the pages. Karen Horton, Project Manager, made far more of an intellectual contribution to this Study Guide than we could have reasonably expected, and her continuing advice and encouragement throughout the project are greatly appreciated.

Finally, we are extremely grateful to all of those important people in our lives—family, friends, students—who put up with our temporary neglect while we focused on the creation of this Study Guide.

Warren Burggren
Denton, Texas
January 2005

Contents

Experimentation and Research in the Biological Sciences

Introduction to Experimentation and Research

Experimentation, or actively altering the course of an activity, process, or event to gauge the potential outcome of that alteration, seems to be as innate to humans as breathing or eating. We have always tinkered with the world around us. In the early days of *Homo sapiens*, or humans, the tinkering may have had immediate, practical purposes—does the wood from the tree with the heart-shaped leaves burn hotter and longer in a cooking fire than the wood from trees with needles? Does an arrowhead with carefully sculpted edges bring down a deer more reliably than a simple triangular arrowhead? Finding answers to these questions by trying different approaches and learning from the outcomes was often a matter of life or death, so it is hardly a wonder that the tendency for "experimentation" appears to have been naturally selected for in human genes.

Modern humans carry out experimentation in the highly organized format we call scientific research. Certainly, humans also continue to carry out some "life or death" experimentation and research, such as the ongoing search for an HIV vaccine, or better

screening procedures for various types of cancers. Such research is called applied research because of the hope that there may be an immediate application of the results of such experiments. Much of the current experimentation in biomedical research is considered to be applied research. However, a great deal of experimentation and research is considered to be basic research (sometimes called curiosity-driven research), which provides basic scientific knowledge and a foundation for applied research.

Descriptions of both basic and applied research form a central, integrated theme in Scott Freeman's textbook, *Biological Science*. Rather than just asking you to memorize "facts," *Biological Science* offers important insights into the generation of the fascinating information you will be learning in this course, and should leave you with a new appreciation for how scientific information is gathered.

Here in **Chapter 1**, we will consider the process of experimentation and research that leads to the creation of new data in an attempt to answer questions posed by scientists. In **Chapter 2**, we will consider how data are tabulated and graphed for the most efficient communication of the patterns that they portray.

Key Concepts of Experimentation

Setting up and carrying out successful experiments depends on careful organization and thought. Let's discuss a few important concepts of experimentation:

- Hypothesis testing
- Independent and dependent variables
- Experimental controls
- Precision and accuracy
- Consistent treatment of experimental populations

Hypothesis Testing

In **Chapter 1** of *Biological Science*, Scott Freeman discusses the creation of a hypothesis—his example stems from a question about why giraffes have long necks (see p. 10 of your text). A hypothesis is a statement that is proposed to explain a certain set of events, which can be tested and proved or disproved. You might alternatively think of a hypothesis as a theory waiting to be put to the test. *Biological Science* poses the question, "Why do giraffes have long necks?" One hypothesis to explain this observation is: "The long necks of giraffes enable them to reach food that is beyond the reach of other mammals." In other words, when a scientist poses a hypothesis as a statement, he or she is initially "taking a position" when trying to answer the question. The scientist may have no idea what the answer to the question may be, but still suggests a possible explanation in the form of a hypothetical, positive statement.

Once a hypothesis is put forth, the next step is to design an experiment to test that hypothesis. The experimental design should be such that the experiment's results will either prove the hypothesis to be valid—the results support the proposed explanation—or the results will disprove, or falsify, the hypothesis. Proving a hypothesis is often a matter of gathering so much evidence in support of a hypothesis that any reasonable person would believe it to be proven. In reality, falsifying a hypothesis is often easier. The philosopher Karl Popper (1902–1994) explained the concept of falsifying hypotheses with an example about ravens. If researchers propose the hypothesis, "All ravens are black," then they might set out to collect as many black ravens as possible. After collecting hundreds

of black ravens, they may believe the hypothesis to be true (that is, to be "proved"). However, the researchers can't be sure that there isn't a white raven out there somewhere! In contrast, if they set out to find a *white* raven and succeed, then the hypothesis that "all ravens are black" is completely and efficiently falsified. The researchers might then suggest the new hypothesis that "All ravens are black except for a very, very small population of white ravens." Falsifying a hypothesis leads to the creation of a new and better hypothesis that takes new results and observations into account. When a hypothesis has undergone numerous modifications and can no longer be falsified, then scientists may permit themselves to refer to the hypothetical statement as a fact. When a fact or body of facts has stood the test of time, it may become a law.

YOUR TURN

In *Biological Science*, Scott Freeman describes many interesting experiments performed by biologists. Turn to page 218 in **Chapter 11** (How Do Genes Work?). Here, Freeman describes an experiment from the early 1940s by Avery, MacLeod, and McCarty that investigates the source of hereditary material. Can you figure out what the original hypothesis proposed by the authors might have been? Try doing the same for the experiments by Candace Galen on Rocky Mountain wildflowers that Freeman describes on pages 422 and 423 of **Chapter 21** in the textbook.

Independent and Dependent Variables

Virtually all experiments, as well as the tables and graphs that present the information these experiments have produced, possess independent and dependent variables. It is crucial that you understand the difference between them. The independent variable is typically the variable that can be controlled, or fixed, by the scientist. For example, consider a plant biologist who is interested in the adaptations of a certain species of cactus to the dry conditions of the Mojave Desert. (You should pay particular attention to this example, because we will use it repeatedly in this and the next few chapters.) The biologist wonders whether the amount of water available to cactus seedlings affects their growth rate. To test this, she might take cactus seeds from native plants and grow cactus seedlings in a green-house (Figure 1.1).

Here are the experimental conditions: All the seedlings are exposed to a constant 18 hours of light. The amount of water that each pot receives is controlled as part of the experiment. Different pots of cactus seedlings are given different amounts of water, and the amount of growth stimulated by each watering regime is carefully measured and recorded each week. The amount of water given to the seedlings is the independent variable, because the amount of water the seedlings receive is fixed and independent of any physiological response the seedlings might have to the water. Their growth in response to watering is the dependent variable because the amount they grow more or less *depends* on the water they receive during the experiment.

The design of an experiment can be simple, as in this case involving cactus seedlings, or can be very complex (for example, simultaneously varying water, nutrients, light, and genetic stock of the cactus and tracking any effects on seedling growth—an experiment with four independent variables). Nevertheless, in every experiment, you should be able to identify a dependent variable (the variable that unpredictably changes) and one or more independent variables (controlled as part of the experimental design). Once you have sorted out the meaning of dependent and independent variables, it will be easier to understand how tables and graphs present their information.

Find a few descriptions of experiments in *Biological Science*, and see if you can identify the independent and dependent variables in each.

Experimental Controls

Every student taking this course has carried out an experiment—both formally and informally. Your formal experience might include experiments in other college or high school science courses. Informally, think of the experiments you carry out every day. For example, have you ever tried to study less and still get the same grade, or tried to study more to get a better grade? Have you ever fiddled with a recipe to see if you can improve it? Have you ever checked to see whether you get better gasoline mileage with one brand of gas over another? These examples of real-life "tinkering" are actually experiments producing data.

While we all may be experimenters in our own right, what is often lacking in our informal experiments is a built-in control. A control is the portion of the experiment in which all possible variables are held constant, contrasting the portion of the experiment in which the variable(s) is altered to observe the resulting effects. The presence of a control ensures that any effects observed in the experimental portion of the experiment can be attributed to the altered variable. For example, you can change a recipe for cookies by adding more chocolate chips, but you

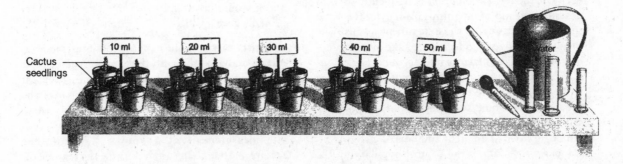

Figure 1.1. In this experiment carried out by a plant biologist, five groups of cactus seedlings are grown in a greenhouse and provided with different amounts of water each week.

don't know if the cookies taste better due to the extra chips or because you are especially hungry when you taste them. What this experiment needs is a *control*, which would involve baking one batch of cookies using the normal recipe side-by-side with another batch using more chocolate chips, then tasting the two batches at the same time.

Many otherwise good experiments can suffer from a lack of adequate controls. Controls help a scientist draw valid conclusions from the observed results of an experiment, which is why trained scientists are scrupulously careful to create and include controls with every experiment. For example, a team of researchers trying to test a new drug they are developing will test the drug on one experimental population of animals or patients while simultaneously giving a second, equivalent population a "placebo," or inactive drug treatment, as the control. Any effects subsequently observed in the experimental population can then be attributed to the drug being tested.

To help develop the concept of controls a little further, let's return to our example about cactus seedlings. The average amount of water that falls in the seedlings' natural habitat during the period when cactus seeds might sprout is about 5–10 millimeters per week (mm/week). This translates into about 20 milliliters (ml) of water per week, allowing the experimenter to give some of the seedlings the same amount of water that they would normally receive in the wild. These plants constitute the control population for this experiment.

The experimental populations are those that receive more or less water than is given to the controls. (By the way, the term *population* is frequently used to describe a group of organisms, usually of the same species.) By setting up the experiment in this way, the plant biologist can be sure to take any extraneous or unexpected factors into account. For example, perhaps the amount of nutrients or light that all of the plants receive in the greenhouse is different than the amount that they would normally receive in the wild. The biologist is still able to test the single effect of different amounts of watering by comparing all data to the control population. Returning to our example of changing a cookie recipe, this is similar to baking two batches of cookies—one the control batch and one the experimental batch.

Even if you are ravenous when you taste the two different populations of cookies, you still can make a comparison between the two recipes.

All of the experiments that Scott Freeman mentions in *Biological Science* involve careful use of controls, enabling them to survive scientific scrutiny to become landmark studies.

YOUR TURN

Precision and Accuracy

We often use the terms *precision* and *accuracy* fairly interchangeably; yet in the context of scientific experiments, they mean two quite different things. Let's consider a vegetable scale at the grocery store. Is it accurate, precise, or both? Let's say you place a bunch of broccoli on the scale and observe that it weighs 2.0 pounds. You then weigh the same bunch of broccoli five more times, and it weights 2.0 pounds each time. In this case, we can say that the scale is "precise" because it repeatedly gives the same weight for the same object. But let's also assume that the broccoli *actually* weighs 1.7 pounds. In this case, the scale—while precise because it repeatedly gives the same weight of 2.0 pounds—is not accurate. It would be accurate only if it indicated that the broccoli weighed 1.7 pounds, and it would be accurate *and* precise only if it repeatedly gave a weight of 1.7 pounds (Figure 1.2).

Scientists carrying out experiments must be both precise and accurate with their measurements. For this reason, experiments involving measurement with instruments (and most do) also involve the process of calibration. This means that an instrument is adjusted until it reliably produces the accurate measurement of a known standard. For example, rather than simply trusting the reading on a thermometer, a scientist will compare the temperature it reads against a known temperature (or a previously calibrated thermometer). Simple mercury

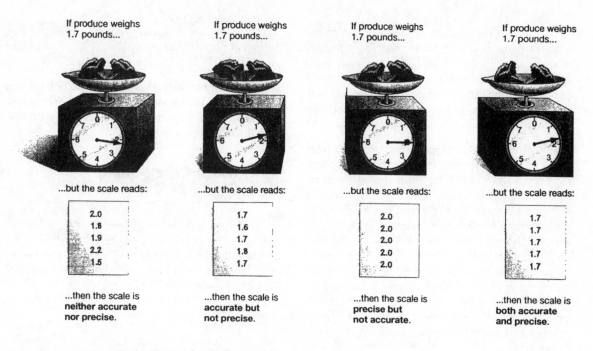

Figure 1.2. A measuring device such as a grocery scale can be (1) neither precise nor accurate, (2) accurate but not precise, (3) precise but not accurate, or (4) both precise and accurate.

thermometers can't be adjusted, but electronic ones can be adjusted to read the correct temperature. Ultimately, the accuracy of data generated depends on how carefully the scientist has calibrated the measurement devices used to collect data. The data on cactus seedling growth, for example, will be useful only if the ruler used for measuring plant height is accurate, the beaker used for measuring water is accurate, and the measurements and water amounts can be carried out precisely.

Consistent Treatment of Experimental Populations

When performing experiments, it is important to ensure that all experiments are carried out in exactly the same way every time. Otherwise, results may be due to experimental error and may not accurately reflect the results of the experiment. For example, in the case of the cactus seedlings, if some plants are measured in the morning and some in the late afternoon, wilting in the late afternoon sun

might cause some plants to measure smaller than they actually are. If some plants are watered on Mondays and some on Tuesdays, some measurements taken Monday afternoon would be made on recently watered, non-wilted plants, while some plants measured will be dehydrated and wilted. Any height differences will be due to the day on which the measurements were taken rather than to how much water they received weekly.

YOUR TURN

Let's imagine that you are carrying out an experiment to see if studying for an extra night each week will improve your grade point average. For the first two weeks, you study on Tuesday night in the quiet library: but in the following two weeks, you study on Friday night in your dorm room with the sounds of loud music and parties all around you. When you tally up the scores from all of your quizzes, exams, and assignments for the month, you find that your GPA remains unchanged. Is there something wrong with your experimental design, or does studying just not matter?

Setting Up an Experiment

Scientists may spend weeks, months, or even years planning an experiment, and they may face limitations due to lack of manpower, funding, or equipment. Some experiments are relatively simple to set up and require little maintenance. Our plant biologist's experiment with cactus seedlings, for example, is fairly straightforward, requiring only that the biologist obtain the seedlings, plant them, water them regularly according to the experimental design, and record their growth. In contrast, research into the effects of a new drug on the heart rate of animals and humans may require the combined efforts of numerous graduate students, postdoctoral fellows, research scientists, and clinical staff, as well as large amounts of expensive equipment—and make take several years to complete. Yet, despite great differences in effort and expense to carry out research experiments, the processes of generating a hypothesis, testing the hypothesis by performing experiments, and analyzing the data are common to all good scientific experimentation.

Try setting up your own hypothetical experiment using this scenario: Major Oil Company needs to build roads through some wilderness areas in their search for fresh oil fields. These roads will cross some streams that are important migratory routes for spawning salmon as they swim upstream. Thus, Major Oil needs to install culverts under the new roads as they are built to allow the water to flow under them. The project's chief engineer wants to install culverts with a 24-inch diameter, which will produce a water velocity of about 25 centimeters per second (cm/sec). The biological consultant hired by Major Oil fears that the fish in the stream won't be able to swim upstream against such a strong current and thinks they should use more expensive culverts with a 36-inch diameter, producing a water velocity of

only 10 cm/sec. Can you design an experiment that would resolve this conflict? Construct a formal hypothesis. Think of the materials and supplies you would need, assuming that all costs will be covered by Major Oil. What would be the experimental procedure? What type of data would you collect? How would you present it to the engineer to make your case?

How Much Data Should Be Collected?

You are probably starting to realize that biological research can be an elaborate and time-consuming enterprise. Poorly planned experiments can result in either an insufficient amount of data for accurate analysis and robust conclusions or the collection of more data than necessary, which is a waste of both time and money.

Let's return to the experiment with cactus seedlings. How many seedlings do you think the plant biologist should use? Should she set up just five flower pots with one seed each, giving each pot a different amount of water? This would be the easiest design, as well as simplest experiment to care for. Should she set up 20 pots? How about 100 or 500 pots? What criteria should she use to decide the size of her experiment?

All scientific data—especially biological data—has inherent variation. The most important sources of this variation are

- **Variation from inherent differences**—natural variation in the samples being examined. For example, some cactus seedlings will grow more rapidly than others simply because of their genetic disposition.
- **Variation from measurement error**—every scientific measurement has some inherent error. Different people measuring one seedling's height will produce different values, and the same person measuring one seedling's height at different times will produce different values. (Remember our earlier discussion of precision and accuracy.)

Carry out a simple experiment to illustrate measurement error. Using a ruler, tape measure, or even the tiles on a floor, measure the length of your normal walking stride. Now estimate the length of a corridor, the width of a small field, or some other distance by counting the number of paces you take as you walk from one end to the other. Do this five times. Ask five of your friends, relatives, or roommates each to estimate the length in question using the same method. Compare the results of your measurements. What would be the most *precise* way to measure this distance? What would be the most *accurate* way to measure this distance? If you owned a land surveying company and your reputation depended on accurate measurements, would you have a team of surveyors carry out the measurements, or would you ask one surveyor to repeat his or her measurements more than once?

Biologists typically spend a lot of time thinking about exactly how many replications—repetitions of an experiment—to carry out. Scientists want to ensure that the measurements they make are as accurate as possible, and one way to ensure accuracy is to take the average of several replicates. Of course, some experiments are relatively easy to replicate; for example, repeatedly measuring the height of a cactus seedling in a flower pot or growing several cactus seedlings rather than one. Consider, however, measuring the compression strength of the bite of a great white shark—and trying to get replicate measurements of the *same* great white shark, as well as several others in the same size range!

A variety of factors affect the design of an experiment with replicate measurements. Assuming that a researcher can manage the practical limitations of time and resources, the most important factor to consider when designing an experiment is the statistical reliability of the data produced. Statistics is a vibrant sub-branch of mathematics, and to discuss the numerous statistical methods used to test the validity of biological research would be beyond the scope of this study guide. However, two simple and useful conventions to learn are the mean and the standard deviation. The mean is the average of all data points collected, and the standard deviation of the mean describes the variation in the data about the calculated mean value in statistical terms.[1] A large variation about the mean (that is, a large standard deviation) indicates the possibility that the data is inconclusive.

To ensure that statistically sound conclusions can be made about our example experiment, the plant biologist might decide to use one seedling in each of five flower pots at every watering amount (for a total of 25 flower pots), as illustrated in Figure 1.3.

This experimental protocol, or plan for the experiment, will allow the biologist to calculate for the five flower pots means values and standard deviations that represent a single watering amount. Without going into statistical detail on the methods used for calculation, taking five accurate measurements (with replicate measurements taken on each seedling) often proves sufficient to determine statistical plausibility.

What do you think are the inherent strengths and weaknesses of using five seedlings to calculate the mean value? (Hint: Murphy's Law also applies to biological research—whatever might go wrong, often does.)

In our example experiment (but not in an experiment involving great white shark bite strength!), it is a relatively small investment of time and money to compensate for error by growing additional seedlings, so the biologist may elect to grow 10 seedlings at each watering level to ensure against the unexpected death of a few seedlings or other possible disasters, such as a dropped flower pot.

---

1. Biologists use several conventions to describe variation—standard deviation, standard error, confidence intervals, and so on. The calculation of each of these can be complex, so a rule of thumb is useful. Generally, if the standard deviations about two means substantially overlap each other, then the two means may not be statistically different.

Analyzing and Interpreting Data

Analyzing data is a crucial step in the experimentation process. It's not enough to simply produce and record data; a scientist must then attempt to make sense of the gathered information to decide whether to accept or reject the hypothesis. You have already undergone similar analytic procedures in your daily life. For example, you may have had a feeling that you are spending more money than you have, a thought which is really an informal hypothesis in disguise. You may then balance your checkbook in an attempt to learn how much money you have, but you shouldn't stop there—you should also compare your monthly balance with your monthly income to see whether you are saving money or falling deeper into debt.

Once a scientific researcher has acquired enough data, he or she will begin to use the individual "trees" to see the overall "forest"; that is, to use the data to recognize broader patterns. Typically, this will be done either by putting the data in tables or by plotting it in graphs. In fact, tables and graphs are such an important part of the data analysis and presentation process that Part One, **Chapter 2** of this study guide is devoted entirely to the topic of scientific data presentation. In a curious way, analyzing scientific data is like solving a word puzzle in which the letters of words are scrambled, and must be rearranged to form pieces of words before whole words begin to take shape. Scientific data alone are like the scrambled letters—a long and tedious process of interpreting the data, comparing it with previous experiments on similar topics, and fitting the outcome into the hypothesis formulated at the beginning of the experiments is required before the scrambled letters begin to resemble words. The truly great scientists of our era are able to put their data in the broadest possible context, so not only do their experiments allow them to accept or reject their own hypotheses, but their conclusions lead to advancements in the field of biology as a whole through insights gained from their experiments.

Publishing Data

The final step in experimentation is publishing the data and conclusions. Scientists receive financial and intellectual support for their work not just to indulge their own personal curiosity, but to encourage the results of their experiments to be distributed throughout the scientific community. Most journals subject submitted articles to rigorous peer review by other scientists in the same area of research. A manuscript is published only after intense scrutiny, and often several rounds of revisions.

There are many outlets for scientific data, most in the form of scientific journals. There are thousands of scientific journals, with titles and topics as broad as *Science* or as focused as *Pediatric Developmental Pathology*. Your institution's library probably has a large section devoted to scientific journals of all kinds. Of course, as electronic publishing on the Web has become commonplace in the last several years, most journals now publish in both traditional paper format and on the Web for subscribers.

YOUR TURN

Go to your library's journal section and look up a biological research journal such as *Journal of Experimental Biology* or *Physiological and Biochemical Zoology*. Find an article about experimentation with "zebra fish." Now, use your Web browser to find that same journal on the WWW, and again try to find an article involving zebra fish. Which format do you find most convenient? Which is easier for finding articles? Which format is better for viewing figures (tables, graphs, etc.)? What do you think will be the future of the electronic vs. paper publishing of scientific information? Ask your biology professor to comment about his or her preference for scientific information.

From the inception of a question posed as a hypothesis through the gathering of resources, the completion of experiments, the analysis and interpretation of data, and finally the publication of results, the overall scientific endeavor can often take years and many thousands of dollars. The next time you are in your library, walk down the aisles of scientific journals and think of the enormous investment of time and resources that has produced our scientific knowledge base. Appreciate the fact that every page of Freeman's *Biological Science* is based on information gathered through careful scientific experimentation.

YOUR TURN

Canadian science commentator David Suzuki once stood outside a grocery store in the dead of the Canadian winter, asking shoppers what role science played in their lives as they left the store with bags containing mangos, pineapples, bananas, and other exotic tropical fruits. Most replied, "None!" Let's consider the same question from a different angle. You have probably recently applied some sort of chemical to your body in the form of de- odorant, cosmetics, soap, shampoo, and so on. Pick one, and try to imagine the scientific experi- mentation that was involved in the creation, pro- duction, and packaging of that product. What do you think of science's role in your own life?

All science, including biological science, depends on the clear, concise presentation of data. Let's turn to that topic in the next chapter.

2

Presenting Biological Data

In Chapter 1 we considered how experiments are set up, carried out, and to some extent how they are analyzed. In this chapter, we will examine ways in which the information produced from such experimental procedures is presented and interpreted.

Introduction to Data

Scientists thrive on data, the information gathered from scientific observation or experimentation. They think about it all the time—what type of data they need, how much data they need, how they can get it, and what it might indicate. Whether they are taking the temperature of the lava from a smoldering volcano or measuring the rate of oxygen consumption in a growing zebrafish embryo, all scientists rely on data.

Data are the currency of science, and learning to interpret the many ways it can be presented will be an important part of your scientific training. But do not think that only scientists deal with data, or that its interpretation is a completely new skill, because you have been handling data in many everyday situations for years! Try to imagine the data you encounter regularly; you might think of the money in your checking account, your credit card balance, or

your grade point average. The data on the monthly statement from your credit card company, for example, tracks the date and amount of your purchases. Interpreting this data involves determining if you are over your limit and budgeting your future credit card purchases to reflect your income.

As our world becomes more technologically advanced, it is becoming increasingly important to understand complex forms of data and to apply the knowledge gained from these data. Drawing accurate conclusions from various forms of data presentation will do more than help you correctly answer questions on your biology tests; the critical thinking and analysis skills you develop will help you excel in other aspects of your education as well as throughout your life.

This chapter introduces the typical ways in which biologists use, present, and interpret data—methods you are also likely to encounter in your classroom and laboratory. Scientists generally present their data in one of several forms:

- Writing
- Tables and graphs
- Simulations and animations

Let's discuss each form in greater detail.

Presenting Data in Writing

You're probably familiar with the saying, "The pen is mightier than the sword," meaning that more can be achieved by effective written communication than by legions of armies. For thousands of years, learned people have recorded information (we might call it data) about their observations of the natural world. For example, Aristotle recorded his observations of the beating heart of a chick embryo, which he exposed by carefully making a small window in the shell of a chicken egg. Socrates left similar written observations of the chick embryo heart, as did Galileo. Since that time, other writings have left a rich legacy of knowledge about the anatomical and physiological development of the chick embryo. These contributions could not be understood without the clear, concise writing that describes the observations—the data—of each writer. Today, writing remains a mainstay of scientific communication. In *Biological Science*, Scott Freeman writes about the great wealth of biological information. While your textbook incorporates a multitude of tables and graphs to support the text, it is apparent that the written word is the primary tool for scientific communication.

You have already used writing to convey scientific information—think of the lab reports, essays, and book reports you have produced so far in your academic career. While you may already understand the components of good scientific writing, Chapter 4 (Part one) of this study guide, "Reading and Writing to Understand Biology," explores these components in considerable detail. It is important to remember that the ability to write clearly is not limited to those who want to become scientists. Those of you studying for careers in business, education, or virtually any other discipline will be required to communicate your ideas and data to others, and your study of the presentation of scientific information will help you successfully communicate all types of data.

Presenting Data in Tables and Graphs

In the previous section, you recalled past lab reports from science courses such as biology, chemistry, and physics. Did you ever use tables or figures to help explain the data you produced? While the pen may be mightier than the sword, you are probably also familiar with the phrase, "a picture is worth a thousand words." A table or a graph is a picture of scientific data. Though you can describe your favorite summer vacation to your friends by talking to them, you may want to take out your photo album as well, using the pictures to supplement your stories. The combination of the pictures and your verbal description provides the most complete communication of your ideas. In the same way, a scientist explaining his or her experiments or an author describing biological principles will almost always resort to a combination of words and visual tools, such as tables and graphs.

Tables and graphs are similar in that they are visual, highly structured ways of looking at data, often displaying independent and dependent variables as discussed in Chapter 1.

Types of Tables

Tables are such a convenient way of communicating data that they can be found all around us. As you read this book, you may have a beverage or a snack close by. Take a minute to look at the table on the packaging. It is probably a relatively simple table showing nutrition information such as caloric content and the amount of fat, cholesterol, sodium, and so on found in the product. While the packaging could describe this information with words, a simple table provides the same information more efficiently. To consider various table formats, let's return to the example of cactus seedlings introduced in Chapter 1. We could describe the experiment in paragraph form, but consider the following table:

Cactus Group (5 pots/group)	Amount of Water Given to Seedlings (ml/week)	Average Seedling Growth (mm/week)
Group 1	10	3
Group 2	20	7
Group 3	30	13
Group 4	40	10
Group 5	50	5

YOUR TURN

Record some of your observations about this table. What does it show you about the experimental design, and what general trends do you notice?

_____

_____

_____

_____

_____

A substantial amount of information about the experiment can be gathered from this table, such as:

- There are five cactus groups represented in the experiment.
- There are five pots in each group.
- A standard watering procedure is shown, in which the most-watered group of seedlings received five times more water than the least-watered group.
- The growth column indicates that all of the seedlings grew to some extent.
- Seedlings receiving a certain amount of water grew the most; seedlings receiving more or less water than this amount did not grow as well.

Tables can be relatively straightforward, as in this example, or relatively complex, with many columns and many rows. Regardless of layout, however, all tables will contain independent and dependent variables.

Types of Graphs

Although tables can be extraordinarily useful to organize data, graphs are also used extensively to communicate this information. If you pick up a newspaper (especially <u>USA Today</u>, which uses more graphics than many other national publications), you will probably find several examples of graphs (look in the "Business" section, where you may find the recent performance of the stock market presented both in a table and a graph).

YOUR TURN

Look through *Biological Science*; how many different types of graphs can you identify?

_____

_____

_____

_____

There are many different types of graphs, each with advantages and disadvantages that suit it for a particular kind of data. Most graphs have another feature in common besides dependent and independent variables; they all have an x-axis, also called an abscissa, and a y-axis, or ordinate. The x-axis runs horizontally along the bottom of the graph, while the y-axis runs vertically (Figure 2.1). Typically, the x-axis is used to plot the independent variable (for example, watering amounts for our seedlings), while the dependent variable (seedling growth) is plotted on the y-axis.

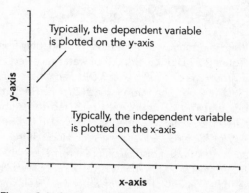

Figure 2.1 Graphs typically have an x-axis, showing the independent variable controlled by the researcher, and a y-axis, showing how dependent variable values change as the independent variable is altered.

Let's identify the specific types of graphs you will encounter in *Biological Science*. We'll continue to use the example of cactus seedlings and growth, because you are now familiar with these data in tabular form.

Scatter Plot of Seedling Growth Data

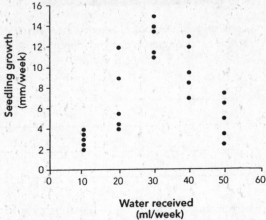

Figure 2.2 A scatterplot is so named because all individual data points are plotted, and may therefore appear to be "scattered" on the graph.

Scatterplots. A scatterplot is aptly named; because each data point is individually represented on the graph (in contrast to graphs in which only the averages of several data points are recorded), it may appear that data are simply scattered on the graph. This scatterplot presents the data from every pot in the cactus seedling experiment (**Figure 2.2**).

YOUR TURN

What information or patterns can you identify from Figure 2.2? Did the amount of water received by a group make a difference? Did each individual plant respond the same in each group? Remember that the control group received the same amount of water as would be received in the wild (20 ml/week); did plants receiving different amounts of water display different responses than the control group? What was the range of weekly seedling growth at 30 ml of water per week, and what was the average response at that level?

_____

_____

_____

_____

Looking at the scatterplot again, you should be able to see that, at each amount of water, there is variation between individual pots. Further, notice that the amount of variation differs for each amount of water, with peak growth when seedlings are given 30 ml of water per week. Scatterplots are very useful in that variation and trends in the data can be examined simultaneously. Sometimes, however, scatterplots are awkward because they can contain too much data, which makes the detection of patterns in the data more difficult. For this reason, scientists sometimes choose to communicate their data using line graphs.

Line Graphs. A line graph does not show individual data points, but instead shows the averages of several data points, connected by one or more lines that best fit the data points. In other words, line graphs appear to "connect the dots" of a scatterplot. The following line graph displays the average seedling growth for each group at each watering amount (**Figure 2.3**).

Line Graph of Seedling Growth Data

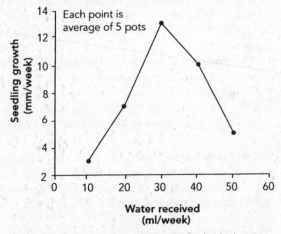

Figure 2.3 A line graph displays either individual or mean values of a dependent variable against the independent variables, and, as the name implies, joins these points with a line to help reveal overall trends.

This line graph clearly shows the pattern that is only hinted at by the scatterplot; namely, that seedlings do poorly with little water, but are also adversely affected by too much water. The graph indicates that an intermediate watering amount, slightly more than the seedlings would receive in their natural habitat, is just right for optimal growth.

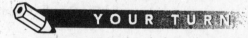

YOUR TURN

Look at some graphs in a printed version of a newspaper or on a news website. Which graph type is more common, scatterplots or line graphs? Look at examples of each—which format works better to communicate the data being presented?

_____

_____

_____

_____

_____

YOUR TURN

Go back to your newspaper or news website. Find a simple line graph with no standard deviations. Without seeing any indication of the variability in the data, can you tell if the trends depicted are "trustworthy?"

_____

_____

_____

While line graphs communicate trends very clearly, they often fail to provide any sense of variation in the data. This problem can be minimized by plotting one of a variety of statistical variables to describe variation in the data. The most common of these variables is the standard deviation of the mean, which is defined in Chapter 1 of this study guide. Figure 2.4 is a replicate of the first line graph we looked at (**Figure 2.3**) with the standard deviation now added. The vertical line drawn through each point represents the amount of variation in the independent variable at each dependent variable value.

Bar Graphs. Scientists often illustrate data with a bar graph in which the height of the bar represents the value of the dependent variable. Bar graphs can be plotted either horizontally or vertically, and there isn't much difference between the two—aesthetics and a sense for which communicates the data more effectively are often key factors when deciding whether to orient a bar graph horizontally or vertically. The two bar graphs presented plot the cactus seedling data in each form (**Figure 2.5**).

Just as line graphs can show variation in data, bar graphs can also be designed to show standard deviations (**Figure 2.6**).

Bar graphs are often most effective when showing the results of an experiment that compares several different variables, or when comparing several different experiments. On a line graph, separate lines often fall on top of one another, which can make data analysis more difficult. A bar graph, however, displays data by using different bars for each dependent variable value.

Let's consider a possible follow-up cactus seedling experiment for our biologist, in which not only does the amount of water differ for each batch of seedlings, but the amount of light given to the seedlings varies—one complete set of pots (5 pots in each of 5 groups) is given 16 hours of light per day, while another set is given 18 hours of light per day. The biologist concludes that the amount of light doesn't make much difference, but she still wants to communicate these findings graphically.

Line Graph of Seedling Growth Data, with Standard Deviations of the Mean

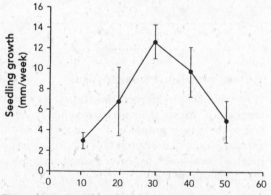

Figure 2.4 Line graphs may also show how the data varies about the plotted mean values, as in this line graph showing the standard deviation of the mean for each mean seedling growth value.

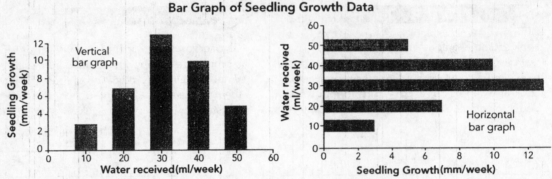

Bar Graph of Seedling Growth Data

Figure 2.5 Bar graphs are often an effective means of showing the relationship between dependent and independent variables. In this figure, the same data set is plotted as a vertical bar graph (upper) and a horizontal bar graph (lower). Which do you think more effectively presents this data?

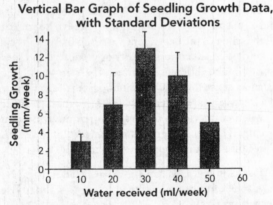

Vertical Bar Graph of Seedling Growth Data, with Standard Deviations

Figure 2.6 As with line graphs, bar graphs can also show variation in the data by displaying standard deviations of the mean.

YOUR TURN

Has the biologist added another dependent variable or another independent variable?

_____

Now let's combine these data using both a line graph and a bar graph to highlight the differences between the two graph types (**Figure 2.7**).

Which do you think is more effective?

Pie Charts. Another common type of graph is a pie graph, or pie chart. As the name implies, a pie chart resembles a pie divided into more than one slice. A pie chart is ideally suited for presenting data expressed in the form of percentages of fractions of a whole. The size of each slice, when appropriately labeled, indicates the relative amount of the graphed variable. For example, a pie chart can effectively present data about the elemental composition of seawater (with each element shown as a percentage) or the fraction of each phylum represented in the total number of species in an ecosystem.

YOUR TURN

Can you think of a good way to present the cactus seedling data using a pie chart?

_____

_____

_____

Putting Your Knowledge to Work

Now that you've learned the common types of graphs, let's assess your understanding by asking you to try a little graphing of your own. For this exercise, you will need your copy of *Biological Science* and some graph paper (or a computer program that allows graphing).

1. Try to find an example of each type of graph in *Biological Science*. For each type, identify the x- and y-axes (except for a pie chart) and the dependent and independent variables.

2. Pick five chapters from *Biological Science* at random. For each chapter, record the total number of graphs, regardless of graph type. Let's consider the chapter number to be the independent variable and the number of graphs per chapter to be the dependent variable. Now, create these graphs:

- Scatterplot, recording each individual data point.
- Line plot, using the average number of graphs for each chapter.
- Two bar graphs, one oriented vertically, the other horizontally.

Which graph do you think is the most effective way to show the data you collected?

3. Refer to the five chapters from *Biological Science* that you used to make the graphs in step 2. Create a worksheet so that you can keep track of the number of each type of graph (line graph, bar graph, etc.) as you flip through these chapters. Using a compass (or a coffee cup or another similar object), draw a circle on a piece of graph paper. Create a pie chart showing the fractions of the total number of graphs represented by each type.

Presenting Data with Simulations and Animations

Some of your favorite sites on the World Wide Web probably contain animations, which use iconic symbols or cartoon drawings that move in a lifelike way. Animations are becoming increasingly important in the communication of scientific data. They are especially effective when describing ongoing processes occurring over a period of time; for example, an animated graph that reveals its data in a series of stages rather than all at once.

YOUR TURN

Your Student CD-ROM that accompanies *Biological Science* includes many animations. For example, look at Activity 52.2 on your CD-ROM, "Habitat Fragmentation." What are the advantages of using animation to communicate this information?

_____

_____

_____

_____

Related to animations are simulations, often in the form of animated graphs or tables, that allow the recreation of a real-life situation. Rather than passively watching animation unfold, you can manipulate data (usually independent variables) in a model to see how the outcome (the dependent variable) changes in response to different data sets. An increasingly important tool for data communication and interpretation, simulations are used in everything from predicting how a forest fire might spread to anticipating a drug's effects on a patient's blood pressure.

Seedling Growth Data from Two Experiments, Plotted as Line and Bar Graphs

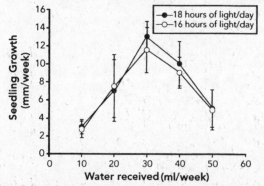
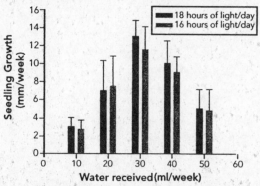

Figure 2.7 When the results of more than one experiment are plotted on a line graph, the data can be difficult to interpret. Sometimes a bar graph (lower), with separate bars representing every dependent variable value, presents the results more clearly than a line graph (upper).

YOUR TURN

Can you think of any times you have used a simulation in your personal life, that is, fed data into a model to see what the outcome would be? You might recall an example involving your personal finances, or perhaps you've calculated the effects of several potential test grades on your grade point average.

_____

_____

_____

_____

As you can see, there are many ways to effectively present scientific information, each with different characteristics that may suit it for specific types of data. Experiments may yield unexpected results, however, and careful interpretation is often necessary before data can be understood. In the next chapter we will explore the challenges of data interpretation, and you'll find techniques to help you understand scientific information and excel in your study of biology.

Understanding Patterns in Biology and Improving Study Techniques

Before you can begin a comprehensive study of life, it is necessary to learn background details about the physiology of life, from large organisms to microbes that cannot be seen without the help of a microscope. This is an immense amount of information, and textbook authors are faced with the challenge of organizing it. How can details about everything from molecules and viruses to ecosystems and evolution be compiled and presented so that they will be understood by even the most inexperienced student? Fortunately, biology is not just a haphazard accumulation of facts. Life is highly ordered and involves many ongoing processes that result in recognizable patterns—some simple, others more complex—that are easily presented in textbooks.

An understanding of detail is crucial to the pursuit of biology, because though individual structures and their functions may not seem important at first, each will play an important role in your discovery of the elaborate patterns found throughout biology. When used to complement your lectures and labs, your textbook is an important tool that will help you master details and discover a rich tapestry of biological patterns. This chapter will help you use the features of Freeman's *Biological Science* to gain full understanding of your course material and of biology as a whole.

 YOUR TURN

To understand how patterns and relationships in biology are organized, go to the section in your library that contains introductory biology textbooks such as Freeman's *Biological Science*. Look at the tables of contents in several different books. Do you see much variation in how the chapters are arranged in each book? What themes are present in every book? Looking at the publication dates of these books, can you see any ways in which the organization of biological information has evolved over the years?

_____

_____

_____

_____

Discovering Patterns in Biology

Both researcher and student are faced with the common challenge of gathering enough specific knowledge about biology to be able to recognize general patterns. Before you can attempt to understand complex biological concepts, you must learn about specific structures and processes that will explain biology in a broader context. In Chapter 1 we considered the pitfalls of "not seeing the forest for the trees," which describes the idea that broad patterns and conclusions can be overlooked when attention is focused solely on detail. Let's apply this metaphor to the study of biology.

Seeing both the Forest and the Trees— Use Details to Understand Concepts

As a beginning biology student, you will notice that there is no shortage of trees! Hundreds of terms, many of which may be unfamiliar, are used to describe organisms, their constituent parts, and the ways they interact. Learning each tree in the forest of biology is a challenge that can be approached in several different ways. Understanding these details is necessary in order to do well in a biology course— for example, learning the basic structure of a plant leaf and the fundamental steps of cellular respiration are prerequisites for studying the more complex process of photosynthesis. Learning these details will be easier and more interesting if you try to understand *why* they are important. As you progress through this course, constantly question yourself about the material to assess your understanding: Why is a plant leaf so structurally complex? What are the functions of its separate components? How do a plant's leaves contribute to its survival? If you keep the *function* of an organism's structural components (the trees, to return to our metaphor) in mind, these details will be easier to learn and will help clarify a broader concept (the forest). As you read your textbook, try not to passively memorize terms; instead, continually ask yourself, "What does this structure do, and how does it relate to the terms and concepts I already know?"

This study guide emphasizes the idea that the material you study will build upon your knowledge from previous chapters and will help you understand future topics. Each chapter in Part Two contains a section entitled "Cross-Cutting Themes," an example of which can be found on page # of this study guide. Its two subsections, "Looking Back—

Concepts from Earlier Chapters" and "Looking Forward—Concepts from Later Chapters," remind you of concepts you already know and indicate connections to concepts you will learn about in chapters to come. If you look for general patterns and broad concepts as you learn the details of biology, you will eventually discover that the shadowy outline of the forest is starting to emerge from your study of the trees.

Look for Hierarchical Relationships

At this point in your academic career, you are probably familiar with outlines. Every outline, whether it clarifies the structure of a college-level essay assignment or describes the layout of an entire book, follows the same basic format. Main ideas are identified by the major headings, and these topics are divided into more specific subheadings. While some ideas stand alone as key concepts, others need the support of more specific facts; key words or phrases can be included within subsections to indicate additional details that will be discussed in the essay or book.

In an outline, facts and ideas are presented in the form of a hierarchy. A hierarchy is a system of categorical groupings that is arranged according to relative importance. A good example of a hierarchy is a book's table of contents, which is essentially a complex outline; its hierarchical structure distinguishes it from an index, in which ideas and facts are listed alphabetically regardless of their relative importance. A good table of contents is more than a guide to help the reader locate information; it also establishes a hierarchical relationship between topics.

It is important to understand the hierarchy of information in *Biological Science* as well as in other textbooks you may encounter. To help you with this concept, a section entitled "Key Biological Concepts" is included in each chapter in Part Two of this study guide. This section presents the main ideas of the corresponding textbook chapter in a hierarchical format, allowing you to easily distinguish pivotal concepts from supporting details based on their relative position. For example, consider this study guide excerpt from Part Two, Chapter 36, on plant reproduction:

36.1 An Introduction to Plant Reproduction

Sexual Reproduction

- *Most plants reproduce sexually. **Sexual reproduction** is based on the reduction division known as **meiosis** and on **fertilization**, the fusion of haploid cells called **gametes**.*
- *Sperm are small cells from the male that contribute genetic information in the form of DNA but few or no nutrients to the offspring. Female gametes are called eggs, which contribute a store of nutrients to the offspring.*

Plant Life Cycles

- *Plants are the only organisms that have both a multicellular form that is diploid and a multicellular form that is haploid. This life cycle is called alternation of generations.*
- *A spore is a reproductive cell that grows into a new individual directly. A gamete is a reproductive cell that must fuse with another gamete before growing into a new individual.*

Asexual Reproduction

- *Asexual reproduction does not involve meiosis or fertilization. It leads to offspring that are genetically identical to the parent plant. However, the salient characteristic of asexual reproduction is efficiency.*

This example introduces several details about plant reproduction, and their relative positions in the outline indicate relationships between them. Based on the structure of the outline, you can determine that there are three important points under the heading of plant reproduction—sexual reproduction, plant life cycles, and asexual reproduction. Each of these topics is supported by specific details; for example, meiosis and fertilization involving gametes are both important components of sexual reproduction. Such details are also included under the topics of plant life cycles and asexual reproduction. Although this information could be discussed in paragraph form, a hierarchical outline allows relationships between topics to be defined clearly by their appearance on the page.

Accept Variability

To understand biology, it is necessary to accept that individual organisms, structures, or behaviors may not always fit into expected patterns. Variability is a natural part of biological science; occasional variations actually emphasize that most organisms match a standard patern. While *Biological Science* concentrates on the typical components of general patterns, it also includes realistic variation. This variation is especially apparent in your book's many graphs, which often show individual data points as well as an overall pattern. Consider this graph from Chapter 22, page 441, which illustrates the relationship between human birthweight and mortality (**Figure 3.1**).

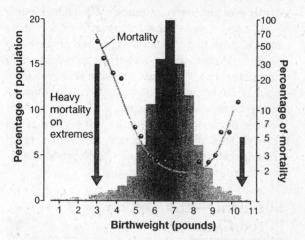

Figure 3.1 The histogram shows the percentages of newborns with various birth weights. The gray dots are datapoints indicating the percentage of newborns in each weight class that died, plotted on the logarithmic scale shown on the right. The gray line is a function that fits the datapoints.

Logically, we might expect to observe lowest mortality at one ideal birthweight, with a greater mortality rate if birthweight is either higher or lower than this optimum. In fact, the actual data do support this hypothesis. However, if you examine the data points representing mortality (in Figure 3.1, the gray points connected by a U-shaped curve) and the actual distribution of body weights at birth (the gray bars), you will see that while there are obvious trends, there is variability among individual data points. In this example, though variability is illustrated by any individual points lying off the indicated curve (drawn from the average of all data points), most points fall fairly close to this line. In some experiments, variability is far more pronounced, requiring sophisticated statistical analysis to identify a pattern in a widely distributed set of data.

Accepting variation will prevent you from being distracted by any data that do not fit the predicted pattern. Variation, though it may appear to disrupt experimental data, is not just a nuisance to be ignored. After all, it is variation among individual organisms that enables a species to change over time, and is therefore the foundation of evolution.

Be Comfortable with the Unknown

In addition to facing the challenges of variability, biology students also must be prepared to encounter uncertainty. New facts emerge and new hypotheses are tested daily, making biology a field that is constantly evolving and reinventing itself. Though much is already known about biology, an even greater amount is waiting to be discovered and understood. *Biological Science* discusses both accepted facts and current theories, distinguishing what is proven from what is conjectured. You may pose questions to your professor and lab instructor that they may not be able to answer, because the answer is not yet known. This uncertainty simply reflects the rapid growth of the field and explains biologists' constant search for greater understanding.

Surviving—and Thriving in the Classroom

As with any course, you will need to thoroughly understand biology in order to succeed on quizzes and tests and achieve your desired grade. Several tools are at your disposal, many integrated directly into your textbook, that can make the learning process easier and more rewarding. Additionally, certain study habits may improve your ability to understand the material you are studying.

Budget Your Time Carefully

The amount of time you spend on your biology course outside of class—studying, preparing for tests, and working on your laboratory assignments—will be reflected in the grade you achieve. Your other classes also require that you take the time to study, read the material, and write papers. How can you be sure that you are devoting enough time to each class?

It is important to be realistic about your time commitments. Everyone in your biology class has different schedules, courseloads, and activities, so it is important to develop a study pattern that suits your life. For example, let's say that your biology course involves two lectures and a lab each week; therefore, you will probably spend at least five hours a week in class for your biology course alone. You may need at least another five hours a week to complete assignments and prepare for exams, meaning that biology alone might take up 10 hours of your time each week. Considering that there are five classes in a typical course load, you may devote 50 hours each week to academics alone, leaving very little time for family, friends, a part-time job, and relaxation! Realistically consider what you can expect to achieve in the time you have available. If you concentrate on what you hope to achieve in college and on your future goals, your courses will become a top priority.

Go to Lectures and Labs

Lectures and labs are not optional; if they were, your course syllabus might say, "Please come to class occasionally." Instead, it probably lists the exact times for all lectures and labs, implying (or specifically stating) that they are mandatory. By attending every lecture and lab, taking careful notes and participating in class while you are there, you can take an active role in the learning process. Going to class will give you the opportunity to question your instructor and classmates about material you may not understand and allow you to gain different perspectives and practical experience. Attending class regularly, therefore, may be the one thing you can do to most improve your performance in a biology course.

Take Good Notes in Class

Attending class is necessary for success, but it is also important to make the most of your class time by taking good notes. Though you may have personal note-taking methods that work well for you, try structuring your notes in the form of an outline to illustrate a hierarchy of ideas. A hierarchy, as introduced earlier in this chapter, arranges concepts according to relative importance. Try to mimic your textbook as you take notes in class, reflecting the hierarchy of facts outlined in the table of contents and in the chapters themselves.

YOUR TURN

Examine the hierarchical format used in these sample notes on Chapter 41, "Gas Exchange and Circulation," and compare it with the format of notes you've recently taken for this class. Are they similar? Can you think of ways to improve your own note-taking technique?

_____

_____

_____

_____

times seem that they are two separate courses. Remembering that the laboratory is designed to support the lecture, and vice versa, will help you understand the material you are studying. Though they may not fall on the same day, the scheduling of laboratories and lectures often coincides so that in a given week, the material discussed in each will be related. For example, in the same week as your lecture on animal diversity, you may have a laboratory examining the morphological diversity of a wide array of animal specimens. Similarly, your lectures on the role of water and ions in an animal's physiology may be accompanied by a laboratory about the structure and function of the mammalian kidney. If you don't immediately see a connection between your lectures and labs, ask your professor or lab instructor to explain how they are related.

Air and Water as a Way of Breathing

Oxygen and Water in Air
- Air is primarily N_2, with 21% O_2 and traces of CO_2.
- Gases move by diffusion.
- Partial pressure is the pressure of gas in a mixture.
  - The partial pressure of air at sea level is about 160 mmHg.
  - Partial pressure on top of Everest is only 53 mmHg because of much higher altitude.

Oxygen and Water in Water
- CO_2 is 30 times more soluble in water than O_2.
- Fish and other aquatic organisms can easily get rid of CO_2, but getting O_2 is a problem.
  - Fish die-offs are almost always due to lack of O_2 rather than buildup of CO_2, while animals breathing air can have trouble getting rid of the CO_2 they produce.
- Gas solubility decreases as water temperature increases.
  - Fish in warm ponds suffer from O_2 deprivation more than in cold ponds, because of the lower solubility of O_2 in warm water.

Even if you do not choose to follow this method precisely, consider modifying your own note-taking technique to include a hierarchy of ideas. This format will help to solidify your understanding of the relationships between concepts.

Integrate Lecture and Laboratory Material

Lectures and laboratories are often taught at different times by different instructors, so it may some-

Take Advantage of Your Textbook

While this study guide is designed to help maximize your understanding of your biology course, it is not meant to replace your textbook. The vast amount of information contained in its chapters makes your textbook an indispensable supplement to your biology lectures and laboratories.

First, familiarize yourself with the book's features—

the table of contents, glossary, index, and so on—as well as with the structure and content of the chapters. Parts of the textbook can be used like a reference book if you need to review topics you recognize from previous courses. You can use the index to locate the page on which a particular term is presented, thus avoiding a tedious search of an entire chapter. The glossary can be a useful tool to test your knowledge: Write down any key terms in your lecture notes or reading assignment, and then define these concepts in your own words and compare your definitions to those offered in the glossary.

Chapter content is carefully designed to introduce and thoroughly explain each topic, because lectures and labs may not always provide enough information to fully understand a concept. For more difficult topics, you may need to refer to the book for additional explanations. A complex subject like genetics, for example, might demand a careful reading of the entire chapter before you can fully understand the topic.

Take Advantage of the Media for Students

Be sure to take advantage of the robust media support package that accompanies your textbook. The Student CD-ROM provides resources to help you visualize difficult concepts, explore complex biological processes, and review your understanding of the material. The CD-ROM media fall broadly into two types—animations and tutorials. The simple animations will help make the one-dimensional art in *Biological Science* come to life. The more complex tutorials will walk you through difficult concepts and include opportunities to test your knowledge.

You'll know when there are animations or tutorials because you'll see a vertical blue tabs on the outside edges of your book pages—see page 5, for example. But if you are not inclined to stop reading in the middle of the chapter, look for the "CD-ROM and Web Connection" at the end of each chapter (under "Chapter Review"). You'll also find a helpful summary there, including an estimate of how long the learning activity takes.

Another feature of the student media is an ability to test your knowledge. The quizzes on the website (www. prenhall.com/freeman/biology) include hints and feedback to help you assess your understanding. You'll also find research tools, including a broad collection of science and research links for the subject areas described in each chapter.

Share Your Knowledge

One of the best ways to learn is to tell others what you know. Find a study partner and meet regularly to discuss material. Take turns explaining topics to each other as thoroughly as possible, using any new terms and concepts you are learning. The partner who is listening can then critique the other's knowledge, identifying any missing information to ensure that both partners thoroughly understand the topic. In this way, you will be able to quickly identify your particular areas of weakness. You will eventually need to explain these concepts on laboratory and lecture exams (especially if these tests involve essay questions); any self-testing you do before an exam will build your confidence and ensure your mastery of the subject.

Though biology can be complex at times, understanding principles of data interpretation and improving your study technique can make learning this subject both challenging and rewarding. Reading effectively and writing to clearly communicate information will also help you in this course, throughout your education, and in the "real world" as well. In the next chapter we will discuss these skills and give you the opportunity to assess your abilities.

4

Reading and Writing to Understand Biology

In Part One, Chapter 2 of this study guide we noted that while scientific information is often conveyed visually with graphs and animations, writing remains the most common and most effective means of scientific communication. This chapter will help you improve your ability to read and interpret information about biology, evaluate your note-taking skills, and learn to convey biological ideas in your own words.

Reading for Understanding

Your success in this biology course (as well as in other classes) may depend on your ability to gather important information from what you read. Do you carefully interpret the material you read? Can you pick out the topic sentence in a paragraph? When you read a description of an experiment, can you identify the hypothesis being tested? Do you have the ability to skim your text when reviewing for a test?

Good readers easily understand written material because they are aware of the structure of the text as they read. When a text is well written, in each paragraph it is easy to identify a topic sentence that defines the main idea. A topic sentence is usually located at or near the beginning of each paragraph and is followed by further support, elaboration, or

explanation. This supporting information may describe examples or analogies, explain how the topic relates to a larger context, or present alternate views to help the reader understand the information.

Certain reading methods can increase the amount of information you gather from the material. When reading something the first time, read carefully, paying attention to any figures you encounter. If you come across a topic you do not understand, refer to your study guide; it may present the same information in different words and may include other examples. Use the glossary or a dictionary to look up the definitions of any unfamiliar words. As you read, stop periodically to ask yourself what you have learned; these mini-reviews will help you remember information later. If you have difficulty understanding a topic, ask your professor, teaching assistant, or another student to explain the subject.

You can use several techniques to improve your understanding and memory of what you read. If you plan to keep your textbook after you complete this biology course, you may want to underline topic sentences and other important points; but remember that underlining is effective only if it is done selectively. If you do not want to mark your book, making notes on sticky pads and attaching them to pages with key information will serve the same purpose.

Note-taking is another effective way to increase the amount of information you retain while reading, but certain methods may work better than others. Take notes only after you have read the material completely and thoroughly understand it. Taking notes as you read may distract you from learning the material and may tempt you to copy the text directly into your notes; you will learn and remember more from writing and reviewing your notes if you use your own words. In the next section of this chapter, "Writing for Understanding," we discuss effective note-taking methods in greater detail.

When you review for a test, look through the chapters you are studying carefully to make sure that you understand the material. Go over the figures in your textbook, testing yourself to see if you can explain each figure without referring to the text. Be sure you can accurately define any boldfaced vocabulary words. As a final review of important topics, re-read your class notes and any notes you took while reading the textbook.

YOUR TURN

Let's practice these reading skills. Read the following passage from Chapter 1, page 4 of *Biological Science* ("*What Is Natural Selection?*"), and underline the topic sentences. (An answer key is at the end of this chapter.)

Natural selection, the process component of the theory of evolution, occurs whenever three conditions are met. The first condition is that individuals within a population vary in their characteristics... Darwin and Wallace had studied natural populations long enough to realize that variation among individuals is almost universal. In wheat, for example, some individuals are taller than others. The second condition is that the variable traits are heritable, meaning that they are passed on to offspring. The third condition is that certain heritable traits help individuals survive better or reproduce more. For example, if tall wheat plants are easily blown down by wind, then shorter plants will tend to survive better and leave more offspring in windy environments.

If all three conditions are met, then a population's characteristics will change over time. In the example just given, populations of wheat that grow in windy environments would tend to become shorter from generation to

generation. A change in the characteristics of a population, over time, is evolution.

To clarify how the process works, consider the origin of the vegetables called the "cabbage family plants." Broccoli, cauliflower, Brussels sprouts, cabbage, kale, savoy, and collard greens are all descended from the same species - the wild plant in the mustard family pictured in Figure 1.3a. To create the plant called Brussels sprouts, horticulturists selected individuals of the wild mustard species with particularly large side buds. In mustards, the size of side buds is a heritable trait. When the selected individuals mated with one another, their offspring turned out to have larger side buds, on average, than the original population (Figure 1.3b). By repeating this process over many generations, horticulturists succeeded in producing a population with extraordinarily large side buds. The derived population has been artificially selected for large buds and barely resembles the ancestral form (Figure 1.3c). Note that during this process, the size of side buds in each individual was set—the change occurred in the characteristics of the population.

Darwin pointed out that natural selection changes the characteristics of a wild population over time, just as artificial selection changes the characteristics of a domesticated population over time. But no horticulturist is involved in the case of natural selection. Natural selection occurs naturally, simply because certain individuals in wild populations have heritable traits that allow them to leave more offspring than individuals without those traits. Evolution, or change in the population over time, is the inevitable outcome of this process.

Take notes on the passage using techniques discussed in this section. (Refer to the answer key for sample notes after you've taken your own.) Look at the passage again and consider the following questions: Is the boldfaced vocabulary word familiar? (If not, look up the definition in the glossary and write it in your notes so you can easily review it later.) Do you recognize the underlined words? (If not, you might want to look them up in a dictionary.) Using only your notes and your memory of the passage, answer the following questions:

- Using the Brussels sprouts example described in the text, explain the three conditions that must be present for natural selection to take place.
- What is the difference between artificial and natural selection?

Thinking about these types of questions as you read will help you understand and remember the content.

_____

_____

_____

_____

Reading and writing are closely related. Taking notes and/or making flashcards as you read helps you understand and remember information and will make it easier for you to study later on.

Writing for Understanding

Taking Notes

The ability to communicate your knowledge to your professor in writing will be essential to your success in this biology course. Don't wait until you get your first test back to assess your writing skills; you can start to practice writing clearly about biology as you take notes and answer the short answer and essay questions in your text and in this study guide. Taking notes on the material you read will help you analyze, understand, and remember the information; answering short answer and essay questions will allow you to consider facts from a different angle and put your thoughts into your own words.

Although it is acceptable to use informal language—incomplete sentences, abbreviations, and so on—when writing notes and flashcards, it is essential to write with clarity, brevity, and logic. Include all important facts, eliminate any extraneous details, and ensure that your notes will still make sense to you months later. (Why are "extraneous details," that you probably won't need to know for quizzes and tests, included in your textbook? These details are meant to support important points in the text. Though it may not appear on exams, this information will help you understand the more complex topics you are required to know.)

If note-taking seems insufficient and you need additional help memorizing information, try making flashcards of important terms and concepts. Write the name of the concept on one side, and use the other side for definitions and examples. Flashcards are convenient because they can be transported easily, enabling you to study even when you may not have your textbook or class notes with you. You can also tape small flashcards around your room in places where they will catch your eye, such as on your mirror or next to the light switch. Once you know a term, replace it with a new flashcard. The benefit of this technique is that you can not only learn new vocabulary and concepts gradually throughout the semester just by walking around your room but also avoid last-minute cramming before tests.

Be sure your notes will allow you to distinguish key concepts from supporting details and examples when you review them later. Consider using the Cornell Note Taking System (you can learn about this method on the Internet; the "Internet References" section at the end of this chapter gives the URL of a helpful website) or another similar method. Though it is a good idea to take notes both in lecture and while reading the textbook, your lecture notes should be more detailed than your text notes. You can always refer to the textbook for further information to support your notes, but lectures are only given once.

YOUR TURN

Let's try reading this passage called "Evolutionary Legacy" from David S. Goodsell's article, "Biomolecules and Nanotechnology" in *American Scientist* 88(3), May-June 2000. First read for understanding, underlining topic sentences as you go, and then take notes about the passage. Be selective, distinguishing main points from supporting examples or details.

_____

The process of evolution by natural selection places strong constraints on the form that biological molecules may adopt. Because genetic information is passed directly from generation to generation, cells must maintain a living line back to the earliest primordial cells. If a cell fails to generate a living descendent, all of its biological discoveries will be lost. This is far more limiting than the technology of our familiar world. If we create machines that don't function, we scrap them and go back to the drawing board. But if a cell takes a gamble and changes a critical machine, it had better get it right the first time or the result will be disastrous.

The picture is not entirely grim, however, as cells have several levels of redundancy within which to develop new machines. First, the plans for a given machine may be duplicated, which allows the duplicate to be modified and ultimately perfected to perform a function different from the original. This is very common in the evolution of life. Hemoglobin, the protein that carries oxygen in our blood, is an example. Our cells contain information for building several different types of hemoglobin. One is optimized for carrying oxygen in the blood of adults, whereas another is found in the blood of a fetus. The fetal hemoglobin has a higher affinity for oxygen, allowing it to capture oxygen from the mother's blood. About 200 million years ago, a gene duplication allowed the fetal hemoglobin to be perfected separately.

Second, biology seldom involves a single cell. A population of cells—billions, trillions—is the biologically relevant entity. Within this population there exists ample room for experimentation. Millions of modifications may be tried, even if most are ultimately lethal. The population will still survive and individuals with rare improvements may grow to dominate in later generations. Human immunodeficiency virus (HIV) shows the benefits of evolutionary change, accelerated so that we can see the effects in months instead of millennia. HIV reverse transcriptase, the enzyme that copies the virus's genetic information, is particularly error-prone. Because of this, the population of viruses within an infected individual contains viruses with all possible single-site mutations—thousands of variants on the wild-type virus. The best of these will dominate, but even the weakest are continually created and recreated in subsequent generations by the low-fidelity copying mechanism. Thus, when an infected individual is treated with anti-HIV drugs, the population has a wide range of different mutants to choose from, some of which may be resistant to the drug: The virus is made more efficient by its very inefficiency. The hallmark of biological evolution is the plasticity provided by mutation and genetic recombination. Within a population, or through genetic duplication within a single cell, a great many variants may be tested and the occasional improvement saved.

Evolution carries with it one important drawback, however: the problem of legacy. Once a key piece of machinery is perfected, it is difficult to replace it or make major modifications without killing the cell. This is particularly true for major molecular processes, such as protein synthesis, energy production and molecular machines. This leads to the remarkable uniformity of all earthly living things when observed at the molecular level. All are built of the same basic components.

Compare your notes with those shown in the answer key at the end of this section. Do your notes mention the same main points as those in the answer key? Do you prefer your notes or the sample notes? You probably find your own notes easier to understand because they are in your own words, phrased in a way that makes sense to you. If you prefer the sample notes, try to figure out why. For example, were your notes too detailed? Were they too brief? Did you miss some main points or fail to make a distinction between major topics and supporting details? Use these observations to identify ways in which you can improve your note-taking skills.

_____

_____

_____

_____

_____

_____

Now that you've improved your note-taking ability, let's discuss more formal writing techniques.

Writing Essays and Other Assignments

Why Is Writing So Important? Many students don't consider writing to be an important part of their science classes. Yet, everyone benefits from good writing skills, including people in science-related professions. A scientist's work is not limited to research conducted in a laboratory; he or she must be able to effectively communicate research goals, procedures, and results in grant applications and scientific papers. As a biology student, you will need to take notes, write laboratory reports, and take tests that may include essay or short-answer exercises. The ability to write clearly and concisely will improve your performance in all these areas.

Evaluating and improving your writing technique will benefit your study of biology in several ways:

• Taking coherent notes in your own words will make studying easier.

- Writing down information will help you recall it later on exams.
- Clearly written short-answer and essay responses will effectively communicate your knowledge.
- Well-written laboratory reports will demonstrate your understanding of the process studied.

Compare these two short-answer responses to the following question: How does what we know about artificial selection support the theory of evolution?

1. Artificial selection is like near to natural selection which darwin said caused evolution and they both are types of selection therefore evolution must occur. When you select something it must be good, so plants and animals change to get better buds or something quickly because their reproduction is helped by it.

2. When humans artificially determine the reproductive success of domesticated organisms with variable, heritable traits, the artificially selected plants or animals show major changes over a relatively short period of time. Thus natural selection could also cause the changes over time (evolution) that have led to the diversity of organisms we find today.

Which answer more effectively communicates the material? Why? Which student would you be more likely to hire to work in your lab?

_____

_____

_____

_____

Elements of Good Writing. We've discussed the many ways in which the ability to write well can be an advantage and explored this idea with examples of short-answer responses. But maintaining clarity and presenting ideas cohesively can be more difficult when completing an assignment requiring a longer response. What can you do to improve your essay-writing technique?

Think First. Before you start writing, be sure you understand your topic (refer to the "Reading for Understanding" section of this chapter) and the writing assignment. Taking the specific demands of the assignment into consideration, decide on a thesis statement; this sentence introduces the topic to be addressed and identifies a particular position that your paper will defend. After you locate examples or supporting arguments to support your thesis, you are ready to make an outline that will provide a layout for the organization of your essay or lab report. Though making an outline may seem unnecessary, it will ensure that your argument is thoroughly developed and can be easily understood by other readers.

Creating an outline is a worthwhile step only if you carefully consider what information it should contain. Whether you prefer to draft detailed outlines or less elaborate ones, certain elements should be included in all outlines:

- A well-defined thesis statement located near the beginning
- Supporting points introduced in a logical order
- Explanations of how supporting details relate to the thesis
- A strong conclusion

You are applying to work with a professor on his or her summer research project. Draft an outline for the cover letter you plan to attach to your resumé and application. See the sample key for a sample.

Write. After completing your outline, use it to write a first draft of your essay. Though it is not necessary to be a perfectionist at this stage—you will edit later—try to create a clear topic sentence for each of your paragraphs. Use active verbs while avoiding passive voice; for example, it is more effective to write, "We inadvertently contaminated our sample," than "Mistakes were made and the sample was contaminated." Be concise, but thoroughly support each of your main points. Consider using metaphors, similes, analogies, and examples to explain your topic; these creative techniques encourage readers to make their own connections, enabling them to understand your position more clearly.

Most important, do not plagiarize. Plagiarism, stealing someone else's ideas and passing them off as your own, is unethical. When you plagiarize, you also sabotage your educational experience by discouraging yourself from identifying and explaining your own ideas and perceptions. This does not mean you cannot use material from outside sources; if you find relevant material on your topic, include it in your essay with a citation acknowledging its source.

Let's take another look at the previous sample essay by David Goodsell and analyze the components of a well-written piece. It can be difficult to examine good writing because while bad writing distracts you as you read, alerting you to problems, good writing may seem almost unnoticeable because it allows you to easily understand the information being discussed.

In the first paragraph, Goodsell introduces his thesis statement—that evolution limits the forms of biological molecules—and explains it with a comparison between the evolution of life and the development of machines. He uses the next two paragraphs to explore ideas that contradict this thesis, suggesting that evolution's limitations are not strong enough to be substantial. He supports this alternate view by modifying his previous analogy between machinery and life to discuss a similarity between the two; an example involving hemoglobin is included for further clarification. Goodsell uses the words *first* and *second* to describe the topics of the second and third paragraphs, ensuring that the reader understands that these two points are meant to support this alternative argument. By presenting a position that opposes the thesis statement, Goodsell uses a popular and effective technique of essay writing. This method assures the reader that the author thoroughly understands the topic and gives the author the opportunity to directly refute arguments that contradict his or her thesis. In the final paragraph, Goodsell explains that his thesis is still relevant despite the conflicting arguments presented in the preceding two paragraphs. He then restates his original thesis and concludes the essay by identifying a situation in which his position is particularly relevant.

Throughout his essay, Goodsell uses clear logic, explaining and supporting his statements appropri-

ately. He employs many techniques of strong writing, such as varying the sentence structure and using repetition—in moderation—to emphasize important points. He also avoids using passive voice. When you consider this example, however, remember that it is an excerpt from a longer article. Had he addressed the same topic in an essay of this length, Goodsell would probably have developed and supported his thesis while spending less time discussing opposing arguments.

How can you, a student writer, be sure that you are applying enough of these writing techniques to your own assignments? While drafting an outline will help you organize information before you begin writing, you may not always remember to vary sentence structure, use active verbs, or support your thesis with examples as you write. This is why editing is such an important step in the writing process.

Edit. The best way to improve a paper you have written is to re-read and edit your work a day after you finish writing. Allowing time to pass before editing will give you a fresh perspective; you will often detect problems that you may not have noticed otherwise. This method may not be an option for procrastinators, however; another, more immediate editing technique involves reading your paper aloud to yourself or a friend. Speaking your words out loud will allow you to catch and eliminate any grammatical errors and awkward phrasing, and you will notice if certain words are used frequently. Avoid using the same word repeatedly by looking up alternative words in a thesaurus. Finally, if your essay is on a computer, use the spell- and grammar-check programs. Misspelled words and poorly constructed sentences will interrupt the flow of your argument and distract your reader.

YOUR TURN

Practice identifying effective writing techniques as you read this passage from Carl Zimmer's article entitled "Do Parasites Rule the World?" (*Discover* 21(8), August 2000). You might first read the article for understanding and then take notes on the passage. Finally, look at the details of sentence structure, observing strengths of the author's writing style.

As scientists discover more and more parasites and uncover the extent and complexity of their machinations, they are fast coming to an unsettling conclusion: Far from simply being along for the ride, parasites may be one of nature's most powerful driving forces. At the Carpinteria salt marsh, Kevin Lafferty has been exploring how parasites may shape an entire regions' ecology. In a series of exacting experiments, he has found that a single species of fluke—Euhaplorchis californiensis—journeys through three hosts and plays a critical role in orchestrating the marsh's balance of nature.

Birds release the fluke's eggs in their droppings, which are eaten by horn snails. The eggs hatch, and the resulting flukes castrate the snail and produce offspring, which come swimming out of their host and begin exploring the marsh for their next host, the California killifish. Latching onto the fish's gills, the flukes work their way through fine blood vessels to a nerve, which they crawl along to the brain. They don't actually penetrate the killifish's brain but form a thin carpet on top of it, looking like a layer of caviar. There the parasites wait for the fish to be eaten by a shorebird. When the fish reaches the bird's stomach, the flukes break out of the fish's head and move into the bird's gut, stealing its food from within and sowing eggs in its droppings to be spread into marshes and ponds.

In his research, Lafferty set out to answer one main question: Would Carpinteria look the same if there were no flukes? He began by examining the snail stage of the cycle. The relationship between fluke and snail is not like the one between predator and prey. In a genetic sense, infected snails are dead, because they can no longer reproduce. But they live on, grazing on algae to feed the flukes inside them. That puts them in direct competition with the marsh's uninfected snails.

To see how the contest plays out, Lafferty put healthy and fluke-infested snails in separate mesh cages at sites around the marsh. "The tops were open so the sun could shine through and algae could grow on the bottom," says Lafferty. What he found was that the uninfected snails grew faster, released far more eggs, and could thrive in far more crowded conditions. The implication: In nature, the parasites were competing so intensely that the healthy snails couldn't reproduce fast enough to take full advantage of the salt marsh. In fact, if flukes were absent from the marsh, the snail population would nearly double. That explosion would ripple out through much of the salt marsh ecosystem, thinning out the carpet of algae and making it easier for the snails' predators, such as crabs, to thrive.

After you read and analyze Zimmer's article, answer the following questions:

After you read and analyze Zimmer's article, answer the following questions:

1. What is Zimmer's thesis?

_____

_____

_____

2. What hypothesis was Lafferty testing?

_____

_____

_____

3. What is the larger context for Lafferty's study?

_____

_____

4. Note the verbs that Zimmer uses. Are they active or passive?

_____

5. Can you identify examples of a simile and a metaphor in this passage?

_____

6. Does sentence structure vary?

_____

7. What other strengths and/or weaknesses can you identify in Zimmer's writing?

 YOUR TURN

Now that you can identify the characteristics of strong writing, take out an essay or lab report you have written recently. Reread it, looking for the elements of good writing described in this chapter. Is your thesis well supported? Which are more common in your paper, active or passive verbs? Does the sentence structure in the essay show enough variation? How might you further edit the piece?

As you can see, there is more to good writing than simply identifying a topic and carelessly listing information about it. The ability to convey an idea or support a position thoroughly in writing is a skill that will improve your grade in this biology course, help you as your education continues, and will be useful throughout your professional career.

References

The following references contain further information about ways to improve writing technique and the fundamentals of language and grammar.

Text References

Goddin, N., and E. Palma, eds. 1993. *Grammar Smart: A Guide to Perfect Usage.* Villard Books.

Kirscht, J., and M. Schlenz. 2002. *Engaging Inquiry: Research and Writing in the Disciplines.* Prentice-Hall.

Learning Express. 1998. *Writing Skills for College Students.* Prentice-Hall.

Pechenik, J. 2000. *A Short Guide to Writing about Biology,* 4th ed. Longman.

Mark, Peter (Roget), and B. Kipfer, eds. *Roget's 21st Century Thesaurus.* Dell Publishing Company.

Strunk, W. Jr., and E. B. White. 2000. *The Elements of Style,* 4th ed. Allyn & Bacon.

B. Internet References

The Cornell Note Taking System
   http://www.dartmouth.edu/admin/acskills/ no_frames/lsg/cornell.html

Study Guides and Strategies, by J. Landsberger
   http://www.iss.stthomas.edu/studyguides/

A Short Guide to College Writing, by J. M. Williams and L. McEnerney
   http://writing-program.uchicago.edu/resources/col-legewriting/

College Tutor Study Guide
   http://www.amelox.com/study.htm

Introduction 4—ANSWER KEY

Reading for Understanding

Topic sentences are highlighted below:

<u>Natural selection, the process component of the theory of evolution, occurs whenever three conditions are met.</u>

<u>If all three conditions are met, then a population's characteristics will change over time.</u>

<u>To clarify how the process works, consider the origin of the vegetables called the "cabbage family plants."</u>

<u>Darwin pointed out that natural selection changes the characteristics of a wild population over time, just as artificial selection changes the characteristics of a domesticated population over time.</u>

Your notes might look like this:

Natural selection = process that causes evolution. Occurs when...

1. Individuals in population vary.
2. Variable traits are heritable.
3. Some heritable traits help individuals survive better, reproduce more.

3 conditions lead to change in a population's traits over time = evolution examples: wheat, artificial selection in mustard family.

Writing for Understanding
Taking Notes

These notes are modeled after the Cornell Note Taking System (Figure 4.1). To learn more about this method, refer to the URL given in the "Internet References" section of this chapter.

Writing Essays and Other Assignments
Why Is Writing So Important?

The second answer communicates the information more clearly. Although the first response contains relevant information, the poor spelling, grammar errors, and misplaced punctuation in this answer are distracting. Its sentences are not structured well, forcing the reader to search for key information and piece it together. To what do the "its" in the last sentence refer? If this was a student answer from an exam, would you be convinced that the student knows what artificial selection is? Careful consideration of this response reveals that the student understands similarities between artificial and natural selection and that evolution is change over time, but a tired or distracted test grader might not take the time to draw these conclusions, deducting points because the response is hard to read.

Elements of Good Writing.

Think First

  I. Thesis: Hire me this summer
    A. Introduce myself—how did I hear about professor?
    B. Professor's research sounds interesting.
    C. Want to get research experience.

 II. I like biology
    A. Have done well in biology classes
    B. Interested in nature: hiking, gardening, animals
    C. Want to major in biology, am especially interested in professor's research area

III. I work hard and am responsible
    A. Held jobs for the past two summers (references available)
    B. Volunteer work at local animal shelter

IV. Conclusion
    A. Hire me this summer
    B. Contact information
    C. Thanks for your consideration

Write and Edit

  1. Zimmer's thesis is that "parasites may be one of nature's most powerful driving forces."

evolution and biological diversity	1. Evolution by natural selection limits the diversity of biological molecules. • All cell lines must descend from progenitor cells. • Machines designs can be scrappped and restarted from scratch. • But cell innovations must succeed the first time or the cell won't leave any descendants.
cell redundancy	2. However, cells do have some redundancy. • The plans (DNA) for some "functions" can be copied, and the copies can be adjusted without fatal results. - e.g. adult vs. fetal hemoglobin
cell populations and modifications	3. Within the whole population of cells, many modifications, even lethal ones, may occur. • Those few cells with changes that are advantageous will increase in frequency in the population over time. - e.g. HIV (high mutation rate causes lots of variants—difficult to kill whole population with drugs.) • Through evolution variants are tested and the rare improvements saved.
evolution limited by legacy	4. Nevertheless, evolution is still limited by history. • Major modifications to important cell equipment almost always kill a cell. • This especially applies to major molecular processes (e.g., protein synthesis). • So, all life is very similar at molecular level.

In <u>Biomolecules and Nanotechnology</u>, David Goodsell argues that despite (1) the occasional functional redundancy within a cell and (2) the large population of cells within which modifications can appear and be tested, evolution is limited by legacy — especially at the molecular level.

Figure 4.1. An example of notes modeled after the Cornell Note Taking System.

2. Lafferty tested the hypothesis that flukes affect snail population dynamics.

3. Lafferty's study was conducted within the context of a marsh ecosystem; it also observed how snail populations affected algae, crab, and bird populations.

4. Zimmer mostly uses active verbs in the present or past tense, for example: "scientists *discover*," "Lafferty *found*," "birds *release*," "eggs *hatch*," and "flukes *castrate*."

5. An example of a simile is "looking like a layer of caviar," and an example of a metaphor is "that explosion would ripple out."

6. Sentence structure varies well; for example, Zimmer begins some sentences with prepositional phrases and others with the subject. Zimmer also uses sentences of many different lengths.

1

Biology and the Tree of Life

A. KEY BIOLOGICAL CONCEPTS

1.1 The Cell Theory

- Scientists Robert Hooke and Anton van Leeuwenhoek were the first to observe cells in the late 1660s (**Figure 1.1**).
- Additional observations made by many researchers confirmed that all organisms are made of cells.

Are All Organisms Made of Cells?

- Yes! Cells are the fundamental building blocks of life, from the smallest single-celled bacteria (200 nm wide) to the largest multicellular organisms (e.g., sequoia trees).
- Cells from different organisms or from different tissues of a multicellular organism may look quite different, but they all consist of membranes that surround compartments with concentrated chemicals dissolved in water.

Where Do Cells Come From?

- Scientific theories usually have two parts: the description of a pattern observed in nature and an explanation of the process that causes the observed pattern.
- The pattern component of cell theory is the observation that all organisms are made of cells.
- An alternative hypothesis that organisms can just appear from nonliving materials (**spontaneous generation**) was disproved by **Louis Pasteur**, who showed that cells do not appear in nutrient broth if an initial cell is prevented from getting into the broth (**Figure 1.2**).
- Bacteria only seem to magically appear in spoiled milk, because bacterial cells blowing in on dust in the air are too small to see with the naked eye.
- Pasteur's experiment was well designed because there was only one difference between his two treatments: the neck shape of the flask. Cells in the air could not get to the broth in the swan-necked flask.
- All cells come from preexisting cells, so single-celled organisms in a population are related to a single common ancestor, and all cells in a multicellular organism also descend from a single ancestral cell (among humans, this ancestral cell is the fertilized egg).

1.2 The Theory of Evolution by Natural Selection

- All life is related; it is descended (with *major* changes along the way) from a common ancestor.
- In 1858, scientists Charles Darwin and Alfred Russel Wallace proposed natural selection as a mechanism for how evolution occurs.

What Is Evolution?

- Species change through time and are related to each other (the pattern part of the theory). Natural selection explains how this change occurs (the process part).

What Is Natural Selection?

- Individuals of the same species living in the same area at the same time are called a **population**. Populations evolve by natural selection when three conditions occur.
- First, individuals in the population must differ from each other for some trait(s).
- Second, some of the variable traits must be **heritable**. In other words, they can be passed on to offspring.
- Third, individuals with certain heritable traits must survive and/or reproduce better than individuals with other traits.
- If all three of these conditions are met, then the population's characteristics will change over time as the individuals with the favorable traits increase in frequency; this is evolution through natural selection.
- Note that the favorable traits depend on the environment. For example, shorter wheat plants may survive and reproduce better in windy environments, but height may not be important in sheltered environments.
- **Artificial selection** of cabbage-family plants is an example of how many varieties can descend from one common ancestor (**Figure 1.3**). In this case, humans determined which plants grew in succeeding generations (thus, how the population's traits changed over time).
- The cell theory and the theory of evolution are two central, unifying ideas of modern biology.

1.3   The Tree of Life

- Cell theory and the theory of evolution by natural selection imply that all species are descended from a single common ancestor such that a family tree of all organisms—the **tree of life**—can be drawn.

Linnaean Taxonomy

- In studying species and their relationships, it is useful to categorize organisms. In 1735, Carolus Linnaeus created the classification system we still use today.
- Each organism is given a two-part name. The first name is the **genus**, which includes one to several closely related species. The second name is the **species**. Individuals that breed together are included in the same species.
- Scientific names are italicized, and genus names are capitalized. Each two-part name is unique.
- For example, the genus name *Canis* is given to wolves and dogs because these species are closely related. But wolves and dogs do not normally interbreed, so they have different species names (*lupus* and *familiaris*).

Taxonomic Levels

- The **family** is a higher taxonomic level that includes one to several related **genera** (the plural of *genus*). Other still higher taxonomic levels are **order**, **class**, **phylum**, and **kingdom** (**Figure 1.4**). Each **taxon** combines related groups of lower (less specific) **taxa**.

How Many Kingdoms Are There?

- Linnaeus proposed that all organisms be placed in the plant or animal kingdom. But the diversity of organisms cannot be accurately separated into the two groups. (Linnaeus' hypothesis about a fundamental pattern in nature has been rejected.)
- Biologists are now working to make **taxonomy** (classification of species) reflect **phylogeny**—the evolutionary relationship between taxa.
- An important division among organisms is that of **eukaryotes** (single- or multicelled organisms having complex cells with a membrane-bound nucleus) versus **prokaryotes** (unicellular and without a nucleus). See **Figure 1.5**.
- Differences between eukaryotes and prokaryotes are much greater than the differences between plants and animals (plants and animals are both multicellular eukaryotes with similar cell structures).
- Another alternative to Linnaeus' two-kingdom system is a five-kingdom system (**Figure 1.6**). Many other hypotheses have also been proposed, but it is unclear which best represents the true phylogeny.

Using Molecules to Understand the Tree of Life

- Carl Woese and colleagues attempted to discover the evolutionary relationships among groups of organisms by studying small subunit RNA, a molecule found in all organisms.
- Small subunit RNA is made of four ribonucleotides abbreviated by letters A, U, G, and C. These ribonuleotides are connected in a linear sequence (**Figure 1.7**).
- Our knowledge of evolution suggests that small subunit RNA has changed over time. Because they share a more recent common ancestor, closely related species should have more similar small subunit RNA than distantly related species do.
- Therefore, small subunit RNA sequences can be analyzed to produce a **phylogenetic tree** showing the probable evolutionary relationships of the organisms studied.

The rRNA Tree

- An analysis of small subunit RNA from many different species produced the tree of life (**Figure 1.8**). You can think of the vertical axis as representing time from the origin of life at the base of the tree to the present at the top.
- There are three major groups of organisms, including the eukaryotes (Eukarya) and two groups of prokaryotes—the Bacteria and the Archaea. Surprisingly, the Bacteria and the Archaea are much more diverse (at the molecular level) than anyone had expected.
- To accommodate this diversity, Woese proposed a new taxonomic level called the **domain**. Each of the three domains (Bacteria, Archaea, and Eukarya) includes several related kingdoms.

The Tree of Life Is a Work in Progress

- Currently, other molecules are being analyzed as well as other types of data. The placement of specific branches is debated, but many major findings from small subunit RNA analysis have been confirmed.

1.4   Doing Biology

- Biologists test ideas about how the natural world works by evaluating the predictions made by alternative hypotheses. Carefully designed experiments are an important tool for hypothesis testing.

Why Do Giraffes Have Long Necks? An Introduction to Hypothesis Testing

- Robert Simmons and Lue Scheepers tested the hypothesis that giraffes evolved long necks (by natural selection) because those with long necks more easily reach food that is unavailable to others. Although this hypothesis seems logical, it must be rigorously tested and evaluated against alternative hypotheses.

The Food Competition Hypothesis: Predictions and Tests

- The hypothesis that giraffes evolved long necks in order to reach high food sources leads to three predictions, each of which can be separately evaluated.
- First, neck length is variable. Previous studies in zoos as well as in nature confirm this prediction.
- Second, neck length is heritable. This prediction is still untested, because breeding experiments in natural giraffe populations are difficult if not impossible. Future zoo data might help scientists to evaluate this prediction.
- Third, giraffes typically feed high in trees, especially when food is scarce. Data do not support this prediction; giraffes feed with bent necks (**Figure 1.9**). So, there may be another, better explanation regarding why giraffes have long necks.

The Sexual Selection Hypothesis: Predictions and Tests

- An alternative hypothesis proposes that giraffes evolved long necks because males with longer necks win more fights than do shorter-necked giraffes, and they can then father more offspring. Data support this hypothesis.

Why Are Chili Peppers Hot? An Introduction to Experimental Design

- Chili peppers are hot to mammals because they contain capsaicin, a molecule that binds to mammalian heat receptors. Birds, however, do not appear to have the same response to capsaicin.
- Chili seeds eaten by an animal may either be destroyed in the digestive tract (seed predation) or voided unharmed with natural fertilizer (seed dispersal).
- The most important fruit- and seed-eating animals in the natural habitat of chili are the cactus mouse and a bird—the curve-billed thrasher. The mouse appears to destroy chili seeds that it eats; thrashers do not destroy the seeds, they disperse them.
- John Tewksbury and Gary Nabham tested the hypothesis that capsaicin in chili peppers is an adaptation that discourages seed predation while not preventing seed dispersal. This hypothesis predicts that only thrashers will eat hot chilies, but both mice and thrashers will eat fruit that does not contain capsaicin.
- A **null hypothesis** expresses the alternative possibility that the explanation offered by the hypothesis does not apply. In this case the null hypothesis predicts that mice and thrashers do not differ in their response to capsaicin and therefore should not differ in their fruit preferences.
- Several cactus mice and curve-billed thrashers were captured and offered fruits from (1) a variety of chili that does not make capsaicin, (2) a hot chili, and (3) hackberries (no capsaicin and do not look like chilies).
- These three fruits were offered to the mice and birds in the same manner, and the amounts of each eaten during a set time interval were recorded (**Figure 1.11**).
- Thrashers and mice ate similar amounts of hackberries; but the birds ate more chili peppers than hackberries (similar amounts of the hot and the mild kind), and the mice ate only a few mild chili peppers and no hot chilies.
- Researchers concluded that capsaicin prevents cactus mice, but not thrashers, from eating chili fruit; the dispersal hypothesis is supported.

- The experiment is well designed: (1) It included a control group (the hackberries) to check for other factors that might influence the results; (2) experimental conditions were controlled to eliminate extraneous variables; and (3) the test was repeated on several individuals to reduce the effects of random variation (sample size).
- Researchers even performed a follow-up experiment to demonstrate that chili seeds are, in fact, destroyed in the mouse digestive system but pass through the thrasher unharmed (**Figure 1.12**).

B. CROSS-CUTTING THEMES

Looking Back—
Concepts from Earlier Chapters
Major themes presented in the Introductory Chapter 1 of this study guide, especially the experimental method, will reappear throughout the text. Further information on some specific topics can be found as detailed here.

Looking Forward—
Concepts in Later Chapters
Cells—Chapter 6 and Unit 2, especially Chapter 7
Evolution of the first cells is discussed in **Chapter 6**, followed by Unit 2 (Chapters 7–11), which covers cell structure and function, cell-cell interactions, some chemical pathways found in many cells, and cell replication (how cells make more cells).

Evolution and Natural Selection—Chapter 4, Unit 5 (Chapters 23–26)
Chapter 4 investigates hypotheses on how first life might have evolved from nonliving macro-molecules. Evolutionary processes in general, including natural selection, are examined in greater detail in Unit 5.

Phylogeny and Taxonomy—Unit 6 (Chapters 27–34)
Chapter 1 introduces Linnean taxonomy (**Figure 1.4**) and the major groups (the five kingdoms) of organisms. The five kingdoms (**Figure 1.6**), and how these groups are related, are described further in Unit 6.

C. DIFFICULT TOPICS

This chapter presents you with some unifying ideas in biology. Many topics introduced in this chapter are covered in greater depth later on in this text. Do not worry about the details of how natural selection works or what rRNA does at this point (or if you do worry, go ahead and look at **Chapters 23 and 15**). However, you should spend some time thinking about how hypotheses are tested in order to discover information about the natural world. Throughout the following chapters, you will be taught biology through explanation of important experiments. If you master the ability to learn about biology through evaluating experiments, you will not have to resort to rote memory because you will *understand* the topic.

D. ASSESSING WHAT YOU'VE LEARNED

(1) Testing Your Knowledge

1. Which statement is *not* part of the cell theory?
    a. All living organisms are made of cells.
    b. All living organisms are multicellular.
    c. Cells arise from preexisting cells.
    d. All cells in an organism are identical.

2. In Pasteur's experiment, what different results would have supported the spontaneous generation hypothesis?
    a. no cell growth in either flask
    b. cell growth only in the swan-necked flask
    c. cell growth in the straight-necked flask
    d. cell growth in both flasks

3. Pasteur's experiment with a swan-necked flask rejected the spontaneous generation hypothesis by demonstrating:
    a. Cells can grow in broth after it has been boiled.
    b. Cells grow in broth unless the broth is sealed off from the air.
    c. Cells grow only in broth exposed to a source of preexisting cells.
    d. Bacteria, but not fungi, can grow in broth.

4. Based on their studies, what novel hypothesis did Charles Darwin and Alfred Russel Wallace present in 1858?
    a. evolution
    b. natural selection
    c. the cell theory
    d. taxonomic classification

5. A group of individuals of the same species living in an area at the same time constitutes a:
    a. community
    b. population
    c. genus
    d. family

6. The _____ is the fundamental structural unit of all organisms.
    a. DNA
    b. tissue
    c. species
    d. cell

7. The theory of evolution claims that:
    a. All species are descended from a common ancestor.
    b. No traits are heritable.
    c. Many species arose independently.
    d. Species remain unchanged over time.

8. Traits that are _____ are passed from one generation to the next.
    a. variable
    b. heritable
    c. prokaryotic

9. Which of the following is a condition needed for natural selection to occur?
    a. Random mating must occur.
    b. The population must be large for natural selection to occur.
    c. Humans select which individuals mate.
    d. Individuals within a population vary in their characteristics.

10. Which of the following is *not* a condition needed for natural selection to occur?
    a. Individuals vary within a population.
    b. Variation within a population is heritable.
    c. Humans apply artificial selection to the population.

d. Certain heritable traits help individuals to better survive and reproduce.

11. _____ was/were domesticated from *Brassica oleracea*, a wild mustard species.
    a. broccoli
    b. cabbage
    c. kale
    d. brussels sprouts
    e. all of the above
    f. broccoli and brussels sprouts

12. Linnaeus' taxonomic system specifies a name unique to each organism. The first part of the name is the organism's _____ and the second part is its _____.
    a. species; family
    b. family; genus
    c. genus; species
    d. genus; order

13. What is the scientific name for modern-day human beings?
    a. Primates
    b. *Homo sapiens*
    c. *Hominidae homo*
    d. *Homo habilis*

14. Modern taxonomy attempts to describe the _____, or historical relationships, among organisms.
    a. variation
    b. natural selection
    c. phylogeny
    d. physical similarity

15. Which of the following kingdoms contain prokaryotes?
    a. Monera and Protista
    b. Fungi and Monera
    c. Monera only
    d. Archaea and Eukarya

16. Use the tree of life (**Figure 1.8**) to determine which of the following is most closely related to humans.
    a. Archaea
    b. Fungi
    c. Ciliates
    d. Land Plants

17. Which statement supports the sexual selection hypothesis of giraffe neck length?
    a. Long necks allow giraffes to feed on higher vegetation that giraffes with shorter necks cannot reach.
    b. Males compete for the opportunity to mate with females by striking each other with their heads. Males with longer necks deliver greater strikes to their opponents than do males with shorter necks.
    c. Long necks help giraffes fight off predators.

18. Which of the following steps is *not* vital to a well-designed experiment?
    a. A control group is included.
    b. All variables other than those being tested are controlled or kept the same between treatments.
    c. Experiments are repeated on many individuals (large sample size).
    d. The steps described in a, b, and c are all important.

(2) Integrating Your Knowledge

(a) Why didn't biologists discover earlier that all life is made of cells?

(b) Human eyesight is variable and heritable. Under what environmental conditions might good eyesight be selected for in humans? Why does eyesight need to be variable and heritable in order to evolve?

(c) If you were a hunter-gatherer without any modern technology, what heritable traits might help you to survive and/or reproduce better?

(d) Can you make up a mnemonic phrase of words starting with D, K, P, C, O, F, G, and S to help you remember the taxonomic groups ordered from largest to smallest?

(e) How does what we know about artificial selection support the theory of evolution by natural selection?

(f) Traditionally, researchers have focused on physical traits to categorize organisms into taxa. What are the advantages to using small subunit RNA sequences in determining the relationships among organisms, and how they should be categorized?

(g) Write out the predictions you would make based on the sexual competition hypothesis, and compare your predictions to the data given to support this hypothesis.

CHAPTER 1—ANSWER KEY

D. Assessing What You've Learned

(1) Testing Your Knowledge

1. b; 2. d; 3. c; 4. b; 5. b; 6. d; 7. a; 8. b; 9. d; 10. c; 11. e; 12. c; 13. b; 14. c; 15. c; 16. b; 17. b; 18. d

(2) Integrating Your Knowledge

(a) Prior to the mid-1600s, microscopes powerful enough to let researchers see cells were not available.

(b) In environments where humans had to get most of their protein by hunting, good eyesight would be selected for if corrective lenses were not available. Traits that do not vary cannot evolve; those traits by definition cannot increase or decrease in frequency, because 100 percent of the population has the same trait. Traits must be heritable in order to evolve, because success based on a specific trait will not lead to an increase in that trait in the next generation if the trait is not heritable.

(c) A keen sense of smell and taste might have helped a hunter-gatherer survive better by improving his or her plant identification skills.

(d) One possible mnemonic is Do Kings Play Checkers or Other Fun Games on Sunday? (Domain, Kingdom, Phylum, Class, Order, Family, Genus, Species)

(e) Artificial selection shows that differential survival and reproduction cause change in a population over time as those heritable traits that are selected for increase in frequency in each succeeding generation.

(f) It is hard to find physical traits shared among bacteria, archaea, plants, and animals; but all of these organisms have RNA. Interpretation of sequence data may also be less qualitative than interpretation of physical traits.

(g) Prediction 1: Males with longer necks win more fights—supported by data.
Prediction 2: Males that win fights have more offspring; this is indirectly supported by data showing that winners of fights gain access to estrous females.
Prediction 3: Neck length is heritable.
Predictions 2 and 3 require further testing to confirm the sexual competition hypothesis as proposed by Simmons and Scheeper.

2

The Atoms and Molecules of Ancient Earth

A. KEY BIOLOGICAL CONCEPTS

- **Chemical evolution** is a hypothesis that explains how complex carbon-containing compounds (and eventually life) could have formed from simpler molecules. This chapter discusses basic chemistry in the context of how chemical evolution could have occurred on ancient Earth.

2.1 The Ancient Earth

Studying the Formation of Planets

- One method scientists use to study planet formation is to create computer programs that simulate this process based on our best knowledge of initial conditions in our solar system. Computer simulation is a powerful tool for studying events that cannot be directly observed.
- Computer simulations support the condensation theory of planet formation; the solar system probably started as dust clouds circling the Sun. Small particles then collide, stick together, make larger bodies, and eventually form planets. Scientists' observation of stars with proto-planets surrounded by dust support this theory.
- The condensation theory predicts that Earth formed through a series of collisions that released heat and kept Earth molten. Collis-

ions continued as Earth cooled, a rock crust formed, water rained down, and oceans were created. During this cooling phase, many volcanoes were active on ancient Earth.

When Did Chemical Evolution Take Place?

- **Radiometric dating** can be used to estimate the age of Earth and when life first appeared.

Atomic Components and Isotopes

- Atoms are composed of negatively charged **electrons** orbiting a **nucleus** made of **protons** (positive charge) and **neutrons** (no charge). See **Figure 2.1**.
- Opposite charges attract and like charges repel. An atom is electrically neutral when the number of protons and electrons are equal.
- Number of protons (the **atomic number**) determines the type of atom or **element**.
- An element always has a specified number of protons, but variations in the number of neutrons can occur and are known as isotopes. Uranium has two isotopes with 143 and 146 neutrons respectively.
- The **mass number** of an atom is the number of protons plus the number of neutrons (electrons have hardly any mass). Thus uranium has a mass number of either 235 or 238, depending on the isotope.

Radioactive Decay

- **Radioactive isotopes** have unstable nuclei that emit particles of radiation (energy) to form new daughter isotopes. This is known as radioactive decay.
- Each radioactive isotope decays at a constant rate quantified as its **half-life**—the time it takes for half the parent isotope to decay into the daughter isotope (**Figure 2.3**). It takes 713 million years for half the atoms in a pure U-235 crystal to decay to lead-207, so the half-life of U-235 is 713 million years.

How Old Is the Earth?

- Rocks containing radioactive isotopes are radiometrically dated by measuring the current ratio of parent to daughter isotopes. The isotope's half-life and the starting isotopic ratio must also be known in order to calculate the rock's age.
- Uranium in cooling molten rock forms pure U-235 crystals, so its starting isotopic ratio is 100 percent U-235.
- The most ancient Earth rocks formed 4.40 billion years ago (or Ga, for "giga-years ago"; **Figure 2.4**).
- Meteorites formed 4.58 Ga, and the Moon formed 4.51 Ga. Earth must be about the same age; but direct radiometric dating is not possible, because Earth was initially molten.
- The oldest sedimentary rocks (made by particles sinking out of water) are from 3.85 Ga; thus, oceans existed then, but may have been present as early as 4.28 Ga.

When Did Life Begin?

- Traces of living organisms in rocks form the **fossil record**. The oldest fossils (from 3.85 Ga) consist of carbon grains that have high levels of ^{12}C relative to other heavier carbon isotopes. Living organisms preferentially take in this lighter ^{12}C from their surroundings.
- Life appears to have formed on Earth around 3.85 Ga, soon after the massive bombardment ended and the oceans were formed.

2.2   The Building Blocks of Chemical Evolution

- Most cells are 96 percent hydrogen (H), carbon (C), nitrogen (N), and oxygen (O).

What Atoms Are Found in Organisms?

- Atoms commonly found in organisms are shown in **Figure 2.5**. The subscript indicates the atomic number (number of protons), and the superscript indicates the mass number for the most common isotope.
- Electrons are the key to understanding how atoms interact. Electron orbitals hold up to two electrons, and **electron shells** are groups of orbitals numbered by distance from the nucleus. Electrons fill the inner shells first (**Figure 2.6**).
- Elements commonly found in organisms have at least one unpaired electron in their outer (**valence**) shell. The number of unpaired electrons present is known as an atom's valence. The presence of unfilled electron orbitals allows the formation of **chemical bonds**, to attach atoms together.

How Does Covalent Bonding Hold Molecules Together?

- Atoms are more stable when each orbital has two electrons. Orbitals with one electron each can overlap so that the nuclei share the two electrons, forming a **covalent bond** (**Figure 2.7**). A **molecule** is formed when two or more atoms are joined by covalent bonds.
- A **nonpolar covalent bond** forms when electrons are evenly shared between two atoms. If one atom holds onto the shared electrons more tightly than the other does, a **polar covalent bond** forms (**Figure 2.8**).
- An atom's affinity for electrons is called its **electronegativity**. Oxygen has a particularly high electronegativity, thus water has two polar covalent bonds; electrons stay closer to the oxygen atom, giving it a partial negative charge ($\delta-$) and the hydrogen atoms partial positive charges ($\delta+$).

How Does Ionic Bonding Hold Molecules Together?

- **Ionic bonds** form when electrons are completely transferred from one atom to another (**Figure 2.9**).
- Charged atoms are called **ions**. An atom that loses an electron becomes positively charged (a **cation**), and an atom that gains an electron becomes negatively charged (an **anion**). Opposite charges attract, and the ionic bonds between cations and anions form salts (e.g., table salt—NaCl).
- Electron sharing in chemical bonds exists in a continuum from equal sharing (nonpolar covalent bonds) to partial sharing (polar covalent bonds) to complete transfer or no sharing (ionic bonds). See **Figure 2.10**.
- Molecules found in organisms mostly have the stronger covalent bonds.

Some Simple Molecules Formed from H, C, N, and O

- The number of unpaired electrons in the valence shell determines the number of bonds an atom can make.
- Thus, C bonds with 4 H to form methane (CH_4), N bonds with 3 H to form ammonia (NH_3), and O bonds with 2 H to form water (H_2O). See **Figure 2.11**.

Double and Triple Bonds

- When there are two unpaired electrons in the valence shell, two nuclei can share four electrons in a double bond (e.g., carbon dioxide, CO_2). A triple bond can form if there are three unpaired electrons (e.g., molecular nitrogen, N_2).

Bond Angles and the Shape of Molecules

- CH_4, NH_3, H_2O, CO_2, H_2, and N_2 are the starting points for chemical evolution. Molecular function is influenced by its shape, so let's examine the shape of these molecules.
- Molecular shape depends on bond angles, which in turn depend on the orbitals participating in the bond. Orbitals are actually the area of space an electron inhabits.
- The four C orbitals with unpaired electrons have a tetrahedral shape, the three unpaired N

orbitals are pyramid shaped, and the two unpaired O orbitals are bent (**Figure 2.12**). Consequently, methane (CH_4) is shaped like a tetrahedron, ammonia (NH_3) like a pyramid, and water (H_2O) is planar and bent.

Representing Molecules (Figure 2.13)

- The **molecular formula** states the numbers and types of atoms in a molecule (e.g., H_2O, CH_4).
- **Structural formulas** show which atoms are bonded together, and they indicate single, double, or triple bonds. Other models show three-dimensional geometry.

Quantifying the Concentration of Key Molecules

- To find the **molecular weight** of a molecule, add up the mass numbers of all the atoms in the molecule.
- One **mole**, or $6.022 \propto 10^{23}$ molecules, has a mass equal to the molecular weight expressed in grams.
- The concentration of a substance dissolved in a liquid (a **solution**) is typically expressed as **molarity** (M)—the number of moles present per liter of solution.

2.3   Chemical Reactions, Chemical Energy, and Chemical Evolution

- A **chemical reaction** occurs when one molecule combines with or is broken into others. The chemical evolution theory proposes that the simple molecules present on ancient Earth reacted with one another to create larger, more complex molecules. Under what conditions could this have occurred?

How Do Chemical Reactions Happen?

- CO_2 (g) + $H_2O(l)$ ← $H_2CO_3(aq)$ describes a chemical reaction between carbon dioxide gas (g) and liquid (l) water (the **reactants**), producing an aqueous solution (aq) of carbonic acid (the **product**). The double arrow means the reaction is reversible and the equation is balanced with the same number of each atom per side.

- **Chemical equilibrium** occurs when forward and reverse reactions proceed at the same rate. Changes in temperature or concentration alter the chemical equilibrium. For example, adding more reactants will cause more product to be made until a new equilibrium is achieved.
- **Endothermic** reactions must absorb heat to proceed; **exothermic** reactions release heat.

What Is Energy? (Figure 2.16)

- **Energy** is the ability to do work or supply heat. **Potential energy** is stored energy, and **kinetic energy** is movement energy. Molecules are always moving; this type of kinetic energy is called **thermal energy**.
- **Temperature** measures thermal energy. If an object is "cold," its molecules are moving slowly.
- When a cold object touches a hot object (with faster-moving molecules), thermal energy (**heat**) passes from the hot to the cold object. Molecules in the cold object heat up and move faster, while molecules in the hot object cool down and move slower.
- One type of energy can change into another type of energy, but it is never created or destroyed. The **first law of thermodynamics** expresses this concept by stating that energy is conserved.
- An electron in an outer shell has high potential energy. As it falls to a lower energy shell, its potential energy is converted to kinetic energy which is, in turn, changed into light or heat as the electron reaches the lower shell. The potential energy lost equals the light or heat energy released.
- Molecules on ancient Earth were exposed to a lot of thermal energy from volcanic eruptions, asteroid impacts, and ultraviolet radiation from the Sun.

Chemical Evolution: A Model System

- Computer models can simulate chemical reactions that might have occurred on ancient Earth. Researchers investigated whether formaldehyde (H_2CO) and hydrogen cyanide (HCN) could have been produced because formation of these molecules is the first step of chemical evolution.
- Formation of these molecules is not spontaneous: Energy input is necessary.

What Makes a Chemical Reaction Spontaneous?

- Reactions are spontaneous when the **Gibbs free-energy change** (EG) is negative (E means change).
- EG = EH – TES, where H is potential energy, T is temperature in degrees Kelvin, and S is **entropy** (a measure of disorder). See **Figure 2.18**.
- Reactions tend to occur spontaneously if the products have lower potential energy (–EH) than the reactants (electrons are held more tightly). These reactions are exothermic—they release heat.
- Reactions also tend to be spontaneous when the products have more entropy (+ES) than the reactants.
- The **second law of thermodynamics** states that in an isolated system, entropy increases over time.
- Reactions that proceed spontaneously (ΔG < 0) are **exergonic**. Reactions that require energy input to occur (ΔG > 0) are **endergonic**. When ΔG = 0, a reaction is at equilibrium and no net change in reactants or products occurs.

Energy Inputs and the Start of Chemical Evolution

- A computer model of chemical reactions among CH_4, NH_3, H_2O, CO_2, H_2, and N_2 molecules included both spontaneous reactions and reactions that occur in the presence of sunlight energy (photons).
- High-energy photons reached ancient Earth because there was little ozone (O_3) in the atmosphere. These photons can knock electrons away from valence shells, breaking apart molecules and forming **free radicals**—highly reactive atoms with unpaired electrons (**Figure 2.19**).

*The Roles of Temperature and
Concentration in Chemical Reactions*

- High temperatures and high concentrations cause more reactant collisions and faster reaction rates.
- As computer models have shown, significant amounts of H_2CO and HCN form under temperature and concentration conditions likely found on Earth about 4 Ga.

How Did Chemical Energy Change during Chemical Evolution?

- Electrons in H_2CO and HCN bonds are farther from the atom nuclei than in the reactant bonds, and thus have more potential energy.
- The formation of H_2CO and HCN is a critical step in chemical evolution because energy from sunlight has been converted to **chemical energy** (potential energy in chemical bonds). This new source of chemical energy makes possible the formation of larger, more complex molecules.

2.4   The Composition of the Early Atmosphere: Redox Reactions and the Importance of Carbon

- Computer models suggest that H_2CO and HCN would form only if the atmosphere included H_2, NH_3, and CH_4 because these molecules trigger **reduction-oxidation** (redox) reactions.
- Volcanic gases (mostly CO_2, N_2, and H_2O) likely dominated the early Earth atmosphere, but H_2, NH_3, and CH_4 were also present in sufficient amounts.

What Is a Redox Reaction?

- In a **reduction-oxidation** or redox **reaction**, one molecule loses electrons (is **oxidized**), and another gains electrons (is **reduced**).
- In redox reactions, one reactant is an **electron donor** and another is an **electron acceptor**. The "loss" of an electron may just mean that the electron moved farther away from the atom's nucleus (**Figure 2.20**).
- As electrons get closer to or farther from a nucleus, their potential energy changes.

- When sunlight energy allows CO_2 (electron acceptor) and H_2 (electron donor) to react, forming CH_2O and H_2O (a redox reaction, **Figure 2.21**), the carbon atom is reduced because the product electrons are closer to the carbon nucleus. This reduction of carbon allows further chemical evolution.

What Happens When Carbon Is Reduced?

- Carbon has four unpaired electrons, allowing it to make four covalent bonds. This flexibility in bonding makes possible an incredible diversity of carbon-based molecules. Almost all molecules made by organisms contain carbon, so molecules with a C–C bond are called **organic molecules** (**Figure 2.22**).

Linking Carbon Atoms (Figure 2.23)

- Organic molecules easily form when reduced carbon compounds (e.g., CH_2O) are heated.
- The potential energy present in reduced carbon compounds allowed the formation of complex organic compounds, some of which are found in organisms today.

Functional Groups

- Groups of H, N, or O atoms (**functional groups**) bonded to C determine the behavior of organic compounds (**Table 2.1**). Carbon provides the structural framework.

2.5   The Early Oceans and the Properties of Water

- Life originated in and is based on water because water is a great solvent (substances dissolve easily in it).

Why Is Water Such an Efficient Solvent? (Figure 2.25)

- The bonds between H and O in water are polar covalent because hydrogen attracts electrons less strongly than oxygen does. The whole molecule is **polar** because the O side of the molecule has a partial negative charge ($\delta-$) and the H atoms carry partial positive charges ($\delta+$).
- Partial charges on the H and O in different H_2O molecules attract each other and form weak **hydrogen bonds**.

- Hydrogen bonds also form between H_2O and other polar molecules or charged substances (**ions**) dissolved in water. These bonds help the substances stay in solution.

How Does Water's Structure Correlate with Its Properties?

Water Is Denser as a Liquid than as a Solid (**Figure 2.26**)

- Hydrogen bonds in ice connect molecules in a crystal-like pattern, whereas there are fewer hydrogen bonds in water, causing water molecules to pack more tightly. Thus, ice is less dense than liquid water, an extremely unusual property. Large bodies of water rarely freeze through because ice floats.

Water Has a High Capacity for Absorbing Energy

- The amount of energy it takes to raise the temperature of 1 gram of a substance 1°C is called its **specific heat**. Water has a very high specific heat because hydrogen bonds must be broken in order for water molecules to move faster (**Table 2.2**).
- Water also has a high **heat of vaporization** (the amount of energy required to change 1 g of liquid to a gas).
- HCN and H_2CO would have rained down from the atmosphere and dissolved in ocean water, where water's temperature-buffering capacity would protect them from energy sources that could have broken them apart.

Acid-Base Reactions and pH

- **Acid-base reactions** are chemical reactions in which one molecule or ion gives up protons (an **acid**), and another accepts protons (a **base**).
- Water can act as both an acid and a base; $H_2O + H_2O \leftarrow H_3O^+ + OH^-$. Pure water at 25°C has a proton concentration, symbolized as $[H^+]$, of $1.0 \propto 10^{-7}$ M. These protons are associated with water molecules as H_3O^+.
- The **pH scale** indicates $[H^+]$: $pH = -\log[H^+]$ (**Figure 2.27**).
- Pure water is neutral and has a pH of 7. Solutions with a pH lower than 7 are acidic (more likely than water to give up protons); those with a pH greater than 7 are basic.

What Was Water's Role in Chemical Evolution?

- Chemical evolution proceeded in the ocean as the Earth's surface was gradually covered by water.
- Earth's oceans contained reduced carbon compounds from redox reactions in the atmosphere and ions dissolved from eroding rocks. Heat caused formation of simple organic molecules, preserved from destructive energy sources by water's high specific heat.

B. CROSS-CUTTING THEMES

Looking Back— Concepts from Earlier Chapters
Evolution—Chapter 1
Having learned that species change over time and that all organisms descend from a single common ancestor, in Chapter 2 you explored the conditions under which this common ancestor (the first life) evolved.

Looking Forward— Concepts in Later Chapters
Chemical Evolution—Chapters 3–6
Chapter 2 sets the stage for and introduces the major players in chemical evolution. **Chapters 3 through 6** continue this story and describe several hypotheses about how life evolved from nonlife.

Redox Reactions—Chapters 9 and 10
Respiration, photosynthesis, and other important metabolic pathways involve a series of redox reactions.

Water Is Special— Chapters 36, 42, 54, and others
Water has some unusual properties that make it essential to life. **Chapters 36** and **42** discuss how plants and animals deal with their water requirements, and **Chapter 54** describes how water cycles through the ecosystem.

**Functional Groups—
Chapters 3, 4, 5, 9, 37, 43, and others**
Functional groups introduced in this chapter are discussed in greater detail later with reference to specific chemical pathways and nutrition.

C. DIFFICULT TOPICS

Chapter 2 is a whirlwind tour of basic chemistry. You need to know these basics in order to understand (1) molecules and chemical reactions that are essential to life, and (2) how life could have evolved from nonlife. You may want to look at an introductory chemistry text to ensure you understand concepts like **electron orbitals**, **potential** and **kinetic energy**, different **bond types**, **pH**, and how **redox reactions** work. You may also want to go back and carefully examine the figures in this chapter. Make a list for yourself, itemizing how the concepts introduced in this chapter relate to our best hypotheses about the environment of ancient Earth and the chemical reactions that could have led to life.

D. ASSESSING WHAT YOU'VE LEARNED

(1) Testing Your Knowledge

1. Two isotopes of an element vary in the number of _____.
   a. electrons
   b. protons
   c. neutrons
2. The number of _____ within an atom never varies among atoms of a specific element.
   a. neutrons
   b. protons
   c. electrons
3. The most ancient fossils are in rocks that formed _____.
   a. 4.4 billion years ago
   b. 3.8 Ga
   c. 4.4 Ga
   d. 65 million years ago
4. 3.5 Ga is about 5 times the half-life for uranium's decay to lead. What ratio of uranium to lead would researchers find in a fossil from 3.5 Ga?
   a. 1 part uranium to 1 part lead
   b. 1 part uranium to 5 parts lead
   c. 1 part uranium to 25 parts lead
   d. 1 part uranium to 32 parts lead
5. An organic molecule always contains one or more atoms of:
   a. nitrogen
   b. sulfur
   c. carbon
   d. oxygen
6. Which group of atoms determines a molecule's behavior?
   a. functional group
   b. carbon skeleton
   c. hydrogen atoms
7. 96% of matter found in living organisms is composed of these four elements:
   a. hydrogen, carbon, oxygen, sulfur
   b. carbon, sulfur, nitrogen, iron
   c. hydrogen, carbon, oxygen, nitrogen
   d. sulfur, hydrogen, oxygen, nitrogen
8. A large boulder is balanced on top of a hill. You give the boulder a push, and it rolls down the hill. This is an example of transferring _____ energy into _____ energy.
   a. kinetic; potential
   b. potential; kinetic
   c. kinetic; thermal
9. Which of these factor(s) determine(s) whether a chemical reaction will occur spontaneously?
   a. amount of entropy (disorder in a group of molecules)
   b. temperature
   c. amount of potential energy differences between products and reactants
   d. all of the above
10. Which phase of matter has the highest entropy?
    a. solid
    b. liquid
    c. gas
11. In a reduction-oxidation reaction, a substance that loses electrons is:
    a. oxidized
    b. reduced
12. Which statement regarding the properties of water is NOT correct?
    a. Water is a good solvent.

b. The oxygen atom in water has a high electronegativity.

c. Water is a polar molecule.

d. Solid water (ice) is denser than liquid water.

13. Substances that give up protons H^+ during acid-base reactions are:

a. isotopes

b. acids

c. bases

d. ions

14. How many protons are present in 1 liter of pure water? (Hint: See definition of molarity.)

a. $1.0 \propto 10^{-7}$

b. $6.022 \propto 10^{23}$

c. $6.022 \propto 10^{16}$

d. No protons are present.

15. In _____ bonds, electrons are shared equally between two atoms.

a. ionic

b. nonpolar covalent

c. polar covalent

d. hydrogen

16. A cation has a _____ charge.

a. positive

b. negative

c. neutral

17. The bonds between the oxygen and hydrogen atoms within a water molecule are _____ bonds.

a. hydrogen

b. ionic

c. polar covalent

d. nonpolar covalent

18. Which of the following were necessary to produce formaldehyde and cyanide, the first reduced-carbon products, on ancient Earth?

a. atmospheric molecules: NH_3, CO_2, CO, N_2, H_2O

b. energy from sunlight

c. energy from heat

d. Answers a and b are correct.

e. Answers a and c are correct.

19. Why do high temperatures speed chemical reactions?

a. Heat increases the concentration of reactants.

b. Heat causes the reactants to move faster and collide more often.

20. Following formation of reduced-carbon compounds, the addition of heat allowed _____ bonds to form.

a. carbon to oxygen

b. carbon to carbon

c. carbon to hydrogen

d. hydrogen to oxygen

(2) Integrating Your Knowledge

(a) Computer models of chemical reactions in the atmosphere of ancient Earth are discussed in this chapter. What assumptions did the programmers make about environment on ancient Earth? Did the results of the computer simulations support or reject the hypothesis regarding chemical evolution?

(b) H, C, N, and O are the most prevalent atoms in organic compounds. Why do these atoms combine to form molecules like H_2O, CO_2, H_2, and N_2?

(c) Take a look at the following reaction: $CO_2(g) + 2\ H_2O(g) \downarrow\ H_2CO(g) + H_2O(g)$. Is EH positive or negative? Is ES positive or negative? How about EG? Will this reaction proceed spontaneously?

(d) A nitrogen atom has 7 protons and 7 neutrons.

What is its atomic number?

What is its mass number?

What is the molecular weight of molecular nitrogen (nitrogen gas, N_2)?

What type of bond holds N_2 together (be specific)?

How much would a mole of N_2 weigh?

Is the bond holding N_2 together stronger or weaker than the bond holding NaCl (salt) together?

(e) Draw H_2CO and HCN, showing the electrons and orbitals involved in the bonds as in **Figure 2.11**.

(f-k) Label the following functional groups:

f) O=P(-O⁻)(-O⁻)(=O) with O⁻

g) HO—C(=O)—R

h) R—SH

i) O=C with R and H (carbonyl: R—C(=O)—H)

j) R—N with two H (amino)

k) R—OH

(l) Some recent evidence shows that ocean temperatures have increased about 1/10th of a degree in the past 50 years. This does not *sound* like much—why is it viewed as strong evidence for global warming?

(m) Compare and contrast hydrogen bonds and ionic bonds.

CHAPTER 2—ANSWER KEY

D. Assessing What You've Learned

(1) Testing Your Knowledge

1. c; 2. b; 3. b; 4. d; 5. c; 6. a; 7. c; 8. b; 9. d; 10. c; 11. a; 12. d; 13. b; 14. c; 15. b; 16. a; 17. c; 18. d; 19. b; 20. b

(2) Integrating Your Knowledge

(a) They assumed that Earth's early atmosphere contained CH_4, NH_3, H_2O, CO_2, H_2, and N_2 exposed to sunlight (this assumption is supported by the best current data about ancient Earth's environment). Their models supported the chemical evolution hypothesis because reduced carbon compounds were made that can combine to form organic molecules when heat is present.

(b) These atoms combine to form these molecules because they have unpaired electrons in their outer orbitals and are not stable as lone atoms, but are stable when they fill their outer orbitals by sharing electrons in covalent bonds.

(c) EH is positive because more bonds are created than broken; the products have more potential energy. ES is negative because there are fewer molecules and thus less entropy in the products than in the reactants. EG must be positive, and this reaction cannot proceed spontaneously—an energy source (e.g., sunlight) must be present in order for it to occur.

(d) The atomic number for nitrogen is 7, and its mass number is 14.
The molecular weight of N_2 is 28. A triple covalent bond holds N_2 together, and a mole of N_2 would weigh 28 g.
Covalent bonds are stronger than ionic bonds. For example, ionic bonds easily dissolve in water.

(e)

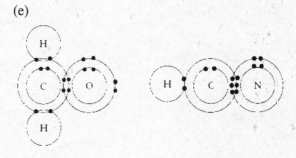

(f) phosphate
(g) carboxyl
(h) sulfhydryl
(i) carbonyl
(j) amino
(k) hydroxyl
(l) A rise in ocean temperatures is strong evidence for global warming because water has such high specific heat. It takes a *lot* of energy to warm all the ocean water even a tenth of a degree.
(m) Both hydrogen bonds and ionic bonds form due to attraction of opposite charges for each other. However, only partial charges (from unequally shared electrons in covalent bonds) are involved in hydrogen bonds, while ions have gained or lost whole electrons. Hydrogen bonds are much weaker than ionic bonds.

3

Protein Structure and Function

A. KEY BIOLOGICAL CONCEPTS

- Alexander I. Oparin and J.B.S. Haldane both proposed the theory of chemical evolution in the 1920s.
- The pattern part of the theory holds that ancient Earth's atmosphere and ocean contained molecules that reacted with one another, forming increasingly complex carbon compounds.
- The process that explains this pattern is the conversion of sunlight and other types of energy into chemical energy (potential energy contained in molecular bonds).
- The four steps of chemical evolution (each requiring energy input) are (1) production of small molecules containing reduced carbon; (2) creation of a **prebiotic soup** in shallow ocean waters as larger organic subunit molecules form; (3) linkage of these organic subunits to make the large organic molecules important in cells today; and (4) evolution of a self-replicating molecule.
- This chapter focuses on proteins: one of the four major classes of large organic molecules.

3.1 Early Origin-of-Life Experiments

- Stanley Miller (1953) was first to simulate chemical evolution in the laboratory in order to experimentally test whether the first stages of chemical evolution would have occurred on ancient Earth.

- He assumed that the early atmosphere had molecules with high free energy that easily participate in redox reactions.
- Methane (CH_4), ammonia (NH_3), and hydrogen (H_2) were placed in a large flask representing the atmosphere. This flask was connected to another flask with a model ocean.
- Miller boiled the "ocean" water to add water vapor to the atmosphere gases. As the vapor cooled, it condensed and rained down, carrying molecules with it through tubing that connected back to the water flask so that the water and molecules dissolved in it circulated through the system (**Figure 3.1**).
- Miller added sparks (simulated lightning) to the "atmosphere" as an energy source.
- Presto! HCN and H_2CO, important precursors for more complex organic molecules, were made along with other more complex organic compounds, including amino acids.

3.2 Amino Acids and Polymerization

- More recent experiments similar to Miller's show that amino acids and other organic molecules easily form under conditions that accurately simulate conditions on ancient Earth (see **Box 3.1**).
- Amino acids are also found in meteorites, lending additional evidence for the formation of a prebiotic soup rich in organic molecules.

The Structure of Amino Acids

- The huge diversity of proteins in living organisms is made from just 20 amino acid subunits. All **amino acids** have a central carbon atom bonded to H, H_2N (an amino functional group), COOH (a carboxyl functional group), and a variable side chain (**Figure 3.4**).
- In water (pH 7), the amino group attracts a proton to form NH_3^+ and the carboxyl group loses a proton to form COO^-. This polarity helps amino acids stay in solution and makes them more reactive.

The Nature of Side Chains

- The 20 major amino acids differ only in the variable side chain or R-group attached to the central carbon (**Figure 3.5**). R-groups differ in their size, shape, reactivity, and interactions with water.
- Nonpolar **hydrophobic** R-groups cannot form hydrogen bonds with water and tend to group together; polar **hydrophilic** R-groups interact readily with water (**Table 3.1**).

Explaining Optical Isomers

- **Isomers** are molecules with the same molecular formula but different structures (**Figure 3.6**).
- **Structural isomers** have the same atoms, but the atoms are attached in different orders.
- **Geometric isomers** differ in the way atoms or groups attach to the two sides of a double bond or ring.
- **Optical isomers** differ in the arrangement of atoms or groups around a carbon atom that has four different groups attached to it.
- All amino acids except glycine have optical isomers—two mirror-image forms.
- Surprisingly, only "left-handed" amino acids are found in cells.

How Do Amino Acids Link to Form Proteins?

- **Polymerization** is the process of linking molecular subunits (monomers) to make **polymers** (macromolecules of linked monomers; **Figure 3.7**).
- A protein is a macromolecule (or polymer) made by connecting amino acid monomers.

- Monomers do not form polymers spontaneously; this is because polymers have less entropy ($-\Delta S$) and higher potential energy ($+\Delta H$) than do unlinked monomer subunits, making ΔG positive.

Could Polymerization Occur in the Energy-Rich Environment of Early Earth?

- Monomers polymerize through **condensation reactions** that release a water molecule as the polymer bond forms. In the reverse reaction, **hydrolysis**, water reacts with the polymer to release a monomer subunit (**Figure 3.8**).
- Condensation reactions require energy input since ΔG is positive for these reactions; but even with energy inputs of heat or electrical discharges, few polymers are produced in experiments like Miller's.
- However, experiments show that polymers on tiny mineral particles like those found in clay or mud are protected from hydrolysis. Chemical evolution of polymers could have occurred in coastal environments.

The Peptide Bond

- Amino acid condensation reactions bond the carboxyl group of one molecule to the amino group on another to form a **peptide bond** (**Figure 3.9**). A chain of linked amino acids (or residues) is a **polypeptide**.
- Side chains stick out from linked amino acids residues (**Figure 3.10a**). Polypeptides are flexible and have directionality (the N-terminus has an unlinked amino group, and the C-terminus has a carboxyl group).
- By convention, polypeptides are shown with the N-terminus on the left and the free carboxyl on the right. Amino acids are numbered starting from the N-terminus (**Figure 3.10b**).
- Peptide bonds cannot rotate, but the bonds to each side of them can, so polypeptides are flexible molecules. **Oligopeptides** have less than 50 amino acids, whereas proteins have 50 or more linked amino acids.

3.3   What Do Proteins Look Like?

- Proteins have incredibly diverse shapes and functions; let us examine the four levels of protein structure.

Primary Structure

- The **primary structure** of a protein is its unique sequence of amino acids. With 20 amino acids, just one protein that is 10 amino acids long has billions of possible amino acid sequences! More specifically, a polypeptide of length n has 20^n possible sequences.
- Because the amino acid R-groups affect a polypeptide's size, shape, chemical reactivity, and interactions with water, just one amino acid change can radically alter protein function (**Figure 3.13**).

Secondary Structure

- Hydrogen bonding between the carboxyl O of one amino acid residue and the amino H of another creates a protein's **secondary structure**. Hydrogen bonding occurs only when a polypeptide is bent such that these functional groups lie close to each other (**Figure 3.14**).
- Usually these hydrogen bonds cause the polypeptide either to coil into an α-helix or fold into a χ-pleated sheet (like the folds in a paper fan).
- Primary structure affects the secondary structures that form; different amino acids facilitate formation of different secondary structures.
- A protein's secondary structure increases its stability; although one hydrogen bond is fairly weak, many hydrogen bonds contribute to each of these secondary structures.

Tertiary Structure

- Amino acid R-groups are largely responsible for a polypeptide's three-dimensional shape or **tertiary structure**. R-groups interact with other side chains or with the peptide-bonded backbone, making the polypeptide bend and fold into a precise shape (**Figure 3.15**).
- R-group interactions include (in order of strength) covalent **disulfinde bonds** between sulfur atoms, ionic bonds, hydrogen bonds, and hydrophilic or hydrophobic interactions with water that cause weak electrical attractions called **van der Waals interactions**.

Quaternary Structure

- Some proteins have several polypeptide subunits; bonding of two or more subunits produces **quaternary structure** (**Figure 3.16**).
- The combined effects of primary, secondary, tertiary, and sometimes quaternary structure (**Table 3.2**) allow for amazing diversity in protein form and function.

Folding and Function

- Polypeptides in solution fold so that hydrophobic side chains group together away from water, and hydrophilic side chains face water. This folding is often spontaneous because the hydrogen bonds and van der Waals interactions stabilize the molecule.
- The folded three-dimensional shape is crucial to protein function. If chemical agents or high temperatures disrupt the hydrogen and disulfide bonds that stabilize the folded shape, a protein is **denatured** and unable to function normally.
- Proteins called **molecular chaperones** help proteins fold correctly in cells. Many of these molecular chaperones are heat-shock proteins, produced after a cell experiences high temperature or other conditions that disrupt protein folding.

3.4 What Do Proteins Do?

Proteins Have Diverse Functions in Cells

- Proteins do most cell work: they defend cells against disease-causing viruses and bacteria, move cells and cell cargo, speed up cell reactions (**catalysis**), carry and receive signals between cells, provide structural support, and transport molecules in and out of cells as well as throughout the body (**Table 3.3**).

An Introduction to Catalysis

- Just because a chemical reaction is spontaneous ($-\Delta G$) does not mean it happens quickly.
- Molecules must collide with the correct orientation and enough energy to break and/or form chemical bonds. Chemical reactions depend on electron interactions, yet electrons repel each other.

- When the kinetic energy of a molecular collision is large enough to overcome electron repulsion, a **transition state** is formed in which a combination of old and new bonds is present.
- **Activation energy** is the free energy necessary to achieve this transition state. Less stable transition states require more activation energy; the higher the activation energy, the slower the reaction rate.
- **Figure 3.19** graphically represents free energy changes during a chemical reaction. ΔG is negative, so this reaction is exothermic and spontaneous. But the activation energy (E_a) needed to form the transition state is high, so this reaction would occur very slowly.
- Chemical reactions are faster at higher temperatutures because molecular kinetic energy is a function of temperature (**Figure 3.20**).
- Chemical reactions also proceed much faster if electrons in the transition state are stabilized by interactions with other ions, atoms, or molecules. A **catalyst** lowers the activation energy of a reaction by stabilizing the transition state (**Figure 3.21**).
- Catalysts are not used up in chemical reactions and do not change ΔG or the energy of the reactants or products. Only E_a, the amount of energy required to achieve the transition state, is affected.
- **Enzymes** are protein catalysts that typically catalyze only one reaction.
- Most biological chemical reactions occur only at meaningful rates in the presence of an enzyme; many enzymes catalyze more than a million reactions per second, thus causing reactions to proceed a million times faster than they would without the enzyme.

How Do Enzymes Work?

- An early model of enzyme action, the **lock-and-key model** (Fischer, 1894), described enzymes as being like locks into which reactant "keys" fit. Inside the enzyme "lock," the reactants or **substrates** participated in the chemical reaction catalyzed by that enzyme.
- This early model correctly identified (1) that enzymes bring substrates together in specific positions that facilitate reactions, and (2) that enzymes are very specific as to which reactions they catalyze because substrates bind only to sites with the appropriate shape and chemical properties.
- The substrate-binding site on an enzyme is called the **active site** (**Figure 3.22**).
- The lock-and-key model has been replaced by the **induced fit** model because enzymes are not rigid like locks; instead, they are flexible and often change shape as substrates bind to the active site.
- Substrate interaction with the enzyme active site causes stabilization of the transition state and a decrease in the activation energy.
- Interactions between the substrate and active-site R-groups include hydrogen bonding, formation of temporary covalent bonds involved in atom transfer, and gain or loss of protons aided by acidic or basic R-groups (**Figure 3.23**).
- Reaction products have a low affinity for the active site and are therefore released from the enzyme.
- Enzyme action has three steps: **initiation**, when reactants are correctly oriented as they bind to the active site; **transition state facilitation**, as interactions between the substrate and active site R-groups lower the activation energy; and **termination**, when reaction products are released from the enzyme (**Figure 3.24**).

Do Enzymes Act Alone?

- Some enzymes require **enzyme cofactors**—atoms or molecules not included in an enzyme's primary structure. These cofactors are either metal ions or small organic molecules called **coenzymes**.
- Enzyme cofactors often bind to the active site and are involved in transition state stabilization.
- Many vitamins are needed for enzyme cofactor production. Vitamin deficiency diseases occur when there are not enough enzyme cofactors to allow normal enzyme function.
- Enzyme regulation is an essential to cell control of chemical reaction rates.
- **Competitive inhibition** occurs when a molecule similar in size and shape to the substrate

competes with the substrate for active site binding (**Figure 3.25a**).

- **Allosteric regulation** occurs when a molecule causes a change in enzyme shape by binding to the enzyme at a location other than the active site. Depending on the way in which the enzyme's shape changes, allosteric regulation either increases or decreases enzyme activity (**Figure 3.25b**).

What Limits the Rate of Catalysis?

- Enzyme kinetics describes how various factors affect enzyme-catalyzed reaction rates.
- **Figure 3.26** shows how the rate of product formation increases linearly for a given increase in substrate concentration at low substrate concentrations, but levels out at high substrate concentrations. Uncatalyzed reactions show linear increases in reaction rate across all ranges of substrate concentrations.
- Initially, each increase in substrate concentration allows the available enzymes to facilitate that many more reactions. But, as the substrate concentration increases, enzyme active sites eventually become filled as fast as they are emptied, and no further increase in reaction rate is possible.
- All enzymes show this type of kinetics, but enzymes with high substrate affinities rapidly achieve their maximum reaction rate as substrate concentration increases, while enzymes with low affinities for their substrate require fairly high substrate concentrations before reaching their maximum rate of reaction.

How Do Physical Conditions Affect Enzyme Function?

- Enzymatic activity is strongly influenced by temperature (effect on enzyme movement) as well as pH (effect on R-group charge and proton- and electron-transfer ability of the active site).
- Most enzymes function best at some particular temperature and pH. These temperature and pH optima usually reflect the environment in which an enzyme functions, because natural selection favors the survival and reproduction of organisms with enzymes that function efficiently (**Figure 3.27**).

- Changes in primary structure that increase enzyme performance at the temperatures and pH it most often experiences help organisms adapt to their specific environments.

How Does ATP Drive Endergonic Reactions?

- Enzymes do not change the potential energy or entropy of reactants or products. How then do endergonic ($+\Delta G$) reactions occur?
- Most of the chemical reactions required for chemical evolution and known to occur inside cells are normally thought of as endergonic reactions. For example, polymers have higher potential energy and lower entropy than do monomers, so polymerization reactions are endergonic.
- Chemical energy, primarily in the form of phosphate groups (PO_4^{3-}) can provide sufficient energy to split one endergonic reaction into a two-step exergonic reaction. In cells, adenosine triphosphate (ATP) typically provides the phosphate (**Figure 3.28**).
- In the first step, ATP transfers one of its three phosphate groups to a substrate, producing an "activated" substrate and adenosine diphosphate (ADP). This reaction is exergonic; the free energy decrease from ATP to ADP more than makes up for the free energy gain by the "activated" substrate.
- In the second step, the "activated" substrate reacts to form the product (e.g., is added to a growing polymer). During product formation, the substrate's phosphate group is released, providing the chemical energy needed to make the reaction exergonic.
- Sometimes the enzyme rather than the substrate gains the phosphate during this two-step reaction coupling.
- In chemical evolution, amino acids in the prebiotic soup could have become phosphorylated by solar, heat, or electrical energy. These phosphorylated amino acids could then have joined to form polypeptides.
- In summary, phosphorylation allows all biological reactions to be exergonic; most of these reactions occur quickly when catalyzed by enzymes.

Was the First Living Entity a Protein?

- A self-replicator would need to catalyze the polymerization of its copy, and enzymes are the most efficient catalysts known. But the first self-replicator likely also needed a copy-making template—something not found in proteins.

B. CROSS-CUTTING THEMES

Looking Back—
Concepts from Earlier Chapters
Chemical Evolution—Chapter 2
The start of evolution, environment of ancient Earth, and basic chemistry needed to understand the chemical reactions that cause chemical evolution are described in Chapter 2.

Chemical Bonds, Chemical Reactions, and Chemical Energy—Chapter 2
Chapter 2 also provided basic background information on how chemical bonds form and how Gibbs free energy (ΔG) determines whether reactions can occur spontaneously.

Looking Forward—
Concepts in Later Chapters
Proteins are the workhorse molecules of cells, so these molecules will appear in many later chapters. Here, a few protein functions are highlighted.

Structural Proteins and Transport Proteins—Chapters 6–8, 13, 36, and others
Proteins form most of the major structures in cells and are responsible for transporting molecules into and out of cells.

Enzymes—Chapters 9, 10, 16, and others
Chapters 9 and 10 focus on some of the most important cell reactions. These reactions are, of course, catalyzed by enzymes. You will learn more about enzyme regulation in these chapters. In Chapter 16 you will learn about how enzymes are involved in transcription and translation—the processes by which cells make enzymes and other proteins.

Other Protein Functions—Chapters 39, 47, and 48 (hormones), Chapter 49 (defense), and others

C. DIFFICULT TOPICS

This chapter gives you lots of information about protein, RNA, and DNA subunits—how they fit together and the chemical reactions involved. Be sure to carefully study the figures in text; you might even want to make your own model of alanine using toothpicks and gumballs in order to get a better understanding of chirality. In the text's discussion of chemical reactions, you are also introduced to the concepts of activation energy (E_a) and transition states.

Sometimes analogies help us understand processes that we cannot directly observe. For exam-ple, the end products of spontaneous chemical reactions typically have less **potential energy** (ΔH) than the starting reactants. Similarly, an object on the floor has less potential energy than an object on the table, because it is closer to Earth's center of gravity. But the object on the table needs **kinetic energy** in order to reach the **transition state** of passing the table edge before it will fall to the floor. If you have many marbles on the table and a fan blowing in the room, air currents may give some of the marbles enough kinetic energy to reach the table edge, at which point they fall to the floor. But if you put the marbles in a box, you have just increased the **activation energy** needed to reach the transition state of reaching the table edge.

As in the example of marbles on a table, even spontaneous chemical reactions may proceed so slowly as to be hardly detectable if the activation energy needed to reach the transition state is high and/or the kinetic energy of the molecules is low. Recall that enzymes increase reaction rates by stabilizing transition states and decreasing activation energy. Reaction rates can also be speeded up by heating the molecules to give them more kinetic energy (is why sugar dissolves in hot water much faster than in cold water).

D. ASSESSING WHAT YOU'VE LEARNED

(1) Testing Your Knowledge

1. Stanley Miller showed experimentally that complex, carbon-containing compounds could form from simple molecules present in early Earth atmosphere and oceans. In Miller's experiment, _____ was the energy source that permitted these complex molecules to form.
    a. heat
    b. light
    c. electricity

2. Amino acids are the building blocks of:
    a. nucleic acids
    b. proteins
    c. carbohydrates

3. There are 20 different amino acids. All amino acids have a(n) _____ functional group and a(n) _____ functional group, but vary in their _____ group.
    a. carboxyl; amino; R-
    b. R-; carboxyl; amino
    c. amino; R-; carboxyl

4. All amino acids except glycine are:
    a. structural isomers
    b. optical isomers
    c. right-handed

5. Which of the following is true of ΔH, $T\Delta S$, and ΔG for polymerization reactions?
    a. ΔH +, $T\Delta S$ +, ΔG +
    b. ΔH +, $T\Delta S$ -, ΔG +
    c. ΔH +, $T\Delta S$ -, ΔG −
    d. ΔH -, $T\Delta S$ -, ΔG −

6. Monomers join to form polymers via:
    a. hydrolysis
    b. condensation reactions

7. The condensation reaction joining two amino acids together forms a:
    a. hydrogen bond
    b. ionic bond
    c. peptide bond
    d. protein

8. The _____ structure of a protein is the sequence of amino acids.
    a. primary
    b. secondary
    c. tertiary
    d. quaternary

9. Which of the following statements regarding proteins is correct?
    a. Alpha helices and beta sheets are forms of a polypeptide's tertiary structure.
    b. Nucleotides join via condensation reactions to form polypeptides.
    c. The words *polypeptide* and *protein* have identical meaning.
    d. Protein structure is correlated with protein function.

10. Which of the following side-chain interactions are important in determining tertiary structure?
    a. covalent bonds
    b. ionic bonds
    c. hydrogen bonds and van der Waals interactions
    d. all of the above
    f. a and c
    g. b and c

11. Which of the following is *not* something proteins do within cells?
    a. cause movement and transport molecules
    b. provide mechanical support
    c. carry genetic information
    d. carry and receive cell signals

12. Which condition(s) will speed the rate of a chemical reaction?
    a. high kinetic energy of reactants
    b. presence of a catalyst
    c. both A and B

13. How do catalysts increase reaction rates?
    a. by stabilizing the transition state
    b. by providing chemical energy
    c. by decreasing ΔG
    d. by increasing the kinetic energy of reactants

14. Further research led to modifications in the lock-and-key model of enzyme action. Which of the following assumptions of the lock-and-key model turn out NOT to be correct?
    a. Enzymes are rigid structures analogous to a lock.
    b. Enzymes bring substrates together so that reactions are more likely to occur.
    c. Enzymes are very specific as to the reaction they catalyze.

15. Enzymes stabilize the transition state by:
    a. providing chemical energy
    b. bringing reactants together in the active site
    c. chemically interacting with the bound substrate
    d. decreasing the activation energy

16. A molecule is a competitive inhibitor of an enzyme if:
    a. It changes enzyme shape by binding to a site other than the active site.
    b. It can bind in an enzyme's active site because it is chemically similar to the substrate.
    c. It competes with the enzyme for substrate binding.

17. At low substrate concentrations, enzyme-catalyzed reaction rates _____ as substrate concentration increases, and at high substrate concentrations, enzyme-catalyzed reaction rates _____.
    a. show an exponential increase; increase linearly
    b. increase linearly; increase linearly
    c. decrease linearly; plateau at a maximum speed
    d. increase linearly; plateau at a maximum speed

18. Which of the following statements is *not* true?
    a. All enzymes function best in the neutral pH range.
    b. Sometimes enzymes have different versions with different temperature and pH optima.
    c. The best enzymes are not affected by temperature or pH.
    d. pH helps enzymes by stabilizing the transition state.

19. Was the first living entity a protein?
    a. Yes, because proteins catalyze reactions and a self-replicator would have had to assemble and polymerize its copy.
    b. No, because proteins cannot be made by chemical evolution.
    c. No, because proteins are not good templates for copy-making.

(2) Integrating Your Knowledge

(a) Look at the structure of the amino acids valine and serine in **Figure 3.3**. Which of these acids is more likely to interact with water, and why?

(b) Explain why glycine is not chiral.

(c) Give three pieces of evidence supporting the hypothesis that amino acids and eventually polypeptides could have formed by chemical evolution on ancient Earth.

(d) In your own words, explain how ATP can drive endergonic reactions.

CHAPTER 3—ANSWER KEY

D. *Assessing What You've Learned*

(1) Testing Your Knowledge
1. c; 2. b; 3. a; 4. b; 5. b; 6. b; 7. c; 8. a; 9. d; 10. d; 11. c; 12. c; 13. c; 14. a; 15. c; 16. b; 17. d; 18. b; 19. c

(2) Integrating Your Knowledge

(a) Serine has a polar OH group; valine just has C and H in its R group. Thus serine is more likely to interact with water.

(b) Glycine's R-group is a hydrogen, which means that the central carbon has two hydrogen atoms attached to it, making its mirror image the same as the molecule itself (like if your hand had two thumbs instead of a thumb and a pinky).

(c) (1) Miller's experiment, and others since then that more accurately simulate ancient Earth conditions, produced amino acids; (2) Meteorites often contain amino acids; and (3) A simulated prebiotic soup (containing amino acids and phosphate) on clay-sized mineral particles can form polypeptides.

(d) ATP provides the energy for endergonic reactions to occur by phosphorylating or "activating" the substrate. This splits one endergonic reaction into two exergonic reactions in which the chemical energy comes from the phosphate groups.

4

Nucleic Acids and the RNA World

A. KEY BIOLOGICAL CONCEPTS

- Key features distinguishing life from nonlife are (1) ability to reproduce and (2) **metabolism**, the ability to acquire certain molecules and control chemical reactions. All living **organisms** have a **cell membrane**, which creates a microenvironment favorable to cell metabolism.
- However, the theory of chemical evolution suggests that the first step in the evolution of life was the formation of a self-replicating molecule. When a descendent of this self-replicator became enclosed in a membrane, it created the first cell.
- Most biologists think the first self-replicator was a molecule of ribonucleic acid (RNA). The Chapter 4 discussion of this **RNA world hypothesis** will also introduce you to nucleic acid structure and function.

4.1 What Is a Nucleic Acid?

- A **nucleic acid** is a polymer of nucleotides in which each **nucleotide** is made up of a phosphate group bonded to a sugar, which is in turn bonded to a nitrogenous base (**Figure 4.1**).
- A **sugar** is an organic compound with a carbonyl group and several hydroxyl groups; the sugar is *ribose* in **ribonucleotides** and *deoxyribose* in **deoxyribonucleotides**—these two sugars differ only in that deoxyribose has an H instead of an –OH on the second ring carbon.
- Nitrogenous bases include adenine (A) and guanine (G) from the purine structural group and cytosine (C), uracil (U), and thymine (T) from the pyrimidine group. The first three are found in both types of nucleotides, but ribonucleotides contain uracil whereas deoxyribonucleotides contain thymine.

Could Chemical Evolution Result in the Production of Nucleotides?

- Chemical evolution simulations have not yet produced nucleotides; sugars and purines are easily made, but researchers think ribose would have had to dominate for nucleic acids to form (the ribose problem), and pyrimidines are not easily synthesized under conditions thought to exist on ancient Earth.

How Do Nucleotides Polymerize to Form Nucleic Acids?

- Nucleotides polymerize to form nucleic acids when a condensation reaction forms a **phosphodiester linkage** between the phosphate group of one nucleotide and the –OH group on the 3≤carbon of another (**Figure 4.2** and **Figure 4.3**).
- **Ribonucleic acid** (RNA) is made from nucleotides containing ribose, and **deoxyribonucleic acid** (DNA) is made from nucleotides containing deoxyribose.

- The sugar-phosphate linkages form the backbone of a nucleic acid; this chain is polar because one end has an unlinked 5≤carbon and the other end has an unlinked 3≤carbon. The nucleotide base sequence creates a nucleic acid's primary structure; it is always written from the 5≤to the 3≤direction.
- In cells, enzyme-catalyzed polymerization occurs after nucleotide free energy has been raised by adding two extra phosphate groups per nucleotide (**Figure 4.4**). These "activated" nucleotides can also form RNA following incubation with tiny mineral particles (**Box 4.1**).
- Consequently, if all the nucleotides were able to form during chemical evolution, they could have polymerized on clay particles to form RNA and DNA.

4.2   DNA Structure and Function

- James Watson and Francis Crick discovered DNA's secondary structure in 1953. Their hypothesis built on the following results from other laboratories.
- Chemists had determined the structure of nucleotides and the nature of phosphodiester linkages.
- Chargaff's rules stated that in DNA the number of purines equals the number of pyrimidines, the numbers of T's and A's are equal, and the number of C's and G's are equal.
- Rosalind Franklin and Maurice Wilkins used X-ray crystallography to calculate distances between atom groups that regularly repeated; they concluded that the structure of DNA is either helical or spiral.
- Analysis of the size and geometry of deoxyribose, nitrogenous bases and phosphate groups led Watson and Crick to conclude that Franklin and Wilkins' 2.0 nm measurement represented the width of the helix and 0.34 nm the distance between bases in the spiral.
- Using models (**Figure 4.8**), Watson and Crick concluded that two antiparallel DNA strands form a double helix with the hydrophilic sugar-phosphate backbone facing the exterior and with purine-pyrimidine pairs of nitrogenous bases on the interior, packed closely together.

- Adenine forms hydrogen bonds with thymine and guanine forms hydrogen bonds with cytosine: this is known as **complementary base pairing** (**Figure 4.9**). The guanine-cytosine pair is held together more strongly than the adenine-thymine pair because this pair forms three, instead of two, hydrogen bonds.
- Watson and Crick's double-helix hypothesis for DNA's secondary structure explains all of Franklin and Wilkins' molecular measurements. There are 10 complementary base pairings for each turn of the helix, and there is a major and minor groove to the double helix (**Figure 4.11**).

DNA Is an Information-Containing Molecule

- The nitrogenous base sequence of DNA stores information required for growth and reproduction of all living cells, and complementary base pairing provides a simple mechanism for DNA replication since each strand can serve as a template for formation of a new complementary strand (**Figure 4.12**).
- But DNA replication requires many enzymes in today's cells—DNA cannot self-replicate.

Is DNA a Catalytic Molecule?

- DNA's stability makes it a reliable store for genetic information; DNA has few exposed chemical groups that could participate in chemical reactions, and the presence of an H instead of OH on the sugar's 2≤carbon makes DNA less reactive than RNA but more resistant to degradation.
- However, stable molecules like DNA make poor catalysts because stabilization of the transition state requires interaction between the substrate and the catalyst. Enzymes, in contrast, have exposed functional groups that interact with substrates.
- Proteins also have incredible structural diversity, allowing interaction with many different types of molecules. DNA, on the other hand, has an extremely simple primary and secondary structure.
- DNA does not appear to be able to catalyze any chemical reaction, thus biologists think that the first life-form was RNA, not DNA.

4.3   RNA Structure and Function

- RNA's primary structure is similar to DNA's, with two differences: (1) RNA contains uracil instead of thymine, and (2) RNA contains ribose instead of deoxyribose. The hydroxyl group on ribose's 2≤carbon can participate in reactions that degrade RNA, making RNA more reactive and less stable than DNA.

- RNA's secondary structure, like DNA's, is due to complementary base pairing between purine and pyrimidine nitrogenous bases. In RNA, two hydrogen bonds form between adenine and uracil and three form between guanine and cytosine.

- However, the bases of RNA typically form hydrogen bonds with complementary bases on the *same* strand: the RNA strand folds over, forming a **hairpin** structure where the bases on one side of the fold align with an antiparallel RNA segment on the other side of the fold (**Figure 4.13**).

- Other secondary structures also occur spontaneously, and all are stabilized by hydrogen bonds.

- Sometimes hairpins and other RNA secondary structures have additional folds or are attached to other RNA strands, giving some RNA molecules tertiary and quaternary structure; this structural complexity produces RNA molecules that vary greatly in their shapes and chemical properties.

RNA as an Information-Containing Molecule

- Due to the predictability of complementary pair bonding, a RNA molecule could theoretically carry the information needed to make a copy of itself. An initial template strand could be used to make a complementary strand that could separate from the template strand and then make a new strand identical to the original template (**Figure 4.14**).

- RNA would be a rather unstable information-containing molecule. But, it may have survived long enough in the prebiotic soup to replicate itself, becoming the first life-form.

Is RNA a Catalytic Molecule?

- RNA's structure is less variable than that of proteins, but it does have some secondary and tertiary structure as well as some chemical reactivity. These qualities allow RNA to stabilize some transition states.

- RNA catalysts are called **ribozymes**, and in at least one single-celled organism (*Tetrahymena*), ribozymes catalyze the hydrolysis and condensation of phosphodiester linkages. Perhaps an RNA molecule could catalyze the reactions necessary for its own replication!

4.4   The First Life-Form

- RNA can both provide a template for copying itself and catalyze the polymerization reaction that links monomers to make the (complementary) copy. These are essential qualities for a self-replicator, so researchers propose that the first life-form was RNA. This is known as the RNA-world hypothesis.

- Because no self-replicating molecules exist today (all were outcompeted by cellular life), researchers test the RNA world hypothesis using laboratory simulations.

- Wendy Johnston and others in David Bartel's lab used artificial selection to create an RNA replicase that could catalyze steps 1 and 2 of Figure 4.14: the addition of ribonucleotides to a growing RNA strand.

- The researchers began by creating many large variable RNA molecules. Then they incubated these molecules with a small RNA template. A few of the large RNAs used themselves as a template and catalyzed, albeit inefficiently, the addition of a few ribonucleotides to the starting template.

- Researchers isolated the round-1 ribozymes, and created the next round of large RNA molecules by copying these ribozymes in a way that caused a few random changes to the primary sequence.

- This step is similar to natural selection in that only the "fittest" molecules produced offspring, but these offspring had a few random changes (mutations). Most of the changes decreased the ribozyme efficiency, but a few improved the ribozyme's catalyst function.

- The above process was repeated 18 times, each time selecting the best ribozymes as templates for the next round. At each step, some ribozymes showed improved catalyst function, and the final population included ribozymes—specifically, RNA replicases—that were pretty good catalysts for the addition of ribonucleotides to a growing complementary RNA strand (**Figure 4.15**).
- These results support the RNA world hypothesis; it appears that researchers may soon be able to produce a self-replicating molecule similar to the one that might have been our world's first life-form.

B. CROSS-CUTTING THEMES

Looking Back—
Concepts from Earlier Chapters
Artificial Selection—Chapter 1
The ribozyme experiment described in Section 4.4 is similar in many ways to other types of artificial selection. Chapter 1 first introduced and defined both artificial and natural selection.

Catalysis and Correlation between Structure and Function—Chapter 3
RNA is able to catalyze some chemical reactions because it shares with proteins some variability in structure and chemical reactivity, permitting it to stabilize the transition state for some chemical reactions, including some reactions key to its potential as a self-replicator.

Looking Forward—
Concepts in Later Chapters
DNA—Chapters 11–20, 24, and others
DNA encodes the genetic information of all cells. Any discussion of genetics therefore necessitates reference to DNA. DNA is equally essential to descriptions of cell reproduction and natural selection (because traits must be heritable in order to be selected).

RNA—Chapters 16 and 34
RNA's main function in extant cells is to use the information contained in DNA to produce proteins. This process is described in **Chapter 16**. **Chapter 34** describes how viruses use RNA.

C. DIFFICULT TOPICS

The ribozyme selection experiment is cutting-edge research; it is hard to understand at first. You might want to reexamine Figure 4.14. Think about why researchers included each of the five steps listed in this figure and about how this process is similar to the artificial selection of cabbage-family plants described in Chapter 1.

D. ASSESSING WHAT YOU'VE LEARNED

(1)   Testing Your Knowledge

1. Which two attributes distinguish life from nonlife?
   a. presence of a plasma membrane and use of DNA to code genetic information
   b. ability to reproduce and presence of a plasma membrane
   c. metabolism and ability to make protein catalysts
   d. ability to control chemical reactions and ability to reproduce

2. The theory of chemical evolution predicts:
   a. reproduction of a life-form is not possible without a cell membrane
   b. a self-replicating molecule evolved before the cell membrane
   c. the first self-replicator was a molecule of DNA
   d. self-replicating molecules exist in nature today

3. Which of these is *not* a component of a nucleotide?
   a. a six-carbon sugar
   b. a phosphate group
   c. a five-carbon sugar
   d. a nitrogenous base

4. Which nitrogenous bases are purines?
   a. cytosine, uracil, and thymine
   b. adenine and guanine
   c. adenine, uracil, and thymine
   d. guanine and cytosine

5. Two major challenges to the theory of chemical evolution are:

a. that living organisms use DNA for their genetic code and proteins as their primary catalysts

b. that no self-replicating molecules exist today, and all organisms are cellular

c. the ribose problem and the origin of pyrimidine bases

d. that RNA cannot catalyze any reactions and proteins cannot serve as templates for replication

6. The condensation reaction joining two nucleotides together forms a _____.
   a. phosphodiester linkage
   b. peptide bond
   c. double helix
   d. pyrimidine

7. An "activated" nucleotide has:
   a. been added to a growing nucleic acid chain
   b. deoxyribose as its sugar
   c. a phosphate group attached to the 5≤ carbon of its sugar
   d. high free energy due to the addition of two phosphate groups

8. The _____ structure of a nucleic acid is the sequence of nitrogenous bases.
   a. primary
   b. secondary
   c. tertiary
   d. quaternary

9. _____ and _____ were the scientists who determined the double-helix structure of DNA.
   a. Rosalind Franklin and Maurice Wilkins
   b. James Watson and Francis Crick
   c. Alexander Oparin and J.B.S. Haldane
   d. Charles Darwin and Alfred Wallace

10. Chargaff's rules state that in DNA, the
    a. # purines = # pyrimidines, the #A's = #G's and the # T's = # U's
    b. # purines = # pyrimidines, the # T's = # A's and the # C's = # G's
    c. sugar-phosphate backbone is formed by phosphodiester linkages and is helical in structure
    d. two strands are antiparallel and have complementary base pairing

11. Which statement about DNA is *not* correct?
    a. DNA contains deoxyribose sugars.
    b. DNA has a double-helix structure.
    c. DNA contains uracil.
    d. DNA has a sugar-phosphate backbone.

12. Franklin and Wilkin's X-ray crystallography measurements of 0.34, 2.0, and 3.4 nm correspond to these DNA measurements:
    a. 0.34 = helix width, 2.0 = distance between bases stacked in spiral, 3.4 = one complete helix turn
    b. 0.34 = helix width, 2.0 = one complete helix turn, 3.4 = distance between bases stacked in spiral
    c. 0.34 = one complete helix turn, 2.0 = distance between bases stacked in spiral, 3.4 = helix width
    d. 0.34 = distance between bases stacked in spiral, 2.0 = helix width, 3.4 = one complete helix turn

13. Each nucleotide has one nitrogenous base: either a purine with a double-ring structure or a pyrimidine with a single ring. Which statement is true regarding complementary base pairing?
    a. Purines always pair with other purines, and pyrimidines always pair with other pyrimidines.
    b. The purine base cytosine (C) always pairs with the purine base guanine (G).
    c. In DNA, the purine base adenine (A) always pairs with the pyrimidine base uracil (U).
    d. Watson and Crick discovered complementary base pairing.

14. DNA is water soluble because:
    a. The exterior-facing sugar-phosphate backbones are negatively charged and hydrophilic.
    b. The nitrogenous bases form hydrogen bonds with water.
    c. The molecule has a hydrophobic interior.
    d. Cells are about 70 percent water.

15. Could DNA catalyze its own replication?
    a. Yes, all living organisms store their genetic information in their DNA sequences.
    b. No, its primary and secondary structures are too simple and unreactive to catalyze any chemical reactions.
    c. Yes, DNA and RNA are structurally and functionally similar
    d. No, DNA cannot serve as a template for its own replication.

16. _____ can both catalyze chemical reactions and carry information to copy itself.
    a. DNA
    b. Proteins
    c. RNA
    d. Nitrogenous bases

17. Which of the following is *not* true about RNA?
    a. RNA primarily functions as a catalyst in modern cells.
    b. RNA is more reactive than DNA.
    c. RNA could theoretically serve as a template for its own synthesis.
    d. RNA is less stable than DNA.

18. Due to complementary base pairing, _____ form(s) a secondary structure called the hairpin loop.
    a. protein
    b. DNA
    c. RNA
    d. sugars

19. Altman and Cech won a Nobel Prize for showing that _____ exist in organisms
    a. ribozymes
    b. mutations
    c. self-replicators
    d. nucleic acids

20. Wendy Johnston and coworkers created RNA molecules that catalyzed:
    a. the formation of hairpin loops
    b. the formation of nucleotides from sugar, phosphate, and nitrogenous base subunits
    c. DNA replication
    d. the addition of ribonucleotides to a growing RNA strand

(2) Integrating Your Knowledge

(a) Fill in the bases that would pair with the following single-stranded DNA sequence.

$3\leq$ C – A – A – T – G – C – T –$5\leq$

$5\leq$ ___–___–___–___–___–___–___ –$3\leq$

(b) List four structural differences between RNA and DNA.

(c) Give three arguments in support of the hypothesis that the first self-replicator was an RNA molecule.

(d) An RNA strand folds into a hairpin structure. One side of the base of the hairpin has the sequence (listed in $5\leq$ to $3\leq$ order): U-U-A-G-U-C-A. What is the nucleotide order on the complementary strand listed in $5\leq$ to $3\leq$ order? Remember that the complementary strand is anti-parallel. A second RNA strand has the sequence C-A-G-G-G-U-C on one side of its hairpin stem. Which hairpin structure will be more stable?

(e) You are told that "Bartle's team copied the selected molecules in a way that introduced a few random changes to their sequence of bases" and that "most of the modified ribozymes worked worse than the original ones." Why then did the research team introduce these random changes? Do you think most natural mutations are also detrimental?

CHAPTER 4—ANSWER KEY

D. Assessing What You've Learned

(1) Testing Your Knowledge
1. d; 2. b; 3. a; 4. b; 5. c; 6. a; 7. d; 8. a; 9. b; 10. b; 11. c; 12. d; 13. d; 14. a; 15. b; 16. c; 17. a; 18. c; 19. a; 20. d

(2) Integrating Your Knowledge
(a) G-T-T-A-C-G-A
(b) RNA differs from DNA in that it (1) has ribose, not deoxyribose, as its nucleotide sugar; (2) uses uracil instead of thymine;

(3) is single stranded, not double stranded; and (4) can have tertiary structure due to the folding of secondary structures into complex shapes (more complex/less stable structure).

(c) Three arguments supporting the hypothesis that the first self-replicator was an RNA molecule: (1) RNA can serve as a template for its own synthesis (although it takes two rounds of replication to get back to the original template molecule), (2) RNA can catalyze some chemical reactions, including phosphodiester bond formation, (3) none of the other macromolecules can do both 1 and 2, and (4) researchers have already successfully created RNA replicases that catalyze the addition of ribonucleotides to a growing RNA strand.

(d) The complementary strand's sequence is U-G-A-C-U-A-A. The second hairpin structure will be more stable because G and C form three hydrogen bonds, but U and A form only two. Thus, the second hairpin stem will have more hydrogen bonds stabilizing it.

(e) Most changes impaired ribozyme function, because it is much easier to break something than to fix it. For example, if you *randomly* made changes to your car engine, most of the changes would cause problems; only a very few would improve the engine's function. The researchers introduced random changes in order to get the very few ribozymes with improved catalyst function. In nature, most mutations are also detrimental.

5

An Introduction to Carbohydrates

A. KEY BIOLOGICAL CONCEPTS

- The term **carbohydrate** refers to single-sugar monomers (**monosaccharides**) as well as many-sugar polymers (**polysaccharides**). These molecules have the generalized chemical formula $(CH_2O)_n$.
- Carbohydrate monomers contain a carbonyl and several hydroxyl functional groups and many carbon-hydrogen bonds.

5.1 Sugars as Monomers

- Sugars provide chemical energy to cells and serve as molecular building blocks; for example, recall that during chemical evolution, ribose had to be present in order for nucleotides to form.
- Monosaccharides, including ribose, are easily made under conditions simulating the prebiotic soup.
- Sugar carbonyl as well as hydroxyl functional groups can participate in many different chemical reactions.
- Monosaccharides vary in the placement of these functional groups: when the carbonyl is at the end of the molecule, the monosaccharide is an aldehyde sugar (*aldose*), but when the carbonyl is in the middle of the sugar's carbon chain, the monosaccharide is a ketone sugar (*ketose*) as shown in **Figure 5.1**.
- The number of carbons in monosaccharides varies. Three-carbon sugars are called **trioses**; five-carbon sugars, like ribose, are called **pentoses**; and six-carbon sugars, like

glucose, are called **hexoses**. Carbons are numbered by starting at the end that is closest to the car-bonyl group.
- Monosaccharides also vary in the arrangement of their hydroxyl groups, so that many sugars have the same chemical formula but different structures. For example, glucose and galactose $(C_6H_{12}O_6)$ are optical isomers (**Figure 5.2**), and only glucose can be directly used by the cell in ATP-producing reactions.
- Sugars in aqueous solution switch back and forth between chain and ring forms, with ring forms predominating. When glucose forms a ring, carbon 1 (the carbonyl carbon) can bond with the fifth carbon's hydroxyl group in two different orientations, resulting in either α-glucose or β-glucose (**Figure 5.3**).
- The combination of different ring forms, isomers, numbers of carbons and carbonyl placements makes possible a huge variety of monosaccharides, all with unique structures and functions.
- Most monosaccharides are easily formed under conditions similar to those in ancient Earth's prebiotic soup. Also, a 3-C ketose and other sugar-like molecules were found on the Murchison meteorite, indicating that they can be synthesized on space debris and could have rained down on ancient Earth.
- It is, however, unlikely that polymerization of monosaccharides occurred during chemical evolution, and researchers still do not understand how ribose came to dominate prebiotic

soup monosaccharides, making possible the synthesis of nucleotides.

5.2   The Structure of Polysaccharides

- **Polysaccharides** form when a condensation reaction between hydroxyl groups of two monosaccharides creates a **glycosidic linkage** (**Figure 5.4**). The simplest polysaccharide is a **disaccharide** of two linked monomers.
- The monomers joined by glycosidic linkages can be identical or different. In contrast to peptide and phosphodiester bonds, glycosidic linkages can form between any two hydroxyl groups, so that the location and geometry of these bonds varies widely.

Starch: A Storage Polysaccharide in Plants

- Plants store sugars as **starch** made of many α-glucose monomers joined by glycosidic linkages between the first and fourth ring carbons (α-1,4-glycosidic bonds). The angle of these bonds causes the monomer chain to coil into a helix.
- Starch is a mix of unbranched and branched helices (amylose and amylopectin respectively; **Table 5.1**). Starch branches when a glycosidic bond forms between carbon 1 of one strand and carbon 6 on another.

Glycogen: A Highly Branched Storage Polysaccharide in Animals

- Animals store sugars as **glycogen**, which is similar to the branched form of starch. Branches (due to α-1,6-glycosidic linkages) occur in glycogen about once every 10 glucose monomers; amylopectin branches occur about once every 30 monomers (**Table 5.1**).

Cellulose: A Structural Polysaccharide in Plants

- Most modern organisms have a protective barrier called a **cell wall** outside their cell membranes, and polysaccharides are major cell-wall components.
- Plant cell walls are mostly made of **cellulose**, a polymer of β-glucose linked by β-1,4-glycosidic bonds (**Table 5.1**). This linkage flips every other glucose monomer upside down,

leading to a geometry that allows extensive hydrogen bonding between adjacent parallel strands of cellulose (**Table 5.1**).

Chitin: A Structural Polysaccharide in Animals

- **Chitin** is found in fungi and algae cell walls and is the primary component of insect and crustacean exoskeletons; although structurally similar to cellulose, chitin is composed of N-acetylglucosamine monomers.
- Chitin monomers, like those of cellulose, are joined by β-1,4-glycosidic linkages in such a way that every other monomer is flipped over, allowing hydrogen bonding between adjacent strands and creating a tough, stiff, protective sheet (**Table 5.1**).

Peptidoglycan: A Structural Polysaccharide in Bacteria

- Bacterial cell walls are primarily composed of **peptidoglycan**, which is made of alternating "M" and "G" monosaccharides joined by β-1,4 glycosidic linkages. Each M monomer is linked to a chain of amino acids, and peptide bonds link the amino acid chains of adjacent strands (**Table 5.1**).
- Some antibiotics kill bacteria by interfering with enzymes that catalyze the formation of the peptide-bond cross-links in peptidoglycan (**Box 5.2**).
- Cellulose, chitin, and peptidoglycan all consist of long parallel strands linked to each other. The resulting structure easily withstands forces that push or pull it, and it functions well as a protective sheet.

Polysaccharides and Chemical Evolution

- Although polysaccharides are important to organisms today, they probably had a minor, if any, role in the origin of life.
- Glycosidic linkages appear to form only in the presence of enzymes, so polysaccharides probably did not form in the prebiotic soup.
- The first life-forms are thought to have been RNA based, but no known ribozyme catalyzes the formation of glycosidic linkages.
- Monosaccharides have fewer functional groups than amino acids do, and polysacchar-

ides have fairly simple secondary structures; that is, they lack the structural and chemical complexity required for catalysis.

5.3   What Do Carbohydrates Do?

- Carbohydrates are important building blocks in the synthesis of other molecules. Recall that ribose and deoxyribose are important nucleotide subunits; and other sugars furnish "carbon skeletons" for the synthesis of many other important molecules, including amino acids.
- Carbohydrates also function to indicate cell identity, store chemical energy, and—as you learned earlier—they form fibrous structural materials that protect cells.

The Role of Carbohydrates in Cell Identity

- Although polysaccharides are unable to *store* information in a way that can be replicated, they do *display* information on the outer surface of cells. **Glycoproteins**—proteins with covalent bonds to carbohydrates—project outward from cell membranes and are essential in cell-cell recognition and cell-cell signaling.
- In multicellular organisms, each cell type has different glycoproteins. Among unicellular organisms, each species displays a different glycoprotein cell coating (**Figure 5.5**).
- Monosaccharide structural diversity makes possible the incredible diversity of glycoproteins necessary to indicate cell identity.

The Role of Carbohydrates in Energy Production and Storage

- Recall that in reduction-oxidation, or redox, reactions, the atom that gains an electron is reduced and the atom that loses an electron is oxidized. In chemical evolution, the reduction of carbon was the key step leading to the creation of complex organic molecules.
- Carbohydrates are good sources of chemical energy because they are good electron donors.

Carbohydrates as Electron Donors

- Today, most sugars are produced via photosynthesis (CO_2 + H_2O + sunlight $\downarrow$ CH_2O +

O_2), a key process that transforms the energy of sunlight into C–H bond chemical energy.
- Carbohydrates have more free energy than CO_2 does, because carbon is reduced during photosynthesis (electrons are closer to the carbon in CH_2O than in CO_2) and the C–H bond electrons are shared more equally and held less tightly than in C–O bonds (**Figure 5.6**).
- Carbohydrates are electron donors in redox reactions that produce ATP, an easily usable form of chemical energy. The oxidation of sugars can be written as CH_2O + O_2 + ADP $\downarrow$ CO_2 + H_2O + ATP. The free energy in ATP drives endergonic reactions and performs all kinds of cell work.
- As CO_2 molecules are reduced (gain electrons), they also gain protons (H^+). The resulting compounds contain many C–H bonds (**Figure 5.7**). The fatty acids have even more C–H bonds, and thus more free energy, than do carbohydrates; both molecules are important sources of chemical energy for cells.

How Do Carbohydrates Store Energy?

- Starch and glycogen are convenient glucose stores because enzymes easily catalyze reactions that release glucose subunits, which can then be used as electron donors in the redox reactions that produce ATP.
- **Phosphorylase** catalyzes the hydrolysis of α-glycosidic linkages in glycogen, and **amylase** catalyzes the hydrolysis of these linkages in starch.

Carbohydrates as Structural Molecules

- The structural polysaccharides—cellulose, chitin, and peptidoglycan (a modified polysaccharide)—all form long strands with bonds between adjacent strands. These strands may then be organized into fibers or layered in sheets to give cells and organisms great strength and elasticity (**Figure 5.8**).
- The β-1,4-glycosidic linkages of these structural molecules, unlike the α-glycosidic linkages discussed earlier, are difficult to hydrolyze. Very few enzymes have active sites that can accommodate their geometry. Molecules with these bonds are thus poor energy sources.

for most organisms, but great structural molecules because they are so difficult to break down. Yet again, structure explains function.

- Structural polysaccharides are essential components of all cell walls. Because polysaccharides were not formed during chemical evolution, the first cell must not have had a cell wall.

B. CROSS-CUTTING THEMES

Looking Back—
Concepts from Earlier Chapters
Chemical Evolution—Chapters 2–4
The start of evolution, environment of ancient Earth, and basic chemistry needed to understand the chemical reactions that cause chemical evolution are described in **Chapter 2**. **Chapters 3 and 4** discuss chemical evolution with respect to proteins and nucleic acids.

Structure-Function Correlation—Chapters 3 and 4
You should be familiar with the ubiquitous correlation between structure and function from the discussion in **Chapters 3** and **4** of this relationship for proteins and nucleic acids. In **Chapter 5** you learn about how the molecular structure of carbohydrates relates to their function.

Looking Forward—
Concepts in Later Chapters
Sugars—Chapters 9, 10, 36, and 43
Chapter 9 describes how organisms break down sugars to release and use the energy contained in their C–H bonds; **Chapter 10** describes how some organisms can use the energy in sunlight to make sugars (photosynthesis). **Chapters 36** and **43** describe the importance of sugars to plants and animals at the organismal level.

Glycoproteins—Chapter 49
Chapter 49 describes the important role that glycoproteins and proteoglycans have in forming a physical shield for a membrane, helping to prevent the invasion of microbial pathogens.

C. DIFFICULT TOPICS

Much of this chapter relates to important structural and functional differences between α- and β-1,4-glycosidic bonds. But the *best* way to get a feel for the structural differences is to make models of each bond type joining two sets of glucose molecules. If you do not have the time and/or materials, draw your own pictures of these molecules (starch, glycogen, cellulose, chitin, peptidoglycan) to improve your understanding of their structural differences. Once you understand the structural differences of the two bond types, make a table summarizing the major structural and functional differences of polysaccharides made from monomers that are joined by these two types of bonds.

	α-1,4-glycosidic bonds	χ-1,4-glycosidic bonds
Bond structure		
Polysaccharides containing this bond and the monomers thus linked		
Cell function		
Relationship between cell function and bond structure?		

D. ASSESSING WHAT YOU'VE LEARNED

(1) Testing Your Knowledge
1. Which functional group(s) is/are found in carbohydrates?
   a. phosphate
   b. carbonyl
   c. hydroxyl
   d. a and b
   e. b and c

2. Which of the following is *not* true of glucose and galactose?
   a. They are optical isomers.
   b. They are pentoses.
   c. Their formula is $C_6H_{12}O_6$.
   d. They are monosaccharides.

3. In α-glucose, the C1 hydroxyl group is: (Refer to Figure 5.3 if necessary.)
   a. gone
   b. turned into a carbonyl group
   c. on the same side of the ring as the C6
   d. on the opposite side of the ring from C6

4. Which of the following statements regarding sugars is correct?
   a. Amino acids undergo condensation reactions to form polysaccharides.
   b. Monosaccharides undergo condensation reactions to form polysaccharides.
   c. Monosaccharides undergo hydrolysis to form polysaccharides.
   d. Glucose is the monomer for all polysaccharides.

5. Which of the following is true of glycosidic linkages?
   a. They can form between hydroxyl groups of any two molecules.
   b. They can form between any two hydroxyl groups on any two carbohydrates.
   c. There is only one type of glycosidic linkages.
   d. They are a type of hydrogen bond.

For questions 6 through 8, Identify the cell-wall polysaccharide for each type of organism using the following lettered choices:
   a. amylopectin   d. glycogen
   b. cellulose   e. peptidoglycan
   c. chitin

6. Bacteria ____
7. Fungi   ____
8. Plants   ____

9. The monomer in chitin is:
   a. α-glucose
   b. β-glucose
   c. amylose

   d. *N*-acetylglucosamine

10. Researchers conclude that polysaccharides played little or no role in the origin of life because:
    a. Ribozymes cannot catalyze the formation of glycosidic linkages.
    b. Polysaccharide monomers are unable to form complementary base pairs.
    c. Polysaccharides cannot catalyze reactions.
    d. All of the above are correct.

11. Carbohydrate cell functions include all of the following *except*:
    a. cell protection
    b. storage of chemical energy
    c. transport of molecules
    d. cell-cell recognition

12. Proteins covalently bonded to carbohydrates are called:
    a. glycoproteins
    b. amyloproteins
    c. cephalosporins
    d. saccharoamides

13. Which of the following is true of redox reactions?
    a. By definition, they involve the transfer of protons (H^+).
    b. All redox reactions involve an electron donor and an electron acceptor.
    c. They occur only in the presence of enzymes.
    d. They occurred only on ancient Earth.

14. In carbohydrates, electrons are:
    a. closer to the carbon atom than they were in CO_2
    b. farther from the carbon atom than they were in CO_2
    c. shared less evenly in C–H bonds than in C–O bonds
    d. more stable in C–H bonds than they are in the C–O bonds of CO_2

15. Fats have more free energy than carbohydrates because:
    a. They are made during photosynthesis.
    b. They have more kinetic energy.
    c. They have more C–H bonds.
    d. They have more C–O bonds.

16. One thing that cellulose, chitin, and peptid-oglycan all have in common is:
    a. hydrogen bonds between adjacent strands
    b. covalent bonds between adjacent strands
    c. α-1,4-glycosidic linkages
    d. β-1,4-glycosidic linkages

(2) Integrating Your Knowledge

(a) What is the difference between an aldose and a ketose?

(b) Identify two similarities and two differences between starch and glycogen.

(c) Why are polysaccharides good at cell identification but poor at information storage?

(d) Do monosaccharides directly provide the energy to do cell work? If not, what does?

(e) Why are only a few enzymes capable of hydrolyzing β-1,4-glycosidic linkages?

CHAPTER 5—ANSWER KEY

D. Assessing What You've Learned

(1) Testing Your Knowledge
1. e; 2. b; 3. d; 4. b; 5. b; 6. e; 7. c; 8. b; 9. d; 10. d; 11. c; 12. a; 13. b; 14. a; 15. c; 16. d

(2) Integrating Your Knowledge

(a) Aldoses have the carbonyl group at the end of the carbon chain, but ketoses have the carbonyl group in the middle of the carbon chain.

(b) Similarities: Both made of glucose monomers, both have α-1,4-glycosidic linkages, both are used to store chemical energy. Differences: Starch is made by plants, glycogen by animals; glycogen is more branched than starch (more α-1,6-glycosidic bonds).

(c) Polysaccharides do not form a template that can be copied and passed on to the daughter cells, so they are a poor molecule for information storage. But polysaccharides are structurally diverse enough to allow each cell type and species to display different polysaccharides.

(d) No, monosaccharides are electron donors in the redox reactions that produce ATP. The ATP directly provides the energy for most cell work.

(e) The structure of the β-1,4-glycosidic linkage does not fit into an enzyme active site very easily, making these linkages very difficult to break.

6

Lipids, Membranes, and the First Cells

A. KEY BIOLOGICAL CONCEPTS

- The **cell membrane** or **plasma membrane** is a layer of molecules, mostly lipids, that surround a cell, separating the life inside the cell from the external environment.
- Development of a plasma membrane around a self-replicator created the first cell. Eventually, cells became able to control their internal environment by keeping harmful compounds out and allowing useful compounds in, greatly increasing the efficiency of chemical reactions performed in the cell.

6.1   Lipids

- Experiments simulating the chemical and energetic conditions of ancient Earth produce several types of lipid. Furthermore, high-magnification pictures (using an **electron microscope, Box 6.1**) of lipid-water mixtures show that some lipids spontaneously form cell-like compartments called vesicles (**Figure 6.2**).

What Is a Lipid?

- Molecules that contain only carbon and hydrogen atoms are called **hydrocarbons**; these molecules are nonpolar and consequently hydrophobic, because electrons are evenly shared in C–H bonds.
- **Lipids** have a significant hydrocarbon component; thus they are nonpolar, hydrophobic organic compounds that do not dissolve in water.

- A **fatty acid** is a hydrocarbon chain bonded to a carboxyl (COOH) group. Fatty acids and isoprene hydrocarbons are the key building blocks for the synthesis of lipids in organisms (**Figure 6.3**).

A Look at Three Types of Lipids Found in Cells (Figure 6.4)

- Amino acids, nucleotides, and sugars are defined by their chemical structure; lipids are defined by their solubility. Thus, lipid structure varies widely.
- **Steroids** (e.g., cholesterol) have a four-ring structure made from isoprene subunits.
- **Phospholipids** consist of a three-carbon glycerol linked to a phosphate group PO_4^{2-} and two isoprene chains (Archaea) or two fatty acids (Bacteria and Eukarya). The phosphate group may also bind to another small organic molecule.
- A **fat**, also known as a *triacylglycerol* or a *triglyceride*, is a glycerol linked to three fatty acids.
- Condensation reactions between glycerol hydroxyl groups and fatty-acid carboxyl groups form **ester linkages** joining the two subunits. Although this linkage is similar to those formed between other macromolecule subunits, phospholipids and fats are not polymers of repeating units (**Figure 6.5**).
- Lipids have many functions in organisms, the most important being the formation of plasma membranes.

The Structures of Membrane Lipids

- Phospholipids and cholesterol are **amphipathic** they contain both a polar, hydrophilic region and a nonpolar, hydrophobic region (**Figure 6.6**). The polar region interacts with water, but the nonpolar end does not, and this "dual sympathy" allows these lipids to form membranes.

6.2   Phospholipid Bilayers

- Phospholipids spontaneously (with no energy input) form **micelles** (tiny droplets) or **lipid bilayers** (two layers) in water so that the polar phosphate "heads" face outward toward water and the hydrophobic fatty-acid "tails" face inward toward other tails (**Figure 6.7**).

Artificial Membranes as an Experimental System

- Lipid bilayers break and form small spherical compartments (vesicles) when shaken. Water is located both outside and inside these spherical bilayers. Because vesicles form spontaneously, any accumulation of phospholipids on ancient Earth would have led to the creation of membrane-bound vesicles.
- Lipid bilayers also form the basic structure of cell membranes.
- **Permeability**, the ease with which substances cross a given barrier, is crucial to plasma membrane function because differentiation between an internal and external environment occurs when certain molecules or ions cross a selectively permeable membrane more easily than others.
- Hypotheses about membrane permeability are tested using artificially formed vesicles called **liposomes** and **planar bilayers**—artificial membranes across a hole in a wall between two aqueous solutions (**Figure 6.8**).
- Researchers investigate membrane permeability by varying either the solute added to one side of the bilayer or the bilayer composition. They then measure the molecule or ion concentrations on both sides of the artificial membrane to determine how fast various solutes cross bilayers under different conditions.

Selective Permeability of Lipid Bilayers

- Experiments with liposomes and planar bilayers show that small, nonpolar molecules move across phospholipid bilayers quickly, but charged substances cross slowly if at all. Thus phospholipids bilayers have **selective permeability**; some molecules cross these bilayers more easily than others (**Figure 6.9**).
- Large or charged compounds cannot pass through lipid bilayers, because they are more stable in an aqueous solution than among the nonpolar, hydrophobic phospholipid tails.
- Water crosses membranes quite rapidly, even though it is polar, because it is such a small molecule.

Does the Type of Lipid in a Membrane Affect Its Permeability?

- Membrane permeability is affected by hydrocarbon chain length and the number of double bonds.
- A hydrocarbon chain with at least one double bond is **unsaturated** because it does not contain the maximum number of hydrogen atoms. A hydrocarbon chain without double bonds is **saturated**; these fats contain more C–H bonds and therefore have more chemical energy than unsaturated fats.
- Double bonds prevent rotation between the bonded carbons, thereby causing a bend or "kink" in the hydrocarbon chain that prevents the close packing of hydrocarbon tails and reduces hydrophobic interactions (**Figure 6.10**). Longer tails, on the other hand, have more hydrophobic interactions than do shorter tails.
- Consequently, bilayers with unsaturated hydrocarbons and/or shorter hydrocarbon tails are more fluid and permeable than those with saturated hydrocarbons and/or longer tails.
- Differences in hydrocarbon fluidity are easily observed in any kitchen: highly saturated fats are solid at room temperature (e.g., butter); those with extremely long hydrocarbon tails form stiff waxes, whereas unsaturated fats are liquid oils at room temperature (**Figure 6.12**).
- Cholesterol also affects membrane permeability (**Figure 6.11**); it fills spaces between

phospholipids, increasing hydrophobic interactions and reducing bilayer permeability.

Why Does Temperature Affect the Fluidity and Permeability of Membranes?

- Experiments with tagged phospholipids show their lateral movement through membranes (**Figure 6.14**). Both temperature and structure (length and saturation) affect how fast phospholipids travel within a membrane (fluidity).
- Cell membranes are fluid at room temperature because the bilayer molecules move relatively fast (in one second, phospholipids can travel the length of a bacterial cell. At lower temperatures the molecules are seen to move more slowly, and the hydrophobic tails pack more tightly.
- Decreased fluidity causes decreased permeability (**Figure 6.13**).

6.3   Why Molecules Move Across Lipid Bilayers: Diffusion and Osmosis

- Dissolved molecules and ions (**solutes**) are in constant random motion due to their thermal energy.
- Molecules or ions that are placed on one side of a phospholipid bilayer generally move toward the other side (**diffusion**). They bump into each other more often in areas of high concentration, and this tends to move them toward regions of low concentration (down their **concentration gradient**).
- Diffusion is spontaneous because it causes an increase in entropy ($+ES$; see **Chapter 2**). Equilibrium (no *net* movement) is achieved when molecules or ions are randomly distributed throughout the solution.
- Diffusion of water is called **osmosis**, which occurs only when solutions of different concentrations are separated by a membrane that is permeable to water but not to the solutes (**Figure 6.16**). Water then moves spontaneously across the membranes toward the solution with the higher solute concentration (i.e., lower water concentration).
- Osmosis can shrink or swell (and even burst) membrane-bound vesicles and cells (**Figure 6.17**).

- If a solution outside a membrane has a higher concentration of solutes than the interior does, and if the solutes are unable to diffuse through the membrane, water will diffuse out of the vesicle (osmosis) and make the vesicle shrivel. Solutions that shrink cells are **hypertonic.**
- **Hypotonic** solutions are less concentrated than the vesicle or cell interior. In this case, if the membrane is not permeable to the solute, water will flow into the vesicle, which in turn will swell and perhaps burst.
- A solution with the same concentration as the inside of a vesicle or cell is **isotonic** and does not affect membrane shape.
- Osmosis and diffusion reduce the differences between solutions on two sides of a phospholipid bilayer.
- Liposomes in the prebiotic soup might have contained self-replicators, but they probably did not provide a significantly different internal environment. But these simple cell-like vesicles could have grown due to the addition of new lipids, and they could have divided when shaken by wave action.

6.4   Membrane Proteins

- Amphipathic proteins can be added into lipid bilayers. Recall that the R-groups of amino acids vary in polarity and thus in their affinity for water. Nonpolar amino acids are stable in membranes due to hydrophobic interactions with the hydrocarbon tails, and polar amino acids are stable next to the polar heads and in the surrounding water (**Figure 6.18**).
- In fact, although phospholipids provide the basic membrane structure, cell membranes contain as much protein as phospholipids do, as measured by mass.
- H. Davson and J. Danielli, in an early model of membrane structure, proposed that proteins coated a pure lipid bilayer. But the more recent **fluid-mosaic model** (S. J. Singer and G. Nicolson) suggests some proteins are inserted within the lipid bilayer, making the membrane a fluid, dynamic mosaic of phospholipids and proteins (**Figure 6.19**).

- **Freeze-fracture electron microscopy** with a scanning electron microscope strongly supports the fluid-mosaic model. The technique involves freezing and fracturing membranes, splitting them in order to view the middle and the surface of the bilayer (**Figure 6.20**).
- Membrane proteins look like mounds sticking out of half of the bilayer, leaving pits in the other half.
- Given the incredible structural diversity of proteins, it is no surprise to learn that they can form tunnels crossing a lipid bilayer, to create a channel or pore. Membrane-spanning proteins are called **integral membrane proteins** or **transmembrane proteins**.
- Other **peripheral membrane proteins** are located on the membrane surface, and they are often attached to the integral membrane proteins (**Figure 6.21**).
- The plasma membrane's interior and exterior surfaces differ greatly in their arrangement of peripheral and integral membrane proteins. These proteins participate in many cell functions, but this chapter focuses on how the integral membrane transports ions and molecules across the plasma membrane.

Systems for Studying Membrane Proteins

- To separate integral membrane proteins from cell membranes, researchers treated the membranes with small, amphipathic molecules called **detergents** (**Figure 6.22**).
- Detergents insert themselves into membranes, and their hydrophobic tails interact with the phospholipid tails, breaking apart the bilayer. Detergents also surround hydrophobic parts of membrane proteins, forming water-soluble detergent-protein complexes.
- The now soluble membrane proteins can be isolated and purified using **gel electrophoresis**, which separates the proteins moving through a gel based on their size and charge (**Box 4.1**).

How Do Membrane Proteins Affect Ions and Molecules?

- Channels, transporters, and pumps all affect membrane permeability.

Facilitated Diffusion via Channel Proteins

- Gramicidin is a membrane peptide that is produced by one type of bacteria to kill other bacteria (this makes it an **antibiotic**). Cells treated with gramicidin lose lots of ions.
- Ions move from areas of like charge to areas of unlike charge as well as from areas of high concentration to areas of low concentration—they move in response to an **electrochemical gradient** (**Figure 6.23**).
- Researchers measured charge flow (electrical current) across membranes with and without gramicidin to test the effect of this peptide on ion flow. In the presence of an electrochemical gradient, current flowed across planar bilayers with gramicidin but not across those lacking gramicidin (**Figure 6.24**).
- Gramicidin makes lipid bilayers permeable to small, positively charged ions (cations) by forming a tunnel—an **ion channel**—through a membrane. Protons (H^+) pass through gramicidin more easily than other larger cations.
- Gramicidin's amino acid sequence and tertiary structure reveal that the molecule forms a hole lined with hydrophilic R-groups. Hydrophobic R-groups on the exterior face of the tunnel interact with the membrane hydrocarbon tails (**Figure 6.25**).
- Membrane channels vary in their structures and functions. **Aquaporins** increase the rate at which water crosses membranes, and **gated channels** open only when bound to a specific molecule or in the presence of a certain electrical charge.
- Although channels allow membranes to exert some control over trans-membrane flow of molecules and ions, these substances always diffuse through channels down their electrochemical gradients. This process is also called **facilitated diffusion** or **passive transport**, since no energy is expended by the cell.
- Passive transport also occurs via ion carriers called **ionophores**, which bind to an ion on one side of a membrane, diffuse across, and then release the ion on the other side (**Figure 6.26**).
- Passive transport, whether by way of channel or ionophore, always occurs down an electrochemical gradient, thus decreasing charge and

concentration differences between the cell's exterior and interior.

Facilitated Diffusion via Transport Proteins

- Glucose is a subunit for important macromolecules and a great energy source, but lipid bilayers are only moderately permeable to glucose.
- Researchers obtain intact cell membranes by bursting red blood cells in a hypotonic solution to form red blood cell "ghosts" (**Figure 6.27**). Ghosts are more permeable to glucose than pure phospholipid bilayers are.
- Researchers isolated a glucose carrier—or **transport protein**—named GLUT-1, which increases membrane permeability to glucose. Glucose enters liposomes with GLUT-1 proteins at the same high rate observed in cell membranes.
- Only right-handed glucose, the form used by cells, is transported by GLUT-1. Researchers think glucose binds to GLUT-1 on the cell's exterior surface, causing a conformational change (just like enzymes change shape during substrate binding) that transports glucose to the cell interior (see **Figure 6.28**).
- Like channels and ionophores, transporters *reduce* differences between the inside and the outside of a cell or vesicle. Facilitated diffusion by these proteins occurs only down an electrochemical gradient.

Active Transport by Pumps

- Cells extant today can also transport molecules or ions *against* their electrochemical gradient; this process requires energy and is called **active transport**.
- Adenosine triphosphate (**ATP**) provides the energy in a form that can easily be used by the protein machine. The **sodium-potassium pump**, Na^+/K^+-ATPase, was the first pump discovered and characterized. This pump uses ATP to transport Na^+ and K^+ and acts like an enzyme in many ways.
- Three Na^+ ions bind to Na^+/K^+-ATPase on the inside of a cell. When ATP donates a phosphate to this Na^+/K^+-ATPase, it then changes shape, releasing the Na^+ ions outside the cell. The ion pump can now bind two K^+, which

are in turn released inside the cell when the phosphate group drops off (**Figure 6.29**).
- This sodium-potassium pump creates a chemical and electrical gradient across the membrane (remember that it returns only two positive K^+ charges for every three Na^+ charges removed). Other pumps are specialized for transport of other ions or molecules.
- Cells use active transport to create an internal environment that significantly differs from the environment outside the cell.

B. CROSS-CUTTING THEMES

Looking Back—
Concepts from Earlier Chapters
The Cell Theory—Chapter 1
Chapter 1 introduced the cell theory as one of the unifying themes of biology: all life is made of cells. In Chapter 6, you learn about the membranes that create a barrier between an organism and the external world and define the cell.

Entropy—Chapter 2
Chemical reactions can proceed spontaneously only when entropy, a measure of disorder, increases and/or potential energy decreases. Now you learn that entropy also determines the movement of molecules and ions across membranes in diffusion and osmosis.

Proteins—Chapter 3
Membrane channels, ionophores, transporters, and pumps are all proteins. You may want to refer to **Chapter 3** to review protein structure in order to better understand how proteins insert into membranes and bind ions or molecules.

Looking Forward—
Concepts in Later Chapters
Cells—Unit 2 (Chapters 7–11), Chapter 22, and other chapters
Because cells are the fundamental units of life, they reappear throughout your text; but the next unit discusses the basics of cell structure, metabolism, and replication (division). Later on, you will learn how cells differentiate to form different tissue types in multicellular organisms.

**Membranes and Membrane Proteins—
Chapters 7–10, 37, 38, 42, 45–47, and 49**
Membranes and their proteins determine what substances can cross the membrane to enter the cell. They thus have a critical role in maintaining homeostasis and transmitting signals to the cell. You'll learn more about these processes in the units on how plants and animals work. In **Chapters 7–10**, you will learn about internal cell membranes and their functions.

C. DIFFICULT TOPICS

One good way to review and test your comprehension of **diffusion**, **osmosis**, **fluidity**, **concentration gradients**, and **electrochemical gradients** is to draw figures like that in **Figure 6.8c** for different experiments you could do. Then explain your expected results based on information given to you in this chapter.

If you put hydrophobic molecules like CO_2 on one side of a lipid bilayer, what would happen? How about if you put glucose on one side? What could you add to the membrane to increase glucose's permeability? In general, how could you modify the composition of the lipid bilayer to make it more or less permeable? What happens when the membrane is impermeable to a solute on one side (but permeable to water)? What happens if you add more water to one side? Ions carry a charge, and opposite charges attract; this is why ions respond to electrical gradients as well as to chemical gradients. What variation on this technique can researchers use to measure membrane permeability to ions?

Answers to all these questions are in the text, but doing your own thought experiments will not only help you to master the information in this chapter but also help you to better understand how researchers test hypotheses about membranes.

D. ASSESSING WHAT YOU'VE LEARNED

(1) Testing Your Knowledge

1. Any organic compound present in organisms that doesn't dissolve in water is a:
   a. fat
   b. lipid
   c. phospholipid
   d. isoprene
2. Any molecule composed of only hydrogen and carbon atoms is a(n):
   a. amino acid
   b. lipid
   c. hydrocarbon
   d. sugar
3. Why are fatty acids called *acids*?
   a. Because they have lots of hydrogens.
   b. Because they form ester bonds with glycerol.
   c. Because the carboxyl group tends to lose a proton in aqueous solution.
   d. Because they like solutions that have a low pH.
4. When phospholipids are placed in water, they spontaneously form:
   a. micelles
   b. lipid bilayers
   c. both A and B
   d. neither A nor B
5. _____ are artificial membrane-bound vesicles.
   a. Phospholipids
   b. Micelles
   c. Lipids
   d. Liposomes
6. Which type of molecules most easily move across a membrane?
   a. hydrophobic molecules, like N_2
   b. large polar molecules, like glucose
   c. ions, like chloride (Cl^-)
   d. small, uncharged polar molecules, like water
7. Why is the cell membrane less permeable to glucose than to glycerol?
   a. Glucose is larger than glycerol.
   b. Glucose contains oxygen; glycerol does not.
   c. Glycerol is nonpolar; glucose is polar.
   d. Glucose is a charged molecule; glycerol is not.
8. _____ fatty acids and _____ hydrocarbon chains increase membrane permeability.
   a. saturated; long
   b. saturated; short
   c. unsaturated; long
   d. unsaturated; short

9. At room temperature, a cell membrane has the consistency of olive oil. Cooling the membrane _____ its fluidity and _____ its permeability.
   a. increases; decreases
   b. decreases; increases
   c. decreases; decreases
   d. increases; increases

10. Spontaneous movement of any molecule from an area of high concentration to an area of low concentration is:
   a. osmosis
   b. diffusion
   c. equilibrium
   d. active transport

11. Distilled water is _____ relative to a solution containing dissolved salts.
   a. hypertonic
   b. hypotonic
   c. isotonic
   d. none of the above

12. A red blood cell placed in distilled water:
   a. remains unchanged
   b. loses water and shrinks
   c. gains water and expands

13. Which of the following techniques are used to visualize proteins in membranes?
   a. freeze and fracture the membrane
   b. scanning electron microscopy
   c. both a and b
   d. gel electrophoresis

14. Which statement best describes the fluid-mosaic model of membrane structure?
   a. The phospholipid bilayer is coated on both sides with hydrophilic proteins.
   b. The phospholipid bilayer contains diverse proteins, including some embedded amphipathic proteins that span the bilayer.
   c. Two layers of proteins are interspersed with phospholipids.
   d. The phospholipids bilayer has only a few proteins associated with it.

15. Detergents are:
   a. amphipathic
   b. hydrophilic
   c. hydrophobic
   d. osmotic

16. Gel electrophoresis moves molecules through a gel medium placed in an electric field. Gel electrophoresis separates molecules based on their:
   a. charge
   b. size
   c. both a and b
   d. neither a nor b

17. The membrane protein gramicidin is used as an antibiotic since it causes cells to lose ions. Gramicidin is an example of a(n):
   a. pump
   b. transport protein
   c. channel protein
   d. gated channel

18. The glucose transporter protein GLUT-1 transports:
   a. all sugars
   b. glucose and other six-carbon sugars
   c. both isomers of glucose (right-handed and left-handed)
   d. the right-handed optical isomer of glucose

19. A protein binds to an ion, diffuses across a membrane, and releases it on the other side. No ATP or other chemical energy is used. This protein is an:
   a. aquaporin
   b. gated channel
   c. ionophore
   d. ion pump

20. Which of the following is *not* true of the sodium-potassium pump?
   a. It becomes phosphorylated by ATP.
   b. It transports sodium into the cell and potassium out of the cell.
   c. It creates an electrochemical gradient across the cell membrane.
   d. It is similar to an enzyme in many ways.

(2)   Integrating Your Knowledge

(a) Look at the structure of fats in **Figure 6.4**. Why don't they form membranes?

(b) Look at labels on butter and cooking oil; compare amounts of saturated and unsatur-ated fats. Use this information to explain why oil is fluid at room temperature but butter is not. Why is butter softer at room temperature than in the refrigerator?

(c) Use your knowledge of how the concentra-tion of solutions outside cells affects cell shape to explain why salting meat is an effective way to preserve it (and keep bac-teria from rotting it).

(d) Pour some oil into a pan or bowl of water. What happens? Does the oil (a lipid) mix with the water? Now add some dishwasher detergent and stir. What happens? See if the detergent has phosphate in it; many do, although phosphates can damage the envir-onment (many states regulate phosphate amounts in detergents).

(e) In this chapter you are told that cholesterol is a steroid. You may have heard that some steroids function as hormones. Based on what you read about steroids in this chap-ter, predict whether steroid hormones can cross membranes easily on their own, or whether they require a transporter to cross the cell membrane.

(f) Compare the permeability scale in **Figure 6.9** with the pH scale shown in **Figure 2.28**. Why is it useful to show these scales as powers of ten?

(g) Gramicidin is more permeable to H^+ than to K^+ since H^+ is much smaller. Draw a graph as in **Figure 6.24**, showing different curves for currents produced by movement of these two ions at various ion concentrations.

(h) Cells often break down glucose as a source of energy. How does this alter the concen-tration gradient for glucose? How does this explain why many cells can rely on trans-porters like GLUT-1 to bring glucose into the cell, when the direction of transport is determined by the electrochemical gradient?

(2) Integrating Your Knowledge

(a) Because fats are not amphipathic.

(b) Oil has more unsaturated fats than butter does. Oil is fluid at room temperature since double bonds in its unsaturated fats make kinks in the hydrocarbon chains and dis-rupt hydrophobic interactions. Butter is softer at room temperature than in the re-frigerator because molecules move slower at lower temperatures, causing hydrocar-bon tails to pack more tightly.

(c) As water leaves their cells due to osmosis, any bacteria reaching the salted meat shriv-el up and die. The salted meat is hyper-tonic.

(d) The oil floats on top of the water and does not mix with the water. When the deter-gent is mixed in, the oil splits up into smal-ler and smaller droplets.

(e) The four-ring hydrocarbon structure of steroids is hydrophobic. Thus, lipid bilay-ers are permeable to steroids.

(f) It is useful to show these scales as powers of ten, because a wide range of permeabil-ities or concentrations can easily be shown in the same figure.

(g) The curves should parallel each other, with the curve for H^+ above the curve for K^+.

(h) As cells use glucose, the concentration of glucose in a cell decreases. This maintains a lower concentration of glucose in the cell than in, for example, the blood. Glucose moves into the cell down its concentration gradient via transporters like GLUT-1.

CHAPTER 6 ANSWER KEY

D. Assessing What You've Learned

(1) Testing Your Knowledge

1. b; 2. c; 3. c; 4. c; 5. d; 6. a; 7. a; 8. d; 9. c; 10. b; 11. b; 12. c; 13. c; 14. b; 15. a; 16. c; 17. c; 18. d; 19. c; 20. b

7

Inside the Cell

A. KEY BIOLOGICAL CONCEPTS

- The cell is the fundamental unit of life; most organisms are unicellular and, even in large multicellular organisms, complex processes originate in the cell. Let's now explore cell structures and their functions.

7.1   What's Inside the Cell?

- Because Bacteria and Archaea are morphologically similar, we first discuss the simpler prokaryotic cells, then the more structurally complex, nucleus-containing eukaryotes.

Prokaryotic Cells

- The prokaryotic **plasma membrane** surrounds one compartment containing the **cytoplasm**, a highly concentrated solution containing all the molecules that make up the inside of the cell (**Figure 7.1**).
- The cytoplasm is typically hypertonic to the outside environment, so water tends to enter the cell by osmosis, causing it to expand. This osmotic pressure is resisted by a tough **cell wall** which, in many prokaryotic species, consists of a carbohydrate-protein complex called peptidoglycan (**Figure 7.2**).
- Cell walls protect cells and give them shape and structure. Outside the cell wall, many bacteria have yet another protective layer consisting of **glycolipids**—lipids with polysaccharides attached.
- The **chromosome**, a large DNA molecule associated with a few proteins, is the biggest cytoplastmic structure. Most prokaryotic species have one circular chromosome in the **nucleoid** region of the cell.
- DNA's nitrogenous base sequence carries the organism's genetic information. A **gene** is a section of DNA containing directions to build one polypeptide or RNA molecule.
- Because the chromosome may be hundreds of times longer than the cell, it is "supercoiled" (**Figure 7.3**).
- Prokaryotes often also contain small supercoiled DNA pieces called **plasmids**. Plasmids are independent mini-chromosomes that usually carry genes needed only under rare environmental conditions. Plasmids can sometimes be passed between cells—even between the cells of different species.
- All prokaryotic cells also contain ribosomes for protein synthesis. Ribosomes have a large and a small subunit and contain both RNA and protein molecules (**Figure 7.4**).
- Some prokaryotes have tail-like **flagella** on their cell surface; these flagella spin around to move the cell.
- Some prokaryotes have membrane-bound storage containers or extensive internal membranes for photosynthesis (**Figure 7.5**). One species even has a simple **organelle**: a membrane-bound cytoplasmic compartment with enzymes specialized for a specific function.
- The inside of prokaryotic cells is supported by a **cytoskeleton** of protein filaments.

- Larger eukaryotic cells have more highly developed cytoskeletons and internal membrane compartments.

Eukaryotic Cells

- The relatively large size of eukaryotes may have evolved as an adaptation for cell predation. However, large size makes it difficult for molecules to diffuse across the entire cell. This problem is partially solved by breaking up the large cell volume into several smaller membrane-bound organelles (**Figure 7.6**).
- The compartmentalization of eukaryotic cells increases chemical reaction efficiency by separating incompatible chemical reactions and grouping together enzymes and substrates.

The Nucleus

- All eukaryotes have a large nucleus surrounded by a double-membrane **nuclear envelope**, supported by **nuclear lamina** proteins that help organize the linear chromosomes.
- Eukaryotic DNA spooled on histone proteins forms **chromatin**, existing as compact supercoiled **heterochromatin** or as long unwound strands of **euchromatin**.
- Ribosome synthesis occurs in the **nucleolus** (**Figure 7.7**).

Ribosomes

- Eukaryotic cytoplasm includes everything inside the plasma membrane except for the nucleus. Many ribosomes are found in the **cytosol**—the fluid part of the cytoplasm.
- As in prokaryotes, eukaryotic ribosomes are made of RNA and protein and have large and small subunits that join to form a molecular machine for protein synthesis (**Figure 7.8**).

Rough Endoplasmic Reticulum

- The rough endoplasmic reticulum (rough ER) is a network of membrane-bound tubes and sacs. It connects to the outer membrane of the nuclear envelope and is studded with ribosomes. These ribosomes make proteins destined for the plasma membrane, the lysosome organelle, or cell secretion.
- Proteins made by rough ER ribosomes move inside the rough ER as they are made and are then folded and otherwise modified in the rough ER **lumen** (**Figure 7.9**).

Golgi Apparatus

- The **Golgi apparatus** is formed by a series of flat membrane sacs (**cisternae**) stacked together. The Golgi's *cis* side ("this side") faces the rough ER and receives products from it; the *trans* side ("across side") faces the plasma membrane and sends finished products to the cell surface in small vesicles (**Figure 7.10**).

Smooth Endoplasmic Reticulum

- Endoplasmic reticulum without ribosomes is called the **smooth endoplasmic reticulum** or **smooth ER**. Fatty acid and phospholipid synthesis, as well as breakdown of hydrophobic toxins, occurs here. The smooth ER may also store calcium ions (Ca^{2+}) that act as cell signals (**Figure 7.11**).
- The smooth and rough ER, together with the Golgi apparatus and the lysosomes, make up the **endomembrane system**. In each case, differences in organelle structure relate to its function.

Peroxisomes

- **Peroxisomes** are single-membrane globular organelles that independently grow and divide (**Figure 7.12**).
- Inside the peroxisomes, oxidation reactions occur, producing hydrogen peroxide (H_2O_2) as a by-product. The enzyme catalase quickly degrades the dangerously reactive H_2O_2 to harmless water and oxygen, so that the cell is not harmed.
- Different types of peroxisomes specialize in different types of oxidation reactions. For example, in plant leaves **glyoxisomes** convert a photosynthetic product into a sugar used to provide energy for the cell.

Lysosomes

- **Lysosomes** vary in size and shape, but are always single-membrane-bound centers for cell storage and/or waste processing (**Figure 7.13**).
- Animal lysosomes have an acid interior (pH 5.0) and contain digestive enzymes called

acid hydrolases that work best at a low pH. These enzymes break macromolecules into their monomer subunits.

- Membrane-surrounded packages sent to lysosomes originate when the animal's cell membrane engulfs a smaller cell or food particle (**phagocytosis**) or when a cell surrounds damaged organelles with a membrane so that they can be digested and recycled in the lysosome (**autophagy**).
- Lysosomes also receive materials via **receptor-mediated endocytosis**; macromolecules outside the cell bind to protein receptors, causing the membrane to fold inward and pinch off, forming a vesicle called an **early endosome**. This membrane-bound vesicle receives digestive enzymes from the Golgi and has its pH lowered as it matures into a **late endosome**, which may eventually turn into a lysosome.
- After macromolecules are hydrolyzed, their subunits are transported out of the lysosome by proteins in the organelle's membrane. These amino acids, sugars, and so on are then used to make new macromolecules.
- Phagocytosis and receptor-mediated endocytosis are two methods by which the cell membrane can pinch off a vesicle to bring outside material into the cell—a process called **endocytosis**. A third type of endocytosis, called **pinocytosis**, brings fluid into the cell. Pinocytotic vesicles do not go to lysosomes.
- Plants and Fungi have very large lysosomes called **vacuoles** that function primarily as water and/or for ion storage and help the cell maintain its normal volume. Sometimes these vacuoles store other molecules (proteins, toxins, pigments) or contain digestive enzymes.

Mitochondria

- **Mitochondria** have two membranes; the inner one is folded into saclike **cristae** that contain the **mitochondrial matrix** (**Figure 7.16**). Most enzymes involved in making ATP (the cell's primary source of chemical energy) are found either in the inner mitochondrial membrane or in the matrix solution.
- ATP production is a mitochondrion's core function.

- Mitochondria grow and divide independently of cell division; they make their own ribosomes and contain their own DNA. Mitochondrial DNA is supercoiled into a small, circular chromosome similar to those found in bacteria.

Chloroplasts

- Most plant and algal cells have **chloroplasts** that, like mitochondria, grow and divide independently, have a double membrane, and contain **chloroplast DNA** in a circular chromosome. Via photosyntheis, chloroplasts convert light energy to chemical energy.
- Inside the chloroplast, hundreds of flat vesicles called **thylakoids** are stacked into **grana**. Everything required for photosynthesis is located in the thylakoid membranes or outside the thylakoids in the chloroplast **stroma** (**Figure 7.17**).

Cytoskeleton

- Eukaryotes have a complex interconnected system of protein fibers that form the **cytoskeleton**. The cytoskeleton gives the cell shape; it also aids cell movement and the transport of materials within the cell.

The Cell Wall

- Fungi, algae, and plants have an outer **cell wall** typically made of carbohydrate fibers in a stiff matrix of polysaccharides and proteins (**Figure 7.18**). This cell wall protects the cell. Some plants produce yet another protective barrier of **lignin**, a tough molecule that makes wood when combined with cellulose.

How Does Cell Structure Correlate with Function?

- An organelle's membrane and enzymes correlate closely with its function (**Table 7.1**).
- Similarly, cell structure (e.g., the type, size, and number of organelles) correlates with cell function; cells that need a lot of ATP have many mitochondria, but cells that export proteins have an extensive rough ER and Golgi apparatus (**Figure 7.19**).

The Dynamic Cell

- Microscopes reveal the basic size, shape, and structure of cells, and differential centrifugation (**Box 7.1**) allows the isolation and chemical analysis of cell components.
- High-magnification microscopes show only "snapshots" of fixed cells, and chemical analyses break up cell parts; but live cells are dynamic, living things with interacting parts and constantly moving molecules.
- Cells take in substances, synthesize molecules, get rid of wastes, and reproduce. Every second, each of your cells hydrolyzes about 10 million ATP, and your enzymes catalyze up to 25,000 or more reactions.
- This rest of this chapter focuses on the dynamic cell by discussing (1) how molecules move in and out of the nucleus and around the cell in an ordered way, (2) how proteins move from ribosomes to their end destination, and (3) how cytoskeletal proteins facilitate movement of cell cargo and the cell itself.

7.2 The Nuclear Envelope: Transport Into and Out of the Nucleus

- The nucleus is a highly organized information center for the cell. It is enclosed by a double membrane called the nuclear envelope, which directly connects to the endoplasmic reticulum; the space between the nuclear envelope's two phospholipid bilayers is continuous with the space inside the ER.
- In the nucleus, the nuclear lamina anchors each chromosome and maintains the overall shape and structure of the nucleus. One easily visible structure is the nucleolus, where ribosomes are produced.
- Molecules shuttle to and from the nucleus through thousands of **nuclear pores** that connect the nucleus with the cytoplasm. More than 50 proteins form the **nuclear pore complex** that extends through both nuclear membranes (**Figure 7.22**).
- Initial research on the nuclear pore complex involved injecting gold particles into cells and determining the location of the gold after various time intervals using electron microscopy. After 2 minutes, some gold was associated with nuclear pores; after 10 minutes, many particles had entered the nucleus.
- Ribonucleic acids are exported out into the cell, whereas nucleotide triphosphates and the proteins involved in RNA synthesis must enter the nucleus; but DNA itself remains inside. The nuclear pore complex selects which particles can enter or exit the nucleus (**Figure 7.23**).

How Are Molecules Imported into the Nucleus?

- **Virus** proteins must enter their host cell's nucleus in order to reproduce themselves; but certain amino acid changes in virus proteins prevent the proteins from passing the nuclear pore.
- The preceding observation led to the hypothesis that proteins destined for the nucleus have a molecular address tag or zip code allowing them to enter the nucleus. Viral proteins have access to the nucleus if they contain this **nuclear localization signal** (**NLS**).
- The NLS was studied using a protein called *nucleoplasmin*, which helps assemble chromosomes. This protein has a globular core surrounded by several long "tails." Radiolabeled nucleoplasmin quickly enters the nucleus after injection into a cell's cytoplasm.
- Researchers separated the nucleoplasmin core from its tails using proteases and radioactively labeled the two components. When injected into cells, isolated tail fragments were transported to the nucleus; but core fragments remained in the cytoplasm. Thus, the NLS must be part of the tail region (**Figure 7.24**).
- Further analysis demonstrated that the NLS, the molecular zip code, is a 17-amino-acid sequence. Other proteins destined for the nucleus have similar localization signals.
- Nuclear localization signals cause proteins to bind to **importins** in the cytoplasm. The protein/importin complex passes through the nuclear pore complex with the help of another protein, called Ran, which undergoes a GTP-binding conformational change causing the release of the protein cargo (**Figure 7.25**).

- Large molecules enter the nucleus only if they have a NLS, and their passage requires chemical energy.

How Are Molecules Exported from the Nucleus?

- Shuttle proteins called **exportins** interact with Ran and GTP to help molecules containing a nuclear export signal exit the nucleus. This process is also highly regulated and requires energy input.

7.3 The Endomembrane System: Manufacturing and Shipping Proteins

- Sequences similar to the NLS target proteins made by ribosomes in the cytosol to the mitochondria or chloroplasts. Other proteins are made in the rough ER and travel through the endomembrane system, where they are further processed before reaching their final destination.
- The secretory pathway hypothesis holds that secreted proteins are synthesized and processed in a multistep pathway involving the rough ER and Golgi apparatus, the dominant organelles in secretory cells.
- Palade and co-workers did **pulse-chase experiments** to determine the movements of newly synthesized proteins within secretory cells. A cell is first given a "pulse"—a high concentration of a labeled molecule for a short time—followed by the "chase"—lots of unlabeled molecules for a longer time period.
- Palade's group used pancreatic cells grown in culture (in vitro) because these cells secrete enzymes and other molecules. First they gave the cells a 3-minute pulse of radiolabeled leucine. This amino acid became part of proteins produced during that 3-minute period, labeling those proteins.
- When cells were killed and examined by electron microscopy immediately after the pulse, all labeled protein was inside the rough ER. After a 7-minute chase, the protein was on the Golgi's *cis* side facing the rough ER. After 17 minutes, the protein was inside the Golgi; and after 80 minutes, the proteins were either in secretory granules on the far side of the Golgi or outside the cell (**Figure 7.27**).

- These experiments showed that secreted proteins are made in the rough ER, travel to the Golgi, and are transported out of the cell in secretory granules.

Entering the Endomembrane System: The Signal Hypothesis

- Gunter Blobel and colleagues proposed the **signal hypothesis**: that as ribosomes make secretory proteins, the first few amino acids act like an address tag that directs the growing polypeptide to enter the ER.
- This hypothesis is supported by the observation that secreted proteins made in a test tube are 20 amino acids longer than are the proteins secreted by cells. Blobel proposed that a 20-amino-acid **ER signal sequence** must direct the proteins to the ER and then be removed before the protein is secreted.
- Blobel's group identified the ER signal sequence. This sequence binds to an RNA-protein complex in the cytosol called the **signal recognition particle (SRP)**. The SRP then binds to a receptor in the ER membrane, the ribosome completes protein synthesis, and the signal sequence is removed (**Figure 7.28**).
- The new protein enters either the rough ER lumen (if it will be shipped to an organelle or secreted) or the rough ER membrane (if it is a membrane protein) and is folded with the help of chaperone proteins.
- Within the ER lumen, carbohydrate side chains are added to proteins (**glycosylation**), producing **glycoproteins** (**Figure 7.29**).

Getting from the ER to the Golgi

- Palade and colleagues proposed that proteins are transported from the ER to the Golgi in vesicles because the area between these two organelles contains many small membrane-bound vesicles. Further tests using differential centrifugation confirmed that these vesicles contained labeled proteins.
- Vesicles containing protein products must bud off the ER, travel to the Golgi, and deposit their contents inside as they fuse with the Golgi's *cis* face.

What Happens Inside the Golgi Apparatus?

- Before exiting on the far side of the Golgi, protein products pass through sequential **cisternae** (Golgi compartments), each containing enzymes for specific glycosylation reactions.
- Current research investigates whether molecules move among cisternae within vesicles or whether new cisternae are created at the *cis* face as older cisternae mature and move toward the *trans* face of the Golgi (vesicle transport vs. cisternal maturation hypotheses, **Figure 7.30**.

How Are Products Shipped from the Golgi?

- Additional molecular zip codes direct Golgi proteins to their final destinations (**Figure 7.31**); proteins with a mannose-6-phosphate side chain bind to receptors in vesicles bound for lysosomes.
- Some proteins are sent to the cell surface in vesicles that fuse with the plasma membrane, releasing their contents to the exterior of the cell (**exocytosis**).

7.4 The Dynamic Cytoskeleton

- Cytoplasm contains a complex network of fibers called the cytoskeleton, which provides structural support for cells. The cytoskeleton is dynamic; it changes in order to alter the cell's shape, transport materials in the cell, or move the cell itself.
- The three types of cytoskeletal elements are actin filaments, intermediate filaments, and microtubules (**Figure 7.32**).

Actin Filaments

- The smallest-diameter cytoskeletal elements are **actin filaments**, also known as **microfilaments**. Actin, a globular protein, polymerizes to form two long, fibrous strands twisted around each other.
- Actin monomers and the filaments made from them are asymmetrical, or polar. Thus, actin filaments grow faster at the plus end than at the minus end because polymerization occurs fastest there.

- Actin filaments are grouped into long fibers or dense networks, which are most often found just inside the plasma membrane. These filaments, linked by other proteins, help define the cell's shape (**Figure 7.33a**).
- Microfilaments constantly grow and shrink as actin subunits are added or removed. Actin also often interacts with **myosin**. A conformational change of myosin's head region uses the chemical energy released by ATP hydrolysis to slide bound actin filaments (**Figure 7.34a**).
- Actin-myosin filament movement causes (1) **cell crawling** when filaments push the plasma membrane into **pseudopodia** and then pull the cell as filaments contract, (2) **cytokinesis** when a ring of filaments contracts to pinch an animal cell in two during cell division, and (3) **cytoplasmic streaming** in plants and fungi (**Figure 7.34b).**

Intermediate Filaments

- **Intermediate filaments** are defined by size instead of by composition (**Figure 7.33b**); all are nonpolar cytoskeletal elements that function as structural support for the cell. For example, keratins in our skin and gut help these cells resist pressure and abrasion, and secreted keratins form our hair and nails.
- **Nuclear lamins** in the nuclear lamina are another type of intermediate filament. Some of these filaments reach through the cytoplasm to the cell membrane and form a flexible skeleton, which contributes to cell shape and holds the nucleus in place.

Microtubules

- Microtubules are large, hollow tubes made of α- and β-tubulin **dimers** (two joined monomers).
- Like actin, microtubules grow or shrink as subunits are added or removed. Microtubules are also polar; the two ends are different, and the molecule is more likely to grow from one end than from the other.
- Microtubules contribute to cell structure and help move the cell itself and materials inside the cell. In animals and fungi, microtubules

radiate from two **centrosomes** during cell division. Each centrosome has a **centriole** that helps organize the microtubules (**Figure 7.35**).

- Plants and other eukaryotes have a **microtubule organizing center** similar to the centrosome in function.

Studying Vesicle Transport

- The giant axon is an extremely large neuron that runs the length of a squid's body. If the squid is disturbed, the axon carries an electrical signal to a muscle band where the neuron releases neurotransmitters that cause the muscle to contract, propelling the animal away from danger.
- Because the giant axon is so large, it is easy to work with. Ronald Vale and co-workers were able to squeeze the cytoplasm out of the cell and observe vesicle transport of the neurotransmitters in this cell-free system.

Microtubules Act as "Railroad Tracks"

- Researchers used video-enhanced microscopy techniques to view cytoskeletal filaments. Neurotransmitter vesicles move along a track that was later determined to be made of microtubules (**Figure 7.35**). Vesicle transport required energy because movement stopped if ATP ran out.
- The movement of vesicles from the ER to the Golgi apparatus also depends on microtubules because movement was abnormal following treatment with a drug that disrupts microtubules. How do these vesicles move through the cell along these microtubule tracks?

A Motor Protein Generates Motile Forces

- Vale and co-workers produced microtubules from purified α- and β-tubulin. They added ATP and transport vesicles isolated by differential centrifugation, but no transport occurred.
- Eventually they found a fraction that allowed vesicle movement. Further purification led to isolation of the **motor protein, kinesin**, which generates vesicle movement by converting the chemical energy of ATP into mechanical work.

- X-ray diffraction of kinesin shows three distinct regions: (1) a head with two globular pieces, each of which can bind to ATP and to a microtubule; (2) a tail that binds to a transport vesicle; and (3) a stalk that connects the head and tail regions (**Figure 7.37**). Different kinesins carry different types of vesicles.
- Researchers propose that the kinesin head region moves a step along the microtubule as it binds and hydrolyzes ATP.

Cilia and Flagella: Moving the Entire Cell

- Flagella are long projections that move cells. Bacterial flagella are made of flagellin and rotate like a propeller, but eukaryotic flagella are made of microtubules and wave back and forth.
- Eukaryotic flagella and cilia are very similar. Eukaryotes with flagella typically have only one or two, whereas those with cilia have many shorter structures (**Figure 7.38**). Many eukaryotes do not have either structure.

How Are Cilia and Flagella Constructed?

- Cilia and flagella have two central microtubules surrounded by nine microtubule doublets (of one complete and one incomplete microtubule each). This 9 +2 structure is called the **axoneme**. Spokes connect doublets to the central microtubules, molecular bridges connect doublets to one another, and each doublet has two "arms" projecting toward an adjacent doublet (**Figure 7.39**).
- The axoneme attaches to the cell at a **basal body**—a structure derived from the centrioles that is important in axoneme growth.

A Motor Protein in the Axoneme

- Ian Gibbons, in the 1960s, isolated *Tetrahymena* axonemes by using a detergent to remove the plasma membrane and then performing differential centrifugation. These axonemes beat only in the presence of ATP, showing that cilia movement requires energy.
- Further experiments with this cell-free system showed that axonemes could not beat or use ATP when protein binding was disrupted. Electron microscopy showed that the "arms" between doublets were missing. These arms are a motor protein called **dynein**, which

changes shape when ATP is hydrolyzed so as to walk along microtubules.

- Cilia and flagella bend instead of elongating because the spokes and bridges constrain movement; this makes the microtubules slide past one another when the dynein arms on just one side of the axoneme walk (**Figure 7.40**).

B. CROSS-CUTTING THEMES

Looking Back—
Concepts from Earlier Chapters
Prokaryotes and Eukaryotes—Chapter 1
Chapter 1 first introduced you to the differences between prokaryotes and eukaryotes and showed how these groups are related on the tree of life. In this chapter you also learned about cell theory—that cells are the basic unit of all life, and all cells come from preexisting cells.

The Cell Membrane—Chapter 6
In the last chapter you learned about the structure and function of the cell or plasma membrane. In this chapter you learn more about internal cell membranes—the endomembrane system. Remember that the structure (phospholipid bilayer with proteins) and function of these membranes is similar to that of the cell membrane.

Looking Forward—
Concepts in Later Chapters
Mitochondria—Chapter 9
You will learn much more about mitochondria structure and function when you read about cell respiration.

Chloroplasts—Chapter 10
You will learn much more about chloroplast structure and function when you read about photosynthesis.

Cell Division—Chapters 11 and 12
During cell division, whole chromosomes need to be moved around the cell. Much of what you learned about intracellular transport in this chapter is thus relevant to your understanding of mitosis and meiosis.

DNA and Genes—Chapters 13–18
Because DNA contains all the information for protein synthesis, these chapters on DNA and genes will aid your understanding of cell function. For example, this chapter mentions that ribosomes are the site of protein synthesis, but the process itself is described in **Chapter 16**.

Prokaryotes—Chapter 27
Most of this chapter describes eukaryotic cell structure and function, but much of life is prokaryotic. You will learn more about these cells in **Chapter 27**.

Microfilaments—Chapters 11, 12, 27, and 43
In Chapter 7, you learned how microfilaments help give shape to the cell. These actin filaments also play active roles in cell division (**Chapters 11** and **12**), pseudopodia motility (**Chapter 28**), and muscle contraction (**Chapter 46**).

C. DIFFICULT TOPICS

Chapter 7 has introduced you to many new terms and research techniques. Consider making note cards for yourself for all these new terms and techniques. If you have trouble memorizing, be sure to give yourself plenty of time before the test. It will be easier if you can study a few note cards at a time. You might even try taping up note cards around your room—above the light switch, on your computer, on your sock drawer, on your mirror. This way you can think about these new terms a little bit each day as you go through your regular daily routine.

Once you know some terms, put up new cards to learn. You can also try explaining vesicle transport to family or friends not familiar with biology. If you can explain it clearly (and correctly) to them, you know you have mastered the material.

D. ASSESSING WHAT YOU'VE LEARNED

(1) Testing Your Knowledge

1. What is the fundamental difference between prokaryotic and eukaryotic cells?
   a. Prokaryotic cells do not have a plasma membrane.

b. Prokaryotic cells do not have a nucleus.

c. Prokaryotic cells do not have ribosomes.

d. Eukaryotic cells never have a cell wall.

2. Which of the following is true of DNA in bacterial cells?

a. All bacterial DNA is part of a chromosome in the cell's nucleoid region.

b. Bacterial DNA usually forms linear chromosomes.

c. Bacterial DNA does not contain genes.

d. Bacterial DNA is supercoiled; full-length uncoiled bacterial DNA cannot fit in the cell.

3. How does compartmentalization into organelles help eukaryotes solve a problem associated with the large size of eukaryotic cells?

a. Compartmentalization reduces diffusion distance and concentrates molecules needed for specific reactions.

b. Compartmentalization makes it possible for eukaryotes to eat bacterial cells.

c. Compartmentalization lets eukaryotes become multicellular.

d. Compartmentalization allows eukaryotes to perform chemical reactions like photosynthesis that are not present in bacteria.

4. The nuclear envelope is:

a. a single membrane surrounding the nucleus

b. continuous with the Golgi apparatus

c. supported by the nuclear lamina

d. made of chromatin

For questions 5–10, match the given functions with the correct organelle:

a. nucleus   e. peroxisomes
b. rough ER   f. lysosomes
c. smooth ER   g. mitochondria
d. Golgi apparatus h. chloroplasts

5. ____ processing and packaging proteins for secretion

6. ____ isolation of dangerous oxidation reactions

7. ____ lipid-processing and phospholipid manufacture

8. ____ ATP production

9. ____ synthesis of plasma membrane proteins

10. ____ hydrolysis of macromolecules

11. Which of the following represents a fundamental difference between plant and animal cells?

a. Plant cells do not have centrioles.

b. Plant cells do not have a Golgi apparatus.

c. Plant cells do not have mitochondria.

d. Plant cells do not have a cell wall.

12. What would happen if an allosteric inhibitor prevented importins from binding to cytoplasmic proteins?

a. No large molecules would enter the nucleus.

b. Chromosomes would leak out of the nucleus.

c. Certain proteins would become more concentrated in the nucleus.

d. Some proteins normally found only in the cytoplasm would appear in the nucleus.

13. Which of these proteins is not likely to be transported into the nucleus of living cells?

a. mRNA for protein synthesis

b. DNA polymerase, the enzyme that catalyzes DNA synthesis

c. RNA polymerase, the enzyme that catalyzes RNA synthesis

d. histones

14. Hundreds of proteins may be specifically localized to the nucleus of cells. What is likely to be true regarding transport of these proteins into the nucleus?

a. There must be an equal number of nuclear pore complexes.

b. There must be hundreds of different receptors, one for each type of protein to be transported into the nucleus.

c. Many different proteins must be transported together as a complex into the nucleus, so only one receptor is required for the group.

d. Proteins specifically localized to the nucleus must have the same or similar nuclear localization signal, so a small number of receptor types are required for transport.

15. The glucocorticoid receptor is a protein found only in the cytoplasm until the hormone glucocorticoid enters the cell and binds to this receptor. The receptor then translocates to the nucleus, where it regulates the transcription of certain genes. If a specific sequence of the receptor is deleted from the protein, the receptor still binds glucocorticoid but does not enter the nucleus. Consider the following additional hypothetical experiments. If the mutant receptor is fused with the tail fragment of nucleoplasmin, the receptor enters the nucleus regardless of whether glucocorticoid is bound. Fusion of the core fragment of nucleoplasmin with the sequence deleted from the glucocorticoid receptor results in accumulation of the core fragment in the nucleus. Which conclusion can be drawn from these observations?
    a. The binding of glucocorticoid to the receptor must expose a nuclear localization signal.
    b. The glucocorticoid receptor does not have a nuclear localization sequence.
    c. Glucocorticoid is the nuclear localization signal for the receptor.
    d. The glucocorticoid receptor enters the nucleus by a mechanism other than nucleoplasmin.

16. What types of questions can a pulse-chase experiment address?
    a. What path does a protein follow as it moves through a cell?
    b. What is the structure of a protein in the cell?
    c. How does the concentration of a protein change?
    d. Items a and b are both correct, but not c.
    e. Items a and c are both correct, but not b.
    f. Items a, b, and c are all correct.

17. The signal hypothesis predicts that:
    a. Nuclear localization signal binds to signal recognition particles in cytosol.
    b. Proteins bound for the endomembrane system have a molecular zip code.

    c. Only proteins destined for the nucleus need a molecular zip code.
    d. Secreted proteins will be 20 amino acids shorter when synthesized in a test tube.

18. It's possible to fuse two different cells so a Golgi apparatus from each is present and remains separate in the same hybrid cell. When such experiments are conducted, proteins that started out in one Golgi apparatus frequently end up in both Golgi complexes. What observation about Golgi function does this result support?
    a. Proteins can be transported among Golgi compartments in vesicles.
    b. Proteins can enter and exit Golgi compartments directly and diffuse through the cytoplasm.
    c. The two Golgi apparatuses fuse.
    d. None of these choices are reasonable.

19. Microtubules, intermediate filaments, and microfilaments are made of _____, respectively.
    a. actin, keratins and other nonpolar filaments, α- and β-tubulin
    b. α- and β-tubulin, lamins, myosin
    c. actin, myosin, α- and β-tubulin
    d. α- and β-tubulin, keratins and other nonpolar filaments, actin

20. The mixing of purified microtubules with transport vesicles and ATP does not result in movement of the vesicles. Why?
    a. Actin is missing.
    b. Kinesin is missing.
    c. Glucose is missing.
    d. Intermediate filaments are missing.

21. Suppose the plasma membrane around a flagellum opens to reveal the axoneme inside. The radial spokes connecting the peripheral microtubule doublets to the central pair are then broken by chemical treatment. ATP is then added. What is the expected observation?
    a. Normal bending of the flagellum
    b. No movement, because the plasma membrane is not present.
    c. The axoneme will elongate.
    d. No movement, because the ability to use ATP is lost.

(2)   Integrating Your Knowledge

(a) List and describe three differences between eukaryotic and prokaryotic cells.

(b) Label the organelles in the diagrams on page 97 of this study guide.

(c) Why did researchers inject the radio-labeled nucleoplasmin tails and cores into different cells?

(d) What is the difference between smooth and rough ER? Why does it make sense that rough ER is more important than smooth ER is in the synthesis and distribution of proteins?

(e) Why is the chase an important step in these pulse-chase experiments?

(f) What two properties made the squid giant axon a useful system for studying intracellular transport?

(g) Cell transport can be likened to boxcars moving cargo along railroad tracks; what cell structures make up the tracks, boxcars, cargo, and fuel?

CHAPTER 7—ANSWER KEY

D.   Assessing What You've Learned

(1)   Testing Your Knowledge

1. b; 2. d; 3. a; 4. c; 5. d; 6. e; 7. c; 8. g; 9. b; 10. f; 11. a; 12. a; 13. a; 14. d; 15. a; 16. e; 17. b; 18. a; 19. d; 20. b; 21. c

(2)   Integrating Your Knowledge

(a) Eukaryotes have nuclei and organelles, and they are bigger than prokaryotes.

(b) For animal cell: (a) centrioles, (b) lysosome, (c) nuclear envelope, (d) nucleolus, (e) chromatin, (f) rough ER, (g) ribosomes, (h) peroxisome, (i) smooth ER, (j) Golgi apparatus, (k) mitochondrion, (l) cytoskeletal element, and (m) plasma membrane. For plant cell: (n) cell wall, (o) chloroplast, (p) nuclear envelope, (q) nucleolus, (r) chromatin, (s) rough ER, (t) ribosomes, (u) smooth ER, (v) Golgi apparatus, (w) vacuole, (x) peroxisome, (y) mitochondrion, (z) plasma membrane, and (aa) cytoskeletal element

(c) If radiolabeled tails and cores had been injected into the same cell, the researchers would not have been able to determine which had entered the nucleus by monitoring radiation—because the cores and tails themselves cannot be detected, only the radiation.

(d) Rough ER has ribosomes attached, and smooth ER does not. Because ribosomes are the site of protein synthesis, it makes sense that rough ER would be responsible for the synthesis and distribution of many cell proteins.

(e) Without the chase, radioactive proteins would have been found throughout the endomembrane system during the later time periods, and it would have been impossible to determine whether the radioactivity observed in the ER and Golgi was due only to new proteins or whether some previously made protein remained in the ER and Golgi.

(f) The squid giant axon is large enough to allow extrusion of the cytoplasm to make a cell-free system, and because it moves lots of neurotransmitter, it performs a great deal of intracellular transport.

(g) The tracks are the microtubules, the boxcars are the kinesins, the cargo is cell vesicles (or chromosomes during cell division), and the fuel is ATP.

(g) Radiolabeling, pulse-chase, green florescent protein, subcellular fractionation/ differential centrifugation, and cell-free extracts (e.g., the squid giant axon)

8

Cell-Cell Interactions

A. KEY BIOLOGICAL CONCEPTS

- Cells constantly interact with their environment. In multicellular organisms, the cells' external environment consists primarily of other cells. This chapter focuses on how cells communicate and cooperate with each other in multicellular organisms.

8.1   The Cell Surface

- The plasma membrane's phospholipid bilayer has *peripheral* proteins attached to its surface and *integral* proteins within it. Proteins crossing the entire membrane are called transmembrane proteins. Some of them attach to the cytoskeleton inside the cell and/or to materials outside the cell (**Figure 8.1**).

The Structure and Function of an Extracellular Layer

- Most cells secrete extracellular material that glues them to other cells, defines cell shape, and/or defends the cell from external threats.
- Almost all extracellular material has a "fiber composite" structure consisting of long cross-linked filaments surrounded by a stiff ground substance. The rods or filaments protect against stretching forces, while the ground substance protects against compression (**Figure 8.2**).
- Different groups of organisms use different molecules to form this extracellular fiber composite, and many extracellular layers are flexible as well as sturdy.

The Plant Cell Wall

- The extracellular material secreted by plant cells first forms a **primary cell wall** consisting of long strands of cellulose. This cross-linked polysaccharide, made by enzymes in the plasma membrane, is bundled into filaments called microfibrils, which form a criss-crossed network that becomes filled with a gelatinous polysaccharide ground substance.
- **Pectins** are common gelatinous polysaccharides found forming the cell-wall ground substance. Pectins are hydrophilic and keep the cell wall moist by attracting lots of water. These gelatinous carbohydrates are made in the rough endoplasmic reticulum (ER) and Golgi apparatus and are then secreted to the extracellular space.
- Plant cell shape is determined by the primary cell wall. Normally, the solute concentration is higher inside the cell than outside, so that water enters the cell by osmosis, filling up the cell volume and pushing the plasma membrane against the cell wall.
- **Turgor pressure** is the force with which the filled cell pushes against the cell wall. In young cells, enzymes called *expansins* break cross-links, allowing microfibrils to slide past one another. Turgor pressure then forces the cell wall to elongate, allowing cell growth.
- Mature plant cells may secrete a **secondary cell wall** inside the primary cell wall. The structure of this secondary cell wall varies widely depending on the cell's function.

- Cells that form wood have a secondary cell wall with lots of **lignin**, a complex substance that is especially stiff and strong.
- Plant cell walls are dynamic and may be reinforced or broken down according to the needs of the plant.

The Extracellular Matrix in Animals

- The fiber composite used by animal cells is called the **extracellular matrix** (**ECM**). Like plants, animals use gelatinous polysaccharides as a ground substance (made either by membrane proteins or in the rough ER). Animals, however, use protein fibers instead of polysaccharide filaments.
- **Collagen** is the most common ECM protein fiber (**Figure 8.3**). It and other protein fibers are synthesized in the rough ER, processed in the Golgi apparatus, and secreted via exocytosis. These proteins are more elastic than cellulose or lignin and form a flexible extracellular layer.
- In animals, the composition and amount of ECM varies widely among various cell types.
- Regardless of its exact composition, the ECM provides structural support for the cell.
- The ECM typically also helps cells stick together. Cytoskeletal actin filaments connect to transmembrane **integrins**, which bind to ECM **fibronectins**, which bind to collagen. These protein-protein attachments directly link the cytoskeletons within cells to the ECM (**Figure 8.4**).
- ECM breakdown can cause major problems (e.g., cancer metastasis and other diseases; see **Box 8.1**).

8.2   How Do Adjacent Cells Connect and Communicate?

- Cells, even unicellular organisms, constantly interact with other cells. For example, some bacteria in your mouth secrete a hard polysaccharide-rich substance that accumulates outside their cell walls. This **biofilm** attaches cells to the surface on which they live.
- **Dental plaque** is a biofilm encasing a multi-species group of oral bacteria (**Figure 8.5**).

Other biofilms are secreted as protection by a single-species group of bacteria.

- Unicellular organisms live together and communicate with each other, but cell-cell physical connections are the basis of **multicellularity**. In multicellular organisms, similar cells with similar functions form **tissues**, which may combine to form an **organ** specialized for one biological function.
- Multicellularity has evolved several times (**Figure 8.7**); and each multicellular group has its own type of cell-cell connections and communication. Many of these cell-connecting structures were revealed by transmission electron microscopy.

Cell-Cell Attachments

- Plant cell walls are glued together by gelatinous pectins and other molecules that form the **middle lamella** (**Figure 8.8**). When enzymes break down this layer, the surrounding cells separate.
- A layer of gelatinous polysaccharides also exists between cells in many animal tissues. Transmembrane proteins called integrins connect cell cytoskeletons to the ECM, and some tissues have additional proteins that more strongly connect neighboring cells.
- These stronger protein attachments are especially found in **epithelia**, the tissues covering external and internal body surfaces.

Tight Junctions

- **Tight junctions**, common in cells lining your stomach and intestines, are a type of cell-cell attachment whereby proteins in adjacent cell membranes line up and bind to one another, stitching the two cells together to form a watertight seal between the two plasma membranes (**Figure 8.9**).
- The exact structure and function of tight junctions depends on the type of epithelia in which they are found. Tight junctions can separate and re-form, allowing immune system cells to pass between epithelial cells.

Desmosomes

- Epithelial cells and some muscle are held together by **desmosomes** made of proteins that link the cytoskeletons of adjacent cells. These proteins bind to each other and to the proteins that anchor cytoskeletal intermediate filaments (**Figure 8.10**).

Selective Adhesion

- In the early 1900s, experiments by H. V. Wilson showed that when adult sponges are treated with chemicals that cause cells to separate from each other, the resulting mixed-up mass of cells reconnects with cells of the same tissue type when the chemicals are removed. This is known as **selective adhesion**.
- Cells from two different sponge species randomly mixed together after treatment with the same chemicals sort themselves into groups containing cells from only one species and tissue type (**Figure 8.11**). This suggests that cell-cell connections are species- and tissue-specific.

The Discovery of Cadherins

- Each major cell type has its own cell adhesion proteins. The **cadherins**, which include desmosome proteins, are a group of cell adhesion proteins found in plasma membranes that bind only to other cadherins of the same type, causing cells of the same tissue type to bind together.
- One experiment to identify cell-adhesion proteins started with the isolation of membrane proteins from a certain cell type. Pure preparations of each protein were then injected into rabbits, one at a time, causing the rabbit to produce **antibodies**—proteins that bind specifically to a part of the injected protein.
- Once researchers had a large collection of antibodies, they added one antibody at a time to mixtures of dissociated cells and observed whether the cells could reconnect normally. Antibodies that prevented cell-cell attachment must bind to adhesion proteins, blocking their normal binding function (**Figure 8.12**).
- The cell-type specificity of cadherins explains how animal cells selectively attach to other animal cells within the same individual.

Cell-Cell Gaps

- Direct connections between tissue cells allow cells to work together.

- Plant cells communicate via cell-wall gaps called **plasmodesmata**, where the plasma membranes and cytoplasms of two cells connect such that smooth ER actually runs between one cell and the next (**Figure 8.13a**).
- Like the nuclear pore complex, plasmodesmata also contain proteins that regulate protein traffic. Some proteins passing through plasmodesmata help coordinate the activity of neighboring cells.
- In most animal tissues, **gap junctions** connect adjacent cells (**Figure 8.13b**). These membrane holes are lined by specialized proteins that match up to form a channel between two cells. Small molecules flow through these channels and can help coordinate activities of the two connected cells.
- Plant and animal cells usually retain their own organelles and large macromolecules but share most or all of the small molecules found in their cytoplasms. Cell-cell gaps allow efficient communication among tissue cells so that the tissue can act as an integrated whole.

8.3 How Do Distant Cells Communicate?

- Chemical signals travel throughout animals and plants to target cells, carrying information from one tissue or organ to another. This intercellular signaling involves four steps: signal reception, signal processing, signal response, and signal deactivation.

Signal Reception

- Intercellular signals are typically small molecules, present in minute concentrations, that significantly affect target cell activity. These chemical messengers are called **hormones**, and they are **ligands** that bind to a specific site on a receptor molecule.
- The functions and chemical structure of plant and animal hormones vary widely (**Table 8.1**). However, all play a role in coordinating cell activity in response to information from outside or inside the body.
- Proteins that change their conformation or activity when a hormone binds to them are **signal receptors**.

- Steroid hormone receptors are located inside cells because steroids are lipid soluble and can easily diffuse through the plasma membrane, but most signal receptors are found in the plasma membrane.
- There are many types of receptors, each of which is found only in certain cell types.
- Receptors are dynamic and may increase or decrease in number over time. For example, receptor numbers tend to decline following prolonged hormonal stimulation.
- Receptors can be blocked. Beta-blockers and other drugs act by binding to and blocking certain receptors.
- In all cases, a change in receptor structure indicates that a signal has been received.

Signal Processing

- Steroid hormones directly initiate cell response; they enter a cell, bind to a receptor protein, and are transported to the nucleus as a hormone-receptor complex that directly alters gene expression (**Figure 8.15**).
- Hormones that cannot diffuse across the plasma membrane rely on a complex series of events called a **signal-transduction pathway**. **Signal transduction** refers to conversion of the hormone signal from its original extracellular chemical form to a new intracellular form.
- Signal transduction occurs at the plasma membrane and typically involves either G-proteins or receptor protein kinases.

G-Proteins

- **G-proteins** are peripheral membrane proteins located inside the cell. They are closely associated with transmembrane signal receptors and are named for their ability to bind guanosine triphosphate (GTP) and guanosine diphosphate (GDP).
- G-proteins are turned on, or *activated*, when they bind GTP; they automatically turn themselves off, or are *inactivated*, when they hydrolyze GTP, forming GDP.
- For example, the plant hormone gibberellic acid 1 binds to plasma membrane receptors of aleurone layer cells. The hormone receptor's conformational change activates a G-protein

by causing it to release GDP and bind GTP. The G-protein then splits into two parts, one of which activates a nearby enzyme in the plasma membrane (**Figure 8.16a**).
- G-protein-activated enzymes catalyze the production of small nonprotein signaling molecules called **second messengers**. These intracellular signals spread the hormone message throughout the cell.
- In the example given earlier, the G-protein activated by giberellic acid 1 activates enzymes that cause the release of calcium ions (Ca^{2+}) from the smooth ER. The same G-protein also activates an enzyme that catalyzes the synthesis of cGMP in the cytoplasm. Ca^{2+} and cGMP are both common second messengers.
- Other common second messengers include diacylglycerol (DAG), inositol triphosphate (IP_3), and cyclic adenosine monophosphate (cAMP). These second messengers are all small molecules that diffuse rapidly, spreading the signal throughout the cell.
- Cells can quickly produce large quantities of second-messenger molecules following each hormone-receptor binding event, so this signal transduction amplifies the original signal.

Receptor Tyrosine Kinases

- **Receptor tyrosine kinases** are transmembrane proteins that form dimers after binding with a hormone signal. These kinases then become activated by phosphorylating each other. This phosphorylated form next activates yet other enzymes by phosphorylating them (**Figure 8.16b**). These enzymes phosphorylate yet more enzymes and so on, creating a **phosphorylation cascade** that activates many enzymes.
- Both G-protein and receptor tyrosine kinase signal transduction converts an easily transmitted extracellular message into a greatly amplified intracellular message that carries information throughout the cell, inducing the cell response.

Signal Response

- Like steroid hormones, some second messengers and phosphorylation cascades cause changes in a cell's gene expression. In other cases, second messengers or phosphorylation cascades cause the activation or deactivation of a specific enzyme (or other protein) that critically alters target-cell activity.

Signal Deactivation

- Turning off cell signals is just as important as turning them on. For this reason, cells have automatic mechanisms for signal deactivation. This allows the cell to remain sensitive to future hormone signals.
- Activated G-proteins can hydrolyze GTP, producing GDP and a free phosphate. This reaction changes the G-protein's conformation, deactivating it so that it no longer activates the second-messenger-producing enzyme.
- Second messengers are then either degraded (for example, phosphodiesterase deactivation of cAMP and cGMP) or pumped back into storage (Ca^{2+}).
- Phosphorylation cascades are turned off by *phosphatases*, enzymes that remove phosphate groups from proteins. These phosphatases are always present in cells, ready to quickly shut off phosphorylation cascades as soon as hormone stimulation stops.
- The end result of cell sensitivity to hormonal signaling is an integrated whole-organism response to changing conditions both within and outside of the multicellular organism.

B.   CROSS-CUTTING THEMES

Looking Back—
Concepts from Earlier Chapters
Enzyme Function—Chapter 3
Chapter 3 described how enzymes function. Hormone-receptor interactions are similar to enzyme-substrate interactions; hormones bind at a specific site, causing a conformational change to the receptor, similar to the way in which substrates often cause a conformational change in the enzyme upon binding to the active site. **Chapter 8** also introduces you to some important groups of enzymes: kinases, phosphatases, and phosphodiesterases.

The Cell Membrane and
Its Associated Proteins—Chapter 6
Cadherins, gap junction proteins, hormone receptors, and G-proteins are just some of many membrane proteins mentioned in this chapter. Remember that the cell membrane is dynamic—these proteins are always moving around, and their motion ensures that activated G-proteins interact with the enzymes they activate.

Looking Forward—
Concepts in Later Chapters
Tissues and Organs—Chapters 35 and 41
Chapter 8 describes how cells physically connect to form tissues and organs. You will learn more about tissue and organ structure and function in **Chapters 35** and **41** (plant and animal form and function respectively).

Chemical Signals in Plants and
Animals—Chapters 39, 47, and others
You may want to come back to this chapter later when you read about specific hormones used as chemical signals in plants and animals.

C.   DIFFICULT TOPICS

Antibodies are an important research tool for the cell biologist. The experiments shown in Figure 8.12 are quite complex. If you did not completely understand these experiments after one reading, consider the following analogy.

You are told that antibodies are proteins that specifically bind to a section of another protein (animals make antibodies that bind to foreign proteins as part of their immune response). Pretend that the proteins isolated from a specific cell type and injected into the rabbit are like various-shaped objects that you place on a table for a five-year-old to use in making play-dough molds. After awhile, you have many molds of an AA battery, a Lego, some children's scissors, and various other objects. Let's pretend that these molds, when left out, eventually find and bind to the object they fit with.

For your first experiment (Experiment 1), you leave out the battery molds and discover that they bind to all the batteries in the house, causing your mini flashlights to go out, etc., having no effect on Lego attachment. For

Experiment 2, you put away the battery molds and take out the Lego molds. Now your mini-flashlights work, but the Legos won't connect anymore, because they are all covered with the molds. If you put the Lego molds away and take out the scissors molds (Experiment 3), the Legos and flashlights work, but kids in the house are no longer able to cut paper.

In Experiments 1 and 3, the molds/antibodies did not affect Lego attachment because the objects used to make the molds are not involved in that process. In Experiment 2, Lego attachment no longer works, because the mold/antibody blocks the attachment/binding site. Taking this analogy one step further, you are told that cadherin proteins are cell-type specific. This could be discovered by noting that the molds that block regular Legos in some houses have no effect on Duplo Legos found in other houses. Each house/cell type has its own attachment proteins that are blocked only by their specific molds/antibodies.

D.   ASSESSING WHAT YOU'VE LEARNED

(1)   Testing Your Knowledge

1. Extracellular material can function to: (pick all that apply)
   a. glue cells together
   b. actively transport nutrients into cell
   c. protect cells
   d. define cell shape
   e. catalyze chemical reactions

2. A fiber composite structure contains
   a. cross-linked rods or filaments and contractile elements
   b. a ground substance and cross-linked rods or filaments
   c. a frame structure and contractile elements
   d. a frame structure and a ground substance

3. The primary cell wall of plants is made of
   a. pectins and expansins
   b. cytoskeletal elements and peripheral proteins
   c. cellulose and gelatinous polysaccharides
   d. lignin and collagen

4. Cellulose is made by _____; and pectins are made _____.
   a. enzymes in the plasma membrane; in the rough ER and Golgi apparatus
   b. the rough ER and Golgi apparatus; in the smooth ER.
   c. in the smooth ER; by free ribosomes
   d. the Golgi apparatus; by enzymes in the plasma membrane

5. Turgor pressure occurs when:
   a. solute concentration is the same inside the cell and outside, so that water pushes on the cell wall
   b. solute concentration is higher inside the cell than outside, so that water leaves the cell by osmosis
   c. solute concentration is lower inside the cell than outside, so that water enters the cell by osmosis
   d. solute concentration is higher inside the cell than outside, so that water enters the cell by osmosis

6. Which of the following statements are true about the animal cell extracellular matrix (ECM)? Choose *all* that apply.
   a. Composition and amount of ECM varies widely among different cell types.
   b. Protein-protein attachments link the cell cytoskeleton to the ECM.
   c. The ECM helps maintain cell turgor pressure in animal cells.
   d. Collagen is the most common ECM protein fiber.

7. Only multicellular organisms:
   a. live together and communicate with each other
   b. have organelles
   c. have extracellular layers
   d. have long term cell-cell physical connections

For questions 8 to 10, use the following options to identify where each structure is found.
   a. between plant cell walls
   b. between the plant plasma membrane and cell wall
   c. between epithelial cells
   d. between all animal cells

8. ____ tight junctions

9. ____ middle lamella

10. ____ desmosomes

11. What evidence suggests that cells have selective adhesion?
    a. Electron micrographs show cell-cell attachments.
    b. Randomly mixed sponge cells sorted themselves by tissue and species.
    c. Chemical treatments are unable to separate same tissue-type sponge cells.
    d. Rabbits produce antibodies to injected membrane proteins.

12. Plant cells communicate via cell-wall gaps called _____.
    a. desmosomes
    b. tight junctions
    c. gap junctions
    d. plasmodesmata

13. Which of the following are shared by cells connected by gap junctions? Pick all that apply.
    a. ions
    b. RNA
    c. amino acids
    d. smooth ER
    e. nucleotides

14. All hormones are _____.
    a. steroids
    b. proteins
    c. ligands
    d. receptors

15. High blood glucose causes secretion of insulin. This hormone causes glucose to enter cells, where it can be converted to fat or glycogen. If someone has chronically high blood glucose and high insulin levels, what would likely happen to their insulin receptors over time?
    a. The number of receptors might decrease over time.
    b. More receptors would be synthesized.
    c. The receptors might become more sensitive to the hormone.
    d. Nothing would happen to their insulin receptors.

16. G-proteins are activated when _____.
    a. they bind to a hormone
    b. they bind to ATP
    c. they bind to GTP
    d. they bind to a second messenger

17. What happens after receptor tyrosine kinases bind with a hormone signal?
    a. They form dimers.
    b. They catalyze their own phosphorylation.
    c. They phosphorylate other enzymes.
    d. all of the above
    e. none of the above

18. How are receptor tyrosine kinase signals deactivated?
    a. Phosphodiesterases remove phosphate groups from activated proteins.
    b. Phosphatases remove phosphate groups from activated proteins.
    c. Membrane pumps return calcium ions to storage.
    d. cAMP and cGMP are converted to inactive AMP and GMP.

(2) Integrating Your Knowledge

(a) List similarities and differences between plant and animal extracellular material.
(b) Compare the functions of tight junctions and desmosomes.
(c) Why are steroid receptors located inside cells, whereas most other hormone receptors are located on the plasma membrane?
(d) Describe what happens during each of the four intercellular signaling steps for a hormonal response involving G-proteins.
(e) Why is signal deactivation important? What would happen if a signal was not deactivated?

CHAPTER 8—ANSWER KEY

D. Assessing What You've Learned

(1) Testing Your Knowledge
1. a, c, & d; 2. b; 3. c; 4. a; 5. d; 6. a, b, & d; 7. d; 8. c; 9. a; 10. c; 11. b; 12. d; 13. a, c, & e; 14. c; 15. a; 16. c; 17. d; 18. b

(2) Integrating Your Knowledge

(a) *Similarities*: Both use gelatinous polysaccharides as a ground substance, and both function as structural support. *Differences*: They have different cross-linked fibers (polysaccharides vs. collagen), and animal cells are more flexible.

(b) Tight junctions and desmosomes both serve as physical cell-cell connections. But tight junctions form a watertight seal between membranes (desmosomes do not), whereas desmosomes extensively link the cytoskeletons of adjacent cells.

(c) Steroid hormones are lipid soluble and can diffuse through the phospholipid bilayer, but most other hormones cannot easily cross the plasma membrane.

(d) *Signal reception*: A hormone binds to a receptor associated with a G-protein, causing a conformational change in the receptor. *Signal processing*: The hormone receptor's conformational change causes the G-protein to bind GTP; this activated G-protein in turn activates other enzymes that cause the production (or release) of second messengers. *Signal response*: The second messenger causes either a change in gene expression or the activation (or deactivation) of a specific enzyme or enzymes. *Signal deactivation*: The G-protein hydrolyzes its bound GTP, inactivating itself, and the second messenger is either degraded or pumped back into storage.

(e) Signal deactivation is important because if signals were not turned off, cells would be unable to respond the next time a signal is sent. Automatic signal deactivation allows cells to remain responsive.

9

Cellular Respiration and Fermentation

A. KEY BIOLOGICAL CONCEPTS

- Hydrolysis of **adenosine triphosphate (ATP)** provides the chemical energy that powers most cell work. ATP is the major energy currency of the cell.
- Energy to make ATP comes from oxidation of sugars and other reduced compounds. This energy is used to add a third phosphate to adenine diphosphate (ADP), making ATP.
- This chapter introduces you to **metabolism**— the chemical reactions that occur in cells.

9.1 An Overview of Cellular Respiration

- Carbohydrates and fats are highly reduced molecules containing a lot of chemical energy, but this energy is not immediately available to do work. First, it must be converted into ATP.
- ATP is continually made and hydrolyzed, so at any one time a cell contains only enough ATP for a few minutes of work. The chemical reactions that produce the ATP are among the most fundamental and universal of all cell processes; some of the enzymes involved originated 3.5 billion years ago.
- Glucose, made by photosynthesis, is a key intermediary in cell metabolism. Non-photosynthetic organisms obtain glucose by decomposing or eating photosynthetic species. Cells use glucose to build fats, carbohydrates, and other compounds; and cells recover glucose by breaking down these molecules (**Figure 9.1**).

- Metabolic pathways called cellular respiration and fermentation break down glucose and produce ATP to provide the cell with a constant source of chemical energy.

The Nature of Chemical Energy and Redox Reactions

- In cells, electrons are the most important source of chemical potential energy. Electron potential energy is a function of its position relative to other electrons and to positive charges in the nuclei of nearby atoms.
- Electrons in ATP have high potential energy because phosphate groups contain four negative charges (**Figure 9.2**). These negative charges repel each other. Less electrical repulsion occurs in the hydrolyzed products ADP and P_i. Thus ATP has more potential energy than ADP, and ATP hydrolysis releases energy (is exergonic; note that the entropy of the products is higher than that of ATP).
- Under standard conditions of temperature and pressure, 7.3 kilocalories (kcal) of energy is released per mole of ATP hydrolyzed. Cells use this energy to do cell work.
- Sometimes the phosphate group released during ATP hydrolysis is transferred to a protein. This reaction is also exergonic because there is less electrical repulsion between the phosphate and the protein than between the phosphate and the other two phosphates of ATP.

- When the two negative charges of a phosphate are added to a protein, the protein's electrons change configuration, typically causing a change in the protein's shape (**Figure 9.2c**). This movement in response to phosphorylation (or the addition of an entire ATP) is responsible for much cell work.

Using Redox Reactions to Produce ATP

- ATP production requires a lot of energy. This energy comes from reduction-oxidation (redox) reactions.
- When molecules are reduced, they often gain a proton (H^+) along with the electron. Thus, reduced compounds typically have many C–H bonds. These bonds have high potential energy because the electrons are not held very tightly.
- When molecules are oxidized, they often lose a proton with the electron. The resulting oxidized molecule has many C–O bonds. Oxygen holds electrons tightly, so its electrons have low potential energy.
- Glucose is highly reduced. Burning, an uncontrolled oxidation reaction, releases the glucose's energy as heat (kinetic energy): $C_6H_{12}O_6 + 6 O_2 \downarrow 6 CO_2 + 6 H_2O + $ energy. The carbon atoms of glucose are oxidized to form carbon dioxide, and the oxygen atoms in O_2 are reduced to form water. Electrons are transferred from glucose to oxygen, and protons follow.
- The complete oxidation of 1 mole of glucose releases 686 kcal of energy. In cells, glucose is slowly oxidized in a series of redox reactions that allow cells to use much of the released energy to make ATP.
- Cell respiration is a three-step process in many organisms: (1) glucose is broken down to pyruvate; (2) pyruvate is oxidized to CO_2; and (3) reduced compounds from steps 1 and 2 are oxidized to make ATP.

Processing Glucose: Glycolysis

- **Glycolysis,** a series of 10 chemical reactions that occur in the cytoplasm, is the first step in glucose oxidation. Glycolysis is an ancient pathway; almost all species use the same 10 chemical reactions.

- During glycolysis (**Figure 9.4a**), glucose is broken down into two 3-carbon molecules of pyruvate. The potential energy released is used to phosphorylate some ADP. Nicotinamide adenine dinucleotide (NAD^+) is also reduced to make NADH (**Figure 9.3**), an **electron carrier** that readily donates electrons to more oxidized molecules.

The Krebs Cycle

- In the presence of O_2, the pyruvate of most cells enters a sequence of reactions known as the Krebs cycle, where each pyruvate is oxidized to form three molecules of CO_2 (**Figure 9.4b**). In eukaryotes, Krebs cycle enzymes are found inside mitochondria.
- As each pyruvate is oxidized, some of the released energy is used (1) to reduce NAD^+ to NADH; (2) to reduce flavin adenine dinucleotide (FAD) to $FADH_2$ (another electron carrier); and (3) to phosphorylate ADP to make ATP.

Electron Transport

- The high potential energy electrons carried by NADH and $FADH_2$ participate in a series of redox reactions. These electrons are passed down a series of molecules in an **electron transport chain** located in the inner mitochondrial membrane of eukaryotes (and in the cell membrane of some prokaryotes).
- At each step, electrons lose potential energy, and the energy released is used to pump protons to the outside of the membrane, forming a strong electrochemical gradient. ATP synthase uses the proton gradient energy to phosphorylate ADP as protons cross back into the mitochondrial matrix.
- The final electron acceptor at the end of the electron transport chain is O_2, which gains protons along with electrons, forming water. This completes the oxidation of glucose.
- The combination of glycolysis, Krebs cycle, and electron transport chain is called **cellular respiration**. This term can refer to any process that produces ATP using a reduced compound as an electron donor, an electron transport chain, and an electron acceptor.

Methods of Producing ATP

- ATP synthase uses the proton gradient set up by the electron transport chain to power ATP production. This method of ATP production is linked to oxidation of NADH and $FADH_2$, so it's called **oxidative phosphorylation.**
- **Substrate-level phosphorylation** occurs when ATP is produced by the enzyme-catalyzed transfer of a phosphate group to ATP from an intermediate substrate.
- Throughout glycolysis and the Krebs cycle, the free energy of the compounds involved gradually decreases. Larger drops in free energy are associated with ATP synthesis; smaller drops are associated with the production of NADH and $FADH_2$ (**Figure 9.5**).
- In most organisms, if no oxygen is available to act as the electron acceptor, cell respiration cannot occur (a few rare organisms use other electron acceptors). Instead, **fermentation** reactions take place that allow glycolysis to continue (**Figure 9.6**). In eukaryotes, fermentation produces ethanol or lactic acid.

9.2 Glycolysis

- Hans and Edward Buchner discovered glycolysis by chance in the late 1890s. They added sucrose to yeast extracts and found that the sugar was broken down and fermented, producing alcohol. This discovery showed that cell metabolism could be studied in vitro (outside the cell).
- The Buchners and others soon found that the fermentation reactions lasted much longer when inorganic phosphate was added, that fructose bisphosphate (with two phosphate groups) was one intermediate, and that all intermediate compounds are phosphorylates; only glucose and pyruvate lack phosphate.
- Glycolytic reactions stopped if the yeast extract was boiled. This suggested the involvement of enzymes, because enzymes are inactivated by heat. In fact, each step of glycolysis is catalyzed by a different enzyme.

A Closer Look at the Glycolytic Reactions

- Differential centrifugation techniques were used to verify that all glycolytic reactions oc-cur in the cytosol. **Figure 9.7** shows all 10 reactions of glycolysis.
- In the first reaction, ATP is hydrolyzed to phosphorylate glucose, forming glucose 6-phosphate. Glucose-6-phosphate is then rearranged, creating fructose-6-phosphate, and a second ATP is used up to make fructose-1,6-bisphosphate. These three steps are called the energy-investment phase of glycolysis.
- In the second half of glycolysis, the energy investment is more than paid off. The sixth reaction reduces 2 NAD^+ (one per three-carbon intermediate), the seventh reaction produces 2 ATP, and the last reaction produces another 2 ATP. The net yield is 2 NADH, 2 ATP, and 2 pyruvate per glucose.
- In glycolysis, ATP is produced by substrate-level phosphorylation, because enzymes catalyze the direct transfer of phosphates from a phosphorylated intermediate to ADP.

How Is Glycolysis Regulated?

- High levels of ATP inhibit the enzyme **phosphofructokinase**, which catalyzes the highly exergonic step 3 of glycolysis (Figure 9.7). The products of steps 1 and 2 can be used in other metabolic pathways, but the product of step 3 (fructose-1,6-bisphosphate) can be used only in glycolysis. Step 3 is thus a good point for regulation.
- Regulation of phosphofructokinase by ATP is unusual in that ATP is one of the substrates for this enzyme. Addition of a substrate typically increases the reaction rate. However, because ATP is also one of the end products of glycolysis, it makes sense that cells with lots of ATP do not need to make more.
- Inhibition of an enzyme by the product of a reaction sequence is called **feedback inhibition** (**Figure 9.8**). ATP inhibition of phosphofructokinase allows cells to stop glycolysis when ATP is plentiful and conserve glucose stores for another time.
- Phosphofructokinase has two binding sites for ATP: an active site, where ATP is hydrolyzed, and a **regulatory site** (**Figure 9.9**), where ATP acts as an allosteric regulator. Binding of ATP to the regulatory site changes

the enzyme's conformation in a way that significantly decreases enzyme activity.

9.3   The Krebs Cycle

- Biologists discovered eight small **carboxylic acids** (R-COOH) involved in redox reactions that produced CO_2 and appeared to be associated with glucose metabolism. Adding any of these molecules to cells caused an increase in cell respiration, but the added molecules were not used up.

- **Hans Krebs** realized that the carboxylic acid reactions could occur in a cycle tied to the breakdown of pyruvate (**Figure 9.11**). This hypothesis was supported by the observation that citrate (the cycle's start point) formed when oxaloacetate (the cycle's endpoint) and pyruvate were added to cells.

- This cycle of redox reactions is known as the Krebs cycle.

Converting Pyruvate to Acetyl CoA

- Experiments with radioactive carbon isotopes confirmed that the cycle occurred as Krebs had proposed.

- The reactions occur in cytoplasm of bacteria and archaea and mitochondria of eukaryotes.

- Mitochondria have two membranes. The inner membrane is folded into saclike **cristae**. Most Krebs cycle enzymes are in the **mitochondrial matrix**—the aqueous area inside the cristae (**Figure 9.12**).

- In eukaryotes, pyruvate made by glycolysis must be transported from the cytoplasm to the mitochondria. Pyruvate crosses the outer membrane through small pores and is actively transported (an energy-using process) into the mitochondrial matrix by an inner membrane protein called the pyruvate carrier.

- Inside the mitochondria, pyruvate reacts with **Coenzyme A (CoA)** to produce **acetyl CoA**. Coenzyme A is a common enzyme cofactor that transfers acetyl (–COCH₃) groups to substrates, and acetyl Co A is the compound that reacts with oxaloacetate to form citrate.

- This reaction between pyruvate and CoA is catalyzed by a three-enzyme complex called **pyruvate dehydrogenase**, which is found in the inner mitochondrial membrane. Pyruvate dehydrogenase requires five cofactors including the three B-complex vitamins.

- The reaction between pyruvate and CoA results in the reduction of NAD^+ to NADH and the oxidation of one of pyruvate's carbons to CO_2. The remaining 2 carbons form the acetyl group transferred to CoA. When acetyl CoA reacts with oxaloacetate to form citrate, CoA is released (**Figure 9.13**).

- This reaction between pyruvate and CoA links glycolysis and the Krebs cycle.

- In the Krebs cycle, the two remaining carbons from pyruvate are oxidized to CO_2 (**Figure 9.14**).

- The complete oxidation of one pyruvate produces 4 NADH (1 from reaction with CoA and 3 from Krebs cycle), 1 FADH₂, and 1 **guanosine triphosphate (GTP)**, which is easily converted to ATP.

How Is the Krebs Cycle Regulated?

- When ATP levels are high, pyruvate dehydrogenase is phosphorylated and catalytic activity is inhibited (due to a shape change). High concentrations of acetyl CoA and NADH also inhibit this enzyme complex by increasing the rate at which it is phosphorylated. This is another example of feedback inhibition.

- Feedback inhibition also occurs at two points in the Krebs cycle. In one case, NADH binds to the enzyme's active site as a competitive inhibitor; in the other case, ATP binds to an allosteric site.

- Pyruvate dehydrogenase is activated by NAD^+, CoA, and adenosine monophosphate (AMP). Thus the Krebs cycle is slowed when ATP and NADH are abundant and speeded up when ATP and NADH supplies are low (high AMP and NAD^+; see **Figure 9.15**).

- Natural selection has favored organisms that regulate enzyme activity, thereby conserving glucose, pyruvate, and acetyl CoA when no new ATP or NADH is needed.

What Happens to All of the NADH and FADH₂?

- Two pyruvates are formed for each glucose molecule that enters glycolysis. After each pyruvate has been fully oxidized in the Krebs cycle, the cell has produced 10 NADH, 2 FADH$_2$, and 4 ATP (**Figure 9.16**):

$$C_6H_{12}O_6 + 10 \text{ NAD}^+ + 2 \text{ FAD} \downarrow$$
$$6 \text{ CO}_2 + 10 \text{ NADH} + 2 \text{ FADH}_2 + 4 \text{ ATP}$$

- The four ATP were made by substrate-level phosphorylation and can be used for cell work, the CO$_2$ is a waste gas that you exhale when you breathe, and NADH and FADH$_2$ are used to make more ATP in a reaction that could be written as:

$$\text{NADH} + \text{FADH}_2 + O_2 \downarrow$$
$$\text{NAD}^+ + \text{FAD} + H_2O + \text{more ATP}$$

- In the 1960s, researchers finally discovered how the high-potential-energy electrons carried by NADH and FADH$_2$ were eventually donated to O$_2$, producing H$_2$O in a process that allows ATP production.

9.4   Electron Transport and Chemiosmosis

- Eukaryotes oxidize NADH in their inner mitochondrial membranes. These membranes, isolated by differential centrifugation, can oxidize NADH, but the mitochondrial matrix and intermembrane fluid cannot. In prokaryotes, oxidation of NADH occurs in the cell membrane.
- Molecules that switch back and forth between a reduced and an oxidized state during respiration are also found in the inner mitochondrial membrane. How are these molecules involved in cell respiration?

Components of the Electron Transport Chain

- Molecules in the inner mitochondrial membrane that are involved in the oxidation of NADH and FADH$_2$ make up the **electron transport chain** (**ETC**).
- Most ETC molecules are proteins that contain distinctive chemical groups (flavins, iron-sulfur complexes, heme groups, or copper atoms) that are easily reduced or oxidized in redox reactions.
- **Ubiquinone**, also called **coenzyme Q** or **Q**, is a carbon-containing ring attached to a long hydrophobic tail of isoprene subunits. Q is lipid soluble and easily moves through the

mitochondrial membrane, whereas all but one of the ETC proteins are anchored in the inner mitochondrial membrane.

- ETC molecules differ in their electronegativity—their tendency to become reduced or oxidized. Electrons proceeding down an ETC are held more and more tightly (have less and less potential energy) as they are passed from molecule to molecule (**Figure 9.17**). Each reaction releases a small amount of energy.
- Researchers used poisons to confirm the existence of ETCs. Poisons that inhibit a specific ETC protein cause ETC molecules that receive electrons before the poisoned protein to stay in a reduced state, whereas those "downstream" of the poisoned protein remain oxidized.
- NADH donates electrons to a flavin-containing protein at the top of the chain, but FADH$_2$ donates electrons to an iron-sulfur protein that passes electrons directly to Q. Oxygen is the final electron acceptor in eukaryotes, and the total difference in potential energy from NADH to oxygen is 53 kcal/mol.

The Chemiosmotic Hypothesis

- No enzymes could be found that directly used the energy released by ETC redox reactions to make ATP. Then Peter Mitchell proposed an indirect connection between the redox reactions and ATP production.
- Mitchell suggested that the electron transport chain pumps protons from the matrix to the intermembrane space. This would cause an electrochemical gradient favoring the movement of protons back into the matrix. A protein in the inner membrane could then use this **proton-motive force** to make ATP.
- Mitchell's idea is known as the **chemiosmotic hypothesis** because he called the production of ATP via a proton gradient **chemiosmosis**. The following experiment supported his hypothesis.
- Researchers made vesicles from artificial membranes containing bacteriorhodopsin (a protein that uses light energy to pump protons) and an ATP-synthesizing enzyme found in mitochondria. When lit, bacteriorhodopsin pumped protons out of the vesicles, creating a

proton gradient, and ATP was produced inside the vesicles (**Figure 9.18**).

- The ETC pumps protons, creating a proton gradient, and ATPs are produced when protons flow down their electrochemical gradient back into the mitochondrial matrix through a protein called ATP synthase.

How Is the Electron Transport Chain Organized?

- Electron transport chain proteins are organized into four complexes. Three of these complexes pump protons. Lipid-soluble Q and the protein **cytochrome *c*** transfer electrons between complexes (**Figure 9.19**). In complexes I and IV, protons pass directly through the electron carriers, but their exact route is unknown.

- Q's interaction with complex III is better understood; when Q accepts electrons from complex I or complex II, it also gains protons. Q then diffuses through to the outer side of the membrane, where its electrons are transferred to complex III and its protons are released into the intermembrane space.

The Discovery of ATP Synthase

- Efraim Racker observed that inside-out mitochondrial membrane vesicles had large stalk- and knob-shaped proteins on their surfaces. The knobs fell off if shaken or treated with urea.

- Isolated knobs hydrolyzed ATP, and vesicles with stalks and no knobs could not make ATP even though proton transport occurred normally. This suggested that the stalk part acts as a proton channel.

- Racker hypothesized that the knobs were enzymes that, depending on the conditions, can either hydrolyze or synthesize ATP. This hypothesis was supported by the discovery that when the knobs were added back to knobless vesicles, the vesicles regained their ability to synthesize ATP.

- The knob-and-stalk protein is now called **ATP synthase**. Protons flow through the F_o unit (the stalk), causing a rod connecting the two units to spin. As the F_1 unit (the knob) spins with the rod, its subunits are deformed in a way that catalyzes the phosphorylation of ADP to ATP (**Figure 9.20**).

Oxidative Phosphorylation

- ATP production via the combination of proton pumping by electron transport chains (to create a proton-motive force) and the ATP synthase action is called **oxidative phosphorylation**. ATP synthase produces about 26 of the 30 ATP molecules produced per glucose molecule during cell respiration (**Figure 9.21**)!

- All eukaryotes and many prokaryotes use oxygen as the final electron acceptor of electron transport chains. These species use **aerobic respiration**. Many other prokaryotes, especially those in oxygen-poor environments, use other electron acceptors in **anaerobic respiration**.

- Oxygen, with its high electronegativity, is the best electron acceptor because the potential energy of its electrons is very low, creating a large energy difference between NADH and O_2 electrons. This large potential energy difference allows the generation of a large proton-motive force for ATP production.

- Cells that use other electron acceptors cannot pump as many protons per NADH; thus they cannot make as much ATP per glucose. These cells grow more slowly than aerobic respirators do. Aerobic respirators typically outcompete anaerobic respirators in environments with available oxygen.

- If oxygen or other electron acceptors are temporarily unavailable, NADH electrons have no place to go. The ETC stops, yet cells must have NAD^+ to produce ATP from glycolysis. What do cells do?

9.5 Fermentation

- **Fermentation** occurs when pyruvate or a molecule derived from pyruvate accepts electrons from NADH instead of oxygen. Fermentation pathways convert NADH back to NAD^+ and allow glycolysis to continue producing some ATP via substrate-level phosphorylation (**Figure 9.22**).

- The molecule formed from the reduction of pyruvate is often excreted from the cell as waste.
- In humans, pyruvate accepts electrons from NADH in **lactic acid fermentation** when muscles metabolize glucose faster than oxygen can be supplied. Lactate and NAD^+ are produced.
- Yeast deprived of oxygen use **alcohol fermentation**, in which pyruvate loses CO_2 and is converted to acetylaldehyde, which accepts an electron from NADH to produce ethanol and NAD^+.
- Other fermentation pathways also exist, and some bacteria and archaea exclusively use fermentation (e.g., in your small intestine) to break down glucose.
- Because the electron acceptors in fermentation are much more reduced than oxygen, only 2 ATP are produced per glucose— compared to about 30 ATP per glucose in aerobic respiration. Thus organisms never use fermentation if an appropriate electron acceptor is available for cell respiration.
- Organisms that can switch between fermentation and aerobic cell respiration are called **facultative aerobes**.

9.6 How Does Cellular Respiration Interact with Other Metabolic Pathways?

- Cell metabolism involves thousands of different chemical reactions. Types and amounts of molecules inside cells are constantly changing (**Figure 9.23**).
- **Catabolic pathways** involve the breakdown of molecules and the production of ATP, whereas **anabolic pathways** result in the synthesis of larger molecules from smaller components.
- Together, these pathways allow cells to survive, grow, and reproduce, supplying chemical energy (ATP) and carbon-containing building blocks for synthesis of the molecules necessary for life.

Processing Proteins and Fats as Fuel

- Cells use enzyme-catalyzed reactions to break down glycogen, starch, and most simple sugars into glucose or fructose, which can enter the glycolytic pathway.
- Fats can be broken down into glycerol, which can be phosphorylated to form an intermediate of glycolysis, and acetyl CoA, which enters the Krebs cycle.
- Carbon compounds remain after proteins are broken down into amino acids and amino groups are removed and excreted. These carbon compounds can then be converted into pyruvate, acetyl CoA, and other intermediates of glycolysis or the Krebs cycle.
- Macromolecules are broken down in different ways, but glycolysis and the Krebs cycle are central to ATP production in these catabolic pathways (**Figure 9.24**). For ATP production, cells first use carbohydrates, then fats, and finally proteins.

Anabolic Pathways Synthesize Key Molecules

- Molecules found in carbohydrate metabolism are used in anabolic metabolic pathways too (**Figure 9.25**).
- About half of our amino acids can be synthesized from Krebs cycle compounds.
- Acetyl CoA is the carbon building block for fatty-acid synthesis.
- The second compound in the glycolytic pathway can be used to synthesize a key intermediate in production of nucleotides for RNA and DNA synthesis.
- Pyruvate and lactate can be used to synthesize glucose when sufficient ATP is present to power these reactions. Glucose can be converted to glycogen (in animals) or starch (in plants) for storage.

B. CROSS-CUTTING THEMES

Looking Back—
Concepts from Earlier Chapters
Natural Selection—Chapter 1
The pathways discussed in this chapter are all a product of natural selection. For example, enzyme regulation likely evolved due to the advantages it confers on an organism. Similarly, we are reminded that selection is always relative to the environment; use of oxygen as an electron

acceptor is highly advantageous in environments with oxygen, but is not beneficial in anoxic environments.

ATP Provides Energy for Endergonic Reactions—Chapters 2, 3, and 7

Chapters 2 and **3** introduce you to the idea that cells perform chemical reactions, and that endergonic reactions are necessary to create the macromolecules essential to life. Most endergonic reactions in cells are made possible by coupling these reactions with ATP hydrolysis. In **Chapter 7** you learn that ATP provides the energy for intercellular transport; in **Chapter 9** you learn how cells make ATP.

Membranes—Chapter 6

In many ways, the structure and function of mitochondrial membranes are similar to the structure and function of the cell membranes described in **Chapter 6**.

Looking Forward—
Concepts in Later Chapters

Many later chapters discuss cell work powered by ATP hydrolysis.

Photosynthesis—Chapter 10

In the next chapter, you learn about another ATP-producing process that uses an electron transport chain—photosynthesis.

Bacteria and Archaea—Chapter 27

Bacteria and archaea are groups that show incredible metabolic diversity. **Chapter 9** just touches on this diversity by mentioning that many of these species use anaerobic respiration.

C. DIFFICULT TOPICS

The chemiosmotic hypothesis can be difficult to understand. This is partly why some researchers opposed this theory for so long. But it really just proposes a slightly different type of energy transformation than those we are more familiar with. For example, in **Chapter 2** you learned how potential energy can be converted to kinetic energy. In the electron transport chain, the potential energy of electrons is simply converted to the potential energy in an electrochemical gradi-

ent. Consider an analogy. The electrochemical proton gradient can be compared to the water and gravity gradient observed in a dam. Cells use redox reactions and the potential energy of electrons to pump the protons in the same way a water pump could burn fuel to pump water across a dam. Then the protons flow back across the membrane through ATPase, and the energy in the proton gradient is converted into energy used to phosphorylate ATP. Similarly, water flowing through a dam can be used to turn turbines that then produce electricity.

D. ASSESSING WHAT YOU'VE LEARNED

(1) Testing Your Knowledge

1. Why do reduced molecules tend to have a lot of C–H bonds?
   a. Reduced molecules hold electrons tightly and C–H bonds hold electrons tightly.
   b. Reduced molecules like glucose can't have any oxygen atoms.
   c. Carbohydrates and fats have lots of C–H bonds.
   d. Reduced molecules usually gain protons along with electrons.

2. What is the purpose in having several steps in glycolysis or the Krebs cycle rather than a single step from glucose and oxygen to carbon dioxide and water?
   a. The multistep approach is the only way to convert glucose to carbon dioxide.
   b. The multistep approach increases the amount of potential energy in the reaction.
   c. The multistep approach makes better use of potential energy in the reaction.
   d. The multistep approach increases the amount of heat produced in the reaction.

3. Which of the following is accomplished during the reactions of glycolysis?
   a. Glucose is converted to water.
   b. Glucose is converted to CO_2.
   c. Glucose is converted to pyruvate.
   d. Glucose is converted to ATP.

4. One purpose of the Krebs cycle is to:
   a. oxidize NAD+
   b. reduce NAD+

c. oxidize NADH

d. reduce NADH

5. Which of the following is true regarding ATP synthesis in prokaryotes?
   a. They oxidize NADH on the cell membrane.
   b. They are unable to use oxygen.
   c. They do not make ATP.
   d. They use mitochondria.

6. In the Buchner experiment, why did boiling the yeast extract prevent the conversion of ethanol from glucose?
   a. Phosphate required for the reactions was destroyed.
   b. Sucrose was destroyed.
   c. Proteins were denatured.
   d. Yeast cells were killed.

7. Why is a different enyzme involved in each step of glycolysis?
   a. Each step occurs in a different subcellular location.
   b. Each step occurs in a different cell.
   c. Each step involves a different change in potential energy.
   d. Each step involves a different chemical reaction.

8. Nearly every living organism uses glucose as a nutrient source of energy. Why?
   a. Glucose is the only molecule capable of providing the energy to produce ATP.
   b. The ability to harvest energy from glucose appeared very early in biological evolution.
   c. Glucose contains more potential energy than any other nutrient molecule.
   d. The structure of glucose is very similar to ATP.

9. Which of the following chemical reactions does not occur during glycolysis?
   a. hydrolysis of ATP
   b. phosphorylation of ADP
   c. reduction of NAD^+
   d. oxidation of NADH

10. If glucose is labeled with ^{14}C, what molecule will become radioactive as glycolysis and the Krebs cycle are completed?
    a. water

b. carbon dioxide

c. ATP

d. NADH

11. At the end of the Krebs cycle but before the electron transport chain, the oxidation of glucose has produced a net yield of:
    a. 3 CO_2, 5 NADH, 1 $FADH_2$ and 2 ATP
    b. 6 CO_2, 10 NADH, 2 $FADH_2$ and 4 ATP
    c. 6 CO_2, 10 NADH, 2 $FADH_2$ and 6 ATP
    d. None of the above are correct.

12. What would have been the result of Krebs' experiment if the Krebs cycle were linear instead of circular?
    a. Oxaloacetate + pyruvate would have produced citric acid.
    b. Oxaloacetate + pyruvate would have produced acetyl CoA.
    c. Oxaloacetate + pyruvate would have produced no products.
    d. Oxaloacetate + pyruvate would have produced glucose.

13. If oxygen is labeled with ^{18}O, what molecule will become radioactive as glycolysis and the Krebs cycle are completed?
    a. water
    b. carbon dioxide
    c. ATP
    d. NADH

14. Cyanide poisons an electron transport chain protein with less free energy than Q. Which of the following would occur during cyanide poisoning?
    a. Q would stay reduced and stop moving protons across the mitochondrial membrane.
    b. Q would stay oxidized and stop moving protons across the mitochondrial membrane.
    c. Q would not be affected, but the proton gradient would dissipate.
    d. The electron transport chain would stop functioning because the cell would run out of NADH.

15. Why does NADH donate electrons to the beginning of the electron transport chain, whereas $FADH_2$ donates electrons to the middle of the chain?

a. FADH$_2$ is more rapidly oxidized than NADH.

b. NADH has more potential energy than FADH$_2$.

c. FADH$_2$ has more reducing potential than NADH.

d. NADH has more electrons to donate than FADH$_2$.

16. Glycolysis occurs in the _____; the Krebs cycle occurs in the _____; and the ETC occurs in the _____.

a. mitochondrial matrix; cytosol; mitochondrial inner membrane

b. cytosol; mitochondrial matrix; mitochondrial inner membrane

c. cytosol; mitochondrial inner membrane; mitochondrial intermembrane space

d. mitochondrial outer membrane; mitochondrial intermembrane space; mitochondrial matrix

17. If a cell is treated with a drug that inhibits ATP synthase, the pH in the mitochondrial matrix will:

a. increase

b. decrease

c. not change

18. If oxygen is removed from a human muscle cell, the concentration of lactate will:

a. increase

b. decrease

c. not change

19. Glucose can be converted to fatty acid, but fatty acid cannot be converted to glucose. What does this result suggest?

a. Conversion of glucose to fatty acid is preferred energetically.

b. Conversion of fatty acid to glucose requires too much input of energy.

c. Glucose and fatty-acid metabolic pathways do not share a common intermediate.

d. Conversion of acetyl CoA to pyruvate does not occur significantly in a cell.

20. An example of feedback inhibition is:

a. the effect of AMP on pyruvate dehydrogenase

b. the effect of NADH on coenzyme Q

c. the effect of high concentrations of ATP on phosphofructokinase

d. the effect of FADH$_2$ on phosphofructokinase

(2) Integrating Your Knowledge

(a) Why did the fermentation reactions observed by the Buchners last longer when inorganic phosphate was added?

(b) What will happen to the glycolysis pathway if a cell runs completely out of ATP?

(c) Explain why it is advantageous for the inner mitochondrial membrane to be highly folded.

(d) A compound called dinitrophenol (DNP) can be used to poke holes in the inner mitochondrial membrane. What would this do to ATP synthesis by oxidative phosphorylation?

(e) Why wasn't it enough for Racker to show that ATP synthesis stopped when the knobs were removed? That is, why was his explanation more convincing after he added back the isolated knobs, causing the vesicles to regain the ability to synthesize ATP?

(f) What would happen if you followed a recipe for making wine, but then bubbled oxygen through your wine container instead of letting it sit undisturbed?

(g) What method of ATP production would you expect most organisms to use in a deep-sea environment with low oxygen? On land?

(h) When phosphofructokinase is inactivated by high levels of ATP, what can the cell do with its glucose?

CHAPTER 9—ANSWER KEY

D. Assessing What You've Learned

(1) Testing Your Knowledge

1. d; 2. c; 3. c; 4. b; 5. a; 6. c; 7. d; 8. b; 9. d; 10. b; 11. b; 12. c; 13. a; 14. a; 15. b; 16. b; 17. a; 18. a; 19. d; 20. c

(2) Integrating Your Knowledge

(a) Inorganic phosphate is one of the substrates for glycolysis. Thus glycolysis and fermentation stop when no more phosphate is available.

(b) If a cell ran completely out of ATP, it could not perform glycolysis because the first and third steps require ATP input before ATP is made in the final steps.

(c) The folds provide more surface area—more area for electron transport chain components and ATPases.

(d) When DNP puts holes in the inner mitochondrial membrane, no ATP can be produced by ATPase because the proton gradient disappears.

(e) Something else that was responsible for ATP synthesis might have been knocked off along with the knobs.

(f) You would produce carbon dioxide, water, and lots more yeast, but little if any alcohol, because the yeast would use aerobic respiration instead of fermentation.

(g) *Deep sea*: anaerobic respiration; *on land*: aerobic respiration

(h) It can make glycogen (if animal) or starch (if plant), or use the glucose in RNA and DNA synthesis.

10

Photosynthesis

A. KEY BIOLOGICAL CONCEPTS

- **Photosynthesis** is the ability to convert energy from sunlight into chemical energy. Photosynthetic organisms are called **autotrophs,** organisms that make all their own food. Organisms that derive their food from other organisms are called **heterotrophs**.
- Photosynthesis is the ultimate food source for most organisms, and the evolution of this metabolic pathway is one the great events in the history of life.

10.1 What Is Photosynthesis?

- Photosynthesis occurs in the green parts of plants. It requires sunlight, carbon dioxide, and water and produces oxygen. The overall reaction is CO_2 + $2H_2O$ + light energy $\rightarrow$ $(CH_2O)_n$ + H_2O + O_2, where $(CH_2O)_n$ stands for carbohydrate. Usually glucose ($C_6H_{12}O_6$) is the carbohydrate made:

  $6CO_2$ + $12H_2O$ + light energy $\downarrow$
  $$C_6H_{12}O_6 + 6H$$
- Photosynthesis allows organisms to convert sunlight energy into carbohydrate bond chemical energy.

Photosynthesis: Two Distinct Sets of Reactions

- Cornelius van Niel studied photosynthesis in purple sulfur bacteria. These grow even when starved of sugars if exposed to sunlight and hydrogen sulfide (H_2S). Sulfur (S), not O_2, is the by-product of carbohydrate production for these cells: CO_2 + $2H_2S$ + light energy $\downarrow$ (CH_2O) + S.

- Because CO_2 is used in the reaction just described but no O_2 was made, van Niel's result suggested that the O_2 released in most photosynthesis must come from H_2O, not CO_2.
- This hypothesis that photosynthetic O_2 comes from H_2O is supported by the observation that isolated chloroplasts make O_2 when exposed to sunlight even if no CO_2 is present. Researchers also showed that algae and plants exposed to water containing the heavy isotope of oxygen, ^{18}O, emitted $^{18}O_2$ gas. Clearly the O_2 released in photosynthesis comes from H_2O, not CO_2.
- Melvin Calvin fed $^{14}CO_2$ to algae, identifying the order in which products became ^{14}C-labeled. He found that $^{14}CO_2$ was used to make carbohydrates even in the dark. This light-independent part of photosynthesis, the **Calvin cycle**, reduces CO_2 and produces sugar.
- In photosynthesis the **light-dependent reactions** produce O_2 from H_2O; the **light-independent reactions** produce sugar from CO_2.
- As H_2O splits and forms O_2, electrons are released and transferred to NADPH, an electron carrier similar to NADH. These electrons are later used in the Calvin cycle reduction of CO_2. ATP made by the light-dependent reactions is also used in the Calvin cycle to make sugars (**Figure 10.1**).

117

The Structure of the Chloroplast

- Photosynthesis occurs in green plant parts—specifically, in small green organelles called chloroplasts. Chloroplast membranes release O_2 when exposed to sunlight.

- Leaf cells contain from 40 to 50 **chloroplasts** (**Figure 10.2**). These organelles have two membranes and are filled with vesicle-like structures called **thylakoids**, which are often found in stacks called **grana**. The surrounding fluid is called the **stroma**, and the space inside the thylakoids is called its **lumen**.

- Thylakoid membranes contain lots of **chlorophyll** pigment. **Pigments** absorb some wavelengths of light and transmit others. Chlorophyll absorbs blue and red light and transmits green light. Pigment makes the green color of plants, algae, and many bacteria.

- Chloroplasts derive from colorless **proplastids** found in embryonic plant cells and rapidly dividing mature plant tissue. The proplastids mature into chloroplasts, leucoplasts, or chromoplasts (**Box 10.1**).

10.2 How Does Chlorophyll Capture Light Energy?

- Different types of electromagnetic radiation correspond to different **wavelengths** (**Figure 10.4**); the part of the **electromagnetic spectrum** that humans see is called **visible light** and encompasses wavelengths from 400 to 710 nanometers (nm). Shorter wavelengths (e.g., blue, UV) have more energy than longer wavelengths (e.g., red, infrared).

- Light has particle-like as well as wavelike properties. Packets of light are called **photons**, and each photon or wavelength has a specific amount of energy. This energy can be absorbed by pigments.

Photosynthetic Pigments Absorb Light

- Photons can be absorbed, transmitted, or reflected by molecules. White light is a mix of all visible wavelengths; when pigments absorb some, but not all wavelengths, they appear to us to be the color of the reflected or transmitted wavelengths.

- Pigments can be extracted and separated via a technique called **paper chromatography** (**Figure 10.5**). Raw extract blotted on the chromatography paper is separated into components as a solvent travels up the paper, carrying along the extract molecules at different rates based on their size and/or solubility.

- Leaves contain several pigments. Researchers can cut out the region of chromatography paper with one pigment and determine the wavelengths it absorbs using a spectrophotometer. A graph of light absorbed versus wavelength gives the **absorption spectrum** for a given pigment (**Figure 10.6**).

- The major leaf pigments are the **chlorophylls** (chlorophyll *a* and chlorophyll *b*), which absorb red and blue light and transmit green light; and the **carotenoids**, which absorb blue and green light and thus appear yellow, orange, or red.

- Algae on a slide lit with a spectrum of colors produce oxygen (measured by bacterial growth) in areas exposed to blue and red light. These wavelengths make up the **action spectrum** for photosynthesis. Chlorophylls absorb red and blue wavelengths, so they must be the primary photosynthetic pigments.

What Is the Role of Carotenoids and Other Accessory Pigments?

- Carotenoids are **accessory pigments** that absorb light and pass the energy on to chlorophyll.

- The two classes of carotenoids found in plants are the **carotenes** (e.g., β-carotene in carrots, **Figure 10.8a**) and the **xanthophylls** (e.g., zeaxanthin in corn). These molecules are similar and are both found in chloroplasts. The accessory pigments give deciduous trees their fall colors when chlorophyll degrades.

- Carotenoids absorb wavelengths of light not absorbed by chlorophyll, thus extending the range of wavelengths that can be used in photosynthesis. Carotenoids also stabilize free radicals produced by electromagnetic radiation, protecting chlorophyll from degradation.

- **Flavonoids** protect chlorophyll and other plant molecules from destructive radiation, in this case by absorbing high-energy ultraviolet (UV) light.

The Structure of Chlorophyll

- Chlorophyll *a* and *b* have similar structures but slightly different absorption spectra. Land plants have about three chlorophyll *a* per chlorophyll *b*. Both chlorophylls have a long tail of isoprene subunits and a large ring structure (the "head") containing a magnesium atom where light is absorbed (**Fig. 10.8b**).

When Light Is Absorbed, Electrons Enter an Excited State

- If the energy of a photon striking chlorophyll is similar to the energy difference between two electron shells in the chlorophyll head, the photon's energy is absorbed and an electron is "excited"—it is raised to a higher energy shell where it has greater potential energy (**Figure 10.9**).
- The energy difference between electron state 0 and state 1 in chlorophyll equals the energy of a red photon; the energy difference between state 0 and state 2 equals the energy of a (higher energy) blue photon. Intermediate energy levels (e.g., those of green light) cannot easily be absorbed by chlorophyll.
- Ultraviolet wavelengths have so much energy that they can eject electrons from pigment molecules, while infrared wavelengths carry so little energy that they merely heat atoms a bit (increase their movement).
- Electrons in high energy states may fall back to their ground state, causing **fluorescence** as they release heat and photons. Isolated chlorophyll fluoresces when exposed to UV light.

How Do the Chlorophyll Molecules in Leaves Work?

- In chloroplasts, chlorophyll shows hardly any fluorescence. To determine if all the chlorophyll molecules in a photosynthetic cell use the photon energy they absorb to drive photosynthesis, Emerson and Arnold measured oxygen produced by a green alga in response to light flashes of increasing intensity.
- At low light intensities, increased intensity increased the rate of photosynthesis; but the photosynthetic rate (measured as oxygen production) leveled out earlier than expected—at a level indicating that many chlorophyll mol-

ecules were not participating in photosynthetic oxygen production.

- Researchers concluded that chlorophyll molecules work in groups (**Figure 10.11**). In the thylakoid membrane, 200–300 chlorophyll molecules and accessory pigments group together in complexes called **photosystems**. The two plant photosystems have both an antenna complex and a reaction center.

The Antenna Complex

- When **antenna complex** electrons absorb photon energy and become excited, they pass the energy, but not the electron, to a nearby chlorophyll molecule. The original electron falls back to its ground state; but the energy transfer excites the nearby chlorophyll molecule's electron. In this way, energy is transmitted from chlorophyll to chlorophyll until it reaches the reaction center (**Figure 10.12a**).

The Reaction Center

- Excited electrons are finally passed to an electron acceptor in the **reaction center**. The reduction of the electron acceptor completes the transformation of electromagnetic energy into chemical energy (**Figure 10.12b**).

10.3   The Discovery of Photosystems I and II

- Researchers found that green algae had similar photosynthetic responses to far-red (700 nm) and red (680 nm) light. But the rate of photosynthesis was **much** higher (more than double) when algae were exposed to both wavelengths at once (**Figure 10.13**). This result was called the enhancement effect.
- Robin Hill and Faye Bendall proposed that green algae and plants have two types of reaction centers: **photosystem II** and **photosystem I**. Photosynthesis is more efficient when both photosystems work together—thus the enhancement effect.
- Many bacteria have photosystems similar to those found in algae and plants, and several bacteria have only one of the two photosystems. This allowed researchers to study each photosystem in isolation before trying to understand how they work together.

How Does Photosystem II Work?

- Photosystem II is similar to the single photosystem found in purple nonsulfur and purple sulfur bacteria. When the antenna complex transmits energy to this photosystem's reaction center, a high-energy electron is donated to **pheophytin**—a molecule similar to chlorophyll that lacks magnesium in its head region.

Electrons from Pheophytin Enter an Electron Transport Chain

- When an excited electron from chlorophyll binds to (and reduces) pheophytin in a reaction center, chlorophyll is oxidized. Pheophytin then passes the electron to an electron transport chain similar to that found in mitochondria.
- Photosystem II's electron transport chain, like that in the mitochondria, contains cytochromes and quinones. Electrons participate in redox reactions and are gradually stepped down in potential energy.
- **Plastoquinone** (**PQ**), a small hydrophobic molecule, receives electrons from pheophytin and transports them across the thylakoid membrane to a cytochrome complex. Plastoquinone carries protons from one side of the thylakoid membrane to the other along with the electrons (**Figure 10.14**).
- This proton transport increases the proton concentration inside the thylakoid, causing a thousandfold difference in proton concentration (thylakoid interior at around pH 5; exterior at pH 8). This proton motive force then drives ATP production.

ATP Synthase Uses the Proton Motive Force to Phosphorylate ADP

- As in the mitochondria, protons diffuse down their electrochemical gradient through ATP synthase, causing a conformational change that drives the phosphorylation of ADP. The capture of light energy by photosystem II to produce ATP chemical energy is called **photophosphorylation**.
- In purple sulfur bacteria, cytochrome donates its electron back to the reaction center; but in cyanobacteria, green algae, and plants, the cytochrome complex of photosystem II donates its electrons to photosystem I.

Photosystem II Obtains Electrons by Oxidizing Water

- Cyanobacteria, algae, and plants produce oxygen because part of photosystem II "splits" water to replace its lost electrons: $2\ H_2O \downarrow 4\ H^+ + 4\ e^- + O_2$. Because oxygen is produced as a by-product, organisms that split water during photosynthesis are said to perform **oxygenic photosynthesis**.
- Purple sulfur and purple nonsulfur bacteria cannot oxidize water. They perform **anoxygenic photosynthesis**.
- Photosystem II is the only known protein complex able to oxidize water in this way, and chemists have been unable to synthesize catalysts that perform this feat. We do not yet completely understand exactly how photosystem II splits water.
- All oxygen we breathe originates from splitting of water in photosystem II. Addition of O_2 to the atmosphere by oxygenic photosynthesis had a huge impact on history of life.
- Oxygen is toxic to anaerobic organisms. In addition, organisms that evolved the ability to use oxygen as an electron acceptor in cell respiration were able to make much more ATP than were organisms using other electron acceptors. Consequently, aerobic organisms became dominant on our planet.

How Does Photosystem I Work?

- Heliobacteria, which produce NADH from NAD^+ in the presence of sunlight, were used to study photosystem I. In cyanobacteria, algae, and land plants, this photosystem reduces $NADP^+$, a phosphorylated version of NAD^+, to yield NADPH, which also functions as an electron carrier.
- Excited electrons from the reaction center of photosystem I are passed down an electron transport chain of iron- and sulfur-containing molecules to **ferredoxin** (**Figure 10.15**).
- The enzyme ferredoxin/$NADP^+$ oxidoreductase (also called $NADP^+$ reductase) transfers a proton and two electrons from ferredoxin to $NADP^+$, forming NADPH. The entire photosystem is based in the thylakoid membrane.

The Z Scheme: Photosystems I and II Work Together

- Robert Hill and Fay Bendall proposed the **Z scheme**, depicted in **Figure 10.16**.
- A photon excites an electron in photosystem II's chlorophyll and the energy is transmitted to the reaction center, where chlorophyll P680 (which most easily loses electrons when absorbing wavelengths of 680 nm) passes an excited electron to pheophytin.
- The electron's potential energy is gradually lowered through redox reactions with quinones and cytochromes, and the released energy is used by plastoquinone to transport protons across the thylakoid membrane. ATP synthase uses the proton-motive force to phosphorylate ADP, creating ATP.
- At the end of photosystem II's electron transport chain, a small protein called **plastocyanin** (PC) gains an electron from the cytochrome complex and carries it back across the thylakoid membrane to photosystem I. In this way, plastocyanin physically links photosystem II and photosystem I.
- A chlorophyll called P700 in the photosystem I reaction center loses electrons most readily upon absorbing wavelengths of 700 nm. The electrons enter an electron transport chain, are eventually passed to ferredoxin, and are then used to reduce $NADP^+$ to NADPH.
- Photosystem II's electrons are replaced by the splitting of water, and photosystem I's electrons are replaced with the electrons that plastocyanin shuttles from photosystem II.
- The Z scheme explains the enhancement effect; photosynthesis is more efficient when both 680 nm and 700 nm wavelengths are available, allowing both photosystems to run at maximal rates.
- Photosystem I sometimes transfers electrons to the electron transport chain of photosystem II to increase ATP production instead of using them to reduce $NADP^+$; this is called **cyclic photophosphorylation** (Fig. 10.17).
- Surprisingly, photosystem I is most commonly found in the exterior unstacked thylakoid membranes, while photosystem II is most common in the interior stacked membranes of grana (**Figure 10.18**).

10.4   How Is Carbon Dioxide Reduced to Produce Glucose?

- Photosystems I and II can produce ATP and NADPH only in the presence of light. But the reactions that produce sugar from carbon dioxide are light independent. Although these reactions can occur in the dark, they require the ATP and NADPH produced by the light-dependent reactions.
- Melvin Calvin led the research team that documented the intermediate compounds produced as carbon dioxide is reduced to sugar.

The Calvin Cycle

- Calvin's group fed radioactively labeled carbon dioxide ($^{14}CO_2$) to green algae and then isolated and identified product molecules containing ^{14}C. By waiting different amounts of time between adding the $^{14}CO_2$ and killing the algae (the pulse-chase technique), they determined the order in which compounds are produced during the reduction of CO_2 to sugar.
- Product molecules from algae extract were isolated using paper chromatography and identified by laying X-ray film over the chromatography paper. Radioactively labeled products created a black spot on the X-ray film (**Figure 10.19**).
- Cells killed right after exposure to $^{14}CO_2$ contained a radio-labeled 3-carbon compound called 3-phosphoglycerate (3PG), which is an intermediate in glycolysis. Because Calvin's pathway produces sugar, and glycolysis breaks glucose down, it makes sense that the pathways share some intermediates.
- A five-carbon compound called ribulose bisphosphate (RuBP) reacts with CO_2 to produce a six-carbon compound that splits in half to form two molecules of 3PG.
- The cycle that reduces CO_2 to produce 3PG (and eventually glucose) is now known as the Calvin cycle. This cycle of reactions occurs in the chloroplast stroma (**Figure 10.20**).
- The Calvin cycle has three phases: (1) *Fixation phase*: CO_2 reacts with RuBP, producing two 3PG molecules. The attachment of CO_2 to an organic compound is called **carbon fixation**. (2) *Reduction phase:* 3PG molecules

are phosphorylated by ATP and reduced by NADPH to produce **glyceraldehyde 3-phosphate (G3P)**. Some G3P is drawn off to make glucose. (3) *Reduction phase:* The remaining G3P is used in reactions that regenerate RuBP.

- Most Calvin cycle reactions also occur as part of glycolysis and/or another metabolic pathway, but the reaction between CO_2 and RuBP is unique to the Calvin cycle.

The Discovery of Rubisco

- Arthur Weissbach and colleagues purified proteins from spinach extracts and then tested whether they could catalyze the incorporation of $^{14}CO_2$ into 3PG. Eventually they isolated a carbon-fixing enzyme called ribulose 1,5-bisphosphate carboxylase/oxygenase (**rubisco**), which is very common in leaf tissue.
- Rubisco is cubical and has four active sites for CO_2 fixation (**Figure 10.21**). Despite its importance, rubisco is slow and inefficient. Oxygen competes with CO_2 at the active site—this apparently **maladaptive** trait may date back to the enzyme's evolution at a time when little O_2 was present on Earth.
- When O_2 and RuBP react in rubisco's active site, one product undergoes reactions that use up ATP and produce CO_2 (**Figure 10.22**). These reactions are called **photorespiration** because they consume O_2 and produce CO_2.
- Photosynthesis rate declines in photorespiration. When cell CO_2 concentration is high and O_2 concentration is low, carbon fixation is favored over photorespiration.

How Is Carbon Dioxide Delivered to Rubisco?

- **Stomata** are the leaf structures where gas exchange occurs. They consist of two **guard cells** that change shape when leaf CO_2 concentration is low during photosynthesis to create an opening called a **pore**. Atmospheric CO_2 then diffuses into the extracellular fluid and on into cells down its concentration gradient (**Figure 10.23**).
- Unfortunately, open stomata also cause leaf water loss because the atmosphere is usually much drier than the leaf interior. Thus, when

conditions are hot and dry, many plants must keep their stomata closed to prevent water loss, causing photosynthesis to stop.

C_4 Photosynthesis

- Research groups using Calvin's pulse-chase approach found that in sugar cane (Kortschack's group) and corn (Karpilov's group), $^{14}CO_2$ was initially incorporated into four-carbon organic acids instead of into 3-PG (**Figure 10.24**). This is now known as **C_4 photosynthesis**.
- In C_4 plants, CO_2 is added to three-carbon compounds by **PEP carboxylase** in **mesophyll cells** near the leaf surface. The four-carbon organic acids produced by this carbon fixation travel to **bundle-sheath cells** surrounding the plant's **vascular tissue**, where water and nutrient transport occurs.
- Rubisco is found in the bundle-sheath cells of C_4 plants where the four-carbon organic acids release CO_2, initiating the Calvin cycle (**Figure 10.25**; model initially proposed by Hatch and Slack).
- The C_4 pathway is mostly found in plants that live in hot, dry habitats. This pathway uses energy, but increases the concentration of CO_2 in the cells where rubisco catalyzes the start of the Calvin cycle. In this way, C_4 plants are able to limit photorespiration.

CAM Plants

- Some plants live in hot, dry environments where they keep their stomata closed all day. These plants open their stomata at night to temporarily fix CO_2 to organic acids using a pathway called **crassulacean acid metabolism (CAM)**.
- The organic acids made at night by CAM plants are stored in the central vacuoles of photosynthesizing cells. During the day, CO_2 is released from the stored organic acids and used by the Calvin cycle.
- Thus, C_4 plants stockpile CO_2 in cells where rubisco is not active, and CAM plants store CO_2 at a time when rubisco is not active. Both C_4 and CAM plants use energy to increase the CO_2 concentration when/where rubisco is active, minimizing photorespiration

when they must keep their stomata closed (**Figure 10.26**).

What Happens to the Sugar That Is Produced by Photosynthesis?

- G3P molecules produced by the Calvin cycle are often used to make glucose and fructose, which can combine to form sucrose in the cytosol (**Figure 10.27**). Water-soluble sucrose is easily transported throughout the plant; it can be broken down to fuel cell respiration or be converted to starch for storage.

- In rapidly photosynthesizing cells, where sucrose is abundant, glucose is temporarily stored as starch in the chloroplast. Because starch is not water soluble, it is broken down at night and used to make more sucrose for transport throughout the plant.

- Intermediates in the production of glucose from G3P also occur in glycolysis and can be used to fuel ATP production via glycolysis and cellular respiration if necessary.

- Sugars and starch from photosynthesis provide food for herbivore growth and reproduction when herbivores eat plants and for carnivore growth and reproduction when carnivores eat herbivores. Photosynthesis is thus the ultimate source of energy for almost all cell growth and reproduction on Earth.

B.   CROSS-CUTTING THEMES

Looking Back—
Concepts from Earlier Chapters

Redox Reactions—Chapter 2

Chapter 2 introduced you to redox reactions and the concept of potential energy. Understanding photosynthesis and other metabolic pathways requires application of these concepts.

Use of Radioactive Isotopes—Chapters 7 and 9

Chapter 7 described "pulse-chase" experiments involving the use of radioactive isotopes to discover how molecules are transported throughout the cell. **Chapter 9** mentions that radioactive isotopes were used to discover intermediate compounds in the Krebs cycle. Here in **Chapter 10** you learn how isotopes were used in research

identifying the metabolic pathways involved in carbon fixation.

Electron Transport Chains—Chapter 9

Mitochondria also use an electron transport chain to convert the energy of high potential energy electrons to more easily usable chemical bond energy (e.g., ATP).

Looking Forward—
Concepts in Later Chapters

Oxygen Produced by Photosynthesis—Chapter 27

Photosynthetic bacteria (some similar to those used in experiments described in Chapter 10) were responsible for producing the oxygen that shapes our current world. Find out more about these organisms in **Chapter 27**.

C_4 and CAM Pathways for Carbon Fixation—Chapter 29

Chapter 29 gives you more information about green plants that use the C_4 and CAM strategies for carbon fixation.

Water and Nutrient Transport in Plants—Chapter 36

Plants must take up water to replace water lost via stomata and transport fructose and other nutrients throughout the organism. These transport processes are detailed in **Chapter 36**.

Global Warming and CO_2 Levels—Chapter 54

You'll learn more about global warming and rising CO_2 levels in **Chapter 54**, on ecosystem ecology.

C.   DIFFICULT TOPICS

Three new major metabolic pathways are described in this chapter: photosystem I, photosystem II, and the Calvin cycle. Connections between these and other metabolic pathways are also discussed. One good strategy for learning this information is to make your own lists of the order in which events occur, drawing flowcharts depicting the connections between pathways.

For example, you might start a list describing photosynthesis as follows: (1) Photon hits chlorophyll and excites electron; (2) excited electron is passed to other chlorophyll molecules until it reaches a reaction center; (3) at the reaction center, the electron reduces an electron acceptor. Another study strategy is to look at the text figures and describe out loud to yourself the names of structures and what happens where. In assessing the importance of each step along a pathway, think about what would happen if that part of the pathway failed.

D. ASSESSING WHAT YOU'VE LEARNED

(1) Testing Your Knowledge

1. How would the experiments of van Niel with purple sulfur bacteria have been different if oxygen came from carbon dioxide during photosynthesis?
   a. Elemental sulfur would be produced from carbon dioxide and hydrogen sulfide.
   b. Oxygen would be produced from carbon dioxide and hydrogen sulfide.
   c. Oxygen would be produced from carbon dioxide and water.
   d. Elemental sulfur would be produced from carbon dioxide and water.

2. If green plant cells are incubated with ^{18}O-labeled carbon dioxide, what molecule will become radioactive as the cells are exposed to light?
   a. water
   b. oxygen
   c. ATP
   d. sugar

3. If green plant cells are incubated with ^{18}O-labelled water, what molecule will become radioactive as the cells are exposed to light?
   a. carbon dioxide
   b. oxygen
   c. ATP
   d. sugar

4. Why is it possible for the Calvin cycle to occur in the dark?
   a. The Calvin cycle uses energy stored previously during the light-dependent reactions of photosynthesis.
   b. It is not possible for any part of photosynthesis to occur without light.
   c. No products of the light-dependent reactions are involved in the Calvin cycle.
   d. The Calvin cycle obtains energy from sugars made previously by the light-dependent reactions of photosynthesis.

5. What two high-energy molecules produced during the light-dependent reactions of photosynthesis are used to drive the Calvin cycle?
   a. ATP and oxygen
   b. NADPH and carbon dioxide
   c. ATP and NADPH
   d. ATP and carbon dioxide

6. What is/are the function(s) of accessory pigments in plants?
   a. They extend the range of wavelengths a plant can use to drive photosynthesis.
   b. They protect plants from the damaging effects of electromagnetic radiation.
   c. Both a and b are functions of accessory pigments in plants.
   d. Accessory pigments have no function in plants.

7. In the dark, plants _____ oxygen.
   a. produce
   b. consume
   c. neither produce nor consume

8. Which of the following factors is *not* involved in determining whether light is absorbed by a molecule?
   a. the intensity of the light
   b. the wavelength of the light
   c. the potential energy levels of electron shells in the molecule
   d. the chemical structure of the molecule

9. What is the purpose of chemical reduction of an electron acceptor in the photosynthetic reaction center?
   a. It allows the energy of absorbed light to be trapped and converted to chemical energy.

b. It adjusts the energy level of chlorophyll electrons to match the energy of light illuminating the leaf.

c. It changes the wavelength of chlorophyll fluorescence so that it will not interfere with absorption of light.

d. It provides electrons to be excited by the absorbed light.

10. What is the evidence for two photosystems?

a. Microscopy reveals two colors of chloroplast within the same leaf cells.

b. The combination of light at 680 and 700 nm is much more effective at stimulating photosynthesis than at either wavelength alone.

c. Chlorophyll absorbs light of two different wavelengths: blue and red.

d. Two different high-energy molecules are produced during the light-dependent reactions of photosynthesis: ATP and NADPH.

11. In an oxidation-reduction reaction, potential energy decreases as:

a. electrons are passed from an oxidized molecule to another oxidized molecule

b. electrons are passed from a reduced molecule to another reduced molecule

c. electrons are passed from an oxidized molecule to a reduced molecule

d. electrons are passed from an reduced molecule to an oxidized molecule

12. What happens to the ATP synthesis rate if the pH of the thylakoid lumen decreases?

a. The rate of ATP synthesis will increase.

b. The rate of ATP synthesis will decrease.

c. The rate of ATP synthesis will not change.

13. In an experiment conducted by Jagendorf, isolated thylakoids were incubated in an acidic solution at pH 4.0 until the pH was equilibrated across the thylakoid membrane. The thylakoids were then transferred to a buffer at pH 8.0 with ADP and inorganic phosphate. ATP was synthesized. Did this experiment require light to generate the ATP?

a. Yes; ATP synthase requires high-energy electrons from photosystem II.

b. No; ATP synthesis is dependent only on the presence of a hydrogen ion gradient and does not require light directly.

c. Yes; the hydrogen ion gradient is generated via transfer of electrons in photosystem II.

d. No; ATP synthesis is part of the light-independent reactions of photosynthesis.

14. Electrons excited by absorption of light in photosystem II are transferred to plastoquinone and so must be replaced. The replacements come from:

a. photosystem I

b. water

c. oxygen

d. cytochrome

15. Electrons excited by absorption of light in photosystem I are transferred to iron-sulfur electron acceptors and must be replaced. The replacements come directly from:

a. photosystem II

b. ATP

c. NADPH

d. water

16. The biochemical objective of photosystem I is to:

a. phosphorylate ADP

b. hydrolyze ATP

c. oxidize NADPH

d. reduce $NADPH^+$

17. Six ATP are used for production of one three-carbon sugar (glyceraldehyde-3-phosphate) from RuBP and carbon dioxide in the Calvin cycle. Nevertheless, the Calvin cycle actually requires nine ATP to function. Why?

a. Three additional ATP are used to phosphorylate carbon dioxide for the next cycle.

b. Three additional ATP are used to reduce $NADPH^+$ to NADPH.

c. Three additional ATP are used to regenerate RuBP.

d. Three additional ATP are hydrolyzed to keep sufficient heat available for the reaction to continue.

18. Rubisco is an important enzyme because:
    a. It catalyzes the splitting water, which provides electrons for photosynthesis.
    b. It catalyzes carbon fixation.
    c. It catalyzes the reaction that allows CAM plants to store CO_2.
    d. It is one of the fastest enzymes known.

19. Why is it critical for plants to maintain a high concentration of carbon dioxide in the leaves?
    a. It helps to prevent photorespiration.
    b. It is the only substrate for rubisco.
    c. Oxygen cannot be produced without it.
    d. It is necessary for regeneration of RuBP from glyceraldehyde-3-phosphate.

20. How does the sequestering of carbon dioxide in CAM plants help them to survive?
    a. It keeps carbon dioxide away from places in the leaves where it would be toxic.
    b. It allows carbon dioxide to be gathered and used at different times of the day.
    c. It allows the light-dependent and light-independent reactions to occur at different times of the day.
    d. It allows the plants to produce sugars in the winter when leaves are dead.

(2) Integrating Your Knowledge

(a) Why are plants green?
(b) Fill in the blanks: Plants have _____ cells that open to form a pore on the leaf surface. This structure is called a _____. In C_4 plants, PEP carboxylase is found in _____ cells located near the surface of the leaves. Other C_4 cells that contain rubisco and surround vascular tissue in the interior of the leaf are called _____ cells. All these leaf cells contain photosynthetic organelles called _____. The internal membranes of these organelles form flat, vesicle-like structures called _____, which form stacks called _____.

(c) You learned in this chapter that purple sulfur bacteria were commonly used in research on photosystem I, and that helio-bacteria were commonly used to research photosystem II. Why are these organisms better models for studying these processes than trees?

(d) In paper chromatography, molecules are separated based on their _____ and _____.

(e) Explain how photosystems I and II are similar and how they differ.

(f) Which photosystem's electron transport chain is more similar to that in the mito-chondria, and why?

(g) Why is rubisco considered an inefficient enzyme?

(h) Name the two pathways that allow some plants to stockpile CO_2 for later use. How do the pathways differ?

CHAPTER 10 ↵ ANSWER KEY

D. Assessing What You've Learned

(1) Testing Your Knowledge

1. b; 2. d; 3. b; 4. a; 5. c; 6. c; 7. b; 8. a; 9. a; 10. b; 11. d; 12. a; 13. b; 14. b; 15. a; 16. d; 17. c; 18. b; 19. a; 20. b

(2) Integrating Your Knowledge

(a) Plants are green because they contain lots of chlorophyll; this pigment absorbs red and blue light but transmits green light.

(b) guard; stoma; mesophyll; bundle sheath; chloroplasts; thylakoids; grana

(c) Some advantages to working with bacteria: They are simpler, faster-growing, and easy to keep lots of in the lab. Trees are more complex, take up a lot of space, and take a long time to grow and reproduce—making them, in many ways, more difficult to work with.

(d) size and solubility

(e) Photosystems I and II are similar in that they both use electrons excited by photons to reduce molecules at the beginning of electron transport chains. Both are located in the thylakoid membrane. On the other hand, photosystems I and II most easily use light of different wavelengths, and they use the potential energy of the excited electrons to produce different compounds (II makes ATP and I makes NADPH). They occur in different parts of the thylakoid membrane, and they replace electrons lost from their chlorophylls in different ways (II splits water and I gets electrons from II).

(f) Photosystem II's electron transport chain is more similar to that in the mitochondria than photosystem I's because photosystem II also creates a proton motive force that is used by ATP synthase to make ATP.

(g) Rubisco is inefficient because: (1) It is slow, and (2) it catalyzes the toxic photorespiration reaction when cell CO_2 concentration is low and O_2 concentration is high. It evolved when there was little O_2 in Earth's atmosphere.

(h) The C_4 pathway allows some plants to stockpile CO_2 in cells next to those with active rubisco; the CAM pathway allows some plants to stockpile CO_2 at night.

The Cell Cycle

A. KEY BIOLOGICAL CONCEPTS

- The cell theory states that all organisms are made of cells and that these cells come from other cells; but how does this cell replication occur? Rudolf Virchow proposed that new cells arise through **cell division**. This hypothesis was confirmed by microscopic observations of early development in **embryos**.

- There are two types of cell division (**Figure 11.1**). Sexual reproduction requires **meiosis**, cell division in which only half of the parent cell's genetic material is passed to the daughter cells (the **gametes**—the sperm and eggs). This type of cell division is described in detail in Chapter 12.

- The second type of cell division is called **mitosis**. Mitosis produces **somatic cells** (all nongamete cells in your body), and daughter cells resulting from mitosis are genetically identical to the parent cell. Growth, wound repair, and **asexual reproduction** are accomplished by this type of cell division.

- More specifically, mitosis refers to the division of a eukaryotic cell's genetic material. It is usually followed by **cytokinesis**, the division of the cytoplasm into two daughter cells.

11.1 Mitosis and the Cell Cycle

- Dyes stain threadlike structures in the nuclei of dividing cells. Observations of their changes during cell division in the salamander embryo were first published by Walther Flemming in 1879 (**Figure 11.2**). Paired threads split apart in a process he called mitosis.

- Roundworm studies revealed that the total number of threads in a cell remain constant through division after division. W. Waldeyer named these threads **chromosomes** in 1888.

- We now know that each chromosome contains a long double helix of deoxyribonucleic acid (DNA) wrapped around proteins. DNA carries the cell's genetic information, and the purpose of mitosis is to evenly distribute this genetic material to daughter cells.

The Cell Cycle

- Growing eukaryotic cells alternate between a dividing phase, known as the mitotic or **M phase**, and **interphase**, when no division occurs. Cells spend most of their time in interphase, during which individual chromosomes are not visible using light microscopy.

- Alternation between M phase and interphase is referred to as the **cell cycle**. One cycle encompasses the sequence of events from a cell's creation (by division of a parent cell) to its own division into two daughter cells.

- Two key cell-cycle events are the **replication** of the hereditary material and the separation of the copied chromosomes into two daughter cells.

When Does Chromosome Replication Occur?

- In the 1950s, availability of radioactively labeled nucleotides and ability to grow eukaryotic cells in culture (**Box 11.1**) allowed Alma Howard and Stephen Pelc to determine when chromosome replication occurs.

- Howard and Pelc supplied a growing cell culture with radioactive thymidine. After 30 minutes, the radioactive thymidine was washed away. Cells were then spread out and X-rayed. Radioactive thymidine, indicated by black dots, was found only in interphase nuclei (**Figure 11.3**).
- Researchers knew that M-phase lasted about 30 minutes and the entire cell cycle took about a day in these cells. Because only replicating DNA could have incorporated the radioactive thymidine, they concluded that chromosome replication occurs during interphase. This DNA synthesis stage is called **synthesis** (or **S**) **phase**.

Discovery of the Gap Phases

- Interphase includes two gap phases during which no DNA synthesis occurs. G_1 **phase**, the first gap, occurs before S phase. G_2 **phase**, the second gap, occurs after S phase and before mitosis (**Figure 11.4**).
- Researchers repeated Howard and Pelc's experiment, but waited different lengths of time before X-raying cells in order to determine the time lengths of various cell phases.
- G_2 phase lasts 4 to 5 hours because there is a 4- to 5-hour time lag between the pulse of radioactive thymidine (marking the replication of DNA) and the appearance of radioactively labeled thymidine in mitotic nuclei (the start of mitosis).
- S phase lasts 6 to 8 hours because labeled mitotic nuclei can be observed for 6 to 8 hours (**Figure 11.4**). These cells must have been in S phase when radiolabeled thymidine was available.
- Researchers subtracted G_2, S-, and M-phase times from total cell cycle time to conclude that the G_1 phase must last about 7 to 9 hours.
- During gap phases cell organelles replicate, making additional cytoplasm. Cells perform all normal cell functions during G_1 phase.

11.2   How Does Mitosis Take Place?
Events in Mitosis

- The number of chromosomes varies widely among species, but all eukaryotic chromosomes consist of a DNA-protein complex called **chromatin**. The proteins in chromatin are called **histones**. Chromosomes are visible at the beginning of mitosis, as chromatin condenses to form a more compact structure.
- Each of the two DNA strands in a replicated chromosome is called a **chromatid**. **Sister chromatids** join together at the **centromere** and are exact copies of the same genetic information (**Figure 11.7**).
- The two joined chromatids separate during mitosis to form independent chromosomes—one copy for each daughter cell (**Figure 11.8**). Each daughter cell thus has the same genetic information as the parent cell.
- Mitosis (M-phase) is a continuous process, but biologists identify several subphases (prophase, prometaphase, metaphase, anaphase, telophase) based on specific events occurring at different stages in the process (**Fig. 11.9**).
- Keep in mind that mitosis describes cell division in eukaryotes. Bacterial cell division differs significantly (**Box 11.2**).

Prophase

- Already replicated chromosomes become visible in the light microscope as they condense.
- In the cytoplasm, microtubules begin to form the **mitotic spindle**. **Spindle fiber** microtubules attach to the chromosomes. These fibers pull chromosomes into the daughter cells later on in mitosis. During prophase the spindles either form on, or begin to move toward, opposite sides of the cell.
- Spindle fibers radiate from microtubule organizing centers. In plants these centers are associated with the nuclear envelope, and in fungi they are called spindle pole bodies. In animals they contain **centrioles** and are known as **centrosomes**. Centrioles are not required for spindle formation; their function is not known.

Prometaphase

- As chromosomes condense, the nucleolus disappears and the nuclear envelope breaks down. Spindle fibers attach to each sister chromatid at structures called **kinetochores**, located at chromosome centromeres. Kinetochore microtubules now start moving chromosomes toward the middle of the cell.

- In animals, centrosomes continue to move toward opposite poles of the cell.

Metaphase

- Centrosomes reach opposite poles of the cell (in animals) and chromosomes, pulled by kinetochore microtubules, reach the middle of the cell. The imaginary plane where the chromosomes line up is called the **metaphase plate**. The mitotic spindle is now complete.

Anaphase

- Kinetochore spindles shorten, pulling apart sister chromatids to create separate chromosomes that are pulled toward microtubule organizing centers at opposite cell poles. Chromatid separation ensures that each daughter cell receives one copy of each parent chromosome.
- Motor proteins also push apart the two cell poles.

Telophase

- A new nuclear envelope begins to form around each set of chromosomes. The spindle disintegrates, and the chromosomes become less compact and less visible as they decondense. When two independent nuclei have formed, mitosis is complete and **cytokinesis** begins—division of the cytoplasm to form the two daughter cells.

Cytokinesis

- The division of the cytoplasm to form two daughter cells typically occurs immediately after mitosis. But sometimes cells complete mitosis without cytokinesis, to become *multinucleate*.
- Cytokinesis in animals, fungi, and slime molds occurs when a ring of actin and myosin filaments inside the cell membrane contracts, causing the cell membrane to pinch inward and form a **cleavage furrow**.
- Plant-cell cytokinesis occurs as vesicles are transported from the Golgi apparatus to the middle of the dividing cell. These vesicles fuse to form a **cell plate** (**Figure 11.10**).
- Refer to **Table 11.1** for definitions of key mitotic structures and to **Figures 11.10** and

11.11 for depictions of the different mitotic subphases.

How Do Chromosomes Move during Mitosis?

Mitotic Spindle Forces

- Because microtubule length depends on the number of α- and β-tubulin dimers it contains, researchers hypothesized that spindle microtubules shorten as tubulin subunits are lost.
- Fluorescently labeled tubulin subunits were added to prophase or metaphase cells. The whole mitotic spindle could then be seen using a fluorescent microscope.
- After the start of anaphase, biologists marked a region of the spindle with a laser light that stops fluorescence in the exposed region. This "photobleached" area remained stationary as anaphase progressed and the chromosomes moved toward the nonfluorescing region (**Figure 11.12**).
- The researchers concluded that chromosome movement occurs because tubulin subunits are lost from the kinetochore ends. The rest of the microtubules remain stationary.

A Kinetochore Motor

- Kinetochore proteins "walk" down the microtubules as they shorten, similar to the way kinesin "walks" down microtubules during vesicle transport (**Chapter 7**).
- Dyneins and other kinetochore motor proteins appear to detach near the chromosome and reattach to the kinetochore microtubule further down its length, causing the microtubule shortening responsible for pulling chromosomes to opposite cell poles (**Figure 11.13**).

11.3   Control of the Cell Cycle

- Cell-cycle length varies greatly from tissue to tissue within the same individual. G_1 phase is practically eliminated in rapidly dividing cells such as those found in the intestine. Other cells get permanently stuck in G_1 phase (e.g., mature human nerve cells); this arrested stage is called the G_0 state.
- The rate of cell division may also respond to changes in conditions. These variations in cell-cycle length suggest that the cell cycle is

regulated and that regulation varies among cells and organisms.

The Discovery of Cell-Cycle Regulatory Molecules

- Certain chemicals, viruses, or electric shocks can fuse two cells, forming a hybrid cell with two nuclei.
- When two cells in different cell-cycle stages were fused, one of the two nuclei sometimes changed phase. For example, fusion of an M-phase cell with an interphase cell caused the interphase nucleus to enter M phase (**Figure 11.14a**).
- Researchers hypothesized that the M-phase cell cytoplasm contains a regulatory molecule that causes interphase cells to start mitosis. This hypothesis was confirmed by experiments with *Xenopus* frog eggs.
- Maturing *Xenopus* eggs change from an **oocyte** into a mature M-phase egg. These eggs are very large, making it feasible to purify large amounts of cytoplasm. This purified cytoplasm is then injected into eggs to test for the presence of mitotic regulatory factors.
- When *Xenopus* M-phase cytoplasm was injected into G_2-like-phase oocytes, the immature oocyte entered M phase. Cells remained in the G_2-like phase after injection with cytoplasm from interphase cells (**Figure 11.14b**).
- Researchers concluded that some factor is present in the cytoplasm of M-phase cells (but not in interphase cells) that induces mitosis. This factor has now been purified and is called **M-phase promoting factor** (**MPF**). MPF induces mitosis in all eukaryotes.

MPF Contains a Protein Kinase and a Cyclin

- MPF contains two polypeptide subunits. One, a **protein kinase**, catalyzes a protein phosphorylation reaction that transfers a phosphate group from ATP to a target protein.
- Many protein kinases are regulatory molecules because adding a phosphate group usually alters the target protein's shape and activity, either activating or inactivating the protein. However, the MPF protein kinase concentration does not change much throughout the cell cycle.

- The other MPF subunit, a **cyclin**, does change in concentration throughout the cell cycle. The MPF cyclin concentration increases during interphase, then peaks in M phase before decreasing again.
- The MPF protein kinase is active only when it is bound to the cyclin subunit; it is a **cyclin-dependent kinase** (**Cdk**). Thus, when cyclin concentrations are high, more MPF is active and the target proteins are phosphorylated, causing initiation of mitosis (**Figure 11.15a**).
- MPF itself is synthesized in an inactive phosphorylated form. Late in G_2 phase, enzymes remove the phosphates from cyclin, causing the activation of MPF, which then phosphorylates many different types of proteins (**Figure 11.15b**).
- One protein activated by MPF causes chromosomes to condense, initiating mitosis. Another protein activated by MPF catalyzes the degradation of the MPF cyclin. This negative feedback produces the observed oscillations in MPF cyclin concentration.

Cell-Cycle Checkpoints

- Other regulators of the cell cycle include a different cyclin and protein kinase that trigger the G_1-to-S-phase transition, and several regulatory proteins that maintain the G_0 state of quiescent cells.
- Leland Hartwell and Ted Weinert analyzed mutant yeast cells with cell-cycle defects. Some grew abnormally due to mutations that affected regulatory points in the cell cycle. These researchers coined the term **cell-cycle checkpoint** to describe a critical point in cell-cycle regulation (**Figure 11.16**).
- In our bodies, cells without effective cell-cycle checkpoints keep growing and form a **tumor**.
- Three checkpoints have been identified, the most important occurring late in G_1. At each checkpoint, interactions between regulatory molecules determine whether a cell proceeds with division. Unicellular organisms stop at the G_1 checkpoint if there is not enough food.
- Other factors affecting whether cells pass the G_1 checkpoint are (1) *cell size*—cells must be large enough to split into two functional

daughter cells; (2) *social signals*—signaling molecules from other cells in a multicellular organism; (3) **tumor suppressors**—regulatory proteins that can stop the cell cycle.

- For example, tumor suppressor **p53** stops the cell cycle if a cell's DNA is damaged. It can initiate **apoptosis** (programmed cell death). Defects in this protein can lead to cancer.
- In general, checkpoints function to ensure that a cell is healthy before DNA can be replicated and cell division initiated.
- Another checkpoint exists between the G_2 and M phases. For example, cells with incomplete or inappropriate chromosome replication arrest at this stage. It appears that DNA damage can block the dephosphorylation and activation of MPF.
- A checkpoint also occurs in metaphase. Cells arrest during M phase if the chromosomes are not properly attached to the mitotic spindle. This mechanism prevents incorrect chromosome separation that could give daughter cells the wrong number of chromosomes.

11.4 Cancer: Out-of-Control Cell Division

- Cancer is a common, often lethal disease affecting many humans. At least 200 different types of cancer can affect different organs— or the same organ in different ways. Despite the differences, all cancers derive from cells with cell-cycle checkpoint failures.

Properties of Cancer Cells

- A **tumor** is formed when one or more cells in a multicellular organism begin to divide uncontrollably. **Benign tumors** are noninvasive, but **malignant tumors** are cancerous and can spread throughout the body via the blood or lymph and initiate new tumors.
- Detachment from the original tumor and invasion of other tissues is called **metastasis**. Once a cancer has metastasized, it cannot be cured by surgery alone (**Figure 11.18**).

Cancer Involves Loss of Cell-Cycle Control

- Mature cells often remain in the G_0 state; but if they pass through the G_1 checkpoint, they

continue through mitosis and divide. Biologists thus hypothesized that cancers involve defects in the G_1 checkpoint. To understand how this checkpoint can fail, we must first understand its normal operation.

Social Control

- Unicellular organisms pass the G_1 checkpoint when nutrients are available and cell size is sufficient. Cells of multicellular organisms instead respond to signals from other cells, so that cells divide only when their growth benefits the whole organism. This is known as *social control*.
- Normally, mammalian cell cultures will not grow, even if provided with nutrients, unless **growth factors** are present. These polypeptides or small proteins are released by cells to signal other cells whether or not to grow. Growth factors were first found in **serum**, the liquid that remains after blood clots and the cells are removed from the blood.
- **Platelet-derived growth factor (PDGF)** is a serum growth factor released into the blood by platelets at wound sites. Cells near the wound divide in response to PDGF and help heal the injury.
- There are many different growth factors. Different cells divide in response to different combinations of growth factors, and those particular growth factors must be present in order for the cell culture to grow.
- Cancer cells, however, do not need growth factors in order to divide—they are no longer subject to social control at the G_1 checkpoint.

Social Controls and Cell-Cycle Checkpoints

- Growth factors initiate cell division by triggering cyclin synthesis. The cyclin then activates a cyclin-dependent kinase (Cdk) that activates the S-phase proteins (**Figure 11.19**).
- In some human cancers the G_1 checkpoint cyclin is overproduced, permanently activating Cdk, which then continuously phosphorylates its target proteins. Either the presence of excessive growth factors or cyclin production in the absence of growth factors can cause cyclin overproduction.
- For example, one target protein activated by Cdk is the retinoblastoma (Rb) protein. In

normal nondividing cells, Rb shuts down the cell cycle by binding to and inhibiting the E2F protein. When Rb is phosphorylated, it releases E2F, and E2F stimulates the production of molecules needed in S phase.

- Because Rb in its nonphosphorylated form prevents the transition to S phase, it is a tumor suppressor—defects in the Rb protein lead to continuously active E2F and uncontrolled cell division.

Cancer Is a Family of Diseases

- Many different types of defects can cause the G_1 checkpoint to fail. Most cancers result from multiple defects in cell-cycle regulation. Each type of cancer is caused by a unique combination of errors.

B. CROSS-CUTTING THEMES

Looking Back—
Concepts from Earlier Chapters
Cell Theory—Chapter 1
See **Chapter 1** for further discussion of the cell theory, its importance, and some supporting evidence.

Use of Radioactive
Isotopes—Chapters 7, 9, and 10
Radioactive isotopes are extremely useful research tools. **Chapter 7** described several pulse-chase experiments involving the use of radioactive isotopes to discover how molecules are transported throughout the cell. Other experiments that used radioactive isotopes to discover intermediate compounds in reaction pathways are described in **Chapters 9** and **10**. The use of radioactive thymidine to identify the timing of DNA synthesis is described here in Chapter 11.

Microtubules—Chapter 7
Microtubules were introduced in their role as cytoskeletal elements in **Chapter 7**. Here you discover that microtubules also make up the mitotic spindle, which pulls apart sister chromatids during mitosis. The way that spindle fibers move chromosomes is somewhat similar to the way that cytoskeletal microtubules function in vesicle transport.

Looking Forward—
Concepts from Later Chapters
Meiosis—Chapter 12
Chapter 12 describes meiosis, the type of cell division used in sperm and egg cell production. In many ways, meiosis is similar to mitosis. However, the daughter cells produced by meiotic divisions contain only half of the parental cell genetic information.

Chromosomes—Chapters 13 and 15
Chromosomes, cell structures made of DNA and protein, are responsible for carrying the cell's genetic information. Chapter 11 discusses how copies of each chromosome are provided to each daughter cell during cell division. In **Chapters 13** and **15**, you'll learn more about chromosomes, their importance, and how genes work.

DNA Synthesis— Chapter 14
Chapter 11 describes evidence that DNA synthesis occurs during S phase of the cell cycle. Details of DNA synthesis are given in **Chapter 14**.

C. DIFFICULT TOPICS

One effective way to learn different phases of the cell cycle is to simulate the events, using thread for the microtubules, shoelaces for chromosomes, and a plastic bag for the nuclear envelope. Recite aloud to yourself names of the phases and major events as you act out the process. You may also want to chart out the different cell-cycle phases, listing their major events and any cell-cycle checkpoints. Finally, use your own mnemonic device for remembering the mitotic phases. You could remember interphase, prophase, prometaphase, metaphase, anaphase, telophase, cytokinesis (IPPMATC) as "I Pet Purple Moose At The Circus."

D. ASSESSING WHAT YOU'VE LEARNED

(1) Testing Your Knowledge

1. What is the fundamental difference between mitosis and meiosis?
   a. Mitosis involves two cell divisions, whereas meiosis only involves one.

b. The number of chromosomes doubles in meiosis; it stays the same in mitosis.

c. The amount of genetic material per cell is cut in half in meiosis, but remains constant in mitosis.

d. Meiosis occurs in prokaryotes, and mitosis occurs in eukaryotes.

2. Chromatin is:
   a. the structure that joins sister chromatids
   b. the microtubule organizing center in animals
   c. one strand of a replicated chromosome
   d. the DNA-protein complex that makes up eukaryotic chromosomes

3. Mitotic spindle fibers are composed of:
   a. microtubules
   b. centrosomes
   c. centromeres
   d. kinetochores

4. A cell with its chromosomes lined up across the middle of the cell is in:
   a. prophase
   b. prometaphase
   c. metaphase
   d. anaphase
   e. telophase

5. During mitosis, it is necessary for the nuclear envelope of the parent cell to dissolve. This is accomplished, at least partly, by phosphorylation of proteins associated with the nuclear envelope. If the enzyme responsible for that phosphorylation event is inhibited, at which phase of mitosis are cells likely to arrest?
   a. prophase
   b. prometaphase
   c. metaphase
   d. anaphase
   e. telophase

6. In addition to a pulse-labeling experiment, the length of different phases of the cell cycle can also be measured by labeling cells continuously with radioactive thymine and waiting to see how long it takes for radioactivity to appear in metaphase chromosomes. What phase of the cell cycle would be represented by the time between

the addition of radioactive thymine and the time that radioactivity first appears in metaphase chromosomes?
   a. G_1 phase
   b. S phase
   c. G_2 phase
   d. M phase

7. In cells that are continuously fed radioactive thymine, after radioactivity appears in M-phase chromosomes, the amount of radioactivity increases steadily over several hours and then reaches a plateau. What phase of the cell cycle would be represented by the time between the first arrival of radioactive thymine in M-phase and the time that radioactivity reaches a plateau?
   a. G_1 phase
   b. S phase
   c. G_2 phase
   d. M phase

8. Although the process of chromosome partitioning during mitosis is visible through the light microscope, the process of DNA replication is not. Why?
   a. Chromosomes form only during mitosis.
   b. Chromosomes are visible only after DNA has been duplicated.
   c. Chromosomes are too extended during S phase to be seen by light microscopy.
   d. Chromosomes do not contain protein until mitosis.

9. A certain species of animal has six pairs of chromosomes. How many molecules of DNA do the nuclei of these animals have during G_2 phase?
   a. 6
   b. 12
   c. 24
   d. 48

10. In animals, cytokinesis occurs _____.
    a. when microtubules pull the cell membrane into a cleavage furrow
    b. when vesicles from the Golgi build up to divide the cell along the cell plate
    c. when G_2 phase is complete
    d. when a ring of actin and myosin filaments contracts

11. The experiment where biologists photo-bleached spindle fibers made with fluorescent tubulin showed microtubules shortened from the kinetochore ends. What differ-ent results would indicate that microtub-ules shorten from the microtubule-organ-izing center end?
    a. movement of the chromosomes toward a stationary photobleached area
    b. movement of the photobleached region toward the cell pole
    c. rapid disappearance of the photobleached microtubules
    d. none of the above

12. Some of the earliest experiments in control of the cell cycle were the cell fusion exper-iments of Rao and Johnson. Fusion of cells in M phase with cells in interphase caused the nuclei from the interphase cells to be-have as if they were in mitosis. These re-sults could have been obtained by adding just a part of the M phase cells to the inter-phase cells. What portion of the M phase cell would suffice to cause the interphase nuclei to mimic the process of mitosis?
    a. the nucleus
    b. ribosomes
    c. the endoplasmic reticulum
    d. cytoplasm

13. What would likely happen if you injected cells with a factor that degrades cyclin but does not need to be activated by MPF phosphorylation?
    a. The cell cycle would stop in G_2 phase.
    b. The cell cycle would stop in G_1 phase.
    c. The cell cycle would stop in M-phase.
    d. Cells would divide uncontrollably.

14. Active Cdk phosphorylates proteins that cause:
    a. M-phase initiation
    b. nuclear envelope breakdown
    c. cyclin degradation
    d. all of the above

15. Cells that are too small to successfully produce daughter cells fail to pass:
    a. the G_1 checkpoint
    b. the G_2 checkpoint

c. the metaphase checkpoint
d. all checkpoints

16. Can surgery successfully cure a cancer that has metastasized?
    a. No, all body cells are dividing uncontrollably.
    b. Yes, it could remove all cells with defective cell cycle regulation.
    c. No, cancer cells are no longer localized in one spot.
    d. Yes, if the tumor is benign.

17. Which of the following are true of social control over cell division? (Pick all that apply.)
    a. It occurs only in multicellular organisms.
    b. It acts at all cell cycle checkpoints.
    c. It typically involves the action of growth factors.
    d. The exact mechanism differs in different cell types.
    e. It is studied in cultured cancer cells.

18. What would be the consequence of inacti-vation of E2F?
    a. Cells would be unable to enter the S phase.
    b. Cells would immediately enter the S phase.
    c. Cells would develop cancer.
    d. The Rb protein would become activated.

19. One cause of cancer is mutation of the genes that encode for proteins in the biochemical pathway regulated by PDGF. How could such mutations cause cancer?
    a. If the mutations caused the biochemical pathway to become permanently active, cancer could result.
    b. If the mutations caused the biochemical pathway to become permanently inactive, cancer could result.
    c. Mutation of any gene will probably cause cancer.
    d. If the mutations altered the biochemical pathway so that PDGF stimulated cell division, cancer could result.

(2)   Integrating Your Knowledge

(a) What might happen to cells that repeatedly divided without any gap phases?

(b) Explain why earlier researchers were unable to determine when DNA synthesis occurred.

(c) Why might there be occasions when a researcher observing a mitotic cell under a microscope would have difficulty determining whether the cell is in prophase or metaphase?

(d) In the experiment on spindle shortening, why was labeled tubulin added during prophase or metaphase instead of during anaphase?

(e) How is the kinetochore motor's movement of chromosomes (along spindle microtubules) similar to kinesin's vesicle transport (on cytoskeleton microtubules)? How is it different?

(f) Why did Hartwell and Weinert use yeast instead of bacteria to study cell-cycle regulation?

(g) What is the difference between a benign and a malignant tumor?

(h) Why is there no social control in unicellular organisms?

CHAPTER 11—ANSWER KEY

D. Assessing What You've Learned

(1)   Testing Your Knowledge

1. c; 2. d; 3. a; 4. c; 5. a; 6. c; 7. b; 8. c; 9. c; 10. d; 11. b; 12. d; 13. a; 14. d; 15. a; 16. c; 17. a, c, and d; 18. a; 19. a

(2)   Integrating Your Knowledge

(a) Cells that divided without any gap phases would probably get smaller and smaller and have fewer and fewer organelles with each succeeding division. Eventually they may not have enough organelles or cytoplasm to function effectively.

(b) Earlier researchers could not determine when DNA synthesis occurred, because before radioactive labeled nucleotides became available, there was no way to mark or see when DNA synthesis occurred.

(c) Recall that mitosis is a continuous process. It may thus be difficult to determine whether the centrosomes have fully completed their migration to opposite sides of the cell and whether the chromosomes have fully, or only mostly, moved to the middle of the cell.

(d) Because the mitotic spindle is complete by the start of anaphase, labeled tubulin added in anaphase would not be incorporated into the mitotoic spindle.

(e) Like kinesin, the kinetochore motor appears to "walk" down the spindle microtubules Unlike kinesin, the kinetochore is located at one end of a microtubule that shortens as it loses tubulin subunits. Kinesin instead walks along microtubule "railroad tracks."

(f) Recall that bacteria replicate very differently from eukaryotes. They do not undergo mitosis, because they have circular chromosomes instead of linear chromosomes. Yeast, however, are eukaryotes that divide quickly and live well in culture—two qualities that often make bacteria useful in research.

(g) Malignant tumors are invasive; they are able to spread through the blood or lymph to other parts of the body. Benign tumors are not cancer; they do not spread to other parts of the body.

(h) Natural selection favors unicellular organisms that leave the most offspring. Thus, unicellular organisms should divide whenever enough nutrients are available to allow sufficient growth for cell division (asexual reproduction). Only in multicellular organisms is it favorable to the organism to allow other cells (in the same organism) to influence cell division.

Meiosis

A. KEY BIOLOGICAL CONCEPTS

12.1   How Does Meiosis Occur?

An Overview of Meiosis

- Examine **Figure 12.1** from the perspective of a scientist trying to understand the nature of chromosome function in the **lubber grasshopper**.

- As you begin studying the process of meiosis, it is important to realize that the germinal cell initiating this process contains **two homologues** of each chromosome (**2n or diploid**), and that the **gamete** cell receives only **one homologue** of each chromosome upon completion of meiosis (it will be 1n or **haploid**). In 1902, **Walter Sutton** published his observations of homologous chromosomes in the lubber grasshopper.

- In a diploid organism, what is the origin of each homologous chromosome? How might carrying this "extra" information be valuable? How might it be detrimental?

- Review **Figure 12.3** as an overview of the meiotic process.

- Just before meiosis begins, each of the chromosomes in the 2n parental cells are **replicated**. The newly synthesized chromosomal strands remain attached to the original chromosome from which they were copied until a later stage of meiosis.

- In general, each replicated chromosome looks like the letter X, with two linear **sister chromatids** being attached in the middle at a structure called the **centromere**.

- Examine **Figure 12.4**.

- **Meiosis I** involves the complex mechanism of crossing over in prophase I (described in the next section) and the separation of homologous chromosomes into two separate cells (each are now 1n).

- **Meiosis II** repeats the steps of meiosis I, the key difference being the physical separation of the sister chromatids (X ↓ < >), and the migration of each into a gamete cell that will contain one copy of each chromosome (**1n** or **haploid** gametes).

- **KEY**: Don't confuse the **homologous chromosomes**, which make the precursor cell 2n, with the copies of each that are produced before meiosis begins. As long as the sister chromatids are attached to each other at the centromere, they are considered a single replicated chromosome.

The Phases of Meiosis I

- Meiosis I contains two key events in the generation of genetic diversity from generation to generation.

- The first event is **crossing over** between homologous chromosomes in prophase I, and the second event is the **separation of homologous chromosomes** in anaphase I. These two steps generate incredible genetic diversity in the gametes, and recognizing how each contributes to this diversity should be a key focus in your study of Chapter 12. Both are discussed in detail here, but before we examine them in detail, let's review the steps of meiosis I.

- **Prophase I**—This step of meiosis takes the most time, primarily due to the complexity of crossing over. As it progresses, the chromosomes **condense** into tightly wound balls of DNA (X-shaped, visible with microscopy). Condensation is mediated by small **histone proteins** that bind to the chromosomal DNA and wrap the DNA around their surface. Crossing over occurs in prophase I and involves exchange of long stretches of DNA between homologous chromosomes. This is a key step in recombination of DNA for the production of genetically diverse gametes.
- **Metaphase I**—The condensed chromosomes migrate to the midline of the cell and form a line. Homologous chromosomes are adjacent to each other at the midline. Cytoskeletal proteins, known as microtubules, mediate this migration.
- **Anaphase I**—The second key step in generation of genetic diversity in gametes. The paired homologues are separated from each other, migrating to opposite ends of the cell. Following migration, each pole of the cell contains one complete set of chromosomes.
- **Telophase I**—The chromosomes complete their migration and are enclosed in two newly formed daughter cells (each is 1*n*; confirm by counting centromeres in each cell.

The Phases of Meiosis II

- **Meiosis II** is really just a repeat of meiosis I, with a few important differences. At the start of meiosis II, each daughter cell contains haploid DNA (1*n*). Each replicated chromosome is composed of the two sister chromatids. These chromatids, attached to each other at the centromere, are separated into separate cells during anaphase II. Once the chromatids have segregated into gametes, they are called chromosomes. Because there is no crossing over in prophase II, this step is more rapid than prophase I.

12.2 The Consequences of Meiosis

Chromosomes and Heredity

- A chromosome is a long, linear **macromolecule** composed of deoxyribonucleic acid (DNA). As discussed in **Chapters 13** and **14**, chromosomes hold information in the form of distinct regions of DNA that contain a code, found within the ordering of chemical subunits of the chromosome.
- Each information-carrying DNA region that carries the blueprint for a single **protein** is known as a **gene**.
- Proteins carry out functions within the cell; they form the cellular infrastructure (cytoskeleton), carry out steps in energy production, break down sugar into usable energy, and carry out cellular communication. *Proteins do things*, while genes are blueprints for the proteins.

How Does the Segregation and Distribution of Homologous Chromosomes Produce Genetic Variation?

- The separation (or segregation) of homologous chromosomes in anaphase I generates a great deal of **genetic diversity** in the subsequent gametes.
- Diploid organisms, carrying two **homologues** of each chromosome, divide up the diploid DNA content into two cells at the completion of meiosis I. This process is random, ensuring that each cell gets one replicated copy of each chromosome (a total of 23 in humans). In the production of gametes, the homologous chromosomes **segregate** into different cells at the completion of meiosis I. As each gamete receives one of the two homologues of each chromosome, and there are 23 chromosomes in humans, we are capable of generating 8.4 million (2^{23}) genetically unique gametes. Add to that the additional diversity generated by crossing over in prophase I, and that value increases dramatically.
- **Sexual reproduction** offers incredible diversity to offspring. As we'll see next, organisms invest a great deal of energy into increasing the **genetic diversity** in their offspring. Why is such effort expended to ensure genetic diversity from generation to generation? Doesn't this increase the likelihood of mistakes occurring during gamete formation?

The Role of Crossing Over

- In prophase I, a complex process of **chromosomal recombination** is initiated. Let's focus on one pair of chromosomes, knowing that each pair of homologues will go through the same process.

- The homologous chromosomes each underwent DNA replication before meiosis began, and are attached to the copy at a centromere. The homologous chromosomes come together (**synapsis**), recognizing each other by their matching DNA sequence and forming a tightly associated **tetrad** of chromosomes. *The tetrad contains both homologues and their copies, for a total of four strands of chromosomal DNA.*

- The four strands of DNA associate very closely with each other, and association is between corresponding regions of each chromosome. Crossing over is simply the exchange of sections of DNA from one strand of DNA to another (**Figure 12.6**).

- Two genes that encode the same protein but have small differences in coding information are called **alleles** of each other. There can be hundreds of alleles of a single gene.

- *Crossing over results in a shuffling of gene alleles between homologous chromosomes, and results in unique chromosomes that were not previously present.* Your cells have two copies of each chromosome (one from mom and one from dad). In production of gametes, you will generate chromosomes carrying alleles from your mother and from your father!

- *The regions exchanged are homologous;* that is, they carry the same genes, but subtle allelic difference in the genes contributed by each parent make these crossover products unique from the original chromosomes.

12.3 Why Does Meiosis Exist? Why Sex?

The Paradox of Sex

- Many types of organisms thrive through the process of **asexual reproduction**. A fascinating question in biology focuses on why sex is such a dominant reproductive strategy in multicellular organisms.

- Does the answer lie in the genes? Are organisms that reproduce sexually able to produce more diversity and adaptability in their offspring?

- Maynard Smith predicted that asexually reproducing organisms should reproduce faster and outcompete similar organisms that invest in sexual reproduction. What assumptions of this model might be problematic when you consider the influence of sexual/asexual reproduction on inheritance? What environmental factors might explain the success of sexual reproduction?

When Do Meiosis and Fertilization Occur during the Life of an Organism?

- In the human **life cycle**, haploid cells are found only in the production of gametes. Once the **diploid zygote** is formed by two gametes, mitotic division of this diploid cell generates all cells of the adult.

- Most multicellular organisms exhibit a similar, short-lived haploid phase within their life cycle, though there are very interesting examples of extended haploid growth in some organisms (**Figure 12.13**).

- Consider the life cycle of lower plants that include the mosses (**Figure 12.13c**); unlike most plants, these plants spend most of their life cycle as **haploid gametophytes**. These develop from spores into simple plants with differentiated tissues. Diploid cells are generated when a male gametophye produces sperm and fertilizes eggs produced by the female gametophyte. The **diploid sporophyte** grows while attached to the gametophyte and produces haploid spores when mature. These spores are released to produce more gametophytes.

- So in this example, haploid and diploid stages are both utilized extensively in this plant's life cycle. How might switching from haploid to diploid like this benefit the plant? One hypothesis is that plants carrying harmful mutations are rapidly removed from the population during the haploid stage of the life cycle, because the defective genes are unable to "hide" as a recessive mutation in a diploid cell. We will discuss how genes interact (dominant/recessive) in later chapters.

12.4   Mistakes in Meiosis

How/Why Do Mistakes Occur?

- Meiotic mistakes are an inherent risk in the complex process of generating genetic diversity during gamete production.
- A high percentage of zygotes carry severe abnormalities—extra chromosomes or missing chromosomes (**aneuploidy**), or fragmented chromosomes that were damaged at some point during meiosis. These zygotes typically do not survive to produce viable offspring.

B. CROSS-CUTTING THEMES

Looking Back—
Concepts from Earlier Chapters

The Cell Cycle—Chapter 11

While reading Chapter 12, you'll notice striking similarities between **mitosis and meiosis** and the basic cellular architecture enabling both processes. Examine **Chapter 11** and focus on cell components that are critical to **cellular division**. How might the **cytoskeleton** (filaments and microtubules), **cellular membranes**, and **chromosomes** be prepared for meiotic division? Would **DNA replication** precede this process?

Looking Forward—
Concepts from Later Chapters

Mendel and the Gene—Chapter 13

Mendel worked out much of the meiotic process with no knowledge of what a chromosome really was. The **principles of segregation and independent assortment** both describe what happens to chromosomes in meiosis.

DNA Synthesis—Chapter 14

We'll learn about DNA replication in **Chapter 14**, identifying how signals in the cell activate the deliberate process of copying the **genome** before division can occur.

Early Development—Chapter 21

Meiosis is a key process in **reproduction and fertilization**. In **Chapter 21**, we examine **gametogenesis** in more detail, and follow the gamete through fertilization.

C. DIFFICULT TOPICS

Mitosis versus Meiosis

Mitosis and meiosis share many characteristics—chromosome condensation, alignment of chromosomes at the cellular midline, migration of chromosomes to the poles, and formation of two new cells. What characteristics are observed only in mitosis or only in meiosis? How is meiosis I different from meiosis II?

Carefully examine these processes, focusing on the ploidy (or DNA content) at each stage and on how segregation at anaphase occurs. Some fundamental differences between these processes illustrate the purpose of each process. Mitosis occurs in all of our somatic cells, and it requires the careful transfer of all genetic information from mother cell to the daughter cells. Meiosis expends energy and engages in the dangerous process of rearranging our chromosomes to generate genetic diversity.

Genetic Recombination

Crossing over is a complex process that is often confusing at first. This is especially true when you consider the impact of crossing over on the independent assortment of genes. Diagram the formation of a tetrad and a crossover event between two homologues (use color to make the crossover more visible). Resolve the crossover by illustrating what would then happen to these homologues in anaphase I; notice how all four chromosomes are now unique, each carrying a slightly different DNA content.

Terminology

This chapter introduces many new terms and concepts that are crucial to your understanding of this, and future, chapters. Take time to learn these concepts, and you'll be prepared to move on to more complex concepts in genetics. Here are some terms to carefully examine/define as you study:

aneuploidy	chiasma	chromatids
diploid	gamete	gene
haploid	homologous	life cycle
chromosomes	nondisjunction	out-crossing
(tetrad)	trisomy	zygote
self-fertilization	sexual reproduction	
synapsed chromosomes		

D. ASSESSING WHAT YOU'VE LEARNED

(1) Testing Your Knowledge

1. Which of the following occurs during meiosis II, but does not occur as part of meiosis I?
   a. crossing over between homologous chromosomes
   b. separation of homologous chromosomes
   c. separation of sister chromatids
   d. migration to middle of cell

2. Crossing over:
   a. occurs between sister chromatids on the same chromosome
   b. occurs between chromatids of homologous chromosomes
   c. occurs very rarely
   d. occurs during prophase II of the cell cycle

3. An individual plant that exhibits self-fertilization can produce offspring different from itself primarily as the result of:
   a. mutations that occur during mitosis
   b. independent assortment and crossing over during the formation of gametes
   c. crossing over (only) during the formation of chromosomes
   d. the fact that the number of chromosomes in pollen is always different from the number of chromosomes in egg cells

4. **Figure 12.11** in your textbook demonstrates which of the following?
   a. Organisms produced by sexual reproduction are more variable than those produced by asexual reproduction.
   b. Sexual reproduction is always more successful than asexual reproduction.
   c. Asexual reproduction is always more successful than sexual reproduction.
   d. Asexual organisms generally produce more offspring than sexually reproducing organisms.

5. Robert's father has brown eyes and brown hair; his mother has blue eyes and blonde hair. Assume that the two traits are linked. Robert has a daughter with brown eyes and blonde hair.
   Genetics tests have shown that Robert contributed his mother's hair and eye color chromosome to his daughter. Which of the following would most likely explain his child's hair and eye color? (Assume, for this question, that the daughter's mother's eye-color and hair-color genes are not expressed.)
   a. independent assortment
   b. crossing over of Robert's mother's chromosomes
   c. crossing over of Robert's genes
   d. crossing over of Robert's father's genes

6. Which of the following best describes "alternation of generations," the life cycle employed by most land plants?
   a. A haploid adult produces haploid gametes, which fuse at fertilization and produce a new diploid adult by mitosis.
   b. A diploid cell undergoes to produce a haploid cell that produces a haploid adult by mitosis. The adult then produces haploid gametes by mitosis.
   c. A diploid organism produces haploid cells (spores), and these cells divide by mitosis to produce a haploid organism that produces haploid gametes by mitosis.
   d. A diploid adult produces haploid gametes, which fuse at fertilization and produce a new diploid adult via mitosis.

7. Which of the following would be most affected by a mutation that occurred at a single locus on one chromosome?
   a. a human adult
   b. a land plant
   c. a human embryo
   d. None of the organisms listed here would be affected by a mutation in a single copy of a gene.

8. Which of the following statements is true?
   a. Organisms such as animals, which reproduce sexually, are immune to disease.

b. Because humans reproduce sexually, they are the predominant species on Earth.

c. All bacteria of a given species are genetically identical to each other.

d. Bacteria have the ability to increase their genetic variation.

9. Diagram meiotic crossing over between two adjacent genes on the same chromosome (*A* and *B*, shown in Chart 1). The parental cell entering into meiosis is genotypically *aB/Ab*, and will generate four haploid gametes at the completion of meiosis. Diagram the genetic makeup of the gametes produced by this meiotic process, and compare these gametes to the original parental contributions.

START OF MEIOSIS

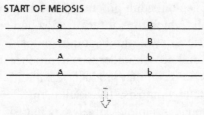

CROSSOVER BETWEEN GENES

GAMETE PRODUCTION

gamete #1

gamete #2

gamete #3

gamete #4

ORIGINAL PARENTAL CONTRIBUTIONS

(2) Integrating Your Knowledge

(a) Compare the ploidy of a diploid (2*n*) cell proceeding through mitosis of meiosis, acknowledging where each change in ploidy takes place.

(b) Describe the two steps in meiosis where most genetic recombination occurs.

(c) Explain nondisjunction as it relates to meiosis; where does this problem predominantly arise?

(d) How might asexual reproduction offer an advantage in some environments?

(e) Why is the freshwater snail (*P. antipodarium*) a good choice for experiments comparing sexual and asexual reproduction?

CHAPTER 12—ANSWER KEY

D. Assessing What You've Learned

(1) Testing Your Knowledge

1. c; 2. b; 3. b; 4. d; 5. c; 6. c; 7. b; 8. d

9. Once the tetrad is formed, strands are exchanged between the homologous chromosomes. If this occurs between the two genes as diagrammed below, then half of the resulting gametes will have a unique combination of alleles relative to each parent.

Chart 1

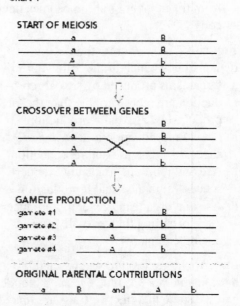

(2)   Integrating Your Knowledge

(a) You can determine ploidy by counting the centromeres; a diploid cell with replicated chromosomes is still diploid, although it has doubled its DNA content.

Mitosis—Before entering into mitosis, the $2n$ cell first replicates its genome. At anaphase, *sister chromatids* are separated and migrate to oppo-site poles of the cell. The cell then divides into two $2n$ daughter cells.

Meiosis—Before entering into meiosis, the $2n$ cell first replicates its genome. At anaphase I, *homologous chromosomes* are separated and migrate to opposite poles of the cell. The cell then divides into two $1n$ cells (sister chromatids are still attached). At anaphase II, *sister chromatids* are separated and migrate to opposite poles of each cell. Then the cell divides into two $1n$ gametes.

(b) **Prophase I**—Homologous chromosomes (replicated) synapse into a tetrad and exchange homologous regions of DNA. This crossing over between chromosomes happens at least once for each tetrad.

Anaphase I—The tetrad separates and homologous chromosomes migrate to opposite poles of the cell. Each centromere remainsintact, with two sister chromatids attached, until anaphase II. The various combinations of homologues generated at the end of mei-osis I offer dramatic diversity to the subse-quent gamete.

(c) The most common site of nondisjunction is meiosis I, during which homologous chromosomes do not properly migrate to opposite poles of the cell. This can occur during meiosis II, but it is less common (see **Figure 12.14**).

(d) Reproduction in the absence of sex is faster, requires less energy, and does not require the fusion of germ cells from two sources. Asexually reproducing organisms can reproduce much more rapidly than sexually reproducing organisms. But, researchers have shown that asexual reproduction offers less adaptability (genetic variability) to the organism.

(e) This snail can reproduce both sexually and asexually. Populations of snails with high numbers of males tend to be reproducing sexually, while asexually reproducing populations have more female snails. The researcher (Curtis Lively) could thus analyze the ratio of males to females in populations of snails that are resistant or sensitive to disease (parasitism).

13

Mendel and the Gene

A. KEY BIOLOGICAL CONCEPTS

13.1   Mendel's Experiments with a Single Trait

What Questions Was Mendel Trying to Answer?

- **Blending vs. acquired characters**—These competing hypotheses of Mendel's time were both severely damaged by data produced from his simple experimentation with pea plants. Review these hypotheses, and the specific results that contradicted them.

- **Incomplete dominance** is a form of blending, with the offspring displaying the phenotype of both alleles. This occurs with some genes, but did not occur in Mendel's analyses with peas. We'll talk more about incomplete dominance later in the chapter.

Garden Peas Serve as the First Model Organism in Genetics

- Consider all of the characteristics you'd want in a **model organism**. If you were setting up the experiments, what category of organism would you select—a plant, insect, worm, fungus, mammal, or primate? How would you select from among all the different possible representatives of each group? What traits would be the most, or the least, valued?

- Mendel chose a common garden pea (*Pisum sativum*), and revolutionized our knowledge of genetics for all eukaryotic organisms. He used the garden pea because:
  1. It is easy to grow.
  2. It is inexpensive; little requirement for specialized equipment.
  3. Its matings can be controlled completely (**Figure 13.1**)
  4. It has easily observed traits (**phenotypes**).
  5. It has a rapid life cycle (seedling to reproductively mature)

Here are some critical terms and concepts from Chapter 13:

- **Allele**—A single gene may have multiple alleles. Each allele is different from all other alleles, with some difference(s) in information content.

- **Dominant/recessive traits**—Most normal or wild-type traits are dominant; mutant traits are recessive. In other words, the mutation is hidden in organisms that carry both a wild-type and a mutant allele (different types of the same gene). Some mutations are dominant to the wild-type allele, but these are unusual.

- **F_1 generation**—The progeny of the parental organisms are called the "first filial" or F_1 generation. These progeny are genetically composed of one copy of each chromosome from the mother and one copy of each chromosome from the father.

- **Genotype**—The collection of alleles comprising the genome (total DNA content) of an organism.

- **Heterozygous**—In somatic cells, individuals carrying two different alleles of the same gene are heterozygous for that gene.

- **Homozygous**—In somatic cells, individuals carrying two copies of the same allele are said to be homozygous for that gene.

- **Parental generation**—In a standard mating, these would be the original parental organisms that donate gametes. In Mendel's experiments, parental plants were true-breeding, indicating several generations of self-fertilization.
- **Reciprocal crosses**—Two separate matings (crosses) performed with reversed male and female contributions. For example, pollen taken from a purple plant fertilizing an oocyte from a white plant (cross 1); and in a second cross, pollen from a white plant fertilizing an oocyte from a purple plant. These crosses are valuable in identifying genes that are located on the sex-determining chromosomes (see the white-eyed fly example, **Figure 13.12**).

Inheritance of a Single Trait (the mono-hybrid cross)

- Gregor Mendel performed many simple crosses, focusing on one trait (such as flower col-or) in experiments known as **mono-hybrid crosses** (**Figure 13.4**). Consider for example, when a wild-type, purple-flowered plant (des-ignated PP) is crossed by a mutant, white-flowered plant (designated pp). If the F_1 prog-eny are all purple, what does that indicate about the relationship between the two alleles?
- You will remember from **Chapter 12** that the product of gamete fusion is **diploid**. Thus the F_1 plant is heterozygous for the flower color gene (Pp), and since the F_1 progeny are all purple, the allele generating purple flowers (P) is **dominant** to the allele generating white flowers (p). The dominant allele is usually written in capital letters, while the **recessive** is written in lowercase letters. *Mendel did not observe blending!*
- Mendel crossed the purple-flowered F_1 prog-eny (or allowed **self-fertilization**), resulting in an F_2 generation. These plants displayed a **3:1 ratio** of purple- to white-flowered plants (¾ of all progeny were purple, and ¼ of all progeny were white). The recessive phenotype reappeared in the next generation, and thus must have been "hidden" in the F_1 plant.
- The segregation of alleles in gamete production is revealed in the ratios of F_2 progeny

phenotypes. Examine the **Punnett square**; it is useful in the analysis of inheritance.
- Mendel described gene segregation in meiosis long before scientists even began to look at chromosomes in the production of sperm or oocytes. Mendel developed a model in which a plant carrying a genotype of Pp generates 50% P gametes and 50% p gametes. Thus, the parental organism carries two determinants for each trait, and these segregate separately in the generation of gametes. This is referred to as Mendel's **principle of segregation**.

Mendel Used Mathematics to Characterize Inheritance (Box 13.1)

- Mendel used probability calculations to predict the ratios produced by specific crosses.
- Using the preceding example, what is the probability of producing a white-flowered (pp) plant if a heterozygous (Pp) plant is allowed to self-fertilize?
- Mendel knew that in order to make this probability calculation, he needed to include the individual probabilities for the contribution of the recessive allele in each gamete.
- The probability for any individual gamete to carry the recessive p allele is .50 (50%). To generate the white-flowered plant, a p oocyte must be fertilized by a p pollen grain. These two independent events must occur simultaneously to generate a homozygous recessive plant; thus, their probabilities are multiplied together (this is called the **product rule**).
- Thus .25 or 25% of the progeny from a self-fertilized heterozygous plant will be white-flowered (pp).
- What is the probability (p) of the same heterozygous F_1 plant generating a genotypically homozygous F_2 plant with respect to the P gene (i.e., either PP or pp)?
- This type of calculation, called an **either/or situation**, requires one additional step. You first use the product rule to determine the individual probabilities for the generation of a PP plant (.25) or a pp plant (.25).
- These independent probabilities are then added together to calculate the probability of generating either one or the other in any given fertilization event (this is called the

sum rule). Thus, .25 + .25 = .50, or 50%. Mendel used simple calculations of this type to make testable predictions about inheritance, and to analyze his data with statistical tests (**Box 13.2**).

13.2   Mendel's Experiments with Two Traits (the Dihybrid Cross)

The Principle of Independent Assortment

- Inclusion of a second gene in a single cross, a **dihybrid cross** reveals the independent assortment of unlinked genes. One example given in the text (**Figure 13.6**) includes the cross of a plant that displays round (*R*), yellow (*Y*) peas by a plant that displays wrinkled (*r*), green peas (*y*). Thus the parental cross is *RRYY* ∝ *rryy*, and the F_1 progeny would be *RrYy*. The F_1 plants display the dominant phenotypes (round, yellow), but the F_2 plants display four combinations of phenotypes. Mendel found traits in these dihybrid crosses to segregate in a 9:3:3:1 pattern. Examine the Punnett square in the figure to confirm that this ratio makes sense.

- By careful analysis of his results, Mendel was able to establish the principle of **independent assortment**. *This principle states that genes segregate independently of each other in the production of gametes* (**Figure 13.6a**). This is true for most genes; but some genes located on the same chromosome do not assort independently, and are described as linked (see **Section 13.4**).

- A simple way to "see" independent assortment is to diagram a simple meiotic division. Without knowing anything about DNA or chromosomes, Mendel was describing chromosomal assortment in meiosis.

- You can predict the genotypes and phenotypes of progeny from specific dihybrid crosses by using the product and sum rules. Compare the results of your calculations with the proportions predicted by a Punnett square.

Using a Testcross to Confirm Predictions

- The **testcross** is a simple analysis used in controlled mating experiments. It allows the investigator to focus on the meiotic events in only one of the two parents.

- A testcross is accomplished by crossing the organism of interest with a true-breeding organism that typically contains only recessive alleles for the genes being studied. Thus, for the genes being studied, the true-breeding organism will produce only one type of gamete.

- Review the example given in text (an *RrYy* plant crossed by an *rryy* plant; **Figure 13.8**). This cross isolates the meiosis of the *RrYy* plant, because only *ry* gametes will be produced by the testcross plant. Thus, the progeny genotype/phenotype reveals what happened during meiosis of the *RrYy* plant. This powerful strategy has many applications in gene analysis.

13.3   The Chromosome Theory of Inheritance

- The **chromosome theory of inheritance** was born out of careful observations of meiosis and the movement of individual chromosomes during the stages of meiosis. Early researchers realized that the reduction in chromosome number during **gametogenesis** (gamete formation) was consistent with Mendel's principle of segregation. They also realized that the movement of homologous chromosomes was consistent with Mendel's principle of independent assortment (**Figure 13.9**), and that the chromosomes likely contained heritable units of information that were passed from generation to generation.

13.4   Testing and Extending the Chromosome Theory

Discovery of Sex-Linked Traits; Discovery of Sex-Linked Chromosomes

- In the early years of observing chromosomes in meiotic cells, scientists struggled with observation of a **heteromorphic** pair of chromosomes that align with each other in anaphase I of meiosis (**Fig. 13.11**). These chromosomes were called heteromorphic since they did not match (different sizes), and all other homologous chromosomes were precisely matched with a chromosome of the same size.

- These heteromorphic pairs represent the sex chromosomes—the X and Y chromosomes.

- It was observed that XY males generate X gametes and Y gametes, and that the composition of the sperm regarding X or Y content determined the sex of any fertilized oocyte.

- Thomas Hunt Morgan, a researcher at Columbia University, utilized the fruit fly (***Drosophila melanogaster***) as a model organism to examine mechanisms of inheritance. He established a line of flies carrying a recessive mutation that resulted in white eyes (rather than red). In monohybrid breeding experiments to examine this trait, Morgan found the expected 3:1 ratio of red-eyed to white-eyed flies in the F_2 generation. However, Morgan found that this trait did not segregate evenly among male and female progeny.

- Examine the reciprocal crosses shown in **Figure 13.12**. Notice that the F_1 and F_2 generations are different depending on which parental fly displays the white-eyed trait. *Note: Whenever the results of reciprocal crosses generate different ratios of males/females that display the trait, the gene is located on a sex-determining chromosome (the X or Y).*

- In pedigree analysis, whenever a phenotype within a pedigree is predominantly associated with males or females, one should suspect that the gene is located on a sex-determining chromosome (the X or Y).

- Review Morgan's data using a **Punnett square**; remember, if the gene is located on the X chromosome, the sex of parental cells used can dramatically influence the phenotypical ratios of the progeny.

- An example—If gene g^+ is located on the X chromosome, then the male genotype may be represented as g^+Y. In gamete production g^+ will segregate with the X chromosome and will be present in 50% of the gametes (carrying the X), while the other 50% of the gametes receive the Y chromosome. Use a Punnett square to analyze reciprocal monohybrid crosses for g. How are the F_1 and F_2 progeny different in these crosses?

What Happens When Genes Are Located on the Same Chromosome?

- When examining dihybrid crosses, all examples of independent assortment we've discussed to this point depend on genes being located on separate chromosomes. What happens when two genes are located near each other, on the same chromosome?

- First, their assortments are not independent of each other; they are physically attached (or **linked**) to each other on that chromosome. Linkage can be identified by segregation patterns that do not show independent assortment of the two genes (something other than 3:1 for a monohybrid, or a 9:3:3:1 for a dihybrid cross).

- This is a difficult concept to visualize; review the meiotic process again, taking into account two genes located on the same chromosome. Those two linked alleles will segregate into the same gamete *every time,* unless a crossover event separates them onto homologous chromosomes. But any gametes that require a crossover event to be produced will be uncommon. Let's look at one example provided in the text (**Figure 13.13**).

- Two genes are utilized to illustrate gene linkage. Genes/alleles included in the example are w^+—red eyes, w—white eyes, y^+—gray body, and y—yellow body. The parental animals in this example donated a w^+y gamete and a wy^+ gamete to generate the F_1 displayed in part (a). Notice that the contributions of the parents are linked on the same chromosomes, and that in meiosis, the only way to generate a w^+y^+ gamete or a wy gamete is for crossing over to occur.

- Therefore, the parental genotype (for w and y) will be the most common genotype found in the gametes produced. Independent assortment will not occur between these two alleles; they are linked. The phenotypes of the F_2 progeny are not 9:3:3:1, but rather are dominated by the parental phenotype (wy^+ and w^+y).

13.5   Extending Mendel's Rules

Incomplete Dominance

- **Mutant alleles** of a single gene are not always clearly dominant or recessive to the **wild-type alleles**. Sometimes the heterozygous organisms display an intermediate phenotype, between the wild-type and mutant phenotypes; this is described as incomplete dominance.

Codominance

- A heterozygous organism, displaying the phenotype of both alleles of a single gene, is said to display **codominance**. Neither allele is dominant or recessive to the other.

Gene Interactions

- There are many examples of genes that influence the phenotype controlled by a second genetic locus.
- Excellent examples include the genes for coat color in mammals. The *B* and *b* alleles control the production of pigment in hair cells, generating either black or brown hair. Thus *BB* and *Bb* animals have black pigments, and *bb* animals have brown pigments.
- But the recessive allele of a second gene (*C*) can completely turn off pigment synthesis. So a *BBcc* animal is white (as are *bbcc* and *Bbcc*), illustrating that *cc* is **epistatic** to *BB* and eliminating its impact on phenotype.

Quantitative (Complex) Traits

- Some traits are influenced by the contribution of multiple genes, and display wide variability (e.g., skin color). These traits are called complex due to the complex pattern of gene and phenotype segregation that is observed in standard crosses.

B. CROSS-CUTTING THEMES

Looking Back—
Concepts from Earlier Chapters
Cell Structure and Function—Chapter 7
As you explore the work of Mendel, review the cell biology behind **phenotype** ("show-type" or form). For example, what aspects of cell biology might result in the loss of specific coloration in a flower? The genes that encode pigment can be **mutated**, and their loss or defect is a very obvious change in flower coloration (abnormal phenotype).

Meiosis—Chapter 12
Without knowledge of the **meiotic process**, it is difficult to follow the key conclusions of

Mendel's work. Review meiosis, examining how gene segregation, independent assortment, and gene linkage fit into the production of gametes. All of Mendel's work was really a characterization of chromosomal behavior in meiosis, though this was not discovered for many years.

Looking Forward—
Concepts from Later Chapters
How Genes Work—Chapter 15
Mendel established the foundations of gene inheritance without knowing what a gene really was! **Chapter 11** examines the basics of **gene structure and function**, revealing details of these heritable units. Each chapter in Unit 3 builds upon Mendel's work, moving toward the utilization of genes in powerful engineering strategies that have revolutionized biology.

Plant Form and Function—Chapter 35
Mendel was able to control mating in plants, fertilizing each cross by hand to ensure that both parents of each cross were known. **Chapter 35** offers additional information on **plant structure** and **plant reproduction**. Mendel also examined traits that include flower color, seed shape and color, and even plant height.

C. DIFFICULT TOPICS

Linkage Analysis
A prominent experimental goal in genetic analysis is to determine the location of a specific gene—first identifying the chromosome and then the region of that chromosome on which the gene is located. Disease-causing genes can sometimes be "mapped" to a specific place by using the medical records of families that carry the disease-causing allele. Finding the gene that causes a specific disease is a difficult task, but the rewards of finding the gene make the incredible effort worthwhile. There are many success stories of scientists identifying disease-causing genes (Huntington's disease, breast cancer). But how does this process work? Scientists use various strategies, one of which is fundamentally described in this chapter—linkage analysis.

Determining genetic linkage involves looking for two genes (or markers) in the genome that do not assort independently of each other. The alleles being studied are often found to segregate together from the parental genome. The easiest way to see linkage is to examine the testcross example. Examine the cross of an *AaBb* organism by an *aabb* organism. What predictions would you make if the genes are unlinked and assort independently of each other? How would linkage between the two genes change the results?

Linkage between *A* and *B* would be revealed by the ratios of progeny produced by this cross. Using these data, scientists can actually determine the distance between two linked genes. As the geneticists gradually refine their assessment of where the gene is located, they can then begin the process of actually finding the gene itself.

D. ASSESSING WHAT YOU'VE LEARNED

(1) Testing Your Knowledge

1. Draw a diagram of a diploid cell with two sets of chromosomes, placing alleles of two genes on the chromosomes as follows: gene **A** on chromosome 1, and gene **a** on chromosome 1≤gene **B** on chromosome 2, and gene **b** on chromosome 2≤Now diagram the 1*n* products of meiosis in these cells. In what four possible ways will these alleles segregate? *Hint:* The key to this segregation is anaphase 1.

2. If an organism is heterozygous for two traits that are linked, how many different genotypes are possible in the gametes? Assume that no crossing over occurs.
   a. 1
   b. 2
   c. 4
   d. 8

3. If an organism is heterozygous for two traits that are linked, how many genotypes are possible in the gametes produced from a single germ line cell if even a single crossing over occurs during meiosis I?
   a. 1
   b. 2
   c. 4
   d. 8

4. Pepper color is a trait that exhibits epistasis. If a pepper plant that has red peppers is heterozygous for pepper color (*Rr*), which of the following represents the phenotypic ratio of the offspring?
   a. unable to determine from the information given
   b. 9:3:3:1
   c. 3:1
   d. 1:2:1

5. A female housefly with red eyes mated with a male housefly with white eyes. Their offspring were:
   90 females with white eyes
   43 males with red eyes
   47 males with white eyes
   Based on this information, which of the following statements is most accurate?
   a. The gene for red eyes must be dominant.
   b. The gene for red eyes is inherited in a 3:1 ratio.
   c. The gene for red eyes must be on the X chromosome.
   d. The gene for white eyes must be on the Y chromosome.

6. As researchers study genes more closely, they are finding that many genes are pleotropic. This means that:
   a. the wild-type allele displays incomplete dominance over mutant alleles
   b. the genes encode multiple unique proteins
   c. the phenotype of mutant alleles depends upon environmental conditions
   d. mutations in these genes generate more than one distinct phenotype (trait)

7. Using Punnett square analysis, examine reciprocal crosses of fruit flies carrying recessive copies of the g^+ gene (as described earlier, the gene is located on the X chromosome). First cross a mutant fly (gY) with a wild-type fly ($g^+ g^+$). Next cross a wild-type fly ($g^+ Y$) with a mutant fly (gg). Do these crosses generate different ratios of F_1 progeny? Calculate the ratio/phenotypes.

Cross 1

Cross 2

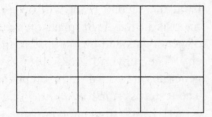

(2)   Integrating Your Knowledge

(a) Mendel's choice of the pea plant (*Pisum sativum*) as his experimental organism was critical to his success. Describe how things might have turned out differently if he had chosen laboratory mice.

(b) Using pea plants, you are analyzing two genes (*A* and *B*) using simple dihybrid crosses. You generate F_1 progeny that are heterozygous at both loci (*AaBb*), and interbreed these plants to generate F_2 progeny. What percentage of the F_2 progeny will carry the genotype *AABB* or *aabb?* Try to answer this question using the Punnett square, then use probability calculations to determine the answer. Do they agree with each other?

(c) Explain the value of a testcross. What advantage is it for a geneticist? Compare a standard dihybrid cross, of *AaBb* by *AaBb*, to a testcross of *AaBb* by *aabb*.

(d) What is a reciprocal cross? In what type of analysis is it particularly valuable?

(e) Allelic codominance is a genetic relationship that Mendel did not observe, but was identified later as an exception and extension to common patterns of inheritance. Give an example of allelic codominance.

(3)   Showcasing Your Knowledge

(a) Walter Sutton and Theodore Boveri hypothesized that the chromosomes they visualized during meiosis were the hereditary determinants predicted by Gregor Mendel 30 years earlier. They utilized light microscopy to observe the behavior of chromosomes over time as meiosis proceeded. Define Mendel's principle of segregation and independent assortment, and the meiotic processes Sutton and Boveri would use to explain these principles (what did they see?).

(b) Define a genetic map. How would a testcross be used to identify the genetic difference between two genes? What is the difference between physical distance and genetic distance?

CHAPTER 13—ANSWER KEY

D. Assessing What You've Learned

(1)   Testing Your Knowledge

1. Four haploid gametes are possible (*AB, Ab, aB,* and *ab*). If the chromosomes assort independently, each of these gametes will be produced in equal proportions.

2. b; 3. c; 4. a; 5. c; 6. d

7. **Cross 1**

	g	Y
g^+	g^+g	g^+Y
g^+	g^+g	g^+Y

Because g^+ is dominant to g, all of the flies will display a normal phenotype.

Cross 2

	g^+	Y
g	g^+g	gY
g	g^+g	gY

All male flies display the mutant phenotype. All female flies are wild type. The reciprocal crosses have different results! The mutant phenotype is associated with males only in the second cross. This gene must be located on the X chromosome.

(2) Integrating Your Knowledge

(a) Laboratory mice are much more difficult to study in large numbers; they reproduce in litters of 5–10. In studying pea plants, Mendel analyzed hundreds to thousands of progeny from each cross and then analyzed the data using statistical analysis. If he was examining only a few progeny, it would have been much harder to establish statistically relevant data.

Peas have many other advantages over mice for Mendel's analysis. Peas are easier to take care of, reproduce more rapidly, display easily identified phenotypes, it is fairly simple to generate true-breeding lines, and they are edible. Mice are important experimental organisms in genetics; but for Mendel's experiments, peas were the better choice.

(b) Determine this answer by making a simple 4 $\propto$ 4 Punnett square like the one in **Figure 13.6b**. The genotypes *AABB* and *aabb* each appear in 1/16 of the progeny. For either *AABB* or *aabb*, the frequency is 1/16 + 1/16 = 2/16, or .125 (12.5%).

Also make this determination by calculating individual probabilities of each genotype. Remember that for unlinked genes in a dihybrid cross, an *AaBb* organism generates an equal distribution of the four possible gametes (25% *AB*, 25% *ab*, 25% *Ab*, and 25% *aB*). Probability of generating an *AABB* plant will require the fertilization of an *AB* gamete by another *AB* gamete. These two independent events must occur simultaneously, thus you will use the **product rule** to calculate probability of this fertilization event.

Probability for *AABB*: (.25) (.25) = .0625 (or 6.25%)

Probability for *aabb*: (.25) (.25) = .0625 (or 6.25%)

Now you can use the **sum rule** to calculate the probability of generating either *AABB* or *aabb*: (.0625) + (.0625) = .125 (or 12.5%).

(c) The testcross allows for careful examination of meiosis in one parent and controls the gametes introduced by the other parent. For example, when crossing an *AaBb* plant by an *aabb* plant, all of the gametes from the second plant will be *ab*. Thus, meiosis in the *AaBb* plant dictates ratios of genotypes and phenotypes in the progeny. The dihybrid cross *AaBb* by *AaBb* generates a 9:3:3:1 ratio of phenotypes. The testcross *AaBb* by *aabb* will generate a 1:1:1:1 ratio of phenotypes.

(d) Reciprocal crosses are two crosses analyzing the same traits, but in each cross the genotype of the parents is reversed. For instance, a *bb* male fruit fly may be crossed by a *BB* female fruit fly. Then the reciprocal cross will be a *BB* male by a *bb* female. These crosses are particularly valuable in identifying sex-linked genes, or genes located on sex-determining chromosomes (X or Y). If reciprocal crosses generate different ratios of progeny genotypes, the gene involved is likely located on a sex chromosome. Review the story of the white-eyed male fruit fly (**Figures 13.12** and **13.13**).

(e) The ABO blood groups in humans illustrate allelic codominance. A, B, and O refer to phenotypes possible from the allelic composition of a single gene (*I*). Three alleles of this gene—I^A, I^B, and *I*—control the composition of cell-surface molecules on red blood cells. If you are genotypically *ii*, there are none of these cell surface molecules; if you are $I^A I^A$, you have the A cell-surface marker; but if you are $I^A I^B$, you have both A and B on the cell surface. These two alleles are codominant relative to each other; the heterozygotes display both phenotypes.

14

DNA Synthesis

A. KEY BIOLOGICAL CONCEPTS

14.1 DNA as the Hereditary Material

Transformation Experiments

- In 1928 Fredrick Griffith utilized the bacterium *Streptococcus pneumoniae* to evaluate its role in human infection and disease. Griffith knew there were **virulent** strains of *S. pneumoniae* that could cause death in the infected individual, and **nonvirulent** strains that were harmless as pathogens.

- What was different about these strains of the microbe? As Griffith examined these strains, he stumbled upon the capacity of one bacterium to alter the other, presumably through the transfer of an information-bearing molecule.

- **Figure 14.1**; the virulent *S. pneumoniae* can be distinguished from the nonvirulent strains simply by growing the cells in a petri dish (virulent = colonies that appear smooth; nonvirulent = rough or bumpy colonies).

- Griffith utilized mice to assay for the virulence of individual samples, injecting the samples into healthy animals and evaluating their tolerance of each sample.

- The discovery that samples containing heat-killed virulent bacteria could somehow transform nonvirulent cells to virulent was exciting. Somewhere within the heat-killed sample were molecules that carried the information for becoming virulent, and the nonvirulent cells were likely taking up those molecules.

- What experiments might you suggest to extend these results in identifying the information-carrying molecule that was transferred between strains? Be imaginative, but be sure to include controls as part of the analysis.

DNA Is the Transforming Factor

- Researchers followed up on Griffith's experiment by attempting to selectively remove macromolecules from the sample of heat-killed, smooth *S. pneumoniae* (**Figure 14.3**). The researchers used enzymes that selectively degrade DNA, RNA, and protein, removing other macromolecules (lipids, carbohydrates) with an extraction step.

- Notice that only the samples treated with a **DNase** (an enzyme that degrades DNA) lost their ability to transform the nonvirulent cells to virulent.

- These experiments strongly supported the concept of DNA being information carrying, altering the virulence of the exposed cells.

- Bacteriophages are made of a protein coat, with DNA inside the phage head structure. This virus attaches to the bacterium and *transfers material into the bacterial cell*, forcing the bacterium to synthesize hundreds of new viral particles. The material carries information, but is the transferred material made of protein or DNA?

- As illustrated in **Figure 14.5**, if viral proteins were radioactively labeled (^{35}S) and DNA (^{32}P), the radioactive DNA will be transferred into the bacterium during infection. Protein is not transferred into the bacterium.

14.2 Testing Early Hypotheses about DNA Replication

- To understand this chapter, you need to review DNA structure, described in **Chapter 4**.

The double helix and the nature of this macromolecular structure are vital review topics.

- Compare the three mechanisms of DNA replication that were proposed following Watson and Crick's discovery of the double-helix structure (**Figure 14.7**).

  1. **Semiconservative replication**—The double helix separates, and each old strand is copied to generate two new chromosomes. Thus *each new chromosome is composed of one strand of old DNA and one strand of newly synthesized DNA.*

  2. **Conservative replication**—During replication the original chromosome is copied but remains unchanged. Both of the original strands would remain at the end of this replication, and the *new chromosome will be completely new.*

  3. **Dispersive replication**—The replication process generates two new chromosomes, with *new and old sections of DNA mixed together randomly.* This model is complex, and would require a great deal of cutting and splicing of DNA strands.

The Meselson-Stahl Experiment

- *Escherichia coli* is a prokaryotic bacterium that divides rapidly, is easy to culture, and is readily available. Many aspects of basic cellular function are shared among all living organisms; and though eukaryotic organisms are generally more complex, the underlying processes are similar. *E. coli* was an excellent **model organism** for early DNA replication studies.

- To allow for experimental separation of newly synthesized DNA from the original template DNA, Matthew Meselson and Frank Stahl grew *E. coli* bacteria in the presence of **"heavy" nitrogen** (^{15}N rather than the normal ^{14}N). Is nitrogen a part of the DNA structure? Where?

- After many generations of growing bacteria in the presence of ^{15}N-containing media, they changed the bacteria to normal ^{14}N media.

- How would this experimental design label DNA differently in each of the three models? Would each generate a unique result? How might you detect this difference?

- Meselson and Stahl separated the DNA by density and found that the semiconservative mode of DNA replication was indicated by their data. Review the data, remembering that the denser the DNA, the further the band will travel downward in the tube (**Figure 14.8**).

- Do the data make sense with your predictions?

The Discovery of DNA Polymerase

- Imagine the difficulties associated with isolating the enzyme responsible for DNA replication. Where would you look for the enzyme? How would you detect its presence in a sample?

- DNA is synthesized from nucleosides (dATP, dGTP, dCTP, and dTTP). The three components of each are a nitrogenous base, ribose, and a triphosphate. Components of the triphosphate moiety are readily labeled with radioactivity (^{32}P).

- Once researchers were able to isolate **DNA polymerase I** from *E. coli*, enzymatic activity detected in an assay was directed by a **DNA template** strand, and the new DNA was an exact copy. Subsequent studies showed that another DNA polymerase, called **DNA polymerase III**, performed the bulk of DNA replication during bacterial cell division while DNA polymerase I performed a key role in the repair of mutation.

14.3 A Comprehensive Model for DNA Synthesis

- DNA polymerase III catalyzes DNA synthesis in the 5≤ to 3≤ direction. This makes more sense when you examine the structure of a nucleotide. New nucleotides are added to the 3≤ end of an existing DNA strand, attaching to the hydroxyl (–OH) on the 3≤ end of ribose.

- On a circular prokaryotic chromosome, DNA replication starts at specific sites and proceeds in both directions from the starting point. This **bidirectional DNA replication** proceeds in each direction with the advancement of a Y-shaped **replication fork** (**Figure 14.11**). The fork originates with the action of **DNA helicase**, an enzyme that opens the double helix to allow enzymes to attach to each strand.

- DNA polymerization requires a primer (a 3≤–OH) on which to add nucleotides. The **primase** enzyme makes short RNA primers to start the DNA synthesis. This RNA is later replaced with DNA.

- How can this replication work if the DNA polymerase works only in the 5≤to 3≤direction? Examine **Figure 14.12**; it displays the continuous **leading strand** and the discontinuous **lagging strand** found at a replication fork. DNA polymerase III enzymes are associated with each replication fork, mediating the leading/lagging strand synthesis reactions.

- Tuneko Okasaki discovered the mechanism of lagging-strand synthesis. The short fragments generated at the site of lagging strand synthesis are known as **Okazaki fragments**.

- Review the stepwise examination of DNA polymerization offered in **Figure 14.13**. Each of the key steps is presented in proper chronology, and should make intuitive sense relative to the process we've introduced up to this point. Remember that this is a summary, and that DNA replication is a complex process, requiring many proteins (some of which have yet to be identified).

14.4 Replicating the Ends of Linear Chromosomes

- The ends of *linear* chromosomes are called **telomeres**, and contain many repeats of a six-base sequence TTAGGG (repeated ~1000 times).

- As the replication fork nears the end of a linear chromosome (like those found in human cells), the lagging strand makes an RNA primer for the end of the chromosome. But there is no way to replace this RNA with DNA, because there is no avail-able primer for DNA synthesis (review **Figure 14.13**, step 4). The RNA is removed, leaving a section of single-stranded DNA (leading strand) at one end of each new chromosome.

- In most dividing cells, this residual single-stranded DNA is degraded and the telomere shortens by that length with each cell division. Telomere length then serves as a record

of cell division, and limits the number of times a somatic cell can divide. But some cells, like germ cells, replicate the telomere completely.

- **Telomerase** is an enzyme that mediates the replication of telomeres, solving the problem of the single-stranded DNA described above. As is illustrated in **Figure 14.15**, telomerase adds more six-base repeats to the end of the leading strand. Then primase makes an RNA primer, and DNA polymerase uses the primer to synthesize the lagging strand sequence.

14.5 Repairing Mistakes in DNA Synthesis

How Is DNA Polymerase Proofread?

- During DNA synthesis, nucleotides are added to the new strand via their ability to form hydrogen bonds with the template nitrogenous base. Thus, using H-bonding interactions, an A in the template will direct the addition of a T to the new strand.

- To find the proofreading enzyme, researchers screened for mutant cells that made lots of mistakes during DNA replication.

- *E. coli* bacteria carrying a mutation in the epsilon (ϵ) subunit of *pol III* were found to make many errors during DNA replication. This subunit mediates **proofreading** during DNA replication by cutting out any nucleotides that were added incorrectly. The enzymatic removal of these nucleotides is termed **exonuclease** activity (**Figure 14.16**).

Mismatch Repair

- If the *pol III* enzyme leaves a mistake behind, the cell has other DNA repair enzymes available to it. *MutS* is one example of a DNA repair enzyme; it identifies **mismatched nitrogenous bases** and replaces the incorrect nucleotide.

- How can *mutS* tell the original strand from the newly synthesized strand? Chemical modifications (**methylation—CH_3**) of DNA are added long after DNA replication. The newly synthesized strand would not have these modifications and would be different from the older DNA strand.

14.6 Repairing Damaged DNA

A Case Study in DNA Repair

- **Xeroderma pigmentosum (XP)** is an inherited disease in humans; it results in sensitivity to UV light, X-rays, and other DNA-damaging treatments. People affected by this disease rapidly develop skin lesions following exposure to sunlight.

- This disease is caused by mutation of one of several **excision repair** enzymes utilized by humans. Diminished excision repair enzymes result in a deficient cellular capacity to repair damaged DNA. As is illustrated in **Figure 14.18**, excision repair involves the removal of a mismatched nucleotide and some adjacent nucleotides, followed by the addition of new, correctly matched nucleotides. This process requires three enzymatic activities; exonuclease, DNA polymerase, and DNA ligase.

- Ultraviolet light can damage DNA by altering nitrogenous bases. A great example is the formation of **thymine dimers**. Adjacent thymines in DNA can actually form a covalent dimer (these are also called **cyclobutane pyrimidine dimers**). See **Figure 14.19** for an illustration of the structure of these dimers, and to examine two key experiments characterizing DNA repair in XP patients.

B. CROSS-CUTTING THEMES

Looking Back—
Concepts from Earlier Chapters

Macromolecules and the
RNA World—Chapter 4
The **double helix, antiparallel strands**, the $5\leq$ and $3\leq$ ends of each strand, and the role of **nitrogenous bases** in DNA structure—these are key concepts in Chapter 14, and their introduction in **Chapter 4** should be reviewed carefully as you begin to study DNA synthesis.

The Cell Cycle—Chapter 11
Finally, in Chapter 14 we begin to truly define the process of **DNA replication** that precedes cell division. This chapter fills in some of the holes from **Chapter 11**. Replication of a cell's genome is no small matter, and performing this task carefully is crucial to the successful transfer of information to the daughter cells.

Looking Forward—
Concepts from Later Chapters
Bacteria and Archaea—Chapter 27
Much of the early work in DNA synthesis and isolation of enzymes in chromosome replication was first identified in **bacteria**. These cells are simple compared to eukaryotic cells, but the underlying mechanisms of cell function are shared. **Chapter 27** offers a critical introduction to bacteria and **prokaryotic cell biology**.

C. DIFFICULT CONCEPTS

One common theme in this chapter is the utilization of mutants to isolate and characterize the mutated gene. Scientists interested in a key cellular function can sometimes predict what a cell lacking that capacity will display as a mutant phenotype. One example from the chapter is the examination of DNA repair enzymes by isolating bacteria that made lots of mistakes during DNA replication. If the mutant phenotype is expected to be lethality, or a dramatic loss of cell division capacity, isolation of these mutants becomes difficult. Then temperature-sensitive mutations might be a better choice, because cells harboring this type of mutation may grow normally at the permissive temperature and display the mutant phenotype only at the restrictive temperature.

Humans afflicted with mutations that weaken the DNA repair mechanisms can display many phenotypes, including light sensitivity, brittle hair and nails, and skin that is easily sloughed off. Classic examples of this type of mutation are **Xeroderma pigmentosum, Cockayne syndrome**, and **trichothiodystrophy (TTD)**. Recently, Vermeulen and colleagues (2001) identified a group of people who suffer the symptoms of TTD whenever they run high fevers.* Upon carefully examining a specific DNA repair gene in these people, the researchers discovered a mutation that results in a **temperature-sensitive** phenotype. This means the mutant enzyme works at normal temperatures, but at slightly elevated temperatures it is unable to function because of its mutation.

Wow! It is easy to think of bacteria or yeast as a cell type that might offer temperature-sensitive mutations to researchers, but this research actually identified them in humans!

_____

*W. Vermeulen, S. Rademakers, N. G. Jaspers, E. Appeldoorn, A. Raames, B. Klein, W. J. Kleijer, L. K. Hansen, and J. H. Hoeijmakers, 2001. A temperature-sensitive disorder in basal transcription and DNA repair in humans. *Nature Genetics* 27(3), 299–303.

D. ASSESSING WHAT YOU'VE LEARNED

(1) Testing Your Knowledge

1. Which of the following best describes the structure of a DNA molecule?
   a. a deoxynucleoside triphosphate with an adenine, thymine, guanine, or cytosine base attached to it
   b. a sugar-phosphate backbone with adenine, thymine, guanine, and cytosine bases projecting toward the inside of the backbone
   c. a sugar-phosphate backbone with adenine, thymine, guanine, or cytosine bases projecting toward the outside of the backbone
   d. a sugar-phosphate backbone made up of adenine, thymine, guanine, or cytosine bases hydrogen bonded to each other

2. The following DNA sequence occurs at the start of a DNA strand. Which of the following sequences would most likely bind to this sequence to initiate DNA replication?

   3≤AATTGCAGATTCA 5≤

   a. 5≤TTAACGTCTAAGT 3≤
   b. 3≤TTAACGTCTAAGT 5≤
   c. 3≤UUAACGUCUAAGU 5≤
   d. 5≤UUAACGUCUAAGU 3≤

3. Which of the following represents the correct order of events as they occur in the process of DNA replication?
   a. helicase opens helix, *pol III* synthesizes RNA primer of leading strand, *pol I* excises primer and fills gap

   b. primase synthesizes primer, *pol I* excises primer, *pol III* elongates primer and synthesizes leading and lagging strands
   c. helicase opens helix, *pol III* elongates primer and synthesizes leading and lagging strands, DNA ligase links fragments of DNA
   d. primase synthesizes RNA primer, *pol III* synthesizes lagging and leading strands, helicase opens helix

4. The classic experiment performed by Alfred Hershey and Martha Chase revealed:
   a. RNA was not the hereditary material.
   b. Microbes could exchange genetic information.
   c. In a viral infection, protein was transferred into the infected cell.
   d. In a viral infection, DNA was transferred into the infected cell.

5. Semiconservative DNA replication involves:
   a. the separation and cutting of template strands; these sections are then copied and recombined to make the new DNA strands
   b. the outward rotation of nitrogenous bases in a double helix, and the generation of completely new double-stranded DNA copy
   c. the separation of strands of DNA, and the synthesis of a new complimentary strand on each old strand
   d. synthesis of an RNA copy of the DNA, and the generation of a new DNA strand using reverse transcriptase

6. Cancer can occur when DNA is damaged. Which of the following statements is true regarding the cause of damage to DNA molecules?
   a. The structure of DNA can be damaged only by environmental toxins such as cigarette smoke, aflotoxins, and other toxins in food.
   b. Excision repair enzymes are able to prevent all mutations in DNA.
   c. Small quantities of ultraviolet radiation have no effect on DNA structure.

d. DNA molecules can break down in the absence of environmental toxins or radiation.

7. Imagine that you are a researcher working on developing a gene therapy for XP patients. Which of the following would be the best "gene" to incorporate into the skin cells of these patients, based on the data shown in Figure 14.19?
   a. a gene for the production of "normal" DNA strands
   b. a gene that would increase the production of DNA polymerase III
   c. a gene that would increase the production of DNA polymerase I
   d. a gene that would increase the production of thymidine in damaged cells

8. Once DNA strands are separated by DNA helicase, _____ block the reformation of a double helix by the template strands.
   a. RNA primer strands
   b. topoisomerase enzymes
   c. single-stranded DNA-binding proteins
   d. RNA polymerase complexes

(2) Integrating Your Knowledge

(a) The Meselson and Stahl experiment focused on discerning between competing models of DNA replication. Describe the results the researchers would have generated if DNA replication proceeded by conservative replication.

(b) What role did Arthur Kornberg play in the characterization of DNA replication?

(c) Detection of DNA polymerase activity in vitro suggested a template-driven process of DNA replication. But what experiments were utilized to show that the template was reliably copied by the DNA polymerase activity?

(d) Explain the need for RNA in the process of DNA replication.

CHAPTER 14—ANSWER KEY

D. Assessing What You've Learned

(1) Testing Your Knowledge

1. b; 2. d; 3. c; 4. d; 5. c; 6. d; 7. c; 8. c

(2) Integrating Your Knowledge

(a) Labeling over many generations with ^{15}N would generate heavy DNA throughout the bacterial culture. Transfer of these cells to ^{14}N media would result in light DNA with all new DNA synthesis. If conservative replication occurred, then density gradient centrifugation would generate only heavy and light DNA bands; no intermediate bands would be observed. Review **Figure 14.8** for illustration.

(b) Kornberg was the biochemist who first isolated DNA polymerase in its active form, characterizing its activity using in vitro assays.

(c) Korberg utilized the DNA polymerase activity to generate copies of a bacterial virus (bacteriophage ΦX174). The copied viral DNA was then assayed for its ability to infect bacterial cells. The DNA polymerase was shown to correctly copy the phage genome into a complete genome that was just as infectious as the original DNA preparation.

(d) DNA polymerases are unable to synthesize DNA from a template strand unless there is an available primer on which to add new nucleotides. RNA polymerases are able to synthesize short RNA primers de novo, and are utilized during DNA replication to prime new DNA synthesis. The RNA primers are later destroyed and replaced with the correct DNA sequence.

15

How Genes Work

A. KEY BIOLOGICAL CONCEPTS

15.1   What Do Genes Do?

The Molecular Basis of Hereditary Diseases

- Archibald Garrod was a physician who first began to think of inherited mutations and how they might influence cellular biochemistry (or **metabolic** processes). Garrod noticed a mutant phenotype that was prominent within a family, and suspected that a mutation was passed from parent to child in the family.
- Garrod called the abnormality an **inborn error of metabolism,** suspecting that the phenotype was the result of abnormal metabolic function. The disease, called **alkaptonuria**, caused affected individuals to display (among other things) black urine.
- How might one defective gene/protein generate abnormal metabolism? Think of metabolic processes as an assembly line, with stages along the line contributing consecutive pieces of the final product (**Figure 15.1**). Individual proteins (enzymes) perform a task at each stage, altering the compound to ultimately generate the final product. The loss of one enzyme results in the accumulation of unfinished product, all stuck at the same stage of processing. This buildup of an intermediate is exactly what happens in alkaptonuria; the accumulated intermediate (homogentisic acid) is disposed of in the urine, turning it black.
- Without knowing it, Garrod was already directing attention toward development of the **one-gene, one-enzyme hypothesis**. An inherited mutation in one gene generates one abnormal protein (or enzyme).

Genes and Enzymes

- Researchers analyzing mutant cell lines of the bread mold, *Neurospora crassa*, utilized biochemistry and genetic analysis to characterize the steps involved in biosynthesis of the amino acid, **arginine**. If you recall the assembly-line example of metabolic pathways described earlier, you'll see that each step in the pathway modifies the substrate molecule, eventually generating the product of the pathway (arginine). These types of experiments led to the *one-gene, one-enzyme hypothesis*.
- Arginine is an amino acid that cells can make for themselves if it isn't in the growth medium. Beadle and Tatum isolated mutant cell lines that were unable to grow without arginine supplementation in the growth medium, indicating that the biosynthetic pathway was not functioning properly.
- *The arginine-dependent cell lines must carry mutations in genes that encode enzymes in arginine biosynthesis!*
- Beadle and Tatum knew some of the details about arginine synthesis, and they knew the identity of some of the intermediates. They thus decided to add these intermediate forms to the growth medium.
- Why? If a single stage in the assembly line is blocked and you introduce an intermediate that occurs downstream from the blocked step, the intermediate will be successfully converted into product (arginine), allowing the cells to grow normally.
- Using this strategy, Beadle and Tatum took individual mutant cell lines and identified where the mutation blocked the pathway. Did

it work? Yes! Review **Figure 15.2** in text to examine the arginine biosynthetic pathway.

15.2 The Genetic Code

- If one gene = one enzyme, then how is the protein encoded within the structure of DNA? DNA is double-stranded, with antiparallel strands and the **nitrogenous bases** adenine (A), cytosine (C), thymine (T), and guanine (G). Francis Crick proposed that the sequence of nitrogenous bases in the DNA strand was a kind of Morse code, defining the protein structure with a linear grouping of the nitrogenous bases (i.e., AGT = amino acid #1). His predictions were fundamentally correct.

RNA as the Intermediary between Genes and Proteins

- RNA is single stranded, and is composed of A, U, G, and C (U stands for uracil; there is no thymine in RNA). Francois Jacob and Jacques Monod first proposed that RNA might serve as intermediaries (messengers) between DNA and protein synthesis machinery (**Figure 15.4**).
- **Ribosomes** are small structures found in the cytosol and associated with the **endoplasmic reticulum** (**ER**). They synthesize proteins within the cell by reading the sequence of messenger RNA (**mRNA**).
- **Jerard Hurwitz** and **J. J. Furth** isolated **RNA polymerase**, the enzyme that synthesizes mRNA by reading DNA sequence. They examined cellular fractions for RNA synthesizing capacity; newly synthesized RNA would be radioactive through incorporation of **radioactive ribonucleotide triphosphates** (not deoxyribonucleotides).
- How does RNA polymerase read the DNA to make the mRNA? In the coding sequence, the RNA polymerase will read a C and add the complementary G to the RNA strand (**Figure 15.5**). One unusual exception is that when the polymerase reads an A in the DNA, a U is added to the RNA (instead of thymine).

How Long Is a Word in the Genetic Code?

- George Gamow predicted that each code word in the genetic code would be three nucleotides long (**3 nucleotides = 1 codon**). How could he make such a prediction? *Key*: Twenty amino acids are used by ribosomes to synthesize proteins.
- Francis Crick (and other key researchers) experimentally revealed that the code is indeed based on three nucleotides for each amino acid. Crick made this discovery by making **deletion/insertion mutations** in viral DNA. He found that the **reading frame** of a gene could be destroyed by mutation, but then restored if the total number of deletions or additions were multiples of three. Review *the fat cat ate the rat* as an illustration of reading frame (also, **Figure 15.7**).

How Did Biologists Crack the Code?

- Identifying the three-nucleotide sequence coding for each amino acid was a difficult proposition. Eventually the genetic code was deciphered, and it is shown in **Figure 15.8**. Each of 20 amino acids used by ribosomes to synthesize proteins have multiple codons that designate their addition to a protein.
- Researchers attempting to break the code found that a polymer of U residues in an RNA strand (5'–UUUUUUUUUUUUUU–3') resulted in the translation of that sequence into a polymer of phenylalanine.
- Thus, if RNA is synthesized as the complementing sequence of the gene encoding DNA, then the codon for phenylalanine must be **AAA**. A piece of the code is revealed! This type of strategy revealed a great deal of the genetic code.
- There is only one **start codon** (AUG), which signifies the start of the protein encoding sequence in an mRNA.
- There are three **stop codons** in the genetic code (UGA, UAA, and UAG).

15.3 The Central Dogma of Molecular Biology

- The **central dogma**: DNA is **transcribed** into RNA, and RNA is **translated** into protein (**Figure 15.10**).
- DNA is a long-term form of information storage, allowing for the stable maintenance

of the information and its passage from generation to generation.

- RNA carries the information from the DNA to the translation machinery (the ribosome). In eukaryotic cells, the DNA and the ribosomes are located in separate cellular compartments.
- RNA is unstable within the cell, lasting only for minutes to hours before it is destroyed.
- The proteins that are synthesized by the ribosomes carry out cellular functions.

Exceptions to the Central Dogma (Essay)

- The tobbaco mosaic virus (TMV) contains no DNA; it is an RNA virus. Upon entering the cell, the viral RNA is directly translated into viral proteins.
- Some RNA viruses enter the cell and are reverse transcribed to DNA as the first step of viral infection. The process of converting RNA to DNA is called **reverse transcription**, and is mediated by an enzyme known as **reverse transcriptase**. This process is discussed in more detail in later chapters.

B. CROSS-CUTTING THEMES

Looking Back—
Concepts from Earlier Chapters
Nucleic Acids and the RNA World—Chapter 4
RNA is more than just a messenger. The capacity of RNA to aid in **protein synthesis**, to have **catalytic activity** (it can cut RNA!), and to carry **complex messages** from the nucleus to the cytosol is truly remarkable. Review this introductory material on RNA to enhance your study of Chapter 15.

Review the structure of **double-stranded DNA**; it will be important to your study of gene structure in this chapter.

Mendel and the Gene—Chapter 13
Phenotype and the influence of mutation on physical characteristics of organisms were crucial to Mendel's work. With an understanding of genes and the central dogma, we can now better understand what phenotype really entails. A **mutant gene** produces a **mutant protein** (or no

protein at all), and the loss of this protein's function generates an abnormal phenotype. The mutant gene really is just an incorrect blueprint for a protein.

Looking Forward—
Concepts in Later Chapters
Transcription and Translation—Chapter 16
In **Chapter 16**, we will examine the process of **protein synthesis**, and learn how the information carried on a messenger RNA (mRNA) is **translated** into a functional protein.

Control of Gene Expression and Cell Differentiation—Chapters 17, 18, and 22
If the **human genome** is composed of 30,000 genes, and every cell in your body contains each of these genes, how are these utilized differently by different cells? Each cell type in the human body acts differently due to **differential gene expression**. The genes expressed in a cell dictate the shape, growth, and function of that cell. Regulating gene expression is central to how we utilize the information in the genome.

C. DIFFICULT TOPICS

This chapter contains some complex experimental data, generated by laboratory strategies that are new to the reader. Anticipate some confusion when first reading through these experiments; but realize that you can make sense of these experiments with a little extra review of the material. The characterization of the arginine biosynthetic pathway offers a good example. This experiment utilizes *Neurospora crassa* (a bread mold), a mutational screen, and finally a carefully designed series of phenotypic analyses. What is the relevance of each step in the experiment? Should you ignore these "minor details" and move on to memorizing the conclusions? No! The experimental strategy is critical to this discussion; work through the details one at a time to ensure that you understand why this experiment worked and what it was designed to accomplish. Read the text, the figure legends, and the study guide closely to ensure that you've thoroughly examined the experiment. Also review background material referenced in previous chapters. The relevance of each of the listed terms is described. *Scientists explore biology through careful experimentation. Recognizing the "how" of important discoveries is critical to your development as a scientist.* Don't give up on the experiments included in the text. They are included expressly to teach each student more than just a set of conclusions about important discoveries, and they offer details of the experiments themselves.

D. ASSESSING WHAT YOU'VE LEARNED

(1) Testing Your Knowledge

1. Archibald Garrod hypothesized that alkaptonuria was:
   a. caused by a DNA mutation
   b. caused by an inherited defect in an enzyme
   c. an example of recessive mutation in a gene
   d. caused by abnormal chromosomal sequence

2. Which of the following serves as a template for making proteins from DNA?
   a. tRNA
   b. mRNA
   c. rRNA
   d. DNA polymerase

3. Which of these describes the function of RNA polymerase?
   a. amplifies the "message" by making multiple copies of an mRNA molecule after it has been transcribed from DNA
   b. converts a protein sequence to mRNA
   c. transcribes DNA to mRNA
   d. translates DNA into protein

4. Which of the following is an *exception* to the "central dogma" of molecular biology?
   a. production of RNA in the nucleus of a eukaryotic cell
   b. double-stranded DNA from a virus inserts into the host genome
   c. single-stranded DNA from a virus inserts into the host genome
   d. single-stranded RNA from a virus is used as a template for DNA that inserts into the host genome

5. In a skin cell, which of the following changes may lead to skin cancer?
   a. a change in a single nucleotide in an mRNA molecule
   b. a change in a single nucleotide in a DNA molecule
   c. a change in multiple amino acids of a single protein molecule
   d. a change in the conformation of an enzyme molecule

6. Consider the following two mRNA molecules. Original RNA sequence:
   acguguaccg
   acguuguacc
   The difference between the two molecules would most likely result in which type of change in the resulting protein molecule?
   a. frameshift mutation, resulting in a change in the amino acid sequence
   b. deletion mutation, resulting in a change in the amino acid sequence
   c. protein would be unchanged
   d. intron

7. George Beadle and Edward Tatum revealed that:
   a. One genetic mutation corresponded to one defective protein (enzyme).
   b. DNA is the heritable material.
   c. RNA is the heritable material.
   d. The arginine biosynthetic pathway is defective in bread molds.

8. Which of the following was considered important evidence that RNA was an intermediary between DNA and protein biosynthetic processes?
   a. RNA is located in the nucleus (with DNA) and the cytosol (with ribosomes).
   b. RNA is synthesized, but is then fairly rapidly degraded.
   c. DNA does not colocalize with ribosomes in the cytosol.
   d. All of the above are correct.

(2) Integrating Your Knowledge

(a) Archibald Garrod linked disease to inheritance, and suggested that heritable units encode enzymes. Describe the clinical evidence he observed that stimulated his hypothesis.

(b) How can a mutation in a gene destroy the activity of the encoded protein (or enzyme)?

(c) To initiate a "mutant hunt," the researcher must first isolate a large group of mutant populations of cells, and then utilize them in further experimentation. Describe the process of isolating arg⁻ populations of *Neurospora crassa*.

(d) George Gamow predicted that the genetic code would contain "words" that were three nucleotides in length. His predictions were based on mathematical considerations rather than experimental data. On what information were his predictions based? Was he correct?

(e) In cracking the genetic code, Marshal Nirenberg and Heinrich Matthaei performed an ingenious yet laborious series of experiments. Which two key experimental designs did they use to decipher individual codons?

CHAPTER 15—ANSWER KEY

D. Assessing What You've Learned

(1) Testing Your Knowledge

1. b; 2. b; 3. c; 4. d; 5. b; 6. a; 7. a; 8. d

(2) Integrating Your Knowledge

(a) Garrod observed a family that had several cases of alkaptonuria, a metabolic disease in which affected individuals accumulate a metabolic intermediate in their body (homogentisic acid). He suspected that the source of the disease was a missing metabolic enzyme, and the deficiency was passed from generation to generation. Garrod linked the heritable units (genes) to enzymes (proteins), and was fundamentally correct.

(b) Enzymes catalyze each step in the biosynthetic pathway, modifying the intermediate forms to eventually generate the desired product. Each enzyme is a protein composed of a chain of amino acids. Changing the amino acid content of an enzyme, by altering the gene on which it is encoded, can have a profound influence on the function of that protein.

(c) A culture of wild-type *N. crassa* would first be mutagenized using chemical treatment or exposure to radiation. The researchers would identify clonal populations of cells that grow in rich medium but cannot grow in medium that lacks the amino acid arginine. Cells unable to grow in the selective medium likely contain a mutation in one of the arginine biosynthesis genes. Review **Figure 15.2** to see how these isolated mutant cell lines can then be utilized to identify the genes involved in arginine biosynthesis.

(d) Gamow was correct. He knew that 20 amino acids were used by ribosomes to build cellular proteins. If the genetic code were only two nucleotides in length, then there would be only 4^2, or 16 codons available. But if they were three nucleotides in length, then the available number of distinct codons would be 4^3, or 64.

(e) The first strategy was to synthesize polymers of RNA; poly(U), poly(A), poly(G), and poly(C). They added these RNA polymers to a mixture of reagents capable of translating RNA into protein. Uracil polymers (UUUUUUUU) resulted in the synthesis of chains of phenylalanine, evidence that the codon for phenylalanine is UUU. The second strategy was to make specific codons and then measure which charged tRNA bound to the ribosome as it read that specific RNA sequence.

16

Transcription and Translation

A. KEY BIOLOGICAL CONCEPTS

16.1　Transcription in Bacteria

- **RNA polymerase** synthesizes mRNA transcript (and other types of RNA transcripts) by reading DNA template at an activated gene.
- An RNA polymerase enzyme attaches **nucleoside triphosphates** together, selecting the correct nucleoside by its hydrogen-bonding capacities and synthesizing the RNA strand in the 5≤to 3≤direction (**Figure 16.1**). How is a deoxynucleoside different from a nucleoside? Review the structural differences. Presence/absence of an –OH (hydroxyl) functional group generates major differences in the characteristics of the polymers (DNA vs. RNA).
- RNA polymerase reads the template strand of the DNA to synthesize RNA.

Initiation: How Does Transcription Begin in Bacteria?

- The prokaryotic RNA polymerase is composed of a core complex (holoenzyme) that has the ability to synthesize RNA. The holoenzyme is made up of four globular proteins that associate together (**Figure 16.2**).
- Initiation of transcription requires an additional protein, called **sigma** (τ), that binds to a holoenzyme not yet associated with DNA. Sigma can "read" DNA and identify the start site of a gene. Thus sigma is critical to the proper initiation of transcription at the correct place on the chromosome, recognizing the DNA sequence that precedes the coding region of a gene (the **promoter**).
- What sequences does the RNA polymerase/sigma complex commonly bind? **David Pribnow** examined promoter sequences in bacterial and viral genomes. He allowed the bacterial RNA polymerase/sigma complex to bind to DNA and then characterized the bound DNA. **Figure 16.3**; Pribnow identified two elements found within most bacterial/ viral promoters: the **–10 box** (**Pribnow box**) and the **–35 box**.
- The –10 box is located 10 bases upstream (in the 5≤direction) from the transcription start site. It contains a 5≤—TATAAT—3≤consensus element.
- The –35 box is located 35 bases upstream from the transcription start site and contains a 5≤—TTGACA—3≤consensus element.
- Sigma identifies the –10 and –35 promoter sites, properly orienting the RNA polymerase core complex for transcription at the gene start site. The polymerase then opens the DNA double helix and begins to catalyze the polymerization of nucleotides, moving down the DNA template as RNA is synthesized.
- Sigma dissociates from the holoenzyme once transcription is successfully initiated.

Elongation and Termination

- Once RNA polymerization is initiated, the enzyme **elongates** the RNA transcript by reading the template strand and synthesizing RNA in the 5≤ to 3≤ direction (moving down the template DNA strand in the 3≤ to 5≤ direction).

- Notice there are grooves in the holoenzyme through which pass the template and non-template DNA strands, nucleotides, and the newly synthesized RNA strand.
- Transcription is terminated when specific sequences, called **transcription termination signals**, are encountered in the DNA template. The process of RNA polymerase dissociating from the DNA can occur by several different mechanisms. One type of termination signal results in the RNA forming a hairpin structure (the RNA folds back on itself to make a short stretch of double-stranded RNA). Formation of the hairpin will "pull" the newly synthesized RNA out of the holoenzyme, terminating additional transcription.

16.2 Transcription in Eukaryotes

- Eukaryotic transcription shares many fundamental characteristics with prokaryotic transcription, but it is more complex and includes some additional processing steps.
- Eukaryotic cells contain three RNA polymerases, though mRNA synthesis is mediated by only one of these, RNA pol II. Thus all protein synthesis in a eukaryotic cell depends on **RNA pol II**. RNA polymerase I makes ribosomal RNA (rRNA); RNA polymerase III is responsible for some rRNA and for transfer RNA (tRNA—see Section 16.4).
- The promoters of eukaryotic cells are highly variable, though many do contain a **TATA box** approximately 30 base pairs from the transcription start site.
- Many small proteins assemble with the RNA pol II to locate promoter elements and to regulate transcriptional activity; these are called **basal transcription factors**. These factors act in a manner that is very similar to the sigma factor of prokaryotic cells.

The Startling Discovery of Eukaryotic Genes in Pieces

- Prokaryotic genomes contain genes in one **continuous** DNA sequence.

- Characterization of eukaryotic genomes has revealed that genes are often **discontinuous**, having been broken into smaller fragments with noncoding sequences between each piece. This was first called the **genes-in-pieces hypothesis** by Philip Sharp, who generated DNA:RNA hybrid molecules by heating double-stranded DNA molecules to denature them (as in PCR), and then mixing the denatured DNA with mRNA strands. The mRNA strands should anneal to the complementary strand of DNA sequence (i.e., the original site of that gene). Sharp discovered that the hybrid molecules frequently contained loops of DNA rather than a continuous annealed structure (**Figure 16.5**). Why? The mRNA must be missing some of the sequence! It must have been removed as the mRNA was processed following transcription! As more scientists began to examine this phenomenon, it became clear that eukaryotic genes are discontinuous in the genome, and that following transcription the gene-encoding fragments in the RNA were spliced together to make the mRNA product.

Exons, Introns, and RNA Splicing

- The coding fragments found in the chromosomal location of a gene are called **exons** (expressed), and the noncoding fragments interspersed between exons are called **introns** (intervening sequences).
- How are the exons spliced together? This complex process is mediated by small nuclear ribonucleoproteins (**snRNPs**, pronounced "snurps"). As shown in **Figure 16.6**, snRNPs identify introns by their sequence at each end (the splice junction site), catalyze the removal of the intron, and splice surrounding introns together. The complex of snRNPs that mediate this process is called a **spliceosome**.

Other Aspects of Transcript Processing: Caps and Tails

- After splicing of the primary transcript, two additional steps of RNA processing are performed in eukaryotic cells—the addition of a **5≤ cap** and the addition of a **poly (A) tail**. (**Figure 16.7**)

- The cap is a modified nucleoside triphosphate that is added to the front or **5≤end** of the transcript. This cap aids mRNA translation by enabling binding of some translation regulatory proteins. It also increases the stability of the mRNA by decreasing the rate at which the mRNA is degraded by cellular RNases.
- The 3≤poly (A) tail is another modification of the RNA transcript that aids mRNA stability in the cell. Over 100 consecutive adenosine nucleosides are added to the 3≤end of the RNA.

16.3   An Introduction to Translation

- Once the mRNA moves to the **cytoplasm**, the information contained within its sequence can be translated by the **ribosome** to generate the functional protein.
- Proteins are synthesized from individual **amino acids**, and each amino acid in the protein is coded for in the mRNA. Each codon (three nucleotides in mRNA) directs the ribosome to add one amino acid to the protein.

Ribosomes Are the Site of Protein Synthesis

- Ribosomes are composed of both protein and RNA, called ribosomal RNA (rRNA).
- Roy Britton used **pulse-chase experiments** to examine whether ribosomes were indeed the site of protein synthesis. *Britton wanted to determine if protein synthesis was occurring at the site of ribosome localization.* He labeled proteins being synthesized for a short period of time by adding a 15-second pulse of radioactive sulfate to the culturing medium of *E. coli* (immediately chased with nonradioactive sulphur). This would result in the incorporation of the radioactive tag with amino acids containing sulphur (methionine and cysteine), and a short window of time in which all proteins being synthesized would be radioactive. But how did Britton detect these proteins? How did he determine that they were associated with ribosomes?
- Ribosomes can be isolated via **density gradient centrifugation**, in which molecules are separated from each other according to their density. Britton found the pulse of radioactivity to localize to fractions containing ribo-somes! As the radioactive pulse was chased with nonradioactive sulphur in the *E. coli* cultures, radioactivity disappeared from ribosomal fractions. This shows that the radioactive sulphur is associated with ribosome activity (the proteins being made), not with the ribosome structure itself.
- Because prokaryotic cells have no nucleus, translation occurs immediately following, or even during transcription (**Figure 16.8a**).
- Eukaryotic cells transcribe RNA in the nucleus, process the RNA to mRNA in the nucleus, and then export the mature mRNA to the cytosol for translation (**Figure 16.8b**).

16.4   The Role of Transfer RNA

- **Adapter-molecule hypothesis**: In imagining how a codon of nucleotides in an RNA molecule could somehow direct the recruitment of specific amino acids, **Francis Crick** (of Watson and Crick) developed one hypothesis. He hypothesized there was an as yet undiscovered adapter molecule that "read" the mRNA codon directly and carried with it an amino acid for protein synthesis. The predicted adapter molecule was discovered (see next item) and named **transfer RNA (tRNA)**.
- Researchers studied protein synthesis by identifying the essential ingredients of active protein synthesis. One previously unknown component was purified from cellular fractions and shown to contain amino acids and RNA. The molecules discovered by this research team were the transfer RNAs.
- A tRNA bound to its corresponding amino acid is called an **aminoacyl tRNA**; the addition of amino acids to tRNAs is mediated by a specific enzyme (**aminoacyl tRNA synthetase, Figure 16.10**). There are many types of tRNAs, each carrying a specific amino acid and recognizing a codon on the mRNA. They are the adapters between the amino acids and the mRNA that Crick had predicted!

Connecting Structure with Function

- It is hard to understand what a tRNA really does during protein synthesis without examining tRNA structure and putting the tRNA, ribosome, and mRNA together in a diagram.

Carefully examine **Figure 16.12**. Notice the 3≤end on which amino acids are added, the **stem-loop structures**, and most importantly the **anticodon** region.

- The anticodon actually forms an **antiparallel association** (by hydrogen bonding) with the codon on the mRNA strand. If the aminoacyl tRNA matches the codon being presented, the amino acid on the 3≤end is presented to the ribosome for addition to the protein being synthesized.

16.5 Ribosomes and the Mechanism of Translation

- Protein is synthesized by the ribosome as it reads the mRNA. This process occurs both in the **cytoplasm** and on the surface of the **endoplasmic reticulum**.
- The ribosome was first isolated using **density gradient centrifugation**. A crude mixture of proteins can be divided into components, separating them by their relative density. Ribosomes are composed of protein and rRNA.
- Density gradient centrifugation was used to isolate intact ribosomes and showed that the *E. coli* ribosome was composed of two major parts, now known as the **50S** and the **30S** subunits (S is a unit of sedimentation density). Each subunit is composed of proteins and rRNA strands. The rRNA found in ribosomes is important to the function and structure of the ribosome; it is never translated.
- Ribosomes **self-assemble**, meaning that if the individual components are placed together, they immediately assemble into a ribosome.
- As shown in **Figure 16.13**, there are several key structural sites in the ribosome for tRNA localization. These are A, P, and E sites; their role in protein synthesis is summarized next.

How Does Translation Occur?

Initiation

- Remember that mRNA contains sequences in addition to an open reading frame (coding for protein); there are noncoding 5≤and 3≤RNA sequences. Thus the ribosome must locate the translation start site correctly to ensure that the correct information is translated into

protein. How would you predict the process might occur? Think back to transcription; how was the correct starting site identified?

- Translation always begins with a **methionine** codon (**AUG** in the RNA sequence). The 30S subunit is recruited to a region just upstream of the translation start site (AUG) by sequences found within the mRNA itself (**Figure 16.14**). Scientists have identified a consensus sequence of nucleotides at the translation start site (the **Shine-Dalgarno sequence**).
- Once the 30S subunit is bound to the mRNA, the anticodon of an **initiator tRNA** (methionine) binds to the AUG. The initiator methionine is unique from other methionine tRNAs because the methionine carried on this tRNA is modified with a formyl group. The resulting methionine is called *N*-**formylmethionine**. Initiator tRNA association is a complex process, mediated by a series of small proteins known as **initiation factors** (**IF**).
- The **50S subunit** binds next, associating with the 30S subunit that is already properly oriented to the mRNA translation start site.

Elongation

- The ribosome-mediated process of protein synthesis can be summarized as regulated formation of peptide bonds between amino acids. *What is a peptide bond?* Do you recall peptide bonds as discussed in **Chapter 3**? If not, review that material to be sure you recognize the structure of amino acid and the reactive groups interacting to form a peptide bond (amine and carboxyl groups). *What characteristics of an amino acid make it unique?* Review the "R" groups or side chains on amino acids; it is their makeup that generates the unique characteristics of each protein.
- Review **Figure 16.15** and the three steps of elongation in protein synthesis:
  1. **Filling the A site**—An aminoacyl tRNA moves into the A site of the assembled ribosome/mRNA complex. The anticodon of the rRNA forms hydrogen bonds with the codon in the A site. The adjacent P site contains the previously added tRNA with its amino acid bound by a peptide bond to the growing polypeptide chain.

2. **Peptide bond formation**—The amino acid in the A site forms a peptide bond with the polypeptide chain, its amine group reacting with the carboxyl group of the P-site amino acid. As this bond is formed, the P-site amino acid breaks its covalent attachment to the P-site tRNA.

3. **Translocation**—The P-site tRNA diffuses out of the complex; the ribosome advances forward one codon (5≤↓ 3≤), transforming the A site to the P site and making room for the next aminoacyl tRNA.

- Energy for elongation is derived from high-energy phosphate bonds in GTP molecules. Small proteins known as elongation factors also aid in the elongation process.

Termination

- When the A site encounters a stop codon, instead of a new aminoacyl tRNA, small proteins (**release factors**) enter the site and initiate release of the synthesized protein.
- Universal stop codons: **UAA, UAG, UGA**.

Post-Translational Modifications

- The completed protein product is often further modified to mediate proper folding and to effect protein function. These modifications can be complex, and vary depending on the specific protein involved.
- A class of proteins that mediates folding by interacting with newly synthesized proteins is called **chaperones**. They are cellular proteins that function to bind to other proteins and enable their proper folding. Many of these proteins are stress activated, expressed in high levels following some type of cellular stress that might threaten proper protein folding. An excellent example is the **heat-shock-activated chaperone proteins**. High temperatures can dramatically influence protein folding; heat-shock proteins protect against this threat.
- Proteins can also be covalently modified as a part of their proper synthesis. These modifications can include the addition of sugar molecules (**glycosylation**) or phosphate molecules (**phosphorylation**). Some proteins are processed by enzymatic **cleavage** of the final protein. Proteinases mediate this process and

may be at the site of action for that protein; or they may be regulated by other signals.

16.6   The Molecular Basis of Mutation

- A **mutation** is any change in an organism's DNA sequence. DNA mutations affect phenotype only when the mutation is expressed (DNA ↓ RNA ↓ protein), and the resulting protein functions abnormally. Not all mutations will affect the protein's ability to function; thus they will not generate a phenotype.
- There are many types of DNA mutations. One of the most common is a **point mutation**, which changes a single nucleotide. These can result from errors in DNA replication (**Figure 16.18**), or exposure to mutagenic toxins.
- **Sickle-cell anemia** is a human disease resulting from a point mutation in the gene for hemoglobin. A T↓ A mutation changes a codon for the amino acid glutamic acid into a codon for valine (**Figure 16.19**).
- Some mutations do not change the amino acid added to the protein (notice that CTT and CTC codons both specify glutamic acid). These are known as **silent mutations**.

B. CROSS-CUTTING THEMES

Looking Back—
Concepts from Earlier Chapters
Nucleic Acids and the RNA World—Chapter 4
Scientists are only beginning to understand the various roles for **RNA macromolecules** in cellular function. The discovery that RNA can be catalytic, functioning in the splicing of RNA strands, was amazing. Recent findings reveal that rRNA in the ribosome catalyzes the formation of **peptide bonds**! In this chapter we introduced several different examples of RNA function in the cell. RNA is an interesting molecule displaying an interesting diversity of activities.

Looking Forward—
Concepts from Later Chapters
Control of Gene Expression— Chapters 17 and 18
The next two chapters focus on the mechanisms controlling gene expression. Regulating gene

expression at the chromosome is much more efficient than regulating RNA processing or ribosome activity. Much of the biology defining a cell type (e.g., a liver cell or a brain cell) is the direct result of **differential gene expression**.

Bacteria and Archaea—Chapter 27

Overuse of **antibiotics** in the United States, and the resulting rise in **drug-resistant bacteria**, are in the news today. Humans utilize molecules that attack critical cellular functions in bacteria (like protein synthesis). Why don't these drugs harm our cells? Study the differences between bacteria and human cells; can you hypothesize how these drugs could be harmless to us?

Viruses—Chapter 34

Many early experiments in RNA polymerase activity were performed using viruses. **Bacteriophage viruses** infect prokaryotic cells, and **adenoviruses** infect eukaryotic cells. Experiments using each of these viruses to characterize RNA polymerase are shown in Chapter 34. Why was viral infection of a cell selected for these classical experiments? Upon entry into the cell, the virus takes over the native RNA polymerase for expression of the viral genes. Thus scientists can examine the relatively few viral genes in their experiments, when compared to the large number of cellular genes. Examine this chapter for a detailed discussion of viral pathogenesis.

C. DIFFICULT TOPICS

X-Ray Crystallography (Figure 16.13)

As we examine ribosomes and protein synthesis, it is important to realize how researchers determine the three-dimensional shape of specific macromolecules. Chapter 4 introduced the process of solving the structure of a protein using X-ray diffraction analysis. Scientists have used this strategy to characterize DNA and protein interactions, protein and protein interactions, and even structural changes occurring with the activation of a specific protein (allosteric regulation).

You can find many crystallography images and resources on the Internet. Explore this experimental strategy with some simple searches. Here are two sites we recommend:

http://ncbi.nlm.nih.gov/Structure/MMDB/mmdb.shtml

http://boatman.med.wayne.edu/~xray/education.html

D. ASSESSING WHAT YOU'VE LEARNED

(1) Testing Your Knowledge

1. What is the maximum number of amino acids that could result from the following mRNA sequence?

   5≤AAUCCGUAAAUGAGACCGUCGAUGAAUUAGCG3≤
   a. 0
   b. 6
   c. 7
   d. 10

2. Which of the following best describes the effect of a mutation in the Trp aminoacyl tRNA synthetase that changed the conformation of the tRNA binding site?
   a. The amino acid Trp would not bind to the aminoacyl tRNA synthetase.
   b. The aminoacyl tRNA synthetase would not be produced.
   c. The Trp tRNA would not be produced.
   d. The enzyme would not be activated.

3. Which of the following are required as an energy source for translation?
   a. mitochondria
   b. aminoacyl tRNA synthetase
   c. GTP and ATP
   d. GTP only

4. What is the best description of the "loops" in the tRNA molecule?
   a. unpaired DNA bases
   b. regions created as the result of hydrogen bonding between base pairs
   c. regions of the molecules that remain as single-stranded RNA due to the absence of complementary base pairs
   d. regions provided for binding to the mRNA codon

5. The tRNA shown in **Figure 16.12** in your textbook depicts the typical "L" shape of this molecule. Which of the following will most likely result if some hydrogen bonds

in the region labeled "hydrogen bonds" in the figure are broken using some type of chemical treatment?

a. The tRNA will not be able to bind the amino acid properly.

b. The tRNA will not have the correct anticodon.

c. The anticodon will remain the same, but the amino acid bound to the tRNA will be different.

d. The tRNA will not fit into the ribosome properly.

6. Refer to **Figure 16.14** in your textbook. Which of the following statements is *not* true regarding translation in bacteria?

a. Translation requires the presence of initiation factors and both the small and large subunits of the ribosome.

b. The peptide bonds between adjacent amino acids are formed within the large ribosomal subunit.

c. The first amino acid in every protein is methionine.

d. Translation proceeds from the 5≤end toward the 3≤end of the mRNA.

7. If there were a mutation in a ribosome that prevented tRNA molecules from moving into the E site, how many amino acids would be found in the resulting peptide chain formed on this ribosome?

a. a maximum of 2

b. a minimum of 2

c. a maximum of 1

d. none

8. Which of the following statements is correct?

a. Protein enzymes catalyze the formation of peptide bonds between adjacent amino acids.

b. Peptide bond formation occurs within the small subunit of the ribosome.

c. Peptide bond formation occurs in the absence of any catalyst.

d. A compound other than a protein enzyme is able to catalyze the formation of peptide bonds.

9. Some proteins can act as infectious proteins, or prions. These proteins have an abnormal conformation and cause other "normal" proteins to change to the mutant form, resulting in a number of disorders, including "mad cow" disease. This process is similar to which of the following normal processes in the cell?

a. transcription

b. translation

c. mRNA processing

d. post-translational modification

(2)   Integrating Your Knowledge

(a) By mixing purified preparations of RNA polymerase and the DNA genome of a T7 virus, scientists were able to estimate the number of promoter elements in this viral genome. How did this experiment work? How were the promoter elements being counted?

(b) The −10 and −35 boxes were identified within DNA fragments that survived the DNase protection analysis. Once the researchers isolated the protected fragments, how were the −10 and −35 regions in these fragments identified as important to RNA polymerase binding?

(c) There are seven different sigma factors in most bacterial genomes. Each sigma (σ) factor plays a critical role in identifying promoters for the RNA polymerase to transcribe. Why would bacteria require seven different sigma factors?

(d) Robert Roeder's research team utilized adenovirus DNA to characterize transcription in eukaryotic cells. They used an in vitro assay, adding purified RNA polymerase to adenovirus DNA, but their early experiments resulted in transcription at random places on the DNA. They suspected that something was missing from the RNA polymerase. How did they alter their experimental strategy to address this problem? Were their changes helpful?

(e) Describe the process of translation initiation in bacteria.

CHAPTER 16—ANSWER KEY

D. Assessing What You've Learned

(1) Testing Your Knowledge

1. b; 2. c; 3. c; 4. c; 5. d; 6. c; 7. a; 8. d; 9. d

(2) Integrating Your Knowledge

(a) This experiment utilized a binding assay, measuring the number of RNA polymerase complexes that bound to each viral genomic fragment. RNA polymerase and the viral DNA (radioactive) were mixed in a test tube, and the mixture was then passed through a nitrocellulose filter that binds to protein. Each viral DNA fragment bound to protein was retained on the nitrocellulose, while unbound DNA passed through the filter. Thus scientists could measure binding simply by measuring the radioactivity found in the filter. The researchers continued to add RNA polymerase to the mixture until all the binding sites on the DNA fragment were completely bound (at a ratio of eight protein molecules to each DNA fragment).

(b) Because several promoters were characterized, the researchers identified DNA sequences shared by many of the promoter regions. These conserved sequences were consistently observed in each promoter analyzed; thus, they appeared to be important to all promoters. These regions were later shown to bind to specific components of the RNA polymerase complex.

(c) Bacteria have sigma factors present in the cell that activate the expression stress response genes. Upon heat shock, for example, a sigma factor that activates heat-shock protein expression is activated. Thus genes that are important for stress response have unique promoters that are recognized only by the stress-activated sigma factors.

(d) Roeder and co-workers hypothesized that their RNA polymerase preparation was "too pure" and lacked a protein that normally helped to identify the proper promoter elements (a eukaryotic sigma factor?). When they repeated the experiments with crude cell extracts, the researchers found that transcription activity was promoter specific! Eventually they purified the component of the crude cell extracts that delivered this specificity.

(e) The ribosome must assemble around the mRNA in order to begin the process of translation. Upstream from the AUG is the Shine-Dalgarno sequence. Initiation factors bind to the Shine-Dalgarno sequence and enable the binding of a ribosomal small subunit (30S) to the region. Next a charged tRNA (aminoacyl tRNA with N-formylmethionine) binds to the AUG with its anticodon sequence. Finally, the large subunit binds and completes the formation of the initiation complex.

17

Control of Gene Expression in Bacteria

A. KEY BIOLOGICAL CONCEPTS

- *Escherichia coli* serves as an excellent model organism for the study of prokaryotic gene regulation. These bacterial cells adapt rapidly to **environmental changes** and can often continue to grow. The adaptation is usually based on altered gene expression, activating genes needed to respond to a specific change in conditions.
- Why not express all the genes all the time? Sets of genes are expressed for a specific cellular function, and most genes are expressed sporadically as they are needed. Continual expression of all genes would be devastating to cellular biochemistry, and would certainly cause the rapid destruction of the cell.
- Regulation of gene expression is fundamental to the identity of a cell, controlling cellular characteristics and physiology. When we discuss eukaryotic gene regulation, you will see how gene regulation controls cell identity in a multicellular organism.

17.1 Gene Regulation and Information Flow

- The central dogma of molecular biology involves several steps that can be regulated before an active protein is produced within the cell (**Figure 17.1**).
  1. **Transcriptional regulation**—Control of gene expression at the chromosome, allowing RNA polymerase binding and activation only under cellular conditions favoring expression of that gene.
  2. **Translational (post-transcriptional) regulation**—Regulatory mechanisms present in cells serve to regulate the stability of mRNA messages, to process a primary RNA transcript to an mRNA, and even to increase the efficiency of translation of that transcript.
  3. **Post-translational regulation**—The completed protein product sometimes undergoes additional processing steps to generate activity. These processing steps include phosphorylation, glycosylation, and cleavage of the protein. See **Chapter 16** for more discussion of these mechanisms.
- These types of regulation occur in both prokaryotic and eukaryotic cells. Prominent within these mechanisms is the regulation of transcription.

Metabolizing Lactose—A Model System

- Sugars are metabolized in cells to generate energy in the form of ATP. The pathway that initiates the breakdown of glucose is called **glycolysis** (breaking down glucose and gathering energetically valuable electrons in the process).
- The fastest and easiest way to feed glycolysis is to import glucose molecules from outside the cell and utilize them in glycolysis. If glucose molecules are not available, the bacterial

cell must convert available molecules to glucose or some other energetically useful molecule. To perform this conversion, enzymes are required. These enzymes are typically expressed only when glucose is unavailable but another convertible molecule is available.

- Lactose is a **disaccharide** (glucose and galactose molecules covalently attached to each other) that can be cleaved to the two constituent **monosaccharides** by an enzyme called **β-galactosidase**.

- In the absence of lactose, no β-galactosidase enzyme is expressed in *E. coli* cells. Lactose is an **inducer** of the β-galactosidase gene (and other genes in the lactose operon).

- What about cells having both lactose and glucose available at the same time? Is one sugar preferred over the other? **Jacques Monod** and colleagues tested the hypothesis that *E. coli* would prefer glucose over lactose (**Figure 17.2**). They measured the expression of β-galactosidase under three conditions: media with only glucose, media with only lactose, and media with both glucose and lactose. They found that β-galactosidase is expressed only when glucose is absent and lactose is present, consistent with Monod's hypothesis.

17.2 Identifying the Genes Involved in Lactose Metabolism

Screening Mutants—Replica Plating and Indicator Plates

- Monod, working with **Francois Jacob**, initiated a hunt for mutant *E. coli* that are unable to metabolize lactose. This strategy allowed the scientists to screen for *E. coli* that are unable to utilize lactose as a source of carbon.

- Bacteria can be easily grown in **petri dishes**. Thousands of cells are spread onto each plate and allowed to grow into clonal colonies. Each cell in a **colony** (a pile of cells) shares the same genetic makeup, because all have grown from a single cell. If the original cell carried a mutation in β-galactosidase, then all cells in the colony will contain this mutation.

- Examine replica plating, displayed in **Figure 17.3**. This technique is used to make "copies" of populations of cells. These copies can then be tested in lactose medium, and researchers will still have copies of each clonal population for analysis in later experiments.

- **Indicator plates** were used in this mutant hunt. Monod and Jacob generated petri dishes containing an o-nitrophenyl-β-D-galactoside **(ONPG) indicator molecule**, with a structure that is similar to that of lactose (two subunits bound by a β-1,4 glycosidic bond). Thus any enzyme that breaks down lactose would be expected to break down ONPG. Once broken down, the ONPG subunits generate a yellow coloration in the media, indicating enzyme activity. The researchers could thus examine thousands of *E. coli* colonies, identifying those that lacked the ability to break down ONPG by the absence of yellow coloration near the colony.

- Monod and Jacob were interested in these mutant cells; they hypothesized that the cells carried mutations in the genes that regulate or mediate lactose utilization.

Different Classes of Lactose Metabolism Mutants

- Monod and Jacob recognized that several enzymes were likely important in the utilization of lactose. They isolated several classes of *E. coli* mutants, each containing a different defect in this process (**Table 17.1**).

1. **β-glycosidic** mutants—Loss of this enzyme activity results in an inability to cleave lactose into glucose and galactose. The gene encoding β-galactosidase was named *lacZ*.

2. **Galactoside permease mutants**—A protein located in the cell membrane mediates the transport of lactose into the *E. coli* cell. In the absence of this enzyme, cells cannot accumulate lactose for cellular metabolism. The gene encoding galactoside permease was named *lacY*.

3. **Constitutive mutants**—This class of mutants has lost the ability to regulate expression of the genes needed for lactose utilization. Unlike wild-type cells, they display ONPG cleavage in the *absence* of lactose, indicating that β-galactosidase is being expressed continuously (constitutive expression). One mutation that resulted in the

constitutive expression of β-galactosidase was found in a gene named *lacI*. The "I" is for the phenotype where an inducer (lactose) is unnecessary for the expression of β-galactosidase. This makes perfect sense because the wild-type lacI⁺ protein **represses** gene expression, and the loss of this repression results in abnormal activation of gene expression. Many genes have been named after the phenotype observed when that protein is nonfunctional.

- Further studies revealed that these three genes are located near each other on the bacterial chromosome (**Figure 17.4**). In prokaryotic organisms, genes with a shared purpose (like lactose metabolism) are sometimes controlled by a single promoter; these groups of genes are called **operons**.
- Once scientists were able to generate a genetic map of the chromosomal region near the *lacZ* gene, they found that *lacY* and a third gene—*lacA* (an acetyltransferase)—were all located adjacent to each other.

17.3   The Discovery of the Repressor

- The lactose operon is regulated by **positive** and **negative control** mechanisms. Positive control is when something must bind the promoter to activate expression; negative control is when something bound to the promoter blocks expression of the operon.
- **Leo Szilard** (1950) suggested that Monod's data indicated the lacI⁺ protein was a negative regulator of *lacZ* and *lacY* expression. Thus lacI⁺ was a **repressor** of the lactose operon.
- Monod had shown that *lacI* expression was **constitutive**; in other words, it was always "on" regardless of the presence or absence of lactose in the medium. How then could lacI⁺ release its repression of the operon in the presence of lactose? Szilard proposed that the lacI⁺ protein interacted directly with lactose, and that when lacI⁺ was bound to lactose, it released the promoter element (**Figure 17.5**).
- **Key Research**—To test this hypothesis, the researchers developed an experimental strategy for transferring DNA from one microbe to another (using the bacterial process of **conjugation**). A recipient group of cells carried the

lacI⁻ mutation; they expressed *lacY* and *lacZ* continually, regardless whether lactose was present or absent. When DNA from a *lacI⁺ lacY⁻ lacZ⁻* strain was introduced, the lacI⁺ protein from the introduced DNA restored negative control of the recipient strains' lactose operon. This experiment and others showed Szilard's hypothesis was correct.

Testing the lac *Operon Model*

- A single promoter regulates the expression of all three *lac* operon genes, and that promoter is activated only under specific conditions (presence of lactose, absence of glucose).
- Key hypotheses proposed and experimentally confirmed by Jacob and Monod: (1) operon genes are expressed on a single **polycistronic** mRNA, and their expression is regulated by a single promoter; (2) the *lacI* gene (expressed constitutively by its own promoter) encodes a repressor protein that binds to the **operator** site of the *lac* operon promoter, blocking expression of the operon; (3) lactose is an **inducer** molecule that binds directly to the repressor, causing it to release the operator element and relieving negative regulation of the operon.

17.4   Catabolite Repression and Positive Control

- **Catabolism** is the metabolic breakdown of large molecules to smaller energetically useful molecules. If the cell has plenty of these breakdown products, the catabolic pathway is typically stopped. This inhibition of a metabolic pathway is called **catabolite repression** (e.g., glucose inhibiting the lactose operon).
- The absence of glucose actually helps to activate expression of the lactose operon, via the action of another DNA-binding protein called the **catabolite activator protein (CAP)**.
- Inactivation of the *lacI⁺* repressor allows for some expression of the *lac* operon; but binding by the positive regulator protein, CAP, generates a massive induction of expression.
- CAP is **allosterically** regulated by binding to a molecule called cyclic adenosine monophosphate (cAMP), a modified nucleotide present in high concentrations only when

glucose levels are depleted (**Figure 17.9**). CAP bound to cAMP induces expression of the lactose operon by binding the **CAP site** in the promoter region to recruit RNA polymerase to that site for transcription.

- CAP protein is an example of positive control of an operon; when bound to the promoter region, CAP *increases* operon expression.

- **Figure 17.10** is a good summary of positive and negative regulation of the lac operon.

17.5 The Operator and the Repressor— an Introduction to DNA-Binding Proteins

Finding the Operator

- The **operator sequence** within the *lac* operon promoter is recognized by the repressor protein (lacI⁺). But how was this region of the promoter identified as being important?

- Scientists used an assay called **DNA footprinting** to precisely identify the site of repressor binding. This assay is summarized in **Box 17.2** and **Figure 17.11**. They first radioactively labeled DNA fragments containing the *lac* promoter and mixed these fragments with repressor proteins to allow for repressor binding. They then added an enzyme that degrades DNA, knowing that any DNA that is bound up by the repressor will be protected from digestion. Upon analyzing the products, they identified one DNA sequence that was consistently bound tightly by the repressor protein—the operator sequence.

- The lactose operon promoter actually contains several consecutive operator sequences, each able to bind a repressor protein. Their sequence and organization are summarized in **Figure 17.12**. Notice that the DNA sequence of the operator is symmetrical (contains **dyad symmetry**); this is common in DNA sequences that bind regulatory proteins.

DNA Binding via the Helix-Turn-Helix Motif

- **Domain**—Within a protein, domains are regions that display unique structural characteristics (β-helix, χ-sheet).

- **Motif**—A motif is a conserved domain that is shared by a related group of proteins.

- Many structural motifs are observed in DNA-binding proteins. A common motif observed in the *lac* repressor is called the **helix-turn-helix motif**. This structure fits easily into the major groove of the double helix, and actually "reads" the DNA sequence via hydrogen-bond interactions between amino acids in the protein and nitrogenous bases in the DNA (**Figure 17.13**). The region of the protein that interacts with the nitrogenous bases is known as the **recognition sequence**.

- The active *lac* repressor is a tetramer of four lacI⁺ proteins, each of which can bind to an operator sequence to block the opening of the double helix for transcription. One tetramer is capable of binding two operator sequences, and will cause the DNA between the operators to "kink" or "loop" (**Figure 17.14b**).

- When a repressor interacts with an inducer (lactose or IPTG, an analog of lactose), the inducer binds to central region of the repressor, inducing a change in shape of the repressor tetramer (**Figure 17.14c**). This **allosteric** influence of lactose on the repressor makes the repressor release the operator sequence, relieving negative regulation of the promoter.

B. CROSS-CUTTING THEMES

Looking Back—
Concepts from Earlier Chapters
How Genes Work—Chapter 15
With this chapter, you are adding to your knowledge of how genes work at the level of molecular biology. Regulating the expression of genes is a crucial aspect of their relevance to cell growth and adaptability. Regulation can be simplified to the interaction of proteins with specific sequences in the **promoter** region that precedes the **open reading frame (ORF)**. This interaction can increase or decrease the ability of **RNA polymerase** to transcribe that gene.

Looking Forward—
Concepts from Later Chapters
Pattern Formation and Cell Differentiation—Chapter 22
Cellular differentiation in multicellular organisms is fundamentally due to the careful control

of gene expression. Your introduction into regulated gene expression in this chapter builds a foundation for your future studies of **tissue differentiation, developmental biology**, and even the cause of diseases like **cancer**. Research into how genes are regulated has broad relevance to many areas of biology.

Animal Form and Function—Chapter 38
It is incredible to imagine a simple microbe "making careful decisions" about which genes to express at any given time. How did this come about? Varied environmental pressures *exert a powerful* selection on microbes, and those that survive from generation to generation must display the ability to adapt to change.

C. DIFFICULT TOPICS

Scientists have been examining regulation of the lactose operon for decades. They have identified a complex but fascinating combination of regulatory proteins and DNA regulatory sequences. It is easy to follow how this regulation works when examining one component at a time; but when viewing both positive and negative regulation, things get complicated.

Examine **Figure 17.10** in your textbook, walking through the process that occurs for each of the four conditions listed. Which conditions might result in weak expression on the operon? What interactions must occur to have full activation of the operon? This exercise is vital to study of this operon; ideally, it will resolve any confusion you may have, helping you see how positive and negative regulation of gene expression work together.

D. ASSESSING WHAT YOU'VE LEARNED

(1) Testing Your Knowledge

1. A hypothetical bacterium isolated from a Martian sea uses silicose as its main energy source. This silica-based sugar is one of many sugars present on Mars, and tests have shown that silicose is present everywhere on the planet. Which of the following would be the most efficient type of control for the production of silicase, the enzyme this bacterium uses to metabolize silicose?
   a. constitutive transcription of silicase gene
   b. negative control of transcription of silicase gene
   c. catabolite repression transcription of silicase gene
   d. inducible operon in control of transcription of silicase gene

2. The role of the operator sequence in an operon is to:
   a. orient RNA polymerase at the transcriptional start site
   b. bind a transcriptional repressor protein
   c. bind an transcriptional inducer molecule
   d. encode a transcriptional regulator (positive or negative)

3. The allosteric regulation of protein activity could be described as:
   a. regulation via allelic duplication
   b. an increase in the expression of a protein
   c. inactivation by proteolytic cleavage
   d. a structural change in a protein, resulting in increased or decreased activity

4. The pattern of χ-galactosidase expression illustrated in **Figure 17.2** indicates that:
   a. Lactose is the preferred source of energy in a bacterial cell.
   b. Glucose is the preferred source of energy in a bacterial cell.
   c. Glucose and lactose are utilized equally in bacteria.
   d. χ-galactosidase is critical to glucose utilization.

5. The *trp* operon encodes genes for tryptophan biosynthesis and is regulated by:
   a. tryptophan, the *trp* repressor, transcriptional attenuation
   b. the concentration of glucose, transcriptional attenuation
   c. the cell cycle (growth), the *trp* repressor
   d. tryptophan, the *trp* repressor

6. Which of the following statements is true regarding the reaction shown in **Figure 17.7b**?
   a. χ-galactosidase tightly regulates cellular glucose concentration.
   b. Increasing the concentration of galactose will block χ-galactosidase activity.
   c. Increasing the concentration of glucose will decrease the production of χ-galactosidase.
   d. Increasing the concentration of glucose will induce production of χ-galactosidase.

7. Which of these conditions would initiate the reaction shown in **Figure 17.8b**?
   a. low levels of glucose in the cell
   b. high levels of glucose in the cell
   c. low levels of lactose in the cell
   d. high levels of lactose in the cell

8. DNA footprinting is a valuable technique in the identification of:
   a. open reading frames within genes
   b. the site of specific DNA/protein interactions
   c. genomic loci containing operons
   d. positive regulators of gene expression

(2) Integrating Your Knowledge

(a) List all aspects of the "central dogma" of molecular biology that involve the regulation of gene expression.

(b) In the presence of lactose and glucose, why do *E. coli* cells prefer glucose as a source of energy?

(c) Isolation of cells having mutations in specific types of genes is a common research strategy. Researchers studying utilization of lactose performed a screen for cells with a mutation in the β-galactosidase gene. How was this screen designed?

(d) Define allosteric regulation and give an example.

(e) Catabolite repression is commonly seen in *E. coli*. It involves inhibition of some operons by the presence of a product of that operon's metabolic action. How does catabolite repression work in the lactose operon? Predict other cellular metabolic pathways that might experience catabolite inhibition.

CHAPTER 17—ANSWER KEY

D. Assessing What You've Learned

(1) Testing Your Knowledge
1. a; 2. b; 3. d; 4. b; 5. a; 6. b; 7. a; 8. b

(2) Integrating Your Knowledge

(a) Gene expression is regulated at several places:
   - transcription (regulation of promoter activity)
   - post-transcriptional processing (RNA splicing, mRNA stability)
   - translation initiation (availability of initiation factors)
   - post-translational processing (modifications of proteins enhancing activity)

(b) Glucose can be acted on directly by glycolysis and other metabolic processes to produce energy. Lactose must be converted to glucose. Thus the production of energy from glucose is more efficient and is favored by bacteria.

(c) Cultures of cells were exposed to a mutation-inducing compound and then screened for the ability to cleave a lactose analog (ONPG). Digestion of ONPG by β-galactosidase results in the release of a bright yellow molecule. Thus cells that lack β-galactosidase activity are observed as white cells in the presence of ONPG, even under conditions that activate β-galactosidase expression.

(d) The allosteric regulation of a molecule involves alteration of its structure to regulate its activity. The binding of lactose to the lactose repressor, and cAMP to the CAP protein, are examples of structural changes that regulate the capacity of these proteins to bind DNA.

(e) Catabolite repression of the lactose operon involves binding of CAP to the promoter region. The presence of this protein dramatically enhances gene expression by helping to recruit the RNA polymerase to that site. CAP can bind to the promoter only when it is associated with cAMP, an allosteric regulator of its binding activity. The presence of high concentrations of glucose in *E. coli* cells generates a very low level of cAMP in the cytosol; the operon is thus repressed until the glucose is depleted.

There are many examples of catabolite repression in bacterial cells. One prominent example is a group of operons that mediate amino acid biosynthesis. In the absence of a specific amino acid, the operons are activated; but when the amino acid is readily available, the operon is repressed.

18

Control of Gene Expression in Eukaryotes

A. KEY BIOLOGICAL CONCEPTS

18.1 Mechanisms of Gene Regulation—An Overview

- Regulation of gene expression in eukaryotic cells is more complex than that observed in prokaryotic cells.
- Eukaryotic DNA is tightly associated with **histone** proteins, generating a DNA/protein complex known as **chromatin**. To express a specific gene, the histones at the site of that gene's promoter must be dislodged, a process known as **chromatin remodeling**.
- Eukaryotic gene expression involves **RNA processing**; introns are spliced out of the primary transcript, a 5≤cap is added, and a 3≤tail is added to the mRNA. The exons of some genes can be spliced together in different ways to generate unique protein products, a process known as **alternative splicing**.
- **Figure 18.1** is a summary of the mechanisms that regulate eukaryotic gene expression.

18.2 Eukaryotic DNA and the Regulation of Gene Expression

- **Chromatin** is the complex formed between DNA and proteins in the cell nucleus. Proteins in this complex are called histones, and are positively charged due the high density of **lysine** and **arginine** amino acids in chromatin's structure. DNA is negatively charged due to the phosphate residues in the phosphodiester backbone; therefore, the DNA/histone complex forms spontaneously.

- **Histones** look like a round disk around which the DNA double helix can wrap, protecting DNA and compressing the long molecule. One chromatin complex with DNA wrapped around its structure is called a **nucleosome**. They are visible in preparations of chromatin and can be visualized using an electron microscope (**Figure 18.2**).
- **H1 histone** is important to chromatin structure; it is not part of the disk-shaped complex, but associates with the "linker" DNA between each nucleosome complex.
- The **30-nm fiber** forms as the nucleosomes form a helical structure, wrapping around an axis and further condensing the chromosome. H1 histones interact with each other to mediate the formation of this structure.

Evidence that Chromatin Structure Is Altered in Active Genes

- **Key Research—Harold Weintraub** and **Mark Groudine** tested the hypothesis that transcription of a gene would not occur if it were tightly associated with histones in nucleosome complexes. They tested this by examining genes that are expressed differentially in varied tissues.
1. **χ-globin** is expressed at high levels in **reticulocyte** cells (a type of blood cell).
2. **Ovalbumin** is expressed at high levels in the **female reproductive tract**, but it is not expressed in reticulocyte cells.

- To evaluate any histone association of these two genes in reticulocytes, the researchers employed an assay to degrade DNA that is *not* associated with protein (**Figure 18.3**). The DNase enzyme cannot attach and digest DNA if it is bound tightly to histone proteins.
- Chromatin (purified from reticulocyte cells) is incubated with DNase I and breaks down the DNA not protected by protein association.
- But how can you identify the χ-globin and ovalbumin chromosome fragments specifically? You must **probe** the DNA fragments produced for the intact genes (a Southern blot). The DNase fragments generated from reticulocyte cells contained intact ovalbumin genes and degraded χ-globin genes. The χ-globin gene was not associated with histones, consistent with their hypothesis.
- A second set of evidence was generated in studies with budding yeast (*Saccharomyces cerevisiae*). Mutations in histone genes resulted in abnormal patterns of gene expression, consistent with the hypothesis that chromatin has a key role in regulating gene expression.

How Is Chromatin Altered?

- If histone association with a chromosome occurs only at the site of inactive genes, activation of a gene must somehow be preceded by histone removal. How does this work?
- In the budding yeast (*Saccharomyces cerevisiae*), a group of proteins called the **chromatin-remodeling complex** is associated with the loss of histones from the chromosome. These complexes add acetyl (–CH_2 COO^-) or methyl (–CH_3) groups to the histones. These additions disrupt histone association with the DNA and diminish the positive charge on the histone (**Figure 18.4**).
- **Histone acetyl transferases** (HATs) add acetyl groups to histones to allow gene expression; **histone deacetylases** (HDACs) remove acetyl groups to allow chromatin reformation.

18.3 Regulatory Sequences and Regulatory Proteins

Promoter-Proximal Elements

- Most **regulatory sequences** are found 5≤or upstream of the open reading frame. The **promoter** is the site where RNA polymerase binds to initiate transcription.
- In eukaryotic cells, the **TATA sequence** is found just upstream of the transcription start site. It is bound by the **TATA-binding protein** (or **TBP**). TBP acts in a similar manner to the sigma factor found in prokaryotic cells, locating the promoter for the RNA polymerase enzyme and aiding in the initiation of transcription. Once transcription is initiated, the TBP dissociates (like sigma factor).
- **Key Research—Yasuji Oshima** and co-workers used the budding yeast, *S. cerevisiae*, to first identify an additional **positive regulator** that aids in regulating gene expression. They focused on genes required to utilize galactose; the genes are active only when galactose is present in the growth medium.
- Yeast-cell lines with a mutation in *GAL4* displayed inability to properly activate expression of galactose utilization genes. Protein encoded by this gene was shown to be a positive regulator of galactose utilization genes (similar to the CAP protein in *E. coli*). The GAL4 protein contains a helix-turn-helix motif (**discussed in Chapter 17**), and a **zinc-finger domain** common in DNA binding proteins. But what does the GAL4 protein bind?
- The GAL4 binding site was identified as a short DNA sequence 20 base pairs long and located 5≤of galactose utilization genes (**Figure 18.5**). The binding site is outside of the RNA polymerase binding site (basal promoter) and called a **promoter-proximal element**. The GAL4 protein binds this site and increases transcriptional activity at the adjacent gene.

Enhancers

- **Enhancer sequences** are another type of regulatory sequence, but they are unique due to the ability to influence gene expression from a long distance (up to 100,000 base pairs from the gene). They can be located upstream or downstream of the gene they regulate, or in intron sequences within the gene.

- **Susumu Tonegawa** and co-workers selected an **immunoglobulin**-heavy chain gene for study. They recognized that regulation of this gene would be dynamic, and they focused attention on the regulatory elements they hypothesized to be present in the gene's introns.
- **Key Research**—Review **Figure 18.6**, which summarizes the work of **Tonegawa** and co-workers as they characterized regulatory elements associated with this immunoglobulin gene. They suspected that a region within an intron was influencing the expression of this gene. They selectively removed regions of this intron using restriction enzymes (**see Chapter 19**), and then assayed the remaining gene locus for expression. This selective cutting and reconnecting (ligating) is a fundamental process in the manipulation and cloning of gene sequences.
- Gene expression was measured in this assay using a **Northern blot**. This type of analysis allows the researcher to measure the expression of specific genes by quantifying the amount of mRNA present in a cellular extract. RNA samples are separated on an agarose gel (similar to the Southern blot), transferred to a nylon membrane, then hybridized with a radioactive probe for a specific gene.
- The native and recombinant forms of the immunoglobulin gene were assayed for gene expression. They identified a region in the intron that was required for expression of the immunoglobulin gene; *when it was spliced out, no expression was observed.*
- **Silencers** are similar to enhancers, with varied location and distance to the regulated gene; but they repress rather than activate gene expression.

How Do Enhancers Work?

- **Key Research**—**Julian Banerji** further characterized enhancer sequences, identifying their role in generating a **tissue-specific activation** of gene expression (**Figure 18.8**). They utilized the immunoglobulin gene discussed earlier, revealing that enhancer activity found in the intron was activated only in B cells (where immunoglobulins are made).

- The researchers then spliced the immunoglobulin enhancer sequence into the regulatory regions of the χ-globin gene (normally not expressed in B cells). The χ-globin transgene was now expressed in B cells, revealing the capacity for the enhancer element to activate a tissue-specific gene expression.
- Transcriptional **regulatory proteins** that are found in these differentiated tissues must bind enhancer, silencer, and regulatory elements found on the chromosome. But what are these proteins?

18.4   Transcription Initiation

- Many DNA-binding proteins that regulate the initiation of eukaryotic transcription have been identified.
- **Basal transcription factors** associate with the core transcription complex and the promoter sequence, while **regulatory transcription factors** bind to enhancers and silencers (promoter-proximal elements).
- **Figure 18.10** summarizes the current model of transcription initiation in eukaryotic cells. This model summarizes the work of many researchers. Key to this model is the physical contact between regulatory proteins and the core transcription complex. For this to occur, the DNA must actually fold back on itself, allowing enhancer-binding proteins to come into contact with the core complex.
- Review **Figure 18.10**; initiation of transcription is summarized in the following steps:
Step 1. Regulatory transcription factors recruit the chromatin-remodeling complex and histone acetyl transferases (HATs).
Step 2. Acetylation of histones results in a weakening of histone-DNA interactions and generates "naked DNA" for TFIID binding and transcription initiation.
Step 3. Additional regulatory transcription factors bind enhancers and promoter-proximal elements. These aid in recruiting basal transcription factors to assemble at the promoter, forming the **basal transcription complex** (~10 proteins). The TBP in the complex identifies the TATA box and orients the complex with respect to the transcription start

site. Its binding is the first event in assembly at the promoter sequence (**Figure 18.11**).

Step 4. The RNA polymerase enzyme and associated proteins join the basal transcription complex at the transcription start site to form the **core transcription complex**. Enhancer-binding proteins continue to interact with the large complex through the looping of DNA between their respective locations.

- A key question in the regulation of eukaryotic gene expression focuses on the regulatory mechanisms of these enhancer/silencer-binding proteins. How do they communicate with the core transcription complex?

18.5   Post-Transcriptional Control
Alternative Splicing of mRNAs

- Eukaryotic genes are often broken into exons that are separated by noncoding introns. The primary transcript contains both introns and exons; in the process of post-transcriptional processing, the introns are spliced out to yield the intact open reading frame for translation.

- Some genes exhibit **alternative splicing** patterns, depending on the tissue type or environmental conditions. But what does alternative splicing involve, and how are the protein products different?

- Alternative splicing simply means that the cell splices different exons together, depending on the cell type or condition, generating a unique protein with each splicing pattern. The process of alternative splicing is regulated by cell-type-specific proteins that regulate the splicesome itself.

- One example from text is the mammalian muscle-specific protein **tropomyosin**, encoded in a large gene with 14 exons. Exons that are spliced together depend on the identity of the cell in which the primary transcript is generated (**Figure 18.12**). Smooth muscle cells and striated muscle cells generate different products. There are many examples of alternative splicing; each generates two or more unique proteins from a single gene.

- At least 35% of human genes undergo alternative splicing; though we have only around 40,000 genes, it is anticipated that we express over 100,000 different protein products.

Translational Control

- In the cytosol, additional regulatory mechanisms control gene expression. The stability of mRNAs in cytosol is highly variable. Some are degraded rapidly, allowing for only a short period of translation; others are quite stable. Casein (a milk protein) has a half-life of 1 hour in nonlactating mammary tissue. During lactation, however, this half-life is expanded to 28.5 hours, allowing massive amounts of this protein to be generated.

- Cytosolic factors may also bind mRNAs to block their translation until needed.

- The initiation of translation is mediated by translation initiation factors (TIFs). These can be inactivated by phosphorylation during times of cellular stress, blocking the translation of all but a few transcripts. Viruses can induce this type of response as they try to force the cell to synthesize huge amounts of viral proteins.

Post-Translational Control

- The activity and stability of proteins is tightly regulated by post-translational events that can include protein glycosylation, cleavage, and phosphorylation (introduced in Chapter 16). Glycosylation (attachment of sugar) of proteins is critical to protein folding and structure. Phosphorylation or cleavage of a protein offers a rapid method for activating or inactivating protein activity.

- The signal transducers and activators of transcription (STATs) are an excellent example of post-translational regulation via protein phosphorylation (**Figure 18.13**). In this case the target of the induced signaling cascade is the activation of expression of select genes.

How Does Gene Expression in Bacteria and Eukaryotes Compare?

- The four primary differences: (1) Packaging of DNA in eukaryotes, (2) splicing of mRNA in eukaryotes, (3) complexity of transcriptional control in eukaryotes, and (4) operons in prokaryotes.

18.6   Linking Cancer with Defects in Gene Regulation

- Humans are composed of trillions of cells, working together to produce a healthy organism. Each day many old, damaged, or infected cells are selectively removed from your body and replaced by the mitotic division of healthy neighboring cells. The maintenance of your size and shape requires a dynamic balance of cell death and cell division. **Cancer** is one disease that results when this balance is lost, and a group of cells grows uncontrollably, ignoring the signals sent to control its division.
- Cancer is frequently the result of mutation. Mutation-causing toxins (**mutagens**) often are shown to be **carcinogenic** (cancer causing) and are labeled as **carcinogens**.
- But how is a normal cell transformed into a cancer cell? As you may suspect, cancer often results from mutations in genes that (1) stop or slow the cell cycle or (2) trigger cell growth/division. Genes that stop/slow the cell cycle are known as **tumor suppressor genes**. When cell activity is lost due to mutation (loss-of-function mutations), cells are released from this negative control of the cell cycle. Genes that trigger cell division are called **proto-oncogenes**. Mutation can abnormally induce these genes (a gain-of-function mutation), inducing uncontrolled cell division.

p53—Guardian of the Genome

- **p53** is a critical regulatory protein that has been shown to be abnormal in many types of cancer. Its normal cellular function is to regulate the progression of the cell cycle. Loss-of-function mutations in *p53* are present in half of all human cancers.
- **Key Research—Warren Maltzman** and **Linda Czyzk** revealed that *p53* expression is dramatically activated following exposure to a chemical mutagen.
- *p53* was later shown to act as a **tumor suppressor gene**, arresting cell division following DNA damage (**Figure 18.14**) and allowing the cell to properly repair any damaged genes. The loss of *p53* function allows the cell to replicate DNA despite the presence of mutations, thus dramatically elevating the mutation rate and increasing the likelihood that a cancer-causing mutation may result.

STAT Mutations in Cancer

- **STAT proteins** are **signal transducers and activators of transcription**. These proteins are present in cytosol in a dormant state, and they are activated rapidly following receipt of an extracellular signal (e.g., phosphorylation of STAT). Following activation. these proteins can migrate to the nucleus and activate the transcription of specific genes. Genes activated by STAT proteins are often involved in regulation of cell division.
- Mutant forms of some STAT proteins have been shown to bind DNA in the absence of any external signal, and to activate unregulated cell growth.
- The active form of STAT3 protein is a **dimer** of two activated STAT3 proteins, and formation of the dimer is induced by phosphorylation of a specific amino acid in the protein.
- **Key Research—Jacqueline Bromberg** showed that a mutation inserting a cysteine residue into the dimerization region results in constitutive dimerization of these proteins. The dimerization is stabilized by the formation of a **disulfide bond** between the cysteine residues. This results in a "grow" signal from the STAT3 protein in the absence of any external signal (see **Figure 15.18** for data). But could this type of mutation, abnormally activating the STAT3 signal, cause cancer?
- Cells expressing mutant *STAT3* genes were injected into mice and produced tumors.

B. CROSS-CUTTING THEMES

Looking Back—
Concepts from Earlier Chapters
How Genes Work—Chapter 15
Structures of genes within chromosomes, and the nature of the **central dogma**, are complicated by the revelation of **exons and introns** in eukaryotic DNA. Each gene may produce a

distinct protein, but the information may be broken up into several distinct units.

Control of Gene Expression in Bacteria—Chapter 17

Now that you've examined regulation of gene expression in eukaryotic cells, review **homologous processes of gene regulation** in prokaryotes. In what ways do regulatory mechanisms from these organisms parallel each other? What are key differences between these organisms?

Looking Forward—
Concepts from Later Chapters

Pattern Formation and Cell Differentiation—Chapter 22

Cellular differentiation in multicellular organisms is fundamentally due to the careful control of gene expression. Your introduction into regulated gene expression in this chapter builds a foundation for your future studies of **tissue differentiation, developmental biology**, and even the cause of diseases like **cancer**. Research into how genes are regulated has broad relevance to many areas of biology.

The Immune System in Animals—Chapter 49

In this chapter, we've examined the regulated expression of **immunoglobulins** in **B cells**. The **immune system** is one of the most fascinating examples of cellular differentiation to generate cells with a specific function. Your body is constantly replenishing the immune system from undifferentiated **stem cells** in bone marrow, stimulating their maturation to an array of cells that fight infection.

C. DIFFICULT TOPICS

Processing of mRNA as a form of gene regulation is a recent addition to our understanding of eukaryotic biology. The genes and enzymatic mechanisms that control alternative splicing are not yet well understood. The concept of differentially splicing a primary transcript depending upon cell type and environment is interesting, yet confusing at the same time. How did genes gain this adaptability, enabling a kind of "mix-and-match" strategy of exon combinations? This

has become one of the most important areas of study in eukaryotic gene regulation. Scientists had anticipated that the genome project would reveal humans to have ~100,000 genes, basing their predictions on cellular complexity, mRNA diversity, and genome size. Completion of the genome in 2001 (see **Chapter 20**) revealed that humans have only ~30,000 genes. At first this did not seem to make sense, but subsequent research has shown that alternative splicing is very common in human cells. So, though we have only 30,000 genes, we may still be able to generate 100,000 different mRNAs by differentially splicing exons together. Alternative splicing is far from an interesting side note in eukaryotic gene expression; it is just as important as promoter-proximal elements. Certainly there will be many more surprises as scientists characterize the regulation of gene expression.

D. ASSESSING WHAT YOU'VE LEARNED

(1) Testing Your Knowledge

1. Which of the following DNA sequences is most likely to serve as a restriction enzyme recognition site? (Hint: Find the palindrome!)
    a. GATTAG
    b. CAATTG
    c. AAATTTCCC
    d. CTAGGG

2. Which of the following cells were included as *positive* controls in the experiment described in the previous question?
    a. normal human B cell, control mouse cell, and intact gene in mouse cell
    b. normal human B cell and intact gene in mouse cell
    c. control mouse cell only
    d. normal human B cell only

3. Which of the following is *not* true about eukaryotic regulatory sequences?
    a. Some increase transcription, and others suppress it.
    b. They can function if their 5≤to 3≤ orientation is reversed.
    c. They always have the same function in different cell types.

d. They can function if they are moved to a new location.

4. Which of the following regulatory elements is matched correctly to its chemical composition?
   a. promoter-proximal element; DNA
   b. enhancer; protein
   c. RNA polymerase; DNA
   d. silencers; protein

5. Which of the following is involved in transcription of prokaryotes, but *not* in eukaryotic transcription?
   a. enhancers
   b. RNA polymerase
   c. promoters
   d. operators

6. In smooth and striated muscle cells, DNA for tropomyosin is the same, yet mRNA produced is different. Which of the following is the cause of this phenomenon?
   a. The presence of cell-type specific compounds influences the splicing patterns of the mRNA.
   b. The tropomyosin gene is very large and contains a large number of introns and exons.
   c. Distinct tropomyosin proteins are produced in the two cell types.
   d. A shorter region of the tropomyosin DNA in striated muscles cells is transcribed into mRNA, resulting in a smaller mRNA transcript.

7. If a hormone bound to a cell in order to induce transcription of a given gene by using the STAT pathway, which of the following would most likely occur?
   a. An inactive STAT protein would bind to the enhancer.
   b. An activated STAT protein would bind to the gene, inhibiting transcription.
   c. The STAT proteins would become non-phosphorylated and remain in the cytoplasm.
   d. The STAT proteins would become phosphorylated and move into the nucleus.

8. Which of the following could be a mechanism to help prevent cancer?
   a. increased expression of a protein similar in structure and function to p53
   b. STAT proteins that are easily dimerized
   c. STAT proteins that are constitutively active
   d. moving genes involved in cell growth close to regions of a chromosome that are actively transcribed

9. Which of the following is an application of recombinant DNA technology?
   a. animal cloning
   b. in vitro fertilization
   c. rice plants that produce rice with χ-carotene
   d. using bacteria to control the insect infestation of food crops

10. Which of these changes to the histone structure generates dissociation from DNA?
    a. proteolytic cleavage
    b. acetylation (negative charge)
    c. glycosylation (sugar addition)
    d. all of the above

(2)   Integrating Your Knowledge

(a) Explain how alternative splicing can generate unique protein products from a single gene.

(b) The GAL4 protein offers important clues into the mechanisms of gene regulation in eukaryotic cells (*S. cerevisiae*). What role does this protein play in activating gene expression under conditions where galactose is present? Is there an analogous protein in bacteria?

(c) Summarize what is known about enhancer sequences.

(d) Explain the role of STAT proteins in regulation of gene expression.

(e) How can the mutations in STAT proteins cause cancer?

(3)   Showcasing Your Knowledge

(a) Discuss p53 as an important control protein in cell division, and as a marker for cancer.

(b) In chronological order, write out the steps that activate eukaryotic gene expression. Provide as much detail as possible. What steps parallel specific events in prokaryotic gene expression?

CHAPTER 18—ANSWER KEY

D. Assessing What You've Learned

(1) Testing Your Knowledge
1. b; 2. b; 3. c; 4. a; 5. d; 6. a; 7. d; 8. a; 9. c; 10. b

(2) Integrating Your Knowledge
(a) Higher eukaryotic organisms divide their genes (open reading frames) into small fragments called exons, which are dispersed along the chromosome. The primary RNA transcript for the gene must then be spliced together to restore the continuous reading frame for the ribosome. There are now many examples of genes that splice the exons together differently, depending on cellular conditions. Consequently, exons are added or deleted from the final product depending on the splicing decisions. The result is a dramatically altered protein product.

(b) GAL4 is a zinc-finger, DNA-binding protein that is analogous to the CAP-binding protein utilized in the lactose operon. In the presence of galactose, this protein binds upstream from the transcription start site of select genes and positively regulates activation of expression.

(c) • Enhancers elevate the level of transcription for a specific gene.

• Enhancers can generate a tissue or cell-type specific regulation of gene expression.

• Enhancers can work even if their normal 5≤↓ 3≤orientation is flipped.

• Enhancers can work even if moved from their normal location.

(d) STAT proteins are signaling proteins normally located in the cytosol. These STAT proteins signal the receipt of external growth signals that bind to membrane-associated receptors. The activated receptors mediate phosphorylation of STAT proteins, and the phosphorylated form migrates to the nucleus to activate gene expression. This signaling mechanism is rapid, because the STAT proteins are readily available to carry this signal to the nucleus.

(e) The STAT proteins display an altered structure when phosphorylated, and this structural change is thought to activate the nuclear migration and activation of gene expression. Mutations in STAT proteins that generate similar changes in structure have been shown to generate unregulated cell division (cancer). Jacqueline Bromberg, James Darnell, and their research teams generated this data.

19

Analyzing and Engineering Genes

A. KEY BIOLOGICAL CONCEPTS

19.1 Using Recombinant DNA Techniques to Manufacture Proteins: The Effort to Cure Pituitary Dwarfism

- **Pituitary dwarfism** results from the abnormal production of **human growth hormone (hGH)**, encoded by the gene *GH1*. This disease can be inherited from generation to generation, or it can occur spontaneously.
- The mutant form of the *GH1* gene is **recessive** to the dominant allele; thus humans can carry the mutant allele without displaying any outward abnormal phenotype.
- Humans affected by pituitary dwarfism grow slowly, reaching a maximum adult height of about 4 feet.
- hGH can be purified from pituitary glands dissected out of **human cadavers**. Isolating the hormone in this way is expensive, but it is effective in treating pituitary dwarfism.
- In the 1980s researchers discovered contamination of this hormone supply with the infectious agent that causes **neurodegenerative** disorders (**Kuru and Creutzfeld-Jacob disease**). Kuru was first observed in cannibalistic tribes in **New Guinea**, passing from person to person via consumption of human tissue. The discovery that some patients receiving hGH therapy had acquired the infectious agent caused this source of hGH to be banned from human treatment.

Using Recombinant DNA Technology to Produce a Safe Supply of Growth Hormone

- **Key Research**—Recombinant DNA strategies for producing human growth hormone focused only on **cloning the human gene**, introducing the gene into bacteria (or yeast), and having these **microbes synthesize the hormone**. This supply of hormone would be comparatively inexpensive and disease free.
- **Overview of genetic cloning (Figures 19.1 and 19.4)**—To isolate the *hGH* gene from human cells, researchers first isolated mRNAs from tissue, copied all of these mRNAs into **cDNA** using **reverse transcriptase**, and inserted these cDNA fragments into circular plasmid DNA. The plasmids are then used to carry the cDNA into living *E. coli* cells. Researchers can identify bacterial cells carrying the plasmid by their ability to grow in the presence of antibiotics (antibiotic resistance genes are engineered onto the plasmid). They then screen through all of the bacteria in search of those carrying the *hGH* gene within a plasmid. The scientists must use a DNA probe to examine each colony of cells. Key steps of this process:
  1. **Using Plasmids**—Plasmids are small, circular segments of DNA carried in prokaryotic cells (separate from the chromosome; see Chapter 7). Scientists use these plasmids as **vectors** for DNA cloning and

manipulation; they have engineered many new plasmids just for use in research, typically adding an **antibiotic resistance** gene and other components to the plasmid sequence. Thus simply adding antibiotics to the media can determine the presence of plasmid within a cell; cells without the plasmid will die, while those with plasmid grow normally.

2. **Using Restriction Endonucleases (Figure 19.3)**—Scientists can choose from a large group of enzymes for use in cutting DNA at specific places. These restriction enzymes are made by bacteria to defend against viral infection. The enzymes cut the viral DNA but ignore similar sequences in the bacterial DNA, and the bacteria block those sites with methyl groups. The enzymes are a key tool in molecular biology; they give researchers a method of cutting DNA at specific places. Thus scientists can selectively isolate a specific piece of DNA from a plasmid or other DNA segment. Restriction endonucleases cut at specific **palindrome** sequences in the DNA (they read the same backward and forward). **Figure 19.1** (step 4) requires the utilization of a restriction enzyme to ensure that the cDNA is inserted into the correct site.

3. **Transformation of Bacteria**—Transferring a plasmid into a bacterial cell requires the DNA to cross the cell wall. To do this, scientists use **chemical treatment** or even a **pulse of electricity** to open holes in the cell wall. Once the DNA is inside, the cells are grown in the presence of antibiotics to select for the plasmid.

4. **Nucleic Acid Hybridization**—To probe for the bacterial cells carrying hGH, scientists had to generate a **probe** for the gene. Scientists use many types of probes to search DNA or RNA for specific sequences; but regardless of the type, each probe searches for its target by looking for nucleic acids to which it can **hybridize**. Double-stranded DNA is bound together through hydrogen bonding of its nitrogenous **bases**. DNA probes the hydrogen bond with **complementary sequence** in exactly the same way, and will search through long DNA and RNA sequences to hybridize to its exact match. **Figure 19.4** displays the screening of thousands of individual bacterial colonies with an *hGH* DNA probe. An imprint of each colony is generated on a positively charged filter paper (DNA is negatively charged). The DNA from each imprint is treated to make it single stranded, and it is then permanently attached to the filter. The filter is then probed with the *hGH* probe to identify which colony on the original plate contains the *hGH* gene.

5. Once the correct plasmid-bearing bacteria are identified, the researchers can use the cloned gene (*hGH*) to synthesize large quantities of hormone. This is accomplished by placing the cloned gene under control of a strong bacterial promoter and putting a plasmid carrying this construct into bacteria. Bacterial cells then make a large quantity of the protein (hGH) that can be purified and used to treat patients.

19.2   Analyzing DNA

The Polymerase Chain Reaction

- The **polymerase chain reaction (PCR)** is simply a methodology for amplifying short fragments of DNA from a few copies to billions of copies. Kary Mullis developed PCR in the early 1980s. The strategy exploits a heat-resistant DNA polymerase cloned from a bacterium that grows readily in hot springs (*Thermus aquaticus*).

- The technique is extraordinary due to its simplicity and its value to several areas of genetic analysis: gene cloning, forensic analysis, and rapid analysis of relatedness or ancestry, to name a few.

- A PCR reaction contains just a few key ingredients:
  1. DNA template
  2. Two DNA primers matching the template
  3. dNTPs (nucleoside triphosphates)
  4. Buffer

5. Heat-resistant DNA polymerase (known as *Taq* DNA polymerase)

- A PCR reaction requires about 30 cycles, with each cycle containing time points at three different temperatures:
  1. A denaturing step (95°C)
  2. An annealing (primer binding) step (45–70°C)
  3. An extension step (72°C)

- How do these changes in temperature amplify DNA? Examine **Figure 19.5**, which illustrates several keys to understanding PCR.
  1. Denaturation of the template means to separate the double-stranded DNA to single strands (using temperature).
  2. PCR primers are designed for a DNA target sequence, annealing on each side of the target and offering a 3≤OH group for DNA polymerization.
  3. The primers anneal or bind to opposite strands of the DNA template. Remember if DNA synthesis is always 5≤to 3≤ this process will work only if the primers activate DNA synthesis toward each other (steps 3 and 4).
  4. The extension step allows the thermostable DNA polymerase (*Taq*) to synthesize a new strand of DNA, starting at the 3≤end of each primer.
  5. In one cycle, we've doubled the number of copies of the target sequence. Imagine repeating this for 30 cycles!

- **Key Research—Svante Paabo** and colleagues used PCR to compare DNA from *Homo neanderthalensis* and *Homo sapiens*, testing a hypothesis that these species may have interbred in the past. They first amplified pieces of *H. neanderthalensis* mitochondrial DNA, isolated from ancient bones (estimated to be 30,000 years old). With the same DNA primers, they used PCR to amplify the same regions of *H. sapiens* mitochondrial DNA. They then used dideoxy sequencing on the amplified DNA fragments to compare them (see next section). They found that *H. sapiens* and *H. neanderthalensis* mitochondrial DNA sequence was very different; thus it is unlikely these two species ever interbred.

Dideoxy Sequencing

- Fredrick Sanger developed dideoxy sequencing, a key strategy for DNA analysis (summarized in **Figure 19.6**).

- The procedure is based on dideoxynucleosides. Review the structure of a normal **deoxynucleoside** and a **dideoxynucleoside**. Why is the missing 3≤OH group so important in DNA polymerization? How might the addition of a dideoxynucleoside—to the 3≤ end of the DNA strand being synthesized—affect the polymerization process?

- All polymerization reactions are initiated with a primer that anneals to the region of DNA to be sequenced. Each extension from this primer proceeds for some distance, unless a dideoxynucleoside is added to the chain. Thus, you can "see" where A, T, G, or C nucleotides are added by examining the fragments generated during a series of polymerization reactions containing the dideoxynucleotides.

- Four separate reactions are performed. Why? Remember that the dideoxynucleosides are added differentially to each reaction (one has ddTTP, one has ddCTP, etc.).

- Automated sequencing has made Sanger DNA sequencing easier and faster to perform. It is discussed in Chapter 20.

19.3 Gene Hunting Based on Pedigree Analysis: The Huntington's Disease Story

- **Key Research—Nancy Wexler** and colleagues' search for the mutation behind **Huntington's disease** is an excellent example of how molecular biology has dramatically enhanced research strategies. This disease is an inherited **neurodegenerative disorder** resulting from a dominant mutation. It includes symptoms such as loss of motor function, changes in personality traits, and decreased intelligence. Onset of symptoms is usually at about 40 years of age.

- Once scientists were certain that this disease passed from generation to generation, they set out to use **pedigree analysis** to find the gene that is mutated in these families. But how does this work? How does a pedigree give information about the gene's identity?

- Researchers must first determine the mode of inheritance (dominant or recessive, X-linked or autosomal). By examining the pedigree data, scientists can generate reliable predictions of the mode of inheritance, though human pedigrees often offer a limited amount of information. Examine **Figure 19.8**; what mode of inheritance is consistent with this pattern of inheritance?

    1. In this example, *all affected individuals were born to affected parents* (no "skipping of generations"). There is no example of unaffected parents generating an affected child (which would be consistent with a recessive mutation).

    2. Also, consistent with autosomal mutations, there seems to be fairly equal distribution of affected males and females.

    3. The mutation displays an **autosomal dominant** mode of inheritance.

Using Genetic Markers

- The DNA samples available from affected families can be analyzed with genetic markers to follow the inheritance of specific DNA regions from generation to generation.

- Researchers look for cosegregation (or linkage) of chromosomal markers with the disease phenotype, hoping to find which chromosome, and where on that chromosome, the disease-causing mutation is located (**Figure 19.9**). But to establish cosegregation of the markers and the mutation, researchers must analyze DNA from many members of affected families.

- Genetic markers are simple differences in chromosomes, generally within noncoding regions. Genetic markers used include restriction enzyme cut sites (RFLP analysis) and hypervariable repetitive sequences. Each individual inherits these polymorphic sites from his or her parents, and thus a specific DNA region inherited from a parent can be identified by the genetic markers associated with it (described as a haplotype). Through this type of genotyping, we can "see" a chromosomal region as it is

passed on through several generations. We see that region by identifying its associated marker in individuals within the pedigree.

- **Restriction Fragment Length Polymorphisms (RFLPs)** are differences between chromosomes that are measured as the loss or gain of a restriction enzyme cut site. **Figure 19.10** illustrates RFLP analysis of a chromosomal locus. Digestion of the chromosome at this site generates a specific array of DNA fragments that can be detected by a probe for that DNA region. This procedure works somewhat like the Southern blots described in **Box 19.2**. The presence/absence of restriction enzyme cut sites generates an altered pattern of bands on the resulting filter. Thus individuals with slightly different DNA sequences are revealed through this RFLP analysis.

- Imagine performing this analysis on DNA samples from a family affected by a disease-causing mutation. These analyses might find an RFLP that is *always* present in affected individuals and is *never* present in unaffected individuals! This result may suggest that the disease-causing gene is closely linked to the chromosomal location of that RFLP. Researchers use probability-based statistical analyses to establish linkage between a marker and the specific disease.

- The location of the Huntington's disease gene was localized to **chromosome 4** as scientists established linkage between the disease and an RFLP. But where *exactly* is the gene, and how do you set out to find it?

Pinpointing the Defect

- Once the location of the Huntington's disease gene was narrowed to ~500,000 base pairs on chromosome 4, the next step was to isolate all the genes in this region and examine them for mutations.

- Each mRNA generated from this site was isolated, and the gene sequence was determined from its cDNA.

- **Key Research**—Scientists found one gene within the isolated chromosomal region that was abnormal in people with Huntington's disease. It had an expanded **trinucleotide**

repeat (CAG) at the 5≤(front) end of the coding region. Normal humans have about 11–25 of these repeats, and affected individuals have 40 or more. The codon codes for **glutamine**, and affected individuals have a protein that is inactivated due to the effect that all the extra glutamines have on its configuration.

- The protein encoded by this gene was called **huntingtin**, and the gene was named *IT15*.

19.4 Can Gene Therapy Cure Inherited Diseases in Humans? Research on Severe Immune Disorders

- Gene therapy involves the introduction of a gene that will replace or augment a mutant gene that is causing an abnormal phenotype. Three criteria must be met for this process to be successful.
  1. The wild-type allele for the gene in question must be sequenced and its regulatory sequences understood.
  2. There must be a method for introducing a copy of the normal allele into affected individuals.
  3. The normal allele must dominate a recessive mutant allele, or replace a dominant mutation completely.

How Can Novel Alleles Be Introduced into Human Cells?

- The greatest difficulty in gene therapy strategies is stable introduction of the therapeutic gene into many cells in a human patient.
- The two prominent vectors for introducing these genes into human cells are both **viral vectors**. These viruses have been stripped of genes that might cause damage to the recipient cells, and serve only as a vehicle for delivery of the transgene.
- The gene is transferred into the viral sequence, and the virus is used to infect the patient's cells.
- The two most commonly used viruses are
  1. **Retroviruses**⏤ These contain an RNA genome and convert themselves to DNA upon infection of the host cell. This type of virus **integrates** into the chromosomal

DNA and remains there indefinitely while expressing the new gene (**Figure 19.11**).
  2. **Adenoviruses**—These viruses contain a DNA genome and are a common type of virus that can infect cells in the respiratory tract through inhalation. These viruses do not integrate into the chromosome, and thus are carried in infected cells for only a transient period.

Using Gene Therapy to Treat X-Linked Immune Deficiency

- The disease known as **severe combined immunodeficiency (SCID)** is devastating; it profoundly weakens the immune system. In this disease, white blood cells are unable to properly mature into **T- cells**.
- **Key Research**—A research team led by **Marina Cavazzana-Calvo** developed a strategy to cure one category of SCID, in which a gene called *SCID-X1* (a gene on the X chromosome) is mutated. The gene encodes a **growth factor receptor** called γc that receives a growth-activating signal from outside the cell.
- The team developed a retrovirus carrying the wild-type copy of this gene and tested the retrovirus on cells from dogs and mice.
- The first attempt at using this strategy on humans included two male infants, each affected with the mutant *SCID-X1* gene. No bone-marrow donors were available for either child.
- The research team removed **bone-marrow cells** from each of the boys, and infected the cells with the retrovirus they had constructed (**Figure 19.13**). The cells were then reintroduced into each patient, and gradually each child began to produce his own T cells! Researchers are now carefully watching T-cell production in these young boys to ensure that the transgene continues to provide a normally functioning γc receptor.

19.5 Biotechnology in Agriculture: The Development of Golden Rice

- Most strategies for genetic engineering in agriculture focus on reducing herbivore damage, reducing competition between the crop plant and weeds (making the crop more herbicide resistant), and improving the quality of the food product.

Synthesizing β-Carotene in Rice

- β-carotene is a dietary component that can be rapidly converted to vitamin A. In countries with diets dominated by rice, people commonly suffer vitamin A deficiency. There is a small amount of β-carotene in the rice husk, but it is removed in the polishing step of rice processing.
- **Key Research—Ingo Potrykus** and colleagues set out to make a rice crop that was enriched for β-carotene to relieve this dietary problem.
- As shown in **Figure 19.15**, the synthesis of β-carotene can be accomplished through the conversion of a molecule called geranyl geranyl diphosphate (GGPP) in an enzyme-mediated process. Potrykus recognized that GGPP was present in the endosperm of rice seeds. His team set out to introduce into the rice endosperm the enzymes needed to convert GGPP to β-carotene. The result would be a rice seed with an accumulation of this dietary precursor to vitamin A; but to accomplish this, he would need to make a transgenic plant.

The Agrobacterium Transformation System

- Researchers who do genetic engineering in plants often use a bacterium (***Agrobacterium tumefaciens***) to mediate the process of transferring genetic information into plant cells. This bacterium often infects plants, and as part of the infection process, the bacterium induces growth of a gall (a tumor-like mass) in the plant (**Figure 19.16**).
- The gall-inducing genes are found on a plasmid in the bacterium (Ti plasmid). The plasmid has several components, but the key part is a DNA sequence (T-DNA) that is transferred into the host plant's cells to integrate into the chromosome.
- Researchers recognized the value of DNA being transferred into the plant cell this

way, and have exploited the process to carry other gene sequences along with the T-DNA sequence.
- Potrykus used *Agrobacterium* to carry the β-carotene genes into cells from a rice plant, eventually generating a transgenic plant—now called golden rice (it is yellow from the high concentration of β-carotene).

B. CROSS-CUTTING THEMES

Looking Back—
Concepts from Earlier Chapters
Mendel and the Gene—Chapter 13
Sex-linked traits were introduced in **Chapter 13** with a discussion of inheritance patterns displayed by a **white-eye trait** in *Drosophila melanogaster*. Thomas Morgan was able to make sense of the unusual segregation pattern of X-linked traits in reciprocal crosses. In humans, examining patterns of inheritance over several generations can reveal X-linked traits.

Control of Gene Expression in Eukaryotes—Chapter 18
Restriction enzymes are a key tool in the characterization and recombination of DNA sequences. Microbes make these enzymes as a defense against viral infection (they degrade the viral DNA during infection). The enzymes are key in analyses such as **Southern blotting, RFLP analysis**, and **gene cloning** into vectors.

Looking Forward—
Concepts from Later Chapters
Viruses—Chapter 34
The life cycle of a virus can be quite complicated, sometimes including a period of integration into the host genome (**lysogeny**) before replicating itself and emerging from the host cell. Researchers have harnessed the lysogenic capacity of some viruses to carry genes into the human genome in **gene therapy** strategies.

The enzyme **reverse transcriptase** is produced by **retroviruses** to copy their RNA genome into DNA as part of the infectious process. Reverse transcriptase is an unusual enzyme that has become tremendously valuable in the production of **cDNA** for research.

C. DIFFICULT TOPICS

Imagine how tedious the early research into Huntington's disease really was. Researchers worked together to find regions in the human genome that displayed variability regarding the presence or absence of a specific restriction enzyme cut site. The researchers then would develop a probe for this region that could be used to screen blots of digested DNA. They prepared DNA samples and screened them for each of these polymorphic regions to determine the individual's RFLP makeup. The data for this type of analysis would be complicated to look at on a gel (remember **Figure 19.10**), but eventually each individual would be typed for each locus. However, now what do the researchers do?

Somehow within the data there is linkage between a marker and the disease-causing allele. With Huntington's disease the mutation was dominant, and thus each person with the mutant allele would display the disease eventually (after age 35).

Examine the pedigree displayed in **Figure 19.9**, which displays only the RFLP makeup of each individual at the chromosome 4 marker. Do you see the 100 percent correlation (cosegregation) between Huntington's disease and the C haplotype? There is no correlation with the A, B, or D haplotypes. It is this type of correlation that scientists hope for as they examine *thousands* of individual genetic loci. If a disease were caused by a recessive mutation, would the process of mapping the gene be significantly influenced? Explain.

Today, the methodology is faster, using PCR to amplify polymorphisms and enabling the simultaneous screening of many loci. But even with these improvements, the mapping of a disease gene is truly a daunting task, requiring a great amount of patience and endurance.

D. ASSESSING WHAT YOU'VE LEARNED

(1) Testing Your Knowledge

1. Genetic engineering became possible with the discovery of which of the following compounds?
    a. restriction enzymes and ligase
    b. ddNTP molecules
    c. DNA polymerase
    d. reverse transcriptase

2. Which of the following is the main reason it was necessary to develop a method of using recombinant DNA technology to produce human growth hormone for treating individuals with pituitary dwarfism?
    a. There was no available source of functional growth hormone.
    b. The drug was extremely expensive.
    c. The growth hormone supply led to hereditary diseases.
    d. The growth hormone supply was contaminated with infectious agents.

3. Try reading the gel presented in **Figure 19.6**. Determine the first 10 nucleotides on the 3≤end of the *template* sequence by reading the X-ray film. Which of the following sequences are correct?
    a. 3≤GGA GAT CAA T–5≤
    b. 3≤CCT CTA GTT A–5≤
    c. 3≤GTA TCC AGA G–5≤
    d. 3≤CAT AGG TCT C–5≤

4. The plasmid used in the experiment illustrated in **Figure 19.1** of your text must include which of the following, before being inserted into *E. coli* cells?
    a. a marker gene
    b. restriction recognition site
    c. cDNA synthesized from mRNA of gene being examined
    d. all of these

5. Refer to **Figure 19.4** of your text to answer this question. Which of the following would be the most likely result if step 3 of the procedure were omitted?
    a. The X-ray film would not show any black spots.

b. The X-ray film would show multiple black spots (one for each colony on the plate).

c. Half of the colonies would demonstrate radioactivity on the X-ray film.

d. The results would be identical to step 5 in the figure.

6. Restriction enzymes are formed in bacteria to break down viral DNA. When restriction sites are placed on a human gene, and human DNA and viral DNA with the same restriction site are cut with the restriction enzyme, the "sticky ends" of the two can recombine, and the virus may be used to incorporate the gene into a human chromosome.

Pick the statement that best describes the paragraph above.

a. This is not true; bacteria are not able to cut viral DNA in the way described.

b. The DNA from a human and a virus would not exhibit the same sequence at the restriction site.

c. A virus would not be able to insert genetic material into a human chromosome.

d. This is an example of the utilization of recombinant DNA technology in gene therapy.

7. How is the pedigree of an individual generally constructed?

a. by sequencing the genome of the individual

b. by interviewing the individual regarding the sex and physical traits of his or her family members

c. by sequencing the genes of the individual's parents

d. by analyzing the genetic fingerprint of the individual and of all of his or her family members

8. One example of a disease caused by the expansion of a repetitive DNA sequence (codon repeats) is:

a. leukemia

b. pituitary dwarfism

c. fragile-X syndrome

d. Down syndrome

9. Which of the following is a disadvantage of using biotechnological techniques in agriculture?

a. Genetically modified corn grows poorly and is less nutritious.

b. Transgenic herbicide-resistant plants may transfer their resistance transgene to closely related weeds (through pollination), making them resistant as well.

c. Genetically modified crops are often slightly poisonous.

d. It is expensive to produce seed from transgenic crop plants because the T-DNA impedes seed production.

10. Examine the RFLP blot and the restriction enzyme map shown in **Figure 19.10c**. Columns A–D reveal the banding pattern generated from persons homozygous for this region (i.e., both chromosomes in a diploid person are the same at this site). These four banding patterns designate a **haplotype**, revealing the pattern produced from one site. Using the space below, show the RFLP patterns that would be generated by an analysis of people who are AC or BC for this locus.

Band Size	Individual AC	Individual BC
17.5 kb		
15.0 kb		
8.4 kb		
4.9 kb		
3.7 kb		
2.5 kb		
2.3 kb		
1.2 kb		

(2) Integrating Your Knowledge

(a) Loss of the gene *GH1* results in the type I form of pituitary dwarfism, in which affected individuals are benefited by simply introducing an external source of human growth hormone (hGH). In

SCID, another example from this chapter, T cells are unable to properly mature, even in the presence of the growth-stimulating signal. Why is the introduction of the external signal sufficient in one case but not the other? Explain.

(b) Why is recombinant hGH better as a therapeutic agent than hGH isolated from human cadavers?

(c) In the process of using a DNA probe to detect single-stranded DNA sequences on a filter paper, how does the probe "search" for the target sequences?

(d) What characteristics in a pedigree would suggest to you that a specific mutation was autosomal and recessive?

(e) When compared to retroviral gene therapy strategies, what disadvantages are there for using adenoviruses instead? Advantages?

CHAPTER 19—ANSWER KEY

D. Assessing What You've Learned

(1) Testing Your Knowledge
1. a; 2. d; 3. b; 4. d; 5. a; 6. d; 7. b; 8. c; 9. b;
10. The AC individuals will have all bands of the A haplotype combined with all bands of the C haplotype. The BC individuals will have all bands of the A haplotype combined with all bands of the C haplotype.

(2) Integrating Your Knowledge
(a) Type I pituitary dwarfism is caused by mutation in a gene responsible for the synthesis of hGH. SCID, in the gene therapy example, is caused by a deficient growth factor receptor. Thus, in SCID-X1, the disease results from cells being unable to *receive* the signal, while the pituitary dwarfism is caused by an inability to *send* the signal.

(b) Isolation of hGH from cadavers is very expensive and carries with it the risk of transferring disease (prion disease).

(c) The probes used in Southern blotting, Northern blotting, and the blots used in RFLP mapping are composed of single-stranded DNA that has been labeled in some way (typically with radioactivity). The single-stranded DNA will anneal tightly with any DNA sequences that are composed of the reverse, complementary sequence of the probe sequence. The probe and the target will anneal to one another in a manner similar to double-stranded DNA annealing to itself. Researchers can then localize the probe attached to the blot by exposing an X-ray film to the filter paper.

(d) Recessive mutations often appear in pedigrees, with affected children being born to unaffected parents. Also, recessive mutations typically skip generations within the pedigree, and sometimes may not reappear unless two related individuals have children together. If the sexes of affected individuals in the pedigree are equally distributed, the mutation is likely in an autosomal gene.

(e) Adenoviruses can be inhaled by the patient, and this offers an excellent delivery mechanism for gene therapy strategies involving the lungs and respiratory system. But these viruses don't integrate into the host-cell genome and don't offer long-term transgene expression.

20

Genomics

A. KEY BIOLOGICAL CONCEPTS

20.1 Whole-Genome Sequencing

Recent Technological Advances

- The first complete sequencing of the genome for a prokaryotic cell was completed in 1995 (*Haemophilus influenzae*), and the first eukaryotic genome followed soon after in 1996 (*Saccharomyces cerevisiae*).
- The human genome sequencing project was completed in 2001 (3 billion nucleotides). Genome sequencing data are publicly available through the GenBank website, which now contains ~28.5 billion nucleotides of DNA sequence.
- The capacity for sequencing facilities to generate sequence data has advanced rapidly in recent years. Review **dideoxy sequencing**, a key research tool introduced in **Chapter 19**. The advances still fundamentally use this strategy, but have changed some aspects of detection to speed the process by automation.
- **Figure 20.1** displays the use of fluorescent tags in sequencing. These are attached to the dideoxynucleoside triphosphate molecules (**ddNTPs**), and they label each DNA fragment generated in the sequencing reaction *at the termination event*. Thus if all the labeled ddNTPs (ddATP, ddGTP, ddCTP, ddTTP) are added to a single sequencing reaction that also contains dNTPS, they will occasionally be incorporated as the polymerase synthesizes the complementary strand. With the incorporation of a ddNTP, the polymerization

is arrested, generating a DNA fragment that contains the fluorescent tag on one end.
- The ddNTPs are each labeled with a different colored **fluorescent tag**, allowing researchers to determine which ddNTP addition terminated the reaction (e.g., ddATP = red, ddGTP = blue, etc.).
- Instead of running four separate reactions labeled with radioactivity, the researchers use one reaction and analyze the fragments by their fluorescence. A computer reads the color of each fragment generated in the sequencing reaction (analyzed by size) and automatically displays the DNA sequence found in the template strand.

Which Sequences Are Genes?

- Imagine the task of finding the genes within the genome sequence. What sequences identify the presence of a gene? Computers can search sequence for reading frames with large regions that lack a stop codon. These regions are called **open reading frames** (ORFs) and are usually a gene.
- In bacteria and archaea there are no introns, and genes can be identified by highly conserved promoter sequences associated with a distinct open reading frame.
- In eukaryotic organisms, the process of finding genes is complicated by introns and diverse regulatory sequences. One strategy is first to identify mRNA sequences using a **cDNA library**, and then to hunt for the genes within the genomic sequences. In the process

of generating an mRNA, the introns have been spliced out; but the sequence of exons should match exactly with short sections of the genomic sequence.
- Also, researchers use known genes from other organisms to search for homologous (similar) genes in the organism being studied.

Shotgun Sequencing and Bioinformatics (Box 20.1, Figure 20.2)

- The availability of automated sequencing facilities dramatically speeds up the process of genome sequencing, but researchers still require DNA template and primers to analyze the sequence of specific regions of a chromosome. How does this process work, and what methods improve the speed of this process?
- Chromosomes are very large strands of DNA, and they must be cut into smaller fragments (using **restriction enzymes**) for sequencing. One **vector** for carrying fragments of a chromosome is called a **bacterial artificial chromosome (BAC)**. Vectors are DNA strands used in cloning to carry DNA fragments inside their circular structure. A BAC is unusual in that it can carry very large fragments of DNA (up to 200,000 base pairs).
- What does it mean for a vector to "carry" a piece of DNA? In cloning, scientists use *Escherichia coli* to make many copies of the DNA being studied. They simply place the microbe in a flask of media, and one cell carrying the vector can rapidly grow to billions (each with the vector). Thus we can now isolate the vector from the bacteria, and have many copies to work with.
- In genome analysis, the entire genome of an organism is cut into fragments and transferred into BACs for amplification in bacteria. This collection of BACs is called a **BAC library**.
- The BAC inserts are then analyzed individually. Each is purified, cut into smaller fragments (about 1000 base pairs each), and inserted into another circular vector, called a plasmid. Sequencing reactions are performed on each plasmid, and the continuous sequence of the entire chromosomal fragment carried on the BAC is determined by computer analysis.

- How does the computer *align* these sequences? The computer exploits one detail in this analysis, and it is well illustrated in steps 5 and 6 of **Figure 20.2**. The fragments that are generated **overlap** with each other; upon sequencing, these overlapping sequences are valuable in identifying the correct order of the sequenced fragments.

20.2 Bacterial and Archaeal Genomes

Natural History of Prokaryotic Genomes

- The size of the genome correlates with the complexity of microbial metabolic processes.
  1. *Mycoplasma* bacteria have a very small genome and are unable to grow in the absence of a **host cell**. They parasitize the host cells, growing in their cytosol and absorbing metabolites for their own use. With no host cell, *Mycoplasma* bacteria cannot survive (**obligate parasites**).
  2. Cells with large genomes are much more complex metabolically; they can survive in varying environments by converting available nutrients to the required compounds.
- There is a great deal of genetic diversity among prokaryotes.
- Some genes are found multiple times within the prokaryotic genome. The value of this **redundancy** is not known, but it may be due to slight differences in function or a requirement for a rapid induction of gene expression following activation.
- Many genes or chromosome regions appear to have originated with other microbes. This transfer of large fragments of DNA from one microbe to another is called **lateral gene transfer**.

Evidence for Lateral Transfer

- Two key pieces of evidence for lateral gene transfer are (1) when a section of the chromosome in a species is more similar to genes in distantly related microbes than to those in closely related microbes; and (2) when sections of the chromosome are rich in either GC or AT nucleotides relative to the rest of the chromosome.

- ***Thermotoga maritima*** is a unique bacterial species thriving deep under the sea near hot-spring vents, where the temperature is very hot. In the same environment, some archaea species also grow well. Genome analysis revealed sections of the *T. maritima* genome that closely match the archaea genome. It appears that DNA was transferred from the archaea to the bacteria to aid in this unusual ability to grow in a harsh environment.
- **Pathogenicity islands** have been discovered in some disease-causing strains of *E. coli*. These are chromosome regions that contain genes for aiding in infection; they also allow for localization of the bacterium to unique areas in the gut, and they generate devastating side effects such as severe diarrhea (see Section 20.4). Origins of these loci have been traced to other bacteria, such as *Shigella sp.*, and even to the viruses that infect bacteria.

20.3   Eukaryotic Genomes

Natural History: Types of Sequences

- **Exons** comprise less than 5% of the human genome, and repeated sequences (noncoding) comprise more than 50% of the genome. In prokaryotic cells, usually 90% of the genome is comprised of genes. Why is there such a dramatic difference?
- **Repetitive sequences** in the human genome are due to simple sequence repeats (noncoding), transposable elements, and families of genes.
- Sources of repetitive sequences include the **transposons**. These DNA sequences, found in our chromosomes, occasionally "jump" from one place to another (they are transposable). Our chromosomes have many transposon sequences.
- **Long interspersed nuclear elements**, or LINES, comprise one category of transposons. These elements have a promoter followed by two genes—**reverse transcriptase** and **integrase** (remember reverse transcriptase from our discussion of cDNA libraries).
- Expression of the LINE locus results in synthesis of both proteins; the reverse transcriptase acts on the mRNA to generate a cDNA of the LINE sequence. Integrase then nicks a

new site in a chromosome and mediates integration of the cDNA fragment into that site. So the LINE transposon has moved, leaving a copy behind; a new copy is now present in the genome. See **Figure 20.5** in your text to review the mechanism of transposition.

Simple Sequence Repeats and DNA Fingerprinting

- **Microsatellites** are repeating sequences that are found in the genome with repeats of 1 to 5 bases.
- **Minisatelllites** are repeating sequences found in the genome with repeats of 6 to 500 bases.
- Where do these repeats originate? One prominent hypothesis is that these repeats are produced by an abnormal process called **slipped mispairing** that can occur during DNA replication. Details of that model are beyond the scope of this text, but it can be thought of as a stuttering of the DNA polymerase, accidentally repeating small sequences.
- Repeating sequences are hypervariable, meaning that the number of repeats is highly variable from one person to the next, and even between homologous chromosomes. Thus each person has a unique set of repeat sequences—a genetic "fingerprint" that is unique to that one person.
- The highly polymorphic nature of these repetitive regions can be traced to crossing over during meiosis (**Figure 20.6**). Repetitive sequences would have the capacity to align improperly during prophase I, and if crossing over were to occur in this misaligned area, the chromosomes generated would have an expansion or a contraction of the size of the repetitive region. Examine the figure closely to understand this point.
- DNA fingerprinting is commonly used today in forensic science as authorities investigate and prosecute criminals.
- DNA fingerprinting has many other applications in the analysis of human ancestry, determining relatedness between people.

What Is a Gene Family?

- Genetic loci that are closely related to each other, and physically close to each other on the chromosome, are called **gene families**.

- One example is the **globin gene family**, which is located as a tandem array of genes, each member highly similar to other members of the family (**Figure 20.9**).
- Some members of the gene family are non-functional **pseudogenes**. They are remnants of a functional copy of the gene; but due to mutation, they have lost the ability to be transcribed.

20.4 Comparative Genomics

- As sequenced genomes become available to scientists, these large data sets can be used to compare/contrast different organisms.

Understanding Virulence

- **Virulence** is the tendency of a parasite to harm its host. *Escherichia coli* is a species containing both benign (harmless) and virulent strains. Researchers have used genomics to study the genetic basis of virulence in *E. coli* strain 0157:H7, a common cause of severe food poisoning from undercooked meat. Researchers found 0157:H7 to contain the normal *E. coli* genome with an extra region of chromosome derived from the *Shigella* bacterial genome, which enables the *E. coli* to cause a severe food poisoning. *Shigella sp.* are a common cause of dysentery.
- Researchers hypothesize that a bacterial virus carried this portion of the *Shigella* genome into *E. coli*; this is a valuable example of lateral gene transfer.

Studying Adaptation

- Examples of organisms with genomic modifications to benefit their survival:
  1. *Anopheles gambiae* (a mosquito species) has 58 genes for fibrinogen-like proteins, used to prevent blood clotting at the site of feeding. Fruit flies eat rotting fruit (not blood) and have only 13 genes for fibrinogen-like proteins.
  2. *Plasmodium sp.* are single-celled, parasitic eukaryotes that are responsible for malaria in humans. In the blood, foreign cells are detected by our immune system cells through examination of cell-surface pro-

teins and sugars. Plasmodium cells have between 600 and 800 copies of genes that code for cell surface proteins. This is hypothesized to be a strategy for evading the immune system.
  3. *Prochlorococcus sp.* are prokaryotic bacteria commonly found in the ocean. These microbes are photosynthetic, gathering light to produce cellular energy. Researchers have found a correlation between the number of chlorophyll genes and the quantity of available light for each species' distinct habitat (depth in the ocean). Species living deeper in the ocean have more chlorophyll genes, and a more sensitive light-catching antennae complex.

Gene Number and Alternative Splicing

- **Table 20.1**—A major surprise from genome sequencing is that there is not a clear increase in gene number as cellular complexity increases. Humans have about the same number of genes as fruit flies and roundworms (approximately 30,000).
- One explanation for this confusing result involves alternative splicing of RNA transcripts (discussed in Chapter 18). This mechanism allows for different proteins to be made from a single gene, depending upon how the RNA transcript is spliced together. Early attempts to quantify the number of alternative products for each human gene have generated a value of ~3 alternative products/gene.

20.5 Future Prospects

Functional Genomics and Proteomics

- With all the information contained in the genome, scientists must now begin to methodically analyze how these genes and their encoded proteins work together to generate a living cell. Many strategies are already available to researchers; two dominant strategies are discussed in the text—**DNA microarrays** and **proteomics**.

DNA Microarrays

- Activation of gene expression is an important function of cell differentiation, stress response, and cell division. Scientists have used Northern blotting as a way to detect changes in gene expression, analyzing one gene at a time. DNA microarrays allow researchers to simultaneously measure the expression of every gene in the genome.

- Imagine comparing all of the genes expressed in liver cells, compared to liver cancer cells. Any differences in gene expression are the result of cancerous changes in gene regulation!
- Researchers create microarrays by putting thousands of tiny spots onto a glass slide (or chip), with each spot containing short strands of single-stranded DNA. Then cDNA strands (copied from mRNA, introduced earlier in the chapter) can be used to probe the glass slide. The spots that bind to the cDNA probe reveal the expressed genes within that cell type (**Figure 20.10**).

Proteomics

- Simultaneous analysis of thousands of proteins, in a manner similar to the DNA microarrays, would greatly enhance scientists' ability to characterize how proteins work.
- One type of proteomic assay involves spotting proteins onto a solid matrix (like glass) and then exposing these proteins to a pure preparation of a specific protein or molecule. Detection of protein-to-protein binding and even catalytic activity may then be measured, thus identifying which proteins in the array interact with the introduced protein/molecule.

Medical Implications

- Genome projects offer huge databases of information for studying living organisms. One application of this knowledge is the pursuit of disease prevention/treatment strategies.
- Researchers are using the genome sequence of *Plasmodium* to develop drug targets for proteins found only in the parasitic microbe, not in the human host.
- Vaccines against disease-causing microbes are being developed using genome databases. One example is *Neisseria meningitidis*, which causes meningitis (see "Difficult Topics").

B. CROSS-CUTTING THEMES

Looking Back—
Concepts from Earlier Chapters
How Genes Work—Chapter 15
With information available from genome analysis, we can look back at our introductory examination of genes in a new light. Key findings include the discovery of **families of genes** clustered together in the genome, **pseudogene**s scattered on our chromosomes, and the amazing complexity of **gene structure**.

Analyzing and Engineering Genes—Chapter 19
Dideoxy sequencing is a critical concept to understand, because we build on that technology with an improved fluorescence labeling strategy. The chemistry of Sanger dideoxy DNA sequencing is still the basis of all sequencing used today, though some details have changed.

Looking Forward—
Concepts from Later Chapters
Bacteria and Archaea—Chapter 27
Bacteria display an incredible diversity of genetic sequence, and the evidence that genetic information is exchanged between even distantly related microbes is truly fascinating. Microbes appear to readily gather available genetic information via **lateral transfer of gene**s, sometimes benefiting their ability to adapt and survive.

C. DIFFICULT TOPICS

Rational Drug/Vaccine Design
Once scientists know all the genes of a particular organism, how can they use this information to develop better antibiotics, vaccines, or even medicines that prevent the progression of disease? Remember that the proteins within cells perform tasks and generate structure in a cell. Knowing a gene's DNA sequence allows researchers to predict the amino acid sequence in the protein that would be synthesized by that gene, and offers clues into the gene's structure and function. Examining the genome offers vital insight into the "complete picture" of cellular function. Thus strategies for treating a specific

disease will now utilize genomics data sets to choose better targets for drug action.

One example included in the text discusses the production of a vaccine against a dangerous microbe (*Neisseria meningitidis*) that is one cause of infections in spinal fluid. A vaccine includes inactivated components of an infectious organism, and is injected into a patient to reveal identifiable characteristics of the pathogen to the immune system. Thus as soon as that organism enters the body, the immune system recognizes the danger and rapidly reacts. Genomic analysis of *N. meningitidis* has revealed all of the proteins encoded with that genome, so that researchers can now test all of these proteins for their ability to stimulate the immune system effectively. By understanding this genome sequence, scientists can make a better vaccine, one that will save many lives. The value of sequencing the human genome is undeniable; but certainly we are only beginning to grasp the best strategies for using this information. What applications do you see?

D. ASSESSING WHAT YOU'VE LEARNED

(1) Testing Your Knowledge

1. The fluorescent tag-based dideoxy sequencing is an improvement on radioactive tag-based dideoxy sequencing for the following reasons (choose all that apply):
   a. Automated laser scanners can read results and compile data on a computer.
   b. Radioactive sequencing is inaccurate.
   c. Each ddNTP nucleotide is a different color, and thus the sequence can be determined from electrophoresis of a single sample (four are needed in radioactive labeling).
   d. Fluorescent nucleotides are less expensive than radioactive nucleotides are.

2. Scientists studying the "normal" benign E. coli and the pathogenic E. coli strain O157:H7 have made which of the following determinations?
   a. The O157:H7 strain has fewer genes than the benign strain.
   b. The O157:H7 strain has more introns than the benign strain in key virulence-determining genes.
   c. The O157:H7 strain has the same genetic sequence as the benign strain, so differences in virulence must be the result of post-transcriptional modification of the mRNA of virulence determinants.
   d. The O157:H7 strain has gene sequences that are similar to *Shigella* bacteria that are not found in the benign strain.

3. Lateral gene transfer is indicated when (choose all that apply):
   a. a microbe has the genetic components for anaerobic growth
   b. a microbe has an isolated region with a distinct %GC content
   c. a microbe has a genomic region containing genes that are similar to a distantly related microbe
   d. a microbe has distinct lateral gene transfer motifs in its DNA sequence

4. *Prochlorotrichus* sp. offer an excellent example of:
   a. genomic adaptation
   b. parasitism
   c. microbial conjugation
   d. microbial virulence

5. Transposable elements:
   a. have the ability to replicate themselves and move from one location to another in a genome
   b. occur very rarely in humans
   c. are always active, if present
   d. make up as much as one-half of a bacterium's genetic material

6. Please refer to **Figure 20.5** in your textbook. If there were a mutation in the reverse transcriptase formed in this cell that prevented it from binding to the LINE mRNA, how would the resulting situation differ from that shown in the figure?
   a. The LINE mRNA would not be translated.
   b. A DNA copy of the LINE transposon would not be made.

c. The reverse transcriptase would not be produced.

d. The DNA would not be nicked.

7. Overall gene number and cellular/organismal complexity do not correlate as researchers had predicted. One explanation for this phenomenon is:
   a. our misunderstanding of biological complexity
   b. our inability to identify genes in genome sequence
   c. the importance of alternative splicing in multicellular organisms
   d. the poor quality of genome sequence data

8. The globin gene family shown in **Figure 20.9** in your textbook is thought to have arisen from which of the following processes?
   a. polyploidy
   b. transposable elements
   c. unequal crossover during meiosis
   d. alternative splicing

9. Shotgun sequencing involves:
   a. cutting genomic DNA into small overlapping fragments, sequencing each individually, and using computer software to align the sequences
   b. isolating bacterial artificial chromosomes (BACs) containing huge genomic fragments, and sequencing each using nonspecific "shotgun" primers
   c. generating genomic DNA fragments using a high-pressure gas "shotgun" syringe, and sequencing each fragment generated
   d. radioactive sequencing rather than fluorescent sequencing

10. A "gene family" is:
    a. a group of functionally related genes that are located together and function together (i.e., lactose utilization)
    b. genes transferred together from one microbe to another, through conjugation
    c. most often observed as a pathogenicity island

d. a group of genes that are similar in sequence, and are thought to be duplicated from a common ancestral gene

(2) Integrating Your Knowledge

(a) In your own words, describe the process of aligning DNA sequences from a "shotgun cloning" strategy. Can you think of a good analogy?

(b) In higher eukaryotic cells, genes are often broken up into exons, with introns separating these coding fragments. How do researchers find these genes in the chromosomal sequence when performing genomic analyses?

(c) Using what has been learned from completed prokaryotic genome projects, what key points can now be made about genomes and prokaryotic natural histories?

(d) What evidence is there in a long interspersed element (LINE) that it is a transposon?

(e) Explain the relevance of *E. coli* O157:H7 to evidence for lateral transfer of DNA between prokaryotic organisms.

CHAPTER 20—ANSWER KEY

D. Assessing What You've Learned

(1) Testing Your Knowledge
1. a, c; 2. d; 3. b, c; 4.a; 5. a; 6. b; 7. c; 8. c; 9. a; 10. d

(2) Integrating Your Knowledge
(a) Of course, this will be the student's answer and own analogy. Imagine having 100 freight trains, each 500 cars long, with exactly the same ordering of freight cars. If you were to break each train into fragments of 20–30 freight cars, each separated at different sites, could you examine the order of each of the sections and regenerate the original sequence of cars? Yes! You could read the sequence of each section and then put the sections together into the correct order by aligning sequences with information from overlapping sections. DNA sequencing with "shotgun cloning" uses exactly this

strategy; the large BAC is broken up, sequenced in small fragments, and the original BAC sequence is determined by aligning the smaller sequences.

(b) This task is difficult, and several strategies are employed. The most productive is to search the DNA sequence for matches with known cDNA sequences isolated from cDNA libraries. The small matching regions will reveal the exons within the larger sequence.

(c) • There is a general correlation between the size of a genome and the metabolic capacities of the organism.

• Prokaryotic organisms display an incredible array of genetic diversity.

• Prokaryotic organisms commonly display genetic redundancy (multiple copies of genes).

• Prokaryotes display a high degree of lateral transfer of genetic information.

(d) LINES contain a transcription start site, a reverse transcriptase gene, and an integrase gene. These two genes are known to mediate the transposition of a DNA sequence from one genomic site to another.

(e) Most *E. coli* are harmless to humans—they are even a common inhabitant of the human gut. The *E. coli* O157:H7 strain can be lethal to humans (especially children), causing a dehydrating "bloody diarrhea." Comparative analysis of the genomes of this pathogenic strain of *E. coli* and other nonpathogenic *E. coli* revealed a distinct region in the O157:H7 strain. This region included genes that closely resemble those of a known pathogenic bacterium (*Shigella*). The *Shigella* genes appear to have been transferred into the *E. coli*, giving this strain an enhanced pathogenic potential.

21

Early Development

A. KEY BIOLOGICAL CONCEPTS

21.1 Gametogenesis

Sperm Structure and Function

- **Gametogenesis** is the formation of gametes (sperm and egg) in the reproductive organs.
- The mammalian sperm cell, which has haploid DNA, contains four major compartments (head, tail, neck, and midpiece) and several critical structural components (**Figure 21.2a**).
  1. **Acrosome**—A lysosome-derived structure at the head of the sperm and containing a mixture of hydrolytic enzymes involved in degrading specific components of the egg cell wall. The acrosome is located between the nucleus and the anterior cell wall of the sperm head.
  2. **Centriole**—In the sperm's neck region, the centriole is a small, hollow, cylindrical structure comprised of microtubule fibrils.
  3. **Mitochondria**—Each sperm structure contains many tightly packed mitochondrial organelles within the midpiece region. The capacity for sperm motility depends on ATP production from these mitochondria.
  4. **Flagellum**—A proteinaceous "propeller" structure composed of microtubules, the flagellum generates the thrust necessary for sperm motility to the egg.
- **Pollen grains** serve as plant sperm, and are small enough to be carried by the wind or to powder the extremities of insects. Once these grains come into contact with an ovule, a pollen tube grows from cells in the pollen grain, a single pollen cell then divides mitotically, and the resulting haploid sperm nuclei migrate down the pollen tube to the ovule (**Figure 21.2b**).

Egg Structure and Function

- The mammalian egg cell is much larger than the sperm cell (**Figure 21.3**) and contains a cytosol enriched with nutrients as well as several structural components regulating the processes of fertilization and early development.
  1. **Yolk granules.** Small, dense structures in the egg cytosol, packed with protein/lipid for energy and macromolecular biosynthesis during early development.
  2. **Vitelline envelope.** A fibrous structure surrounding the cell membrane and lying beneath any associated gelatinous layer (egg jelly, zona pellucida, etc.).
  3. **Cytoplasmic determinants.** These protein and mRNA molecules are key regulators of early development in many animals (excluding mammals). Segregation of these molecules during cleavage controls key processes leading to cell differentiation.
  4. **Cortical granules.** These small vesicles are localized to the periphery of the cytosol. They contain enzymes critical to fertilization (see "Blocking Polyspermy" in Section 21.2).

21.2 Fertilization

Fertilization in Sea Urchins

- The sea urchin has proven to be an incredible research system for the characterization of fertilization. Researchers can isolate large quantities of sperm and eggs quite easily.

- As summarized in **Figure 21.4b**, the fertilization of a sea urchin egg involves a complex chain of events.
  1. The sperm is recruited to the egg by an attractant that is present in the **egg coat** and slowly diffuses out into the surrounding seawater. The **attractant** is a small protein (peptide) that is recognized by receptor proteins present in the sperm.
  2. Upon direct contact between the sperm head and the egg cell, the acrosome releases its contents from the sperm head.
  3. This mixture of hydrolytic enzymes degrades the gelatinous egg coat.
  4. At the same time the hydrolytic enzymes are released, a structure is built on the sperm head by the **polymerization of actin** proteins together. This structure is called the **acrosomal process**, and it mediates the fusion of sperm-cell membranes with the egg-cell membranes.
  5. Once the membranes have fused, the sperm nucleus can enter the egg cytosol, eventually fusing with it to generate the zygote with a $2n$ chromosomal content.

Species Recognition in Sea Urchins

- When exposed to a mixture of sperm from different sea urchin species, an egg can be successfully fertilized only by sperm of its own species. If the mechanisms of fertilization are similar between these species, how is it possible for an egg to discriminate between different sperm so effectively?
- A molecule called **fertilizin** on the surface of each sea urchin *egg* binds to sperm cells only from the corresponding species.
- In the 1970s researchers made another breakthrough in the study of species recognition. They isolated a protein on the surface of the *sperm* and named the protein **bindin**. Each species had a slightly unique form of this protein, which bound only to eggs from the same species.
- Kathleen Foltz and William Lennarz tested the hypothesis that fertilizin protein on the egg was the binding partner of bindin on the surface of the sperm. As is summarized in **Figure 21.5**, they used proteases to release cell-surface proteins from sea urchin eggs and

then isolated any of these proteins that could bind sperm or bindin proteins of its own species. Key to this strategy was the prediction that in the collection of proteins released from the egg would be the fertilizin receptor proteins.

- The egg-sperm interaction was species-specific due to the presence of the bindin proteins in sperm and the fertilizin receptors in eggs. These proteins bind tightly to enable fertilization, but are very selective, binding to each other only if each protein originated from the same species (species lock and key).

Blocking Polyspermy

- One key question of fertilization is how only one sperm is successful in fertilizing the egg, despite the presence of hundreds of sperm nearby, all of them intent upon fertilizing the egg.
- A Ca^{2+}-based signal is induced rapidly after fertilization and inhibits fertilization by any additional sperm by generating a **fertilization envelope**. As is summarized in **Figure 21.6**, fertilization results in the release of Ca^{2+} from intracellular stores. The increase in cytosolic Ca^{2+} results in the fusion of cortical granules with the egg-cell membrane. These granules are small, membrane-bound vesicles located near the cell membrane; upon fusion with the membrane, they release their contents into the extracellular matrix.
- What do these cortical granules contain that could inhibit additional fertilization events?
  1. **Proteases** that break down proteins associated with the egg cell wall. These include the receptor protein, and the loss of these structures inhibits receptor binding to additional sperm.
  2. A high concentration of **solutes**; these alter the concentration of solutes at the cell exterior and induce a sudden influx of water from outside the cell (**osmosis**).
- The influx of water generates a water-filled space between the vitelline envelope and the cell membrane, generating a barrier to further fertilization events. The thick matrix of cell-wall materials that have been pushed away from the cell membrane by water influx is called the fertilization envelope.

Fertilization in Mammals

- Mammalian fertilization is unique from that of invertebrates like the sea urchin in several ways, including the absence of the species-specific binding between sperm and egg.
- **Paul Wassarman examined** the proteins that mediate binding between sperm and eggs during mammalian fertilization. He isolated **glycoproteins** (proteins with sugar additions to the structure) from mouse eggs, and identified one (**ZP3**) that binds to sperm heads. ZP3 did not display species-specific binding, as was observed with sea urchin proteins that shared a similar function.

Fertilization in Flowering Plants

- The plant spore germinates upon contact with the top of the female reproductive structures. The pollen then generates sperm and carries them to the ovule with elongation of a **pollen tube** (**Figure 21.7**).
- Attachment and germination of the spore is the result of specific interactions between proteins on the surface of the pollen grain and the female reproductive structures. Similar to sea urchin fertilization, these interactions are species specific.
- The plant ovule is fertilized by two sperm nuclei carried by the pollen tube. This **double fertilization** is key to the development of both the embryo (2*n*) and the **endosperm** tissue that is important to providing nutrition for the young plant.

21.3   Cleavage and Gastrulation

- **Cleavage** refers to the rapid cell divisions that follow fertilization (w/o growth) to generate a spherical **blastula**. **Gastrulation** entails the movements of cells to generate distinct developmental regions within the embryo (**Figure 21.8**).

Controlling Cleavage Patterns

- In 1894, H. E. Crampton established a link between the pattern of cleavage in early development and the establishment of shell-coiling direction later in development. Working with a species that produced a left-handed coil, Crampton was occasionally able to find mutant snails that display a right-handed coil (**Figure 21.9**). Surprisingly, when examining early development in these snails, Crampton discovered that the mutant cells underwent cleavage events with a switched orientation.
- An unusual phenomenon was observed with more extensive experiments. The coil direction of offspring was determined completely by the *mother's genotype*, not the genotype of the offspring. If the mother was homozygous for the recessive allele (*dd*), then all of the offspring displayed left-handed shells. If the mother carried the dominant allele (*DD* or *Dd*), then all of the offspring displayed right-handed shells.
- This **maternal effect** upon embryonic development was due to mRNAs or proteins present in the egg before fertilization, and therefore affects development independent of sperm genotype or the zygote genotype. Another example of maternal effect inheritance is the *bicoid* mutation in the fruit fly (see Chapter 22).
- This is a key example of a **cytoplasmic determinant** regulating developmental processes.

The Role of Cytoplasmic Determinants in Development

- Scientists have successfully generated the **fate map** for a select group of organisms. This map is a careful determination of each cell division event that must occur in the development of a mature organism. These maps are possible only for small organisms with a small number of cells in the adult.
- **Edward Conklin** generated one of the first fate maps known, using the **sea squirt** (*Stylea partita*). These animals are easy to view with a light microscope because they are transparent and lightly pigmented in specific subsets of cells (**Figure 21.10**).
- One example is a band of yellow-pigmented cells in the sea squirt; these cells always differentiated into muscle in the mature larvae. The yellow-pigmented cells contain some type of cytoplasmic determinant or other maternal factor that determines the

developmental fate of cells derived from that cell. Thus cytoplasmic determinants affect cell **differentiation**.

The Role of Cell-to-Cell Signals in Development

- The African clawed frog (*Xenopus laevis*) egg displays an asymmetric distribution of yolk, which aids in defining the **vegetal hemisphere** (yolk-enriched) and the **animal hemisphere** (yolk-poor) of the egg.
- A technique called in situ hybridization was used by researchers to localize specific mRNAs within the egg. If specific mRNAs are critical for the development of select cells during cleavage, then—as Douglas Melton hypothesized—the mRNA would be concentrated in one part of the egg cytosol.
- In situ hybridization is a technique similar to Northern blotting, in which a labeled probe is used to screen unlabeled mRNAs for its complementary sequence. In this case the target mRNAs are located within existing tissue, thus in situ (Latin for "in the normal or natural position"). See Section C, "Difficult Concepts," in this chapter of the study guide; and read **Box 21.1** in the text for a review of this technique.
- Melton identified an mRNA that was concentrated within the vegetal portion of the egg (**Figure 21.11**), and named the corresponding gene *Vg1*. Subsequent researchers identified a second mRNA localized to the vegetal pole, and they named this mRNA *VegT*. During cleavage, cells that receive the *Vg1* and *VegT* mRNAs differentiate into **endodermal** cells (forming the digestive tract of the frog).
- *VegT* is a cytoplasmic determinant. Translation of the VegT mRNA into protein generates a protein that binds DNA and activates endoderm-specific genes.
- Vg1 protein is made in endodermal cells, and it is then secreted out of the cell. Why would this gene segregate into a specific cell type, only to have the protein product secreted? Vg1 is a molecular signal that activates the differentiation of tissues adjacent to the endoderm into the dorsal **mesoderm** (forms the muscle of the frog). The Vg1 protein is a **cell-to-cell** molecular differentiation signal.

Gastrulation

- During **gastrulation**, a complex migration of cells occurs, organizing the embryo and further establishing the key embryonic cell types (endoderm, ectoderm, and mesoderm).
- Review **Figure 21.14** to examine the formation of distinct layers of cells during gastrulation in newt embryos. Early experiments characterizing these processes utilized simple staining of subsets of cells during cleavage, and then followed the cells through gastrulation. This mapping of cell fate has proven to be incredibly useful in the study of early developmental processes.
- Within the blastula are the following:
  1. **Animal pole—ectoderm** (generating skin and nerve cells)
  2. **Vegetal pole—endoderm** (generating gut and organs)
  3. **Central region—mesoderm** (muscle and organs)
- Gastrulation begins with the formation of a blastopore (hole) on the surface of the blastomere. This structure eventually develops into the anus.

Organizing the Major Body Axes

- Hans Spemann and Hilde Mangold sought to identify collections of cells within newt embryos that regulate developmental processes. They transferred groups of cells from one embryo to another and evaluated the influence of that group of cells upon development of the recipient embryo (**Figure 21.16**).
- To facilitate these observations, the researchers chose to utilize two different species of newt, each containing a unique pigmentation, and allowing for the differentiation of cells from different species.
- A key discovery from their work was the identification of a group of cells that, when transplanted into another embryo, induce a duplication of organizational poles within the embryo. The result was a **twinned embryo**; basically two animals emerge from the one embryo, though they remain attached across the abdomen.

- The transplanted cells were named the **organizer** (also called the **Spemann organizer**) due to their ability to coordinate gastrulation.
- The second embryo was found to have cells from both the transplanted tissue and the original embryonic tissue, indicating that the organizer was capable of **recruiting** cells from the existing embryo and inducing their organization into a newt embryo.

Molecules Involved in Organizer Function

- The **gray crescent** is a distinct region of cytoplasm present in the zygote soon after fertilization. This region is located on the opposite side of the embryo from the site of fertilization and develops into the organizer region during cleavage events.
- The organizer is involved in the coordination of key developmental processes and cell migration events, but what signals or mechanisms are responsible for this coordinated embryonic development?
- Organizer function can be revealed by character-izing the mRNAs present during early embry-ogenesis (**Figure 21.17**).
- Researchers **irradiated** frog embryos, inhibiting the formation of the organizer, and then injected selected mRNA sequences to determine if the encoded proteins were able to reinitiate the formation of an organizer region.
- They discovered an mRNA, encoding a protein called **noggin** that upon injection into the embryo could completely rescue the arrested embryo, activating normal developmental processes.
- The noggin protein must be a signal generated within the organizer and responsible for the organizer-associated activation of development.
- Noggin is an additional example of a **cytoplasmic determinant**, a signal present within the zygote that regulates early developmental processes.

21.4   Patterns of Development

- Development from a zygote to a multicellular adult is more than just the result of extensive mitotic divisions. In a tightly controlled manner, groups of cells **differentiate** into **specialized tissues** and **structures**. A central focus of researchers studying developmental mechanisms is how this process of differentiation occurs.

Embryonic Development in **Arabidopsis thaliana**

- *A. thaliana* is a fast-growing, flowering mustard plant that has been used in many research studies of plant biology and genetics (**Figure 21.18**).
- Each plant produces male gametes (**sperm**) and female gametes (**eggs**), and the fertilization of egg produces a **zygote**, which has the capability of developing into an adult plant.
- The first stages of *A. thaliana* development occur in a structure called an **ovule**. The outer shell of the ovule is called the **seed coat**. Inside the seed coat is a layer of cells that is enriched for nutrients, called the **endosperm**. At germination, the endosperm functions as a source of nutrition for the seedling.
- The seedling is composed of **cotyledons** (young leaves), the **hypocotyls** (stemlike structures), and a **root** structure.
- The first mitotic division of the zygote following fertilization generates two cells of different size (see **Figure 21.18c**). The large **basal** cell ultimately generates the suspensor structure.
- The smaller **apical** cell give rises to the plant embryo. The **apical meristem** cells arise from the apical cell, offering the plant a lifelong supply of cells for differentiation into adult tissues (e.g., the root, cotyledon, and hypocotyl cells in early development).
- Three additional differentiated tissues, illustrated in **Figure 21.19**, include the **epidermis, ground tissue**, and **vascular tissue**.

An Invertebrate Model: **Drosophila melanogaster**

- The fruit fly, a key model organism for the study of genetics, has also been studied extensively to characterize its developmental processes. Flies display developmental processes that are dramatically unique from those of plants and vertebrates; but as we shall see, many underlying mechanisms and molecular signals are shared among these organisms.

- The female *D. melanogaster* deposits the fertilized eggs at a moist site with plenty of readily available food (like rotting fruit) for the larva. Maturation occurs independent of any additional contribution by the parents.

- The fertilized egg undergoes a number of mitoses in the absence of cytokinesis. The result is a single-celled embryo with **multiple nuclei**. The cytoplasm is rich in nutrients and is referred to as the **yolk**.

- Eventually each nucleus migrates to the outside of the embryo cell and receives a cell membrane (**Figure 21.20**). Thus the embryo becomes an outer sheet of cells surrounding the original cytoplasm (or yolk).

- **Gastrulation** begins with the formation of a **cleft or furrow** along one side of the embryo. This furrow folds into the center of the embryo and is followed by the formation and movement of several more furrows. The embryo begins to emerge as a series of segments that are distinct from each other, identifying the **anterior and posterior** (head and tail) of the embryo.

- After a period of days, the larva forms a **pupa** and enters into its **metamorphosis**, from which it emerges as a fruit fly. The segments generated during gastrulation are each responsible for the formation of distinct segments of the adult fly.

A Vertebrate Model: The Frog

- The early mitotic steps of frog (*Xenopus laevis*) embryonic development are similar to those we've already discussed. The frog embryo progresses from one large cell to a ball of cells (blastula) through a series of cleavage events (**Figure 21.22**). However, the size of the embryo does not increase as the total number of cells increases from one large cell to hundreds of small cells.

- Gastrulation in frog embryogenesis begins with the formation of the **blastopore** (an invagination of one wall of the blastula). This

site of blastopore formation will eventually be the site of the anus. Thus the anterior-posterior axis of the embryo is established with the formation of the blastopore.

- The **neural tube** is formed from ectoderm located adjacent to the mesoderm band. The ectoderm at this site folds into the neural tube, and is an identifiable marker for the establishment of the dorsal-ventral axis.

- In comparing frogs to flies, the development of **mesoderm** into a series of segments that subsequently differentiate is similar between the two organisms.

Early Development in Humans

- Human eggs are released into the **fallopian tube** by the ovary, and are then available for fertilization. If it is successfully fertilized, the zygote proceeds through the first series of cell division events (cleavage) as it migrates down the fallopian tube toward the **uterus**.

- During these first days after fertilization, the embryo proceeds from one cell to about 100 cells. Between the 8- and 32-cell stages of development, the embryo is called a **morula**; and between the 64- and 128-cell stages, the embryo is called a **blastocyst**.

- Four to five days after fertilization, the embryo is ready to **implant** (attach) into the wall of the uterus (**Figure 21.23**).

- Gastrulation and neural tube formation occur after implantation into the wall of the uterus.

- The placenta forms from embryo-derived and maternally derived cells. The placenta functions to carry nutrients to, and waste from, the developing **fetus**.

Making Sense of Developmental Variation and Similarity

- Why doesn't development progress in a similar manner in all organisms? Shouldn't the process of development be conserved among all multicellular organisms?

- Plants and animals approach development through very different processes that are consistent with the evolutionary distance between these organisms.

- Even when comparing animals, which share the processes of gastrulation and cleavage, we find that they utilize these processes in widely varying ways.

- The following are common themes in development:
  - Axis formation is determined early in development and establishes an important embryonic signal for subsequent developmental processes.
  - In animals, cleavage divides the large maternal cytoplasm, generating smaller cells as the number of cells increases.
  - In both plants and animals, three embryonic tissues are responsible for the development of all adult tissues.

B. CROSS-CUTTING THEMES

Looking Back—
Concepts from Earlier Chapters
Genetics—Chapters 12 and 13
In Section 21.3 there is a discussion of the coiling patterns displayed by snails. Researchers identified a **mutation** that causes a reversal of the direction of coiling during development. Crucial to this discovery was that the **maternal genotype** determined the direction of coiling in the progeny (wild-type or mutant). Therefore, the researchers' analysis of the genetics behind this phenomenon offered important insight into how **cytosolic determinants** regulate early development.

Control of Gene Expression in Eukaryotes—Chapter 18
The regulation of **cadherin expression** is a critical component of development, enabling **selective adhesion** between specific groups of cells and disabling the adhesion between other cells. Beneath these complex developmental processes is the critical regulation of gene expression.

Analyzing and Engineering Genes—Chapter 19
The isolation of genes that regulate and mediate developmental process offers scientists the capacity to perturb these processes in order to better characterize the molecular mechanisms behind development. Using genetic engineering, scientists can produce an abnormal pattern of gene expression, delete the expression of specific genes, and/or observe the influence of mutant proteins on the system. Examine the methodology used to clone the *noggin* gene, and review the experiments used to characterize *noggin* function in frog development.

Looking Forward—
Concepts from Later Chapters
Animal Reproduction—Chapter 48
Key to this chapter are the processes of egg fertilization and embryo development. Examine **Chapter 48**, recognizing the complexity of these processes and the molecular mechanisms that drive development and differentiation forward.

C. DIFFICULT CONCEPTS

In Situ Hybridization Techniques
A critical research tool introduced in this chapter is in situ hybridization, which enables scientists to detect the location of gene expression in an organism. The technique is similar to that used for a Northern blot, whereby mRNA transcripts are detected with a radioactive probe (DNA or RNA) that is complementary to a specific mRNA species (one gene). In Northern blotting, the pool of mRNAs are run on an agarose gel and transferred onto a membrane before the probe is used to screen through the transcripts. In situ hybridizations involve the addition of probe directly to the tissue.

Tissues to be examined using in situ hybridization are "fixed," freezing cellular macromolecules in place with a chemical treatment (formaldehyde). The probe enters into the cells of the tissue being tested and hybridizes to the mRNA present in the fixed cells. The site of probe hybridization is detected by placing an emulsion (a kind of liquid X-ray film) over the tissue, which is typically attached to a slide. The dots are generated as the emulsion is exposed to radioactivity and then developed in a manner similar to that used for photographic film. Researchers can then determine (with proper controls) the localization of gene expression within a tissue sample, or even within an entire organism, depending upon the size of the organism. In **Figure 21.11**, the site of *Vg1* expression within a frog egg is revealed using in situ hybridization. How was this information valuable to scientists studying the process of egg maturation in frogs? Could this information have been gathered in another way?

D. ASSESSING WHAT YOU'VE LEARNED

(1) Testing Your Knowledge

1. Compare and contrast the animal sperm cell to the plant pollen grain (**Figure 21.2**).

_____

_____

_____

_____

_____

_____

_____

2. What three tissues, from which are derived all adult tissues, are present in animal embryos? Define the tissues that are generated by these three tissues in animals.

_____

_____

_____

_____

_____

3. Which of the following is *not* necessary for fertilization to take place?
   a. The sperm and egg must be in the same place at the same time.
   b. The sperm and egg must fuse.
   c. The fusion of the two nuclei must trigger the onset of development.
   d. Intercourse must occur between members of the opposite sex of the same species.

4. In sea urchins, the gelatinous coat that surrounds the egg contains a molecule that attracts sperm. Which of the following is true regarding this phenomenon?
   a. The attractant is a small peptide that diffuses out of the jelly layer into the surrounding seawater.
   b. The attractant is not species specific, and this can result in sperm from one species of sea urchin trying to fertilize an egg from a different species.
   c. Sperm respond to the attractant by swimming toward any area that contains it, regardless of the concentration.
   d. The release of the attractant also triggers the polymerization of actin into microfilaments that form a protrusion, which extends until it makes contact with the egg-cell membrane.

5. Research over the past 70 years has revealed a wide array of mechanisms that block polyspermy in various animal species. In sea urchins, fertilization results in the erection of a physical barrier to sperm entry via the formation of a fertilization envelope. Which of the following are involved in the mechanism that creates the fertilization envelope?
   a. The entry of the sperm causes calcium ions to be released from stored areas inside the egg.
   b. Cortical granules located just inside the membrane fuse with the egg-cell membrane and release their contents to the exterior.
   c. An influx of water separates the envelope matrix from the cell.
   d. The contents of the cortical granules include proteases that digest the

exterior-facing fragment of the egg-cell receptor for sperm.

 e. All of the above apply.

6. Spemann and Mangold located the organizer of embryonic tissue using the experiment shown **Figure 21.16** of your textbook. What conclusions were Spemann and Mangold able to draw from this experiment?

 a. Because the second embryo contained no cells from the host, it had developed completely on its own.

 b. Because both embryos were comprised only of cells from the original host, the host tissue's organizer cells removed the donated tissue.

 c. The transplanted dorsal blastopore lip must be able to alter the fates of host tissues.

 d. The host tissue's organizer cells controlled the transplanted dorsal blastopore lip.

7. In flowering plants, an event known as double fertilization occurs when:

 a. Two sperm enter the ovule; one fertilizes the egg, and the other fertilizes a $2n$ cell to generate endosperm.

 b. Two sperm cells fertilize one normal egg cell to make a $3n$ embryo; the embryo later becomes $2n$ during a unique mitotic division.

 c. Two eggs are simultaneously fertilized by two sperm nuclei.

 d. None of the above apply.

8. Which of the following is true of cleavage?

 a. Cleavage divisions partition the egg cytoplasm during additional cell growth.

 b. These divisions proceed slowly in the initial creation of a multicellular embryo.

 c. When cleavage is complete, the embryo consists of a sphere of cells that is ready to undergo gastrulation.

 d. Cleavage occurs exclusively in plant cells.

(2) Integrating Your Knowledge

 (a) Identify the steps leading to the establishment of dorsal/ventral and anterior/posterior axes during early frog development.

 (b) Why did Bill Smith and Richard Harland use irradiation of zygotes as part of their strategy to identify molecular signals that regulate embryonic cleavage (see **Figure 21.17**)?

 (c) At what point during the development of a frog embryo does expression of the zygotic genome become critical? How was this point experimentally determined by researchers? Do humans display the same timing of activating the zygotic genome?

CHAPTER 21—ANSWER KEY

D. Assessing What You've Learned

(1) Testing Your Knowledge

1. The plant sperm cell is a pollen grain. This structure contains very little of the structural specialization displayed by the animal sperm cell. The pollen grain contains two haploid nuclei, whereas the animal sperm contains only one. The plant pollen grain relies on attachment to the fertile plant and growth of the pollen tube to the egg cell. Two nuclei are delivered! The relevance of these two nuclei involves the formation of endosperm ($3n$ in DNA content) along with the zygote ($2n$).

2. (a) **ectoderm** (generates skin and nerve cells);
 (b) **vegetal pole—endoderm** (generates gut and organs);
 (c) **mesoderm** (generates muscle and organs).

3. d; 4. a; 5. e; 6. c; 7. a; 8. c

(2) Integrating Your Knowledge

 (a) **Dorsal/ventral**—The neural tube is formed from ectoderm located adjacent to the mesodermal band. The ectoderm at this site folds into the neural tube and is an identifiable marker for establishment of the dorsal/ventral axis.

Anterior/posterior—Gastrulation in frog embryogenesis begins with formation of the blastopore (an invagination of one wall of the blastula). This site of blastopore formation will eventually be the site of the anus. Thus the anterior/posterior axis of the embryo is established with formation of the blastopore.

(b) Smith and Harland wanted to identify the signals responsible for formation of the organizer region. They irradiated frog embryos to inhibit the cortical cytosol rotation that always precedes the formation of the organizer. Without reintroduction of a specific, organizer-inducing signal, the embryos would be unable to develop normally. The researchers tested individual mRNAs isolated from other frog embryos and found one mRNA that was able to activate formation of the organizer. They named the gene *noggin*.

(c) Using a potent inhibitor of transcription (α-amanitin), and injecting it into fertilized frog eggs, researchers were able to determine the stage of embryonic development at which zygotic gene expression is required. Surprisingly, frog eggs developed normally to the midblastula transition stage of development before displaying abnormalities. Human embryos require zygotic gene expression much earlier, at the 2-cell stage of development.

22

Pattern Formation and Cell Differentiation

A. KEY BIOLOGICAL CONCEPTS

22.1   Pattern Formation in Drosophila

The Discovery of bicoid

- **Pattern formation** in a developing embryo is the establishment of the three-dimensional, spatial organization that initiates many developmental processes. Cells receive molecular signals that indicate where they are within the embryo, and that information directs gene expression and cell-fate determination.

- The fruit fly, *Drosophila melanogaster*, has been a key model organism for the study of pattern formation.

- Christiane Nüsslein-Volhard and Eric Wieschaus performed a large mutant hunt for genes that influence pattern formation in *Drosophila melanogaster*. They **mutagenized** flies and then examined the fly progeny for defects in development. They looked specifically for mutant flies that generated progeny with abnormal pattern formation.

- The *bicoid* mutation generated a dramatic phenotype, generating larvae that lacked structures on the anterior end (**Figure 22.1**). The normal head/thorax region develops into posterior structures, making the larvae *bicoid*, which means "two-tailed."

- The *bicoid* gene was shown to display **maternal effect inheritance** (**Box 22.1**). This term means that the phenotype in the larvae is due only to the mother's genotype. If the mother lacks a functional copy of the *bicoid* gene, then the embryo will be two-tailed. The

mother is responsible for providing functional *bicoid* mRNAs to the egg, and these cytoplasmic determinants aid in larval development (recall the maternal effect on snail-shell coil direction, introduced in Chapter 21).

What Does bicoid *Do?*

- The *bicoid* gene was isolated via **gene mapping** techniques. Review the examination of **polymorphisms** and **genetic linkage** discussed in **Chapter 13** to illustrate how this process works.

- Once the gene was isolated, researchers could generate probes for localizing *bicoid* mRNA (in situ hybridization) and protein (immunolocalization) within the embryo. In **Chapter 21**, we introduced the localization of mRNA with radioactive probes that complement a specific gene. Antibodies specific for a protein can be used for the same purpose, determining where and when the protein is synthesized in the protein.

- Using in situ hybridization analysis, *bicoid* mRNA was found in the anterior of the egg (**Figure 22.3a**). Remember, this mRNA is from the mother and is present at fertilization as a crucial cytoplasmic determinant.

- Antibodies specific for the *bicoid* protein were used to determine when and where the protein was synthesized (**Figure 22.3b**). Immediately following fertilization, *bicoid* protein is synthesized, and its concentration within the zygote is organized as a gradient. There are very high concentrations of *bicoid* protein in the zygote's anterior and low concentrations at its posterior.

- The **gradient hypothesis** proposed by Nüsslein-Volhard stated that this gradient of *b-icoid* protein was key in establishing anterior/ posterior axes within the embryo. Thus, high concentrations of *bicoid* protein were hypothesized to signal anterior development and low concentrations of *bicoid* protein were hypothesized to signal posterior development.

Testing the Gradient Hypothesis

- To test the gradient hypothesis, Nüsslein-Volhard purified *bicoid* mRNA and injected it into different places in developing embryos. To determine if purified bicoid mRNA was functional when injected into an embryo, the researchers injected the mRNA into the anterior of embryos derived from mutant *bicoid* females. The injected embryos developed normally and did not display the two-tailed phenotype.

- When the researchers injected *bicoid* mRNA into wild-type (nonmutant) embryos, anterior structures always developed at the site of injection. These results strongly supported the gradient hypothesis.

- The *bicoid* protein was shown to act as a **transcription factor** within *Drosophila* embryos, binding DNA and regulating the expression of other genes vital to pattern formation. Identifying what genes are regulated by *bicoid* was vital to unraveling some of the mysteries of early development.

The Discovery of Segmentation Genes

- Many other genes were identified via the mutant hunt conducted by Nüsslein-Volhard and Wieschaus, and each of these genes is vital to early developmental processes. Some of them can be grouped into one of three general categories of genes involved in embryo **segmentation** during *Drosophila* development (**Figure 22.4**).

- **Gap genes**—Loss-of-function mutation in gap genes results in a loss of several consecutive segments in the fly embryo. One key example of a gap gene is named *hunchback*. Mutation of *hunchback* results in embryonic phenotypes similar to those described for *bi-*

coid (lacking segments in the anterior of the embryo). However, *hunchback* functions via embryonic gene expression and is not a maternally derived cytoplasmic determinant like *bicoid*.

- **Pair-rule genes**—Loss of function in pair-rule genes results in the loss of **alternating** segments in the embryo, indicating that these genes are critical to the formation of every other segment formed during embryogenesis.

- **Segment polarity genes**—These genes are critical to formation of components of each segment; the loss of these genes generated embryos lacking portions of each segment.

Do Segmentation Genes Interact?

- Segmentation genes are expressed in a precise order during embryogenesis:

  Gap genes ↓ Pair-rule genes ↓

  Segment polarity genes

- How does this pattern of consecutively activated genes get started? Do the genes activate each other? The maternal cytoplasmic determinant *bicoid* is known to activate expression of *hunchback* (a gap gene whose expression is located in the embryo anterior; similar to the *bicoid* expression pattern), so this is an example of a cytoplasmic determinant activating a member of the gap genes.

- But do the gap genes activate the pair-rule genes? Yes, many of the gap genes code for transcription factors that bind to regulatory sequences in pair-rule gene promoters, activating their expression at a precise time during development. As you might now suspect, members of the pair-rule gene family activate the segment polarity genes.

- This complex array of stepwise gene activation events is called a **regulatory cascade**, and can be quite complex to consider in its entirety (**Figure 22.5**). These waves of gene expression are critical to establishing proper **pattern formation**. They continue through development, eventually being restricted to specific groups of cells as specific tissues begin to form. But how are distinct groups of cells given unique signals during development? In other words, what signals establish the individual segments?

The Discovery of Homeotic Genes

- Edward Lewis (1940s) performed a mutant hunt with *Drosophila melanogaster*, isolating flies that displayed abnormal body patterns in the *adult fly*. Lewis isolated many mutant flies and recognized that many of the mutant phenotypes were due to abnormal development of individual segments. Some mutants had segments that developed into a phenotype that matched an adjoining segment, generating legs where antennae should be, an extra set of legs, or even an extra set of wings.

- This phenomenon of one segment developing incorrectly into the phenotype of another segment is called **homeosis** (like condition).

- The genes responsible for proper identity in each segment were named **homeotic complex**, or **HOM-C** genes. Eight were identified by gene cloning and sequencing, each gene being responsible for the proper development of a specific segment or group of segments in the embryo (**Figure 22.8**).

- All eight of these genes are located on one chromosome. The five genes expressed in the anterior of the embryo are called the *Antennapedia* complex; three that are expressed in the posterior of the embryo are called the *Bithorax* complex.

- The expression of the *HOM-C* genes is sequential, moving in order along the chromosome (*lab* ↓ *AbdB*) and from the anterior to the posterior of the embryo (*lab* is expressed first and in the far anterior of the embryo; *AbdB* is expressed last and at the far posterior of the embryo). Examine **Figure 22.8** closely to follow the parallels between chromosome location and the timing/distribution of gene expression.

- Each of the proteins encoded by the *HOM-C* genes contains a **DNA-binding region**; they are thought to act as transcription factors, activating genes critical for proper development of the relevant segments (e.g., *lab*-activating genes for development of anterior structures). The DNA sequence encoding these regions in each protein is called the **homeobox** (a 180-bp segment within each of the *HOM-C* genes).

- How are the *HOM-C* genes activated during development? There is clear evidence that the segmentation genes play a critical role in activating the proper set of *HOM-C* genes in each segment. Flies carrying mutant segmen-tation genes display abnormal activation of the *HOM-C* genes.

22.2 Pattern Formation in *Arabidopsis*

The Root-to-Shoot Axis of Embryos

- A key model organism for the study of plant biology and development is *Arabidopsis thaliana*, a fast-growing mustard plant. Pattern formation during *Arabidopsis* development, and also later in the life cycle, has been carefully studied to determine which genes regulate this process.

- Gerd Jurgens used a mutant hunt to isolate plants displaying abnormal development of the root-to-shoot axis (**Figure 22.9**). The mutants they isolated can be placed into three groups: **apical mutants** lacking cotyledons, **central mutants** lacking hypocotyl structures, and **basal mutants** lacking hypocotyl *and* root structures.

- Plants displaying the basal mutant phenotype were found to carry a mutation in a gene called *monopterous*, which encodes a putative transcription factor. The expression of *monopterous* is activated by **auxin**, a plant-signaling molecule that is produced in the apical meristem.

- Researchers draw parallels between auxin and *bicoid* as molecules that establish polarity within the embryo through their differential concentration along the axis of the embryo.

Flower Development in Adults

- *Arabidopsis* flowering is initiated from the apical meristem, resulting in the production of four structures, consistently organized in a similar manner. As shown in **Figure 22.10**, the **sepals** are the most external structure, offering protection to the more fragile flower **petals** that enclose the male reproductive organs (**stamens**) and the female reproductive organ (**carpel**).

- In the 1800s, researchers identified mutant *Arabidopsis* plants that resulted in one category of reproductive organ replacing another (reminiscent of our discussion of homeotic genes in *Drosophila*). Years later, Elliot Meyerowitz isolated a group of mutant *Arabidopsis* plants displaying abnormalities in flowering; and he identified the genes responsible for each of the observed phenotypes. The genes he identified can be broken up into three categories, with each category of mutant phenotype displaying the absence of two adjacent whorls. The three phenotypes are pistil/stamen only, sepal/pistils only, and sepals/petals only (**Figure 22.11a**).
- Meyerowitz explained these three categories of genes/mutant phenotypes using the **ABC model** of flowering (**Figure 22.11b**). This model includes three central components:
  1. Each of the three genes regulating flowering is involved in the formation of two adjacent whorls.
  2. Four different combinations of gene products are possible, each generating a unique signal (A, AB, BC, C).
  3. These unique signals are capable of activating different whorls of differentiated meristem tissue: sepal (A), petal (AB), carpel (BC), and stamen (C).
- Each of the three genes was also shown to encode DNA-binding regions in its protein product, sharing functional homology with the homeobox sequences found in the *HOM-C* genes. The plant DNA-binding motifs are known as **MADS boxes**.

22.3 Does the Genetic Makeup of Cells Change as Development Proceeds?

Are Differentiated Animal Cells Genetically Equivalent

- A key question in the study of development and cellular differentiation focuses on the effect of differentiation on the genome—the DNA content of a cell. Is there an irreversible change in the DNA of a cell as it differentiates? A key technology used to test this question has been nuclear transfer experiments, in which the nucleus of a differentiated cell is transplanted into an unfertilized egg. Researchers hypothesized that if there is no change to the DNA content of a differentiated cell, then the transferred nucleus should enable the normal development of a complete individual.
- In the 1950s, scientists were first able to isolate the nucleus of one frog cell and transfer it to an unfertilized frog egg, a process called **nuclear transfer**. Researchers found that the transfer of a nucleus from a blastula- or neurula-stage embryo into an egg was capable of generating a normal frog tadpole.
- However, isolation of nuclei from tadpoles for use in similar experiments was unsuccessful. Something was different about the DNA in these differentiated tissues.
- Ian Wilmut (1997) performed nuclear transfer experiments with sheep mammary gland cells and enucleated sheep eggs. Though the success rate was very low, this process was successful in generating cloned sheep (**Figure 22.13**).
- Tissue differentiation does not involve any irreversible change in DNA content, and it lies primarily within the differential regulation of gene expression in different cell types.

22.4 Differentiation: Becoming a Specialized Cell

- **Differentiation**—the generation of cellular diversity (structure, function) is established through differential gene expression.
- **Morphogenesis**—the assembly of cells into recognizable tissues and organs.
- **Determination**—a cell's irreversible commitment to a particular cell type.
- Chronology within development:

  Pattern formation ↓ Morphogenesis ↓
  Determination ↓ Differentiation

Organizing Mesoderm into Somites

- At the completion of vertebrate gastrulation, the mesoderm is a thin layer of cells located between the endoderm and ectoderm (**Figure 22.14**). As development proceeds, the endoderm forms the **digestive tract** along the ventral side of the embryo. The ectoderm, located along the ventral side of the embryo, folds to form the **neural tube**. At the same time the mesoderm begins to form the **notochord** adjacent to the neural tube, these two structures form together (exchanging molecular signals to regulate each process).

- The fully formed neural tube emits a molecular signal that organizes the neighboring mesodermal cells into blocks of cells known as **somites** (**Figure 22.16**). The neural tube is soon surrounded by somites, mediated by selective adhesion events (mediated by cadherins).

- After cell fate determination occurs, the somites develop into a variety of adult tissues (vertebrae, ribs, body wall, some muscle, and even skin).

Determination of Muscle Cells

- Cells in the somites receive molecular signals from several surrounding tissues (neural tube, ectoderm, and mesoderm). Cells located on the periphery of each somite are destined to become muscle cells.

- Researchers examined the molecular characteristics of myoblasts, somite-derived cells whose cell fate is determined to be muscle. They isolated mRNA from myoblasts, converted the mRNA into cDNA through the process of **reverse transcription**, and reintroduced the **cDNAs** into fibroblast cells (see Section C, "Difficult Concepts"). These researchers sought to identify genes whose expression in the fibroblasts results in the development of muscle-like characteristics. They were hunting for the genes that activate cell-fate determination in the development of muscle cells.

- Researchers discovered one gene that could convert the fibroblasts into muscle-like cells, and they named the gene *MyoD* (myoblast determination).

- *MyoD* codes for a transcription factor that binds to the regulatory sequences of muscle-specific genes and activates their expression. Activation of *MyoD* expression in somites is stimulated by signals from neighboring embryonic tissues, and it results in the development of muscle from some of the somites.

B. CROSS-CUTTING THEMES

Looking Back—
Concepts from Earlier Chapters
Mendel and the Gene—Chapter 13
The cloning of specific target genes from a large genome can be a daunting task. **Chapter 13** introduces how the chromosomal location of a specific gene can be determined using **gene mapping** techniques (the analysis of phenotype and genetic markers through multiple generations). *Bicoid* was cloned through just such a strategy, identifying female flies that produce abnormal progeny and following the **inheritance** of the recessive mutant *bicoid* allele through multiple generations. Eventually, the gene itself was mapped and cloned to enable a variety of additional experimental strategies.

Control of Gene Expression in Eukaryotes—Chapter 18
In Chapter 22 we've examined how vital the regulation of gene expression is in development. Many of the key regulatory molecules in early development are **transcription factors**, binding promoter elements to turn on genes at the proper place in the embryo and at the proper time (review the segmentation genes in *Drosophila*).

Early Development—Chapter 21
The utilization of **in situ hybridization** is a key component in the study of differentiation and morphogenesis. This technique was introduced in earlier chapters, and was carefully examined in **Chapter 21** (see **Box 21.1** and the "Difficult Concepts" section in Chapter 21).

C. DIFFICULT CONCEPTS

Reverse Transcription and cDNA Libraries
Our discussion of the work of Harold Weintraub and colleagues (*MyoD* cloning) introduces an important research tool, the utilization of a **cDNA library** to hunt for genes. The researchers set out to find genes involved in muscle cell determination, and they prepared mRNAs from a collection of myoblasts. Certainly, within the pool of mRNAs are some transcripts that signal and maintain the muscle-specific determination of these cells. But finding the relevant gene is

not easy. It requires a functional screening of all the mRNAs to find those associated with cell determination.

To screen mRNAs, the researchers copied each single-stranded RNA molecule into a copied DNA or cDNA. An enzyme called **reverse transcriptase** is utilized for this purpose. Retroviruses (a category of RNA viruses) use reverse transcriptase to convert their RNA genome into DNA during the process of viral infection. The enzyme can be purified for research purposes, converting mRNAs into cDNAs.

Weintraub and colleagues generated cDNA for each of the mRNAs isolated from myoblast cells and then set out to screen each of these cDNAs for an ability to alter the determination status of fibroblast cells. Their goal was to find, amidst that collection of thousands of different cDNAs, those capable of activating "muscle-like" determination in the fibroblasts. To do this, each of the cDNAs would have to be expressed within a group of fibroblasts to identify any influence of that gene on fibroblast biology. The researchers used DNA plasmids, each carrying a strong promoter, and transferred the cDNAs into their structure. The result is a collection of plasmids with cDNA inserts in their structures, each of which expresses the gene encoded on the cDNA. The plasmids were then transferred into fibroblasts and the plasmid-carrying fibroblasts were analyzed for changes that might indicate muscle determination.

In this instance, the research team was successful in identifying one plasmid carrying the *MyoD* cDNA, which dramatically altered the fibroblasts. The cDNA library had been successfully screened, and a critical insight in cell determination was achieved.

D. ASSESSING WHAT YOU'VE LEARNED

(1) Testing Your Knowledge

1. List the advantages of using *Drosophila melanogaster* as a model organism for identifying the genes and processes responsible for cell-fate decisions in early development.

a. _____

b. _____

c. _____

d. _____

2. If the mother of larvae displaying the mutant *bicoid* phenotype carries only the mutant gene, how was the mother able to develop normally in the first place?
   a. The bicoid phenotype can be suppressed by culturing the embryos at cool temperatures.
   b. The *bicoid* mRNA is a maternal effect gene, donated to an egg by the mother regardless of the egg genotype.
   c. The mother likely also carried a *bicoid* suppressor mutation.
   d. All of the above apply.

3. What was the experimental focus of the screen performed by Nüsslein-Volhard and Wieschaus?
   a. to evaluate the mutagenic properties of varied treatments
   b. to discover the *bicoid* gene
   c. to generate two-tailed fruit-fly embryos
   d. to generate fruit flies with mutations in genes that control fruit-fly development

4. Which of the following is *not* an accurate description of the *bicoid* gene?
   a. Mutation generates larvae without anterior development.
   b. Encoded protein is a DNA-binding protein.
   c. The gene displays a maternal effect mode of inheritance.
   d. A male that is homozygous for this mutation cannot generate viable offspring.

5. Which of the following is true of maternal effect inheritance?
   a. A gene is transcribed in the offspring and affects the phenotype observed in the mother.
   b. A gene is transcribed in the mother and affects the phenotype observed in the offspring.

c. The genes of the mother negate the genes of the father.

d. The genes of the father negate the genes of the mother.

6. What do you call the 180-base-pair DNA sequence that was revealed in each gene in the *Antennapedia complex* and the *Bithorax complex*?

a. homeobox

b. homeotic complex

c. homeosis

d. homeotic genes

7. What is an important advantage of studying the pattern formation of plants?

a. Plants have mutant individuals to study pattern formation.

b. Pattern formation is not limited to the period of early development.

c. Plants have a short life span.

d. Plant formation is very similar to animal pattern formation.

8. When is a cell considered differentiated?

a. when a cell first becomes irreversibly committed to a particular fate

b. when a cell begins its pattern formation

c. when a cell manufactures proteins that are specific to a particular cell type

d. when a cell is part of recognizable tissues or organs

9. What were researchers able to determine by transferring cells from one somite to another?

a. A cell in a somite can become any of the somite-derived elements in the body.

b. Somites are responsible for controlling apoptosis.

c. Somites are transient structures that appear relatively briefly during development.

d. Mesodermal cells in a somite are destined for a variety of structures.

10. Weintraub and co-workers hypothesized that myoblasts contain at least one regulatory protein that commits them to their fate. Their research found the gene that encodes this protein, which they named the protein MyoD. All of the following are true of MyoD *except*:

a. MyoD encodes a transcription factor.

b. MyoD is the gene that regulates activities of *ced-3* and *ced-4* genes.

c. MyoD protein binds to enhancer elements located upstream of muscle-specific genes.

d. MyoD protein activates expression of the *MyoD* gene.

(2) Integrating Your Knowledge

(a) Within the *Drosophila* model of development, what early events establish the signals that initiate pattern formation?

(b) In *Drosophila*, how does the observed pattern of consecutively activated segmentation genes, as illustrated in **Figure 22.4**, get started?

(c) Compare/contrast the terms *morphogenesis, determination*, and *differentiation* as they relate to early development.

(d) How do researchers establish when a somite cell fate is determined?

(e) Describe the methodology used for the cloning of genes involved in muscle-cell differentiation. How were researchers able to assay each isolated gene for the capacity to activate muscle-specific differentiation?

CHAPTER 22—ANSWER KEY

D. Assessing What You've Learned

(1) Testing Your Knowledge

1. (a) *Drosophila* generate large numbers of progeny very rapidly (the time from mating to reproductive maturity is about 3 weeks).

(b) *Drosophila* exhibit complex developmental processes (multicellular organism), serving as a better model than simpler multicellular organisms.

(c) *Drosophila* has well-characterized genetics (**Chapters 12** and **13**).

(d) Fruit flies are easily maintained and bred in a laboratory setting.

(e) Fruit flies can be mutagenized fairly easily (X-rays or chemical mutagens), and flies with mutant phenotype detected by visual inspection of progeny.

(f) Fruit-fly embryos are deposited into a rich nutrient source (in the laboratory, culturing medium), and are therefore easy to isolate for study.

2. b; 3. d; 4. d; 5. b; 6. a; 7. b; 8. c; 9. a; 10. b

(2) Integrating Your Knowledge

(a) The maternally derived cytoplasmic determinant, *bicoid*, is expressed in the anterior of the fertilized embryo and generates a gradient of *bicoid* protein (high concentration in anterior to low concentration in the posterior). This gradient of concentration establishes a polarized pattern of gene expression.

(b) *Bicoid* is a transcription factor, activating the expression of some segmentation genes along its concentration gradient in the early embryo. The key example is the gap gene *hunchback*, whose expression is directly activated by *bicoid*. Then some of the gap genes activate the pair-rule genes, and some of the pair-rule genes activate the segment polarity genes, driving embryonic development forward.

(c) *Morphogenesis* is the migration and assembly of cells into distinct structures/tissues. *Determination* refers to the molecular "commitment" to a cell type, activating cellular changes that result in cellular differentiation. Activation of *MyoD* within myoblasts is an excellent example of cellular determination. Differentiation refers to the generation of diversity in cellular structure and function, and is established through differential gene expression.

(d) As illustrated in **Figure 22.17**, somites may be removed from one site and transplanted to another to determine if the transplanted somite is capable of differentiating into other cell types. Early in somite maturation, each transplanted somite cell is capable of differentiating into any of the somite-derived cell fates. Later during somite development, this is no longer possible, indicating the somite fate is now irreversibly determined.

(e) Harold Weintraub and colleagues isolated mRNAs from determined muscle precursor cells called myoblasts. The mRNAs were reverse transcribed into cDNAs and cloned into expression plasmids. This enables the expression of these cloned fragments in any cell type. They screened through their cDNA clones, transferring each into fibroblast cells and examining the cells for muscle-like differentiation following expression of the cDNA clone. The muscle-like changes within these cells would be identified as changes in cellular morphology and the expression of genes that are active only in muscle.

Evolution by Natural Selection

A. KEY BIOLOGICAL CONCEPTS

- The theory of evolution by natural selection, jointly proposed by Charles Darwin and Alfred Russel Wallace in 1858, is one of the key theories in biology (**Box 23.1**).
- The pattern recognized by this theory is that species are related and change through time. Natural selection is the mechanism that explains this pattern. Modern biologists continue to study evolution in action, amassing yet more evidence in support of this theory.

23.1   The Pattern of Evolution: Have Species Changed through Time?

- Darwin described evolution as "descent with modification"—that change over time produces modern species from ancestral species.

Evidence for Change through Time

- Traces of organisms that lived in the past are called **fossils** (**Figure 23.2**). Those that have been found and described in the scientific literature make up the **fossil record**.
- Most fossils are found in **sedimentary rocks** that form in layers from deposits of sand, mud, or other materials at beaches or river mouths. Fossils from lower sedimentary layers are older than those from upper layers are. These layers are associated with different periods in the **geologic time scale**.
- Researchers can now use radioactive isotopes to date rocks, and they have assigned absolute ages to the various geologic time periods.

Geologic data show that Earth is about 4.6 billion years old; and fossil data show the earliest signs of life in rocks about 3.85 billion years old.

Extinction

- Researchers have discovered many fossils unlike any known living organisms. These species are **extinct**; they no longer exist. More species have gone extinct than exist today, and species have gone extinct throughout Earth's history.
- Darwin interpreted extinction as evidence that species are dynamic and can change.

Transitional Forms

- Fossils often resemble living species found in the same region. This "law of succession" means that extinct fossil species are typically succeeded, in the same region, by similar species. Darwin interpreted this as evidence that extinct forms are the ancestors of modern forms, and species change over time.
- Improvements in the fossil record led to the discovery of many **transitional forms** with traits that are intermediate between older and younger species.
- The whale lineage includes fossil transitional forms that show a gradual change over a 12-million-year period from land-dwelling hippopotamus-like animals to modern whales (**Figure 23.3**). Many other transitions are similarly well documented and provide strong evidence for species change through time.

from terestrial animal to aquatic animal.

227

Environmental Change

- Many rocks and fossils from ancient oceans are found thousands of feet above sea level, suggesting that Earth's topography and environment have drastically changed over time. Our planet and its species are dynamic.
- **Vestigial traits** are functionless structures similar to functioning structures in related species. Biologists interpret these vestigial traits as evidence that trait structure and function change over time (e.g., whale hip bones, human appendix, pseudogenes; **Figure 23.4**).
- Changes in modern species in response to environmental change are also well documented (e.g., evolution of antibiotic resistance in bacteria, herbicide resistance in weeds, etc.). Thus change in species over time continues today and can be measured directly.

Evidence that Species Are Related

- Darwin collected mockingbirds found in the Galápagos Islands on his *Beagle* voyage. The mockingbirds were superficially similar, but different islands had distinct species. Darwin proposed that different island species were similar since they had descended with modification from a common ancestor (**Fig. 23.5**).
- This pattern of similar yet distinct species is commonly found on island groups worldwide. Related species share a **phylogeny** or family tree, and their relationships can be graphically depicted in a **phylogenetic tree**.
- Resemblances between non-island species are widespread; and their structure, development, and genetics can be observed and analyzed.

Structural Homologies

- **Homology**, in biology, refers to resemblance due to common descent (similar traits derived from a common ancestor). **Structural homology** refers to similarity of morphological traits.
- For example, all vertebrates share the same general limb structure (**Figure 23.6a**). It is not logical to use the same underlying pattern for limbs of such differing function, but the similarity is easily explained if all mammals descended from a common ancestor.

Developmental Homologies

- **Developmental homology** refers to similarity in embryo morphology and/or pattern of tissue differentiation.
- Early in embryonic development all **vertebrates** (animals with backbones) have gill pouches and tails (**Figure 23.7**). Biologists hypothesize this pattern exists because the common ancestor of all vertebrates is a fishlike animal with gill pouches and a tail.
- An example of developmental homology in tissue differentiation is that the same group of embryonic cells develops into the jaw of both fish and mammals. Once again, the similarity in trait is explained by the trait's presence in a common ancestor.

Genetic Homologies

- Adult structures develop from embryonic structures and processes, so structural homologies are due to developmental homologies. Both types of homologies are in turn due to **genetic homologies**—similarity in the DNA sequences of genes from different species.
- For example, fruit-fly and human eyes are very different, but the gene that determines where eyes develop is similar in these two groups (**Figure 23.8**). Biologists conclude that a common ancestor of fruit flies and humans had a similar gene that was involved in forming a primitive light-gathering organ.
- Evidence suggests all Earth's organisms descend from a single common ancestor. Thus almost all organisms use the same 64-mRNA codon to specify the same amino acids.
- Similar traits that are not inherited from a common ancestor are said to be **analogous** and are a product of **convergent evolution** (**Box 23.2**).
- Because all organisms share a common ancestor, they also show genetic homology. For example, almost all organisms use the same mRNA codons to specify the same amino acids; and all organisms have a plasma membrane made of a phospholipid bilayer with interspersed proteins, use RNA polymerase to transcribe the information coded in DNA into RNA, use ribosomes to synthesize proteins, use ATP, and so forth.

[handwritten notes at bottom of page:] Convergent evolution: evolution of similar organisms due to adaptations to similar ways of life. → often produces analogous traits. traits in distantly related similar environments and a

- **Homology**, in biology, refers only to similarities due to common descent. But sometimes organisms have similar traits that are *not* inherited from a common ancestor. These are called **analogous traits** or **convergent traits**.

How Do Biologists Distinguish Homology from Analogy?

- Sometimes species inhabiting similar environments have similar traits due to **convergent evolution**. Convergent evolution occurs when natural selection leads to similar solutions to the challenges of a particular habitat.
- For example, icthyosaurs and dolphins show morphological similarities (e.g., streamlined bodies) that their reptilian and mammalian ancestors did not have. Instead, these traits were selected for in both these groups because they are advantageous in fish-eating marine animals that chase down fish in open water (**Figure 23.9**).
- Key to differentiating homology from analogy is determining if the similar structures or genes derive from a common ancestor.
- For example, unlike ichthyosaur and dolphin morphology, the *HOM* genes of insects and the *Hox* genes of vertebrates *are* thought to derive from a common ancestor; they show similar organizations, share a *homeobox* sequence, and code for products with similar functions and patterns of expression. Further, many animals from lineages that branch off between these two groups have similar genes.

Darwinism and the Pattern Component of Evolution

- Darwin and others have convincingly shown that the types of evidence just described are best explained by the theory of evolution, which states that species are related and change through time (**Table 23.1**).
- An alternate theory that species were instantaneously and independently created by a supernatural being (special creation) is not supported by the *scientific* evidence.

23.2 The Process of Evolution: How Does Natural Selection Work?

- The idea of evolution predates Darwin (**Box 23.2**), but Darwin and Wallace were the first to propose a process, natural selection, that satisfactorily explains the observed pattern of descent with modification.
- A population experiences natural selection when these four conditions apply: (1) individuals vary in their traits; (2) some of these variations are heritable; (3) some individuals survive and reproduce better than other individuals; and (4) differential survival and reproduction (**Darwinian fitness**) is influenced by the heritable traits of individuals.
- In other words, some heritable traits help an individual to survive or reproduce better than other individuals. Such traits will, over time, increase in frequency in the population, causing **evolution**—the genetically based change in a population's traits over time.
- **Adaptations** are heritable traits that increase an individual's fitness relative to individuals lacking that trait. Adaptations must be discussed in reference to an organism's environ-ment because traits that increase fitness in one environment may not be advantageous in another environment.
- Note that while natural selection acts on the individual, it is the population that evolves.

23.3 Evolution in Action: Recent Research on Natural Selection

- The theory of evolution by natural selection is testable. Here are some examples of how biologists test for evolution by natural selection in actual populations. Hundreds of other examples are available.

How Did Mycobacterium tuberculosis Become Resistant to Antibiotics?

- *Mycobacterium tuberculosis* causes **tuberculosis (TB)**, which killed up to a third of all adults in big cities in the nineteenth century. TB still kills more adult humans than does any other virus or bacteria.
- Sanitation, nutrition, and antibiotics succeeded in greatly reducing deaths due to TB in industrialized countries between 1950 and 1990. Unfortunately (but predictably), TB is on the rise again as antibiotic-resistant strains evolve and spread (**Figures 23.10** and **23.11**).

- Consider the evolution of antibiotic resistance in a young man who was admitted to a hospital with fever and coughing. Chest X-rays and lung-fluid bacterial cultures confirmed a diagnosis of TB. After more than 33 weeks of antibiotic treatment, his lung fluid and chest X-rays were normal and he was considered cured.
- Two months later, the man was readmitted to the hospital with TB and treated with more antibiotics, including rifampin (a common antibiotic for TB). But new lung-fluid cultures grew rifampin-resistant bacteria. Antibiotics did not kill the bacteria; the man died.
- DNA from the rifampin-resistant bacteria that killed the patient was compared to DNA from the original antibiotic-sensitive bacteria cultured when the man was first admitted to the hospital with TB. The only difference between the two DNA sequences was a single point mutation (C ↓ T).
- This mutation caused a single amino acid change in the *rpoB* gene that codes for a component of RNA polymerase, the protein that transcribes DNA to mRNA.
- Rifampin acts by binding to RNA polymerase in *M. tuberculosis* and preventing mRNA transcription (eventually killing the cell). But because rifampin does not bind effectively to the new mutant RNA polymerase, it cannot kill cells with this mutation.
- Researchers deduced that one or a few cells with this mutation were present in the original population. Cells with this mutation don't usually survive and reproduce as well as cells with the normal *rpoB* gene (the mutant RNA polymerase does not work as well).
- When antibiotics were added to the bacteria's environment, bacteria with the normal *rpoB* gene died; the bacterial population declined so much that the man appeared cured. However, cells with the mutant *rpoB* gene now had a huge selective advantage—they survived better than bacteria without the mutation, reproduced, and eventually became so numerous that they killed their host.
- Natural selection had caused evolution—a change in a population's allele frequencies over time.

- (1) Variation existed in the initial bacterial population. (2) The variation was heritable. (3) Bacteria differed in their reproductive success. (4) Bacteria with certain traits produced more offspring than other bacteria; when rifampin was present, bacteria with the mutant *rpoB* gene were more fit than other bacteria—so the mutant *rpoB* gene is an adaptation to antibiotic-containing environments.
- Currently, **evolution** is most often defined as change in a population's allele frequencies (genetic makeup) over time. In the example given, characteristics of the *M. tuberculosis* population changed over time as the frequency of the mutant *rpoB* gene increased.
- Individual bacterial cells did not evolve; they merely survived or died and reproduced more or less. It was the population that evolved over time due to the effects of natural selection on individual cells.

Why Are Beak Size, Beak Shape, and Body Size Changing in Galápagos Finches?

- Biologists can study natural selection even when the alleles and molecular mechanisms responsible for variation in the population are not known.
- Peter and Rosemary Grant led long-term research on changes in body size and beak size and shape in medium ground finches on Isle Daphne Major of the Galápagos Islands.
- Because these finches live on a small island, researchers could catch, weigh, measure, and mark all individuals. They found that body size and beak size and shape varied among individuals and that variation in these traits was heritable (e.g., parents with deep beaks tend to have offspring with deep beaks).
- In 1977, a severe drought eliminated 84% of the island's finch population. Survivors had much deeper beaks on average than did nonsurvivors. Grant's group hypothesized that individuals with large deep beaks were more likely to survive because they could eat the tough fruits of the *Tribulus cistoides* plant.
- Data show that natural selection led to an increase in the finch population's average beak depth (**Figure 23.12**). Alleles responsible for

development of deep beaks must have increased in frequency in the population because a large, deep beak is an adaptation for cracking large, tough fruits and seeds.

- In 1983 there was much more rain than average, causing lots of plant growth, and finches fed primarily on abundant small, soft seeds. Small finches with small, pointed beaks had particularly high reproductive success during this period, causing yet another change in the population's characteristics (**Figure 23.13**).
- The Grants continue documenting evolution in response to the continuing environmental changes on Daphne Major.

Can Natural Selection Be Studied Experimentally?

- Candace Galen began research on alpine sky-pilots after observing that those in the tundra above the timberline tended to have larger, sweeter-smelling flowers and longer stalks than did those lower down the mountain.
- Skypilots above the timberline are pollinated primarily by bumblebees; flies pollinate most of the plants below timberline. Bumblebees are larger than flies and are attracted to sweet smells, whereas flies tend to prefer smaller, skunky-smelling flowers.
- Galen hypothesized that the different pollinator preferences led to habitat-specific natural selection, favoring individuals with the traits preferred by the primary pollinator. Over time, different selection in the two habitats acted on variation in flower size to produce the observed population differences.
- Galen experimentally tested this hypothesis by establishing two treatment groups: a control group to determine what happens when bumblebees are not agents of natural selection, and an experimental group to determine what happens when bumblebees do cause natural selection (**Figure 23.14**).
- For the control treatment, Galen hand-pollinated randomly selected timberline skypilots with pollen from randomly selected individuals. Skypilots in the experimental treatment were pollinated by bumblebees. Galen collected and germinated seeds from both groups in a greenhouse and planted the resulting seedlings in randomly assigned locations in the timberline habitat.

- After the seedlings matured, Galen compared flower sizes in the two groups. The offspring of bumblebee-pollinated plants had significantly larger flowers than did the offspring of hand-pollinated plants.
- Previous experiments had determined that flower size is heritable, so Galen concluded that the experimental group offspring differed genetically from the control group offspring and that natural selection by bumblebees favors increased flower size.

23.4 The Nature of Natural Selection and Adaptation

Selection Acts on Individuals, but Evolutionary Change Occurs in Populations

- Individuals do not change during natural selection. Those that are selected simply produce more surviving offspring than other individuals do, causing a change in the genetic makeup of the population.
- Darwinian evolution is quite different in this respect from Lamarck's proposed evolution by inheritance of acquired characters (**Box 23.2**). Individuals may change over their lifetime in response to environmental change, but these changes are not passed on to offspring.
- **Acclimation** occurs when an individual changes in response to changes in their environ-ment, but **adaptation** occurs only when a population changes in response to natural selection.

Evolution Is Not Progressive

- Groups that appear later in the fossil record are often more morphologically complex than related groups that appeared earlier, but this tendency is by no means universal. Traits can be lost or gained due to evolution by natural selection.
- For example, populations that become parasitic tend to evolve simpler morphologies (e.g., tapeworms lost their gut). These and other data show that evolution is not progressive. This contrasts with Lamarck's idea that evolution caused progress toward higher levels on a "ladder of life" (**Figure 23.15**).

- According to modern evolutionary theory there is no such thing as "higher" or "lower" organisms. There are only more ancient and less ancient groups with different adaptations that allow the groups to thrive in different environments.

Not All Traits Are Adaptive

- Adaptation is not a perfect process. For example, vestigial traits do not increase fitness; they exist simply because they were present in the ancestral population.
- Other nonadaptive traits include holdovers from structures present early in development and neutral mutations where a change in the DNA sequence does not change the amino acid sequence of the protein encoded by that gene (due to the redundancy of the genetic code).
- Even structures that currently function in organisms may be subject to genetic or historical constraints.

Genetic Constraints

- Many alleles that affect body size also affect other aspects of size. In the finches studied by the Grants, for example, beak depth was correlated with beak width. Thus deep, narrow beaks were not able to evolve even though birds with narrow beaks were better able to twist open the tough *Tribulus* fruits.
- When selection on alleles for one trait causes a correlated but suboptimal change in another trait, the number of possible evolutionary outcomes are limited. This type of constraint is called **genetic correlation**.
- Lack of genetic variation can also constrain evolution because natural selection can work only on extant variation in a population.

Historical Constraints

- Because all traits evolve from previously existing traits, adaptations are constrained by history.
- For example, the bones in our middle ear derive from bones that used to be part of the jaw and braincase of mammal ancestors. Natural selection had to work on the traits available—unlike an engineer, who is free to imagine the best possible way to solve each structural challenge.

B. CROSS-CUTTING THEMES

Looking Back—
Concepts from Earlier Chapters
Evolution and Natural Selection—Chapters 1 and 4
The process of evolution by natural selection was introduced in **Chapter 1** along with some examples of how researchers study evolution. **Chapter 4** discussed the related process of chemical evolution and gave an example of experimental selection. Here in Chapter 23, natural selection and research on its action in modern organisms are described in greater detail.

Inheritance of Traits—Chapters 13 through 16
Chapter 23 emphasizes that only heritable traits can be acted on by natural selection to cause evolution in a population. The genetic basis of inheritance was explained in earlier chapters.

Mutation—Chapter 14
Chapter 23 describes how a single-point mutation led to antibiotic resistance in *Mycobacterium tuberculosis*. For additional information on the molecular basis of mutation, look back to **Chapter 14**.

Transcription—Chapter 16
Chapter 23 introduces you to an antibiotic, rifampin, that can disrupt RNA polymerase's ability to transcribe DNA. See **Chapter 16** for an explanation of DNA transcription in bacteria.

Looking Forward—
Concepts from Later Chapters
Natural Selection and Mutation—Chapter 24
More information on natural selection and mutation can be found in **Chapter 24**, which explores all the possible causes of evolution.

Bacterial Diseases—Chapter 27
Mycobacterium tuberculosis is only one of many disease-causing bacteria. Chapter 27 presents more information on bacterial diseases.

Pollination—Chapters 29 and 40

The second example of evolution in action from this chapter involved selection exerted by bumblebee pollinators on alpine flowers. You can find out more about pollination and plant reproduction in **Chapter 40**. See **Chapter 29** for a review of pollination in the plant life cycle.

C. DIFFICULT TOPICS

Natural selection sounds relatively simple, yet it occurs in so many different settings with such drastically different results. Don't lose sight of the basics when learning the details of individual instances. Instead, try to relate the details to the four steps leading to natural selection: variation, heritability, differential reproductive success, and reproductive success based on those variable heritable traits. Test your understanding by imagining how natural selection might have acted to produce one of your favorite species. What variable traits does that species have? Are those traits heritable? Would those traits affect survival and reproduction? If so, that species may have changed through time as individuals with the successful heritable traits had more offspring than those with less successful traits.

Remember that natural selection acts on the individual: an individual is either selected for or selected against. *On average*, some traits lead to greater success than do other traits (random exceptions will occur; they just won't matter in the long run if the population is large enough—more on this in the next chapter). This process makes the *population* evolve. Individuals do not change their heritable traits, but the frequency of those traits changes in a population over time in response to natural selection on individuals.

D. ASSESSING WHAT YOU'VE LEARNED

(1) Testing Your Knowledge

1. Which of the following is a correct restatement of Darwin's phrase "descent with modification"?
    a. Species change over time, and those that exist today descended from other, preexisting species.
    b. Humans are different from a preexisting ancestor species that lived in trees.
    c. Ancestral species are more primitive and lower down on the phylogenetic tree than descendant species alive today.
    d. Natural selection changes species over time.

2. Which of the following are supported by evidence from the fossil record (choose all that apply).
    a. the concept that species can go extinct
    b. the hypothesis that individuals evolve by acquiring new traits
    c. the idea that life first evolved about 3.85 billion years ago
    d. the existence of transitional forms
    e. the idea that Earth's topography and environment have changed over time

3. The theory of evolution by natural selection is attributed to:
    a. Charles Darwin alone
    b. Charles Darwin and Gregor Mendel
    c. Charles Darwin and Alfred Russel Wallace
    d. Charles Darwin and Jean-Baptiste de Lamarck
    e. None of the above

4. Which of the following is a statement of the law of succession?
    a. Species in the fossil record were succeeded, in the same region, by similar species.
    b. Species in the fossil record were always succeeded, in the same region, by species more closely adapted to the current environment.
    c. Species in the fossil record were succeeded, in the same region, by larger and more complex species.
    d. Species in the fossil record were succeeded, in the same region, by species that were more successful.

5. Why did Darwin document and describe vestigial traits?
    a. Vestigial traits result from divergent evolution. This is an important pattern component of the theory of evolution,

but it is inconsistent with the theory of special creation.

b. Vestigial traits result from convergent evolution. This is an important pattern component of the theory of evolution, but it is inconsistent with the theory of special creation.

c. Vestigial traits indicate that species do not change over time. This contrasts with the theory of evolution, but it is consistent with the theory of special creation.

d. Vestigial traits indicate that species change over time. This is an important pattern component of the theory of evolution, but it is inconsistent with the theory of special creation.

6. An example of a vestigial trait in humans is:
   a. kidneys
   b. cataracts
   c. goose bumps
   d. There are no vestigial traits in humans.

7. In **Figure 23.5a** in your text, which two species are most closely related? Which two species are probably the most genetically, morphologically, and/or behaviorally distinct from each other?
   a. Most related = *N. melanotis* and *N. parvulus*; most distinct = *N. macdonaldi* and *N. trifasciatus*
   b. Most related = *N. trifasciatus* and *N. parvulus*; most distinct = *N. macdonaldi* and *N.melanotis*
   c. Most related = *N. melanotis* and *N. macdonaldi*; most distinct = *N. parvulus* and *N. trifasciatus*
   d. Most related = *N. trifasciatus* and *N. parvulus*; most distinct = *N. melanotis* and *N. macdonaldi*

8. Which of the following is considered Darwin's greatest contribution to science?
   a. Darwin recognized the pattern of evolution (descent with modification).
   b. Darwin recognized the process of evolution (natural selection).
   c. Darwin recognized "The original species."

d. Darwin recognized the heritable basis of traits (genetics).

9. Which of the following is the best definition of Darwinian fitness?
   a. Darwinian fitness is the ability of an individual to survive.
   b. Darwinian fitness is the ability of a population of organisms to persist.
   c. Darwinian fitness is the ability of an individual to stay healthy by eating well-balanced meals and exercising.
   d. Darwinian fitness is the ability of an individual to survive and reproduce.

10. Which of the following is a modern definition of evolution?
    a. Evolution is a change in an individual's ability to survive and reproduce as it grows older.
    b. Evolution is a change in allele frequencies in a population over time.
    c. Evolution is an increase in the fitness of a population.
    d. Evolution is a change in an individual's fitness over time.

11. Bird feathers in warblers and ostriches are:
    a. homologous traits
    b. analogous traits
    c. convergent traits
    d. transitional traits

12. What are the four requirements for evolution by natural selection in a given population?
    a. (i) population varies; (ii) all variation is heritable; (iii) some individuals produce more offspring than others; (iv) populations differ in production of offspring.
    b. (i) individuals in a population vary; (ii) the variation is neutral; (iii) offspring are produced that can survive; (iv) individuals produce offspring.
    c. (i) individuals in a population vary; (ii) variation is heritable; (iii) more offspring are produced than can survive; (iv) all individuals have different alleles.
    d. (i) individuals in a population vary; (ii) variation is heritable; (iii) some individuals produce more offspring than others;

(iv) certain traits lead to greater reproductive success than other traits.

13. Darwin's theory of evolution by natural selection is testable. What does it mean for this theory to be "testable"?
   a. If the theory is testable and correct, then a biologist should be able to observe evolution by natural selection in a real-life population.
   b. If the theory is testable and incorrect, then biologists should not be able to observe natural selection.
   c. If the theory is testable, then biologists should be able to test each postulate of the theory separately.
   d. If the theory is testable, then all of the above are true.

14. Which statement restates the fact that rifampin resistance in *Mycobacterium tuberculosis* is an adaptation?
   a. Rifampin resistance is a heritable trait that increases fitness of individuals compared to susceptible individuals.
   b. Rifampin resistance is caused by a point mutation in the *rpoB* gene.
   c. A mutation occurred in the *rpoB* gene of one cell, and this mutant cell continued to grow and divide.
   d. Drug-resistant strains now account for about 10% of the *Mycobacterium tuberculosis*-caused infections throughout the world.

15. Has evolution by natural selection recently occurred in populations of *Mycobacterium tuberculosis*?
   a. Yes, some people are now resistant to tuberculosis, but before they weren't. The resistance alleles have changed frequency in some human populations.
   b. Yes, the frequency of the adaptive allele for drug resistance has increased over time.
   c. No, evolution has not occurred, because the resistance trait occurs with a frequency of only about 10%.
   d. No, *Mycobacterium tuberculosis* is still the same species.

16. Resistance to a wide variety of insecticides, fungicides, antibiotics, antiviral drugs, and herbicides has recently evolved in hundreds of insects, fungi, bacteria, viruses, and plants. Why?
   a. Mutations are on the rise.
   b. Humans are altering the environments of these organisms, and the organisms are evolving by natural selection.
   c. No new species are evolving, just resistant strains or varieties. This is not evolution by natural selection.
   d. Humans have better health practices, so these organisms are trying to keep up.

17. Why did the Grants observe larger beak depths (in the medium ground finch), on average, following a drought?
   a. Deeper beaks help finches find water.
   b. Larger birds always survive better than smaller birds.
   c. During the drought, natural selection led to increased beak depth in the finch population.
   d. The Grants did not observe deeper beaks after the 1977 drought.

18. Consider the alpine skypilots case study. What is the explanation for large-flowered plants being above timberline while small-flowered plants are below timberline?
   a. Plants above timberline evolve large flower size as they grow older in response to pollination by bumblebees.
   b. Plant populations above timberline have evolved large flower size in response to natural selection by bumblebees.
   c. Plant populations below timberline have evolved small flower size in response to natural selection by herbivores.
   d. A mutation in a temperature regulation gene activated a cold-tolerance response.

19. Which of the following statements is true according to evolutionary theory?
   a. Acclimation can lead to evolution in a population.
   b. Flowering plants are more advanced or "higher" organisms than algae are.

c. Natural selection always creates more complex species from simpler ancestor species.

d. Individuals do not change during evolution by natural selection.

20. Which of the following are examples of evolutionary constraints?
    a. genetic correlation between beak depth and beak width in medium ground finches
    b. lack of genetic variation in some populations
    c. the bones of the mammalian inner ear
    d. all of the above

(2) Integrating Your Knowledge

(a) List the traits that might help some individuals survive or reproduce better than other individuals for each of these organisms: rabbit, turtle, oak tree, bacteria.

(b) Can you think of an analogous structure or shape in two species that are not closely related but live in similar environments (see example in **Figure 23.9**)?

(c) How does the presence of vestigial traits in many species support the theory of evolution?

(d) Why do traits need to be heritable in order for natural selection to occur?

(e) Natural selection acts on the _____, but it is the _____ that evolves.

(f) How did the *M. tuberculosis* researchers establish heritability?

(g) A pattern of repeated speciation events is often found in island chains. What is it about islands that promotes speciation? What would she do to show that evolution of flower scent had also occurred?

CHAPTER 23—ANSWER KEY

D. Assessing What You've Learned

(1) Testing Your Knowledge

1. a; 2. a, c, d, e; 3. c; 4. a; 5. d; 6. c; 7. b; 8. b; 9. d; 10. b; 11. a; 12. d; 13. d; 14. a; 15. b; 16. b; 17. c; 18. b; 19. d; 20. d

(2) Integrating Your Knowledge

(a) *Rabbit*—ability to run fast; *turtle*—shell thick enough to protect against predators, but not too heavy to carry around; *oak tree*—leaves that taste bad to herbivores; *bacteria*—ability to grow and divide quickly

(b) Wings of bats, birds, and pterosaurs; succulent morphology of cacti, euphorbs, and milkweeds found in deserts

(c) Vestigial traits are evidence of common ancestry and change in a population over time.

(d) Traits that are not heritable may increase an individual's ability to survive or reproduce. But because the trait is not passed on to the offspring, it does not change the population frequency for that trait.

(e) individual; population

(f) Researchers identified the DNA mutation that caused the antibiotic-resistant trait. Any trait in a single-celled organism that is coded for by a specific DNA sequence is by definition heritable because DNA carries the genetic information. In multicellular organisms, mutations that occur only in nonreproductive cells (e.g., skin cancer) are not heritable.

(g) Islands promote speciation because the organisms on each island experience different environments and therefore different types of natural selection. In the next chapter you will learn some other reasons why islands can be sites of rapid speciation (genetic drift).

24

Evolutionary Processes

A. KEY BIOLOGICAL CONCEPTS

- Natural selection is the only process that consistently leads to adaptation (increased fitness). However, there are three other processes that can change allele frequencies and cause evolution in a **population**—a group of individuals from the same species that live and breed together.
- The four mechanisms that can cause evolution are natural selection (increases frequency of alleles that improve reproductive success), genetic drift (randomly changes allele frequencies), gene flow (due to migration in and out of a population), and mutation (introduction of new alleles).
- Effects of these evolutionary processes can be compared to a null model, which shows what happens to allele frequencies when none of the evolutionary mechanisms are affecting a population.

24.1 Analyzing Change in Allele Frequencies: The Hardy-Weinberg Principle

- Evolutionary processes are often first analyzed using mathematical models. Model predictions are then tested with experimental or observational data. The model may then be further modified if necessary, and the results used to solve problems in conservation biology or human genetics.
- G. H. Hardy and Wilhelm Weinberg independently developed a mathematical model to calculate what happens to allele frequencies when no evolution is taking place. Biologists compare data to this null model in order to determine whether evolution is taking place in a population.
- Hardy and Weinberg started by considering all the gametes produced in one generation by individuals in a population and placing them all in a **gene pool**. They then calculated what would be observed if the next generation were made by randomly picking gametes from this gene pool.
- The researchers first examined what happens for a particular gene found in a population. Let A_1 and A_2 represent the only two alleles for this gene. The frequency of A_1 is designated p, and the frequency of A_2 is symbolized by q. There are only two alleles, so the two frequencies must add up to 1: $p + q = 1$.
- Three genotypes are possible: A_1A_1, A_1A_2, and A_2A_2. When gametes are randomly picked from the gene pool, the resulting frequency of the A_1A_1 genotype in the new generation is p^2 (the probability of drawing two A_1 alleles $= p \propto p$), and the frequency of A_2A_2 will be q^2. The frequency of A_1A_2 will be $2pq$ (**Figure 24.1**).
- Because all individuals in the new generation must have one of the three genotypes, the sum of the three genotype frequencies must equal 1 (100% of the population): $p^2 + 2pq + q^2 = 1$. When allele frequencies are calculated

for this new generation, the frequency of A_1 is still p and the frequency of A_2 is still q.

- No evolution occurred; this result is called the **Hardy-Weinberg principle**: (1) If allele frequencies are p and q, then genotype frequencies will be p^2, $2pq$, and q^2 for generation after generation; (2) allele distribution under Mendelian inheritance does not change allele frequency.

The Hardy-Weinberg Model Makes Important Assumptions

- The Hardy-Weinberg model assumes that mating is random with respect to the gene being examined and that none of the four evolutionary mechanisms are acting.
- The Hardy-Weinberg principle holds when there is (1) no selection—all parents contribute equally to the next generation; (2) no genetic drift—alleles are drawn from the gene pool in exactly the frequency with which they are found in the population; (3) no gene flow; (4) no mutation; and (5) random mating.

How Does the Hardy-Weinberg Principle Serve as a Null Hypothesis?

- A null hypothesis specifies what we should observe when the hypothesis being tested is wrong. When biologists want to test whether natural selection, nonrandom mating, or some other evolutionary mechanism is acting on a particular gene, the Hardy-Weinberg principle serves as a null hypothesis.
- Whenever biologists observe genotype frequencies that do not conform to Hardy-Weinberg proportions, it indicates that evolution or nonrandom mating is occurring in that population.

MN Blood Types in Humans

- Most human populations have two alleles for the MN blood group. This gene codes for a protein on the surface of red blood cells, and human genotype (*MM*, *MN*, or *NN*) can be determined from blood samples.
- Using a large sample, geneticists can estimate genotype frequency for a population by dividing the number of individuals observed for

each genotype by the total number in the sample. Allele frequencies can then be calculated from the genotype frequencies (**Table 24.1**).

- Allele frequencies can then be used to calculate the expected genotype distribution according to the Hardy-Weinberg principle; if mating is random and no evolution is occurring, genotype frequencies (*MM* : *MN* : *NN*) should equal $p^2 : 2pq : q^2$ (p = frequency of M allele; q = frequency of N allele).
- Statistical tests can determine whether any differences between observed and expected values are so small as to be due to chance (for example, which individuals happened to be included in the blood sample group) or large enough to reject the Hardy-Weinberg null hypothesis for a particular allele in a given population.
- **Table 24.1** shows that in all cases, observed and expected genotype frequencies are almost identical for MN blood groups. Because genotypes occur in Hardy-Weinberg proportions, evolutionary processes do not currently affect MN blood group, and mating must be random with respect to this trait.

HLA Genes in Humans

- The *HLA-A* and *HLA-B* genes code for proteins that help immune system cells recognize and destroy invading bacteria and viruses. Alleles for these genes are codominant, and they code for proteins that recognize slightly different disease-causing organisms.
- Researchers hypothesized that heterozygotes might be more fit than homozygotes because their immune systems would recognize more disease-causing organisms. To test this hypothesis, they determined the genotypes of 124 Havasupai tribe members and used these data to estimate population allele frequencies.
- Researchers calculated the expected genotype frequencies using the Hardy-Weinberg principle and compared the observed and expected frequencies. The Havasupai population has more heterozygotes and fewer homozygotes than expected (**Table 24.2**), so at least one of the Hardy-Weinberg assumptions must be violated for these alleles in this population.

- Mutation, migration, and genetic drift are neg-ligible in this case, so either mating is not random with respect to the *HLA* genotype and/or heterozygous individuals have higher fitness.
- College students can distinguish genotypes at *HLA*-like loci by body odor and are more attracted to the smell of genotypes different from their own. In other research, Hutterite women with the same *HLA*-related alleles as those of their husbands were found to have difficulty getting pregnant and had higher rates of spontaneous abortion.
- Research continues, but it appears that non-random mating, as well as low fitness of homozygous fetuses, may cause the higher than expected frequency of heterozygotes observed in the Havasupai tribe.

24.2 Natural Selection and Sexual Selection

- Natural selection occurs when individuals with certain heritable phenotypes survive and reproduce better than others do. The alleles responsible for this increased reproduction then increase in frequency in succeeding generations, causing evolution.
- The pattern of natural selection documented in the Havasupai tribe is known as **heterozygote advantage**. When heterozygous individuals have higher fitness than homozygous individuals have, natural selection maintains **genetic variation**—the number and relative frequency of alleles in a population.
- Low genetic variation can be dangerous for a population because if the environment changes, alternative alleles that confer high fitness under the new conditions may not be present. If the environmental change is severe enough, the decrease in average population fitness can even cause extinction.
- Different patterns of natural selection exist, each with its own causes and consequences.

Directional Selection

- Directional selection occurs when natural selection increases the frequency of one allele, as in the drug-resistance example (**Chapter 23**). This type of selection reduces population genetic diversity over time.
- If directional selection continues for long enough, the favored alleles eventually reach a frequency of 1.0 (100%) and are said to be *fixed,* whereas those alleles that are no longer found in the population are *lost.*
- Often several genes influence a trait, causing a bell-shaped distribution of population phenotypes. Directional selection for these traits acts on many genes at once and causes the bell curve to shift in one direction (**Figure 24.2**). This type of selection was observed in cliff swallows during a cold snap.
- Directional selection is not necessarily constant throughout a species' range or over time. Sometimes opposing patterns of directional selection on related traits maintain genetic variation in a population (e.g., in cliff swallows, selection for large body size vs. selection for maneuverability).

Stabilizing Selection

- **Stabilizing selection** occurs when individuals with intermediate traits reproduce more than others do; this type of selection maintains intermediate phenotypes in a population (**Figure 24.3**). Stabilizing selection decreases a population's genetic variation over time, but does not change its average trait value.
- Human birth weight is under stabilizing selection because babies of average size survive better than do babies with low or high weights.

Disruptive Selection

- **Disruptive selection** is the opposite of stabilizing selection; it occurs when intermediate phenotypes are selected against and extreme phenotypes are favored. Disruptive selection maintains genetic variation but does not change the mean value of a trait (**Figure 24.4**).
- For example, African seedcrackers with very short or very long beaks survive better than do those with intermediate phenotypes due to the types of food available.

- Disruptive selection can cause speciation if individuals with one extreme of a trait start mating preferentially with individuals that have the same trait.

Sexual Selection

- Mate choice often plays an important role in speciation. **Sexual selection** is selection for enhanced ability to attract mates.
- The Bateman-Trivers theory states that sexual selection typically acts more strongly on males than on females, so that the mate-attracting traits tend to be more elaborate in males. This pattern is explained by the **fundamental asymmetry of sex**—females usually invest more in their offspring than do males.
- Eggs are large and energetically expensive. Females typically produce relatively few offspring during their lifetime, and their fitness is primarily limited by the ability to gain resources necessary to produce and rear young (not by the ability to find a mate). Females should therefore be choosy about the males with which they mate.
- Sperm, on the other hand, are cheap to produce, so males can father many offspring. Male fitness is primarily limited by the males' ability to acquire mates (not by the ability to acquire resources for offspring production). Therefore, males should compete with each other for mates, and sexually selected traits (used in courtship or competition for mates) should be found primarily in males.

Sexual Selection via Female Choice

- Female birds invest a great deal of time and energy in offspring (**Figure 24.5**), and they should be choosy about the males they mate with. In several bird species, females prefer healthy, well-fed males.
- Red and yellow pigments called carotenoids cause bright feather and beak coloration in many bird species. In animals, carotenoids also stimulate the immune system and prevent tissue damage from free radicals. However, animals cannot make carotenoids and must obtain them in their diet.

- Birds with the brightest beaks and feathers must be well fed and healthy, because carotenoids are used in beaks and feathers only when they are not needed by the immune system.
- Research on European blackbirds showed that (1) an immune response to the injection of sheep cells caused a decrease in the brightness of male beaks; and (2) the females preferred bright-beaked males that were fed a diet high in carotenoids to their duller-beaked brothers, which were fed a diet with fewer carotenoids (**Figure 24.7**).
- Other experiments on other bird species also support the conclusion that bright male coloration and other courtship displays indicate the healthiness of the male. Females that choose to mate with such a male usually have offspring with good alleles that help them fight disease and feed efficiently.
- Males that provide resources for egg production or care of young are also preferred mates in many animal species.

Sexual Selection via Male-Male Competition

- Male elephant seals fight over beach patches where females come to bear their pups. Males that win battles obtain larger territories. A **territory** is an actively defended area where the owner has exclusive use—in this case, exclusive access to females inhabiting that part of the beach.
- Males elephant seals are more than four times larger than females. Large male size has evolved because fights for beach territories are almost always won by the larger males. The alleles of large territory-owning males have a fitness advantage because of sexual selection.
- A few male elephant seals father the vast majority of offspring produced each year, so sexual selection is intense in this species and is driven by male-male competition rather than by female choice.

What Are the Consequences of Sexual Selection?

- The Bateman-Trivers theory states that sexual selection is strongest in the sex that invests least in offspring.

- For example, female elephant seals invest much more than males do in their offspring, and the females' reproductive success is limited by energy stores, not by their ability to obtain copulations. In male elephant seals, the number of copulations (which varies greatly) is the main determinant of reproductive success (**Figure 24.9**).
- Because reproductive success usually varies more in males than in females, sexual selection is typically more intense in males. Sexual selection acts differently on males and females, and it can cause the evolution of **sexual dimorphism**: male-female differences for sexually selected traits (**Figure 24.10**).
- Sexual selection is a form of natural selection for the ability to attract mates.

24.3 Genetic Drift

- **Genetic drift** is change in allele frequencies due to chance (**sampling error**). It causes allele frequencies to drift up and down randomly over time.
- Offspring production can be simulated by flipping coins to determine which parent alleles happens to be in the gametes that contribute genetic material to the next generation (**Table 24.3**).
- Just by chance, in the coin flipping you could get more heads than tails, causing one allele to increase in the frequency and another allele to decrease in frequency—evolution by random chance.
- *Genetic drift is random with respect to fitness*. Allele frequency changes are due to chance and are not adaptive.
- *Genetic drift is most pronounced in a small population*. It is easier to deviate significantly from a 50% heads ratio when you flip a coin 10 times than when you flip a coin 1000 times. In the first case, a single coin-flip has a large impact on the heads "allele" frequency. Chance events have a larger impact on allele frequencies in small populations than in large populations.
- *Over time, genetic drift can lead to the random loss or fixation of alleles*. "Lucky" alleles may become fixed, whereas "unlucky"

alleles may be lost, causing a decline in the population's genetic variation.

Experimental Studies of Genetic Drift

- Warwick Kerr and Sewall Wright studied genetic drift in fruit flies using leg-bristle morphology (normal straight bristles versus "forked," bent-looking bristles) as a **genetic marker**—an easily identifiable trait with known alleles.
- Four females and four male flies were placed in each of 96 cages such that the initial frequencies of normal and forked-bristle alleles were 0.5 in each cage. In each succeeding generation, four male and four female offspring were randomly chosen to continue each population.
- After 16 generations, 29 of the 96 populations had only forked leg bristles, and 41 had only normal bristles. In these populations one allele had become fixed.
- Leg bristles do not affect fitness in the lab, migration was not allowed, and mutation rate for the bristle gene is very low; therefore, the observed changes in population allele frequencies must have been due to genetic drift.

Genetic Drift in Natural Populations

- A sampling process occurs during fertilization in every population in each generation, but sampling error typically significantly affects population allele frequency only in small populations. Sampling always occurs during fertilization, but other sampling events can also affect populations.
- Genetic drift is currently a topic of great concern to conservation biologists, because population sizes are reduced due to habitat destruction and other human activities. Further, small populations found on nature reserves or in zoos are especially susceptible to genetic drift.
- Genetic drift can have an effect even in moderately large populations over long periods of time. For example, silent mutations and pseudogenes that do not change gene products have been observed to drift to high frequency or even fixation.

How Do Founder Effects Cause Drift?

- A **founder event** occurs when a group leaves a population, emigrates to a new area, and starts to establish a new population. If the founding group is small, its allele frequencies probably differ from those of the source population. This sampling effect on the new population's allele frequencies is called a **founder effect**.
- Founder events and founder effects are especially common in the colonization of isolated habitats like islands, mountains, caves, and ponds.

How Do Population Bottlenecks Cause Drift?

- A sudden decrease in population size, called a **population bottleneck**, can lead to a **genetic bottleneck**—that is, a sudden reduction in the number of alleles in a population.
- Genetic bottlenecks are commonly caused by disease outbreaks and natural catastrophes.
- Genetic drift often occurs in the resulting small population (for example, the serious vision deficit, *achromatopsia*, that occurred on Pingelap Atoll).

24.4   Gene Flow

- **Gene flow** is the movement of alleles from one population to another. It occurs any time individuals leave one population, join with another, and reproduce.
- Gene flow reduces genetic differences between the source and recipient populations.
- Following the volcanic explosion of Mount St. Helens in the state of Washington, prairie lupine seeds recolonized the ash plain surrounding the mountain. Researchers determined allele frequencies in several new prairie lupine populations, and they observed changes in allele frequencies over time.
- Initially, new populations showed strong founder effects (often only one seed founded a new population), but population allele frequency differences decreased over time. Researchers proposed that populations became more similar as bees caused gene flow by carrying pollen between populations of lupines (**Figure 24.11**).

- Gene flow can increase the average fitness of individuals in a population where genetic diversity was lost due to a nonadaptive process like genetic drift; but gene flow can decrease the average fitness of individuals in a population where natural selection had produced adaptation to a specific habitat.

24.5   Mutation

- Whereas most evolutionary mechanisms reduce genetic diversity, **mutation** restores genetic diversity and creates new alleles. A mutation occurs when copy errors by DNA polymerase cause random changes in DNA sequence that may lead to an altered amino acid sequence.
- Because errors are inevitable, mutation is always adding new alleles into populations at all gene loci.
- Mutation is a random process; because most organisms are well adapted to their current habitats, random changes usually result in **deleterious alleles**—that is, alleles which lower fitness because their gene products do not work as well as do the other naturally selected alleles.
- However, mutation does occasionally produce an advantageous allele. Beneficial alleles created by mutation can then increase in frequency in a population due to the effects of natural selection.

Mutation as an Evolutionary Mechanism

- For a given allele (in humans), there is only about one mutation per 10,000 gametes—that is, mutation rates are too low to significantly affect allele frequencies in and of themselves. Another evolutionary mechanism must act on alleles created by mutation in order for significant evolutionary change to occur.

The Role of Mutation in Evolutionary Change

- Richard Lenski and colleagues used single *Escherichia coli* cells to found many populations of *Ara*+ and *Ara*− bacteria (The *Ara* gene allowed researchers to easily determine relative growth in competition experiments).

E. coli is asexual, so mutation was the only source of genetic variation.

- Cells were grown under identical conditions. Each day some cells from each population were transferred to new growth media. At regular intervals, researchers froze sample cells from each population.

- Natural selection acted on these populations, but no gene flow was allowed, and populations were large enough that genetic drift was unlikely to be important.

- At the end of the experiment, frozen "fossil" cells from previous generations were thawed and grown with cells from more recent generations. Those newer generations produced more cells during these competition experiments, thus demonstrating their adaptation over time to the experimental environment (see **Figure 21.12**).

- Relative fitness increased by irregular intervals over the experiment, for an almost 30% total increase in fitness. Each fitness increase must have been due to a new beneficial mutation that then rapidly increased in frequency through succeeding generations due to natural selection.

- Mutation provides the genetic variation upon which natural selection can act to produce evolution. **Table 24.4** summarizes evolutionary mechanisms and describes their effects on fitness and genetic variation.

24.6   Inbreeding

- Mating in natural populations often is not random. In small, isolated populations, for example, matings between relatives—known as **inbreeding**—may be common. Because relatives are more likely to share alleles, inbreeding violates the Hardy-Weinberg assumptions.

- Inbreeding decreases the frequency of heterozygotes and increases the frequency of homozygotes in each generation (**Figure 24.13**). Although inbreeding affects genotype frequency, it does not change allele frequencies; thus it does not cause evolution.

- **Inbreeding depression** is the decreased fitness that occurs as more deleterious recessive traits are expressed in the phenotypes of ho-

mozygous recessive individuals (see **Figure 24.14**). Homozygotes also have lower fitness for traits that show a heterozygote advantage.

- Inbreeding depression can increase the rate at which natural selection removes disadvantageous recessive alleles by exposing these alleles in the homozygous phenotype.

- Inbreeding and sexual selection both violate the assumption of random mating. Inbreeding affects only genotype frequencies (not allele frequencies); in contrast, sexual selection (a type of natural selection) affects both allele and genotype frequencies, but only for genes involved in mate competition.

- Many species, including humans, have mechanisms to avoid inbreeding. As **Table 24.5** shows, inbreeding depression is also pronounced in humans.

B. CROSS-CUTTING THEMES

Looking Back—
Concepts from Earlier Chapters
Evolution—Chapters 1, 6, 23
This chapter builds on discussions of evolution in earlier chapters by presenting you with processes that produce evolution in a population.

Mutation—Chapters 4, 14, 15, 18
In **Chapter 4** you first learned how some researchers use mutation in selection experiments. In **Chapters 14**, **15**, and **18** you learned more about how mutations occur. In this chapter, the role of mutation as the ultimate source of genetic variation is highlighted. Mutation provides the variation upon which natural selection acts.

Genetics—Chapters 13, 15, 19
Chapters 13 and **15** provide background information critical to your understanding of how allele frequencies change in populations and how genes work to determine organism phenotype. The discussion in Chapter 24 of the Hardy-Weinberg principle shows how researchers study genetics at the population level. **Chapter 19** described how researchers can artificially alter individual (and population) allele frequencies.

Looking Forward—
Concepts from Later Chapters

Evolution—Chapters 25, 26, 51, and others

Because evolution is one of the unifying themes of biology, it reappears throughout your textbook. Larger-scale evolutionary patterns are examined in **Chapters 25** and **26**. **Chapters 27** through **34** discuss the evolution of specific groups of organisms, and **Chapter 51** examines the evolution of behavior.

Adaptation—
Chapters 27–34,
35, 41, and others

As with evolution, you can find mention of adaptations (traits that increase fitness) throughout this textbook. **Chapters 27** through **34** describe the major organismal groups—something that could hardly be accomplished without mentioning the salient characteristics, or adaptations, of each group. **Chapters 35** and **41** have more examples of plant and animal adaptations.

Populations—Chapter 52

Evolution occurs in populations. Although much of this textbook examines other biological levels (species, organism, cell), **Chapter 52** focuses on populations again, in a thorough discussion of their ecology.

C. DIFFICULT TOPICS

Understanding the Hardy-Weinberg equations can be challenging for those lacking an intuitive understanding of the math involved. It may help the student to look at a more concrete example like that given in question (b) of this chapter's "Integrating Your Knowledge" exercise. You could also use two candy colors or two different coins to go through the process of picking gamete alleles and calculating genotype frequencies, as shown in **Figure 24.1**.

Keep in mind that in a large population, if 70% of the alleles in the gene pool are A_1, then 70% of the gametes should be A_1 (under Hardy-Weinberg conditions). But if you try to replicate this on your own, picking blindly, you may get a different percentage in the gametes due to

random chance—in which case you have just shown the effects of genetic drift!

Now think about this process of picking alleles to go into the next generation when the other evolutionary processes are acting. How do the processes affect the allele frequencies? For example, if you are doing this exercise with orange and green candies, and you prefer the flavor of the green ones, you may find that you eat the green ones. The orange alleles survive better and are thus more likely to make it into the next generation. Presto—natural selection!

D. ASSESSING WHAT YOU'VE LEARNED

(1) Testing Your Knowledge

1. What is a population?
   a. A population is any group of many individuals of the same species.
   b. A population is a group of individuals that live in the same area and that regularly interbreed.
   c. A population is a group of interacting species that live in the same area.
   d. A population is two or more groups that regularly interbreed.

Use the following information to answer questions 2 through 4:

It is easy to determine the frequency of homozygous recessive genes (q) because individuals with the homozygous recessive genotype show the recessive phenotype (whereas heterozygotes and homozygous dominants may have the same phenotype). There is some evidence that straight hairlines are recessive and widow's peaks are dominant in humans.

If Hardy-Weinberg assumptions are met for this trait in a human population, and 81% of the population have straight hairlines (A_2A_2), then:

2. q^2 must equal _____.
   a. 0.19
   b. 0.81
   c. 0.90
   d. 1.0
   e. none of the above

3. *q* must equal _____ and *p* must equal

    _____.
    a. 0.9; 0.1
    b. 0.3; 0.7
    c. 0.9; 0.19
    d. 0.5; 0.5
    e. none of the above

4. In the human population, the frequency of homozygous dominants (A_1A_1) must be _____, and the frequency of heter-ozygotes (A_1A_2) must be _____.
    a. 0.81; 0.9
    b. 0.18; 0.01
    c. 0.01; 0.18
    d. 0.25; 0.25
    e. none of the above

5. The Hardy-Weinberg principle acts as a null model because it describes the relationship between allele and genotypic frequencies under conditions where:
    a. None of the four evolutionary forces is acting and mating is random.
    b. New species are differentiating.
    c. Evolution by natural selection increases the fitness of individuals.
    d. Individuals in a population are not mating randomly.

6. Which of the following is *not* one of the Hardy-Weinberg assumptions?
    a. no genetic drift
    b. no mutation
    c. no natural selection
    d. no random mating
    e. no gene flow

7. Which of the following scenarios best illustrates heterozygote advantage?
    a. There are more heterozygous individuals in a population than expected according to the Hardy-Weinberg principle.
    b. For a given locus, heterozygous individuals are more fit than are homozygous individuals.
    c. Individuals in a population are nonrandomly mating without in-breeding depression.

    d. Parents with similar alleles produce more offspring than do parents with dissimilar alleles.

8. Which of the following examples of natural selection in action would tend to increase the genetic variation in the population?
    a. Natural selection causes dragonflies to evolve longer tails.
    b. Alpine skypilots evolve large flowers above the timberline and small flowers below the timberline.
    c. Natural selection simultaneously selects against heavy field mice (they starve during winter) and light field mice (they freeze during winter).
    d. Males with bright red feet produce more offspring than do those with duller feet in one bird species.

9. In mammals, young born from pregnancies that are much longer or shorter than the species average tend to have higher rates of infant mortality. This is an example of _____ selection.
    a. dispersive
    b. directional
    c. sexual
    d. stabilizing

10. Why aren't there any female elephant seals with over 90 offspring?
    a. because males are larger than females
    b. because females must compete for male copulations
    c. because female elephant seals can have at most one pup per year, and they do not live 90 years
    d. because the beaches would get too crowded

11. Evolution by natural selection is a powerful force of change. What keeps organisms from becoming perfectly adapted to their environment?
    a. Environments change, so different alleles become more fit.
    b. Genetic drift causes random changes in allele frequencies.
    c. Mutations introduce new variation into a population.

d. All of the above processes impede, constrain, or prohibit evolution by natural selection.

12. Which statement below most fully characterizes the fundamental asymmetry of sex?
    a. Female fitness is limited most by the ability to get resources for producing eggs and rearing young, whereas male fitness is limited by the ability to attract females.
    b. Female fitness is limited most by the ability to attract males, whereas male fitness is limited by the ability to get resources for provisioning the female.
    c. Female fitness is limited most by the ability to get resources for producing eggs and rearing young, whereas male fitness is limited by the ability to get resources for provisioning the female.
    d. Female fitness is limited most by the ability to attract males, whereas male fitness is limited by the ability to attract females.

13. Genetic diversity is required for natural selection to act, but natural selection can reduce or eliminate diversity. What process can restore genetic diversity to a population?
    a. genetic drift
    b. mutation
    c. sexual selection
    d. stabilizing selection

14. Which of the following biological processes causes adaptation?
    a. mutation
    b. migration
    c. natural selection
    d. genetic drift

15. What is an important consequence of gene flow in natural populations?
    a. Gene flow increases the mutation rate among sedentary organisms.
    b. Gene flow moves individuals from one habitat to another on a seasonal basis.
    c. Gene flow tends to separate allele frequencies among populations.

d. Gene flow tends to reduce genetic differences among populations.

16. In what kind of population is random genetic drift most pronounced?
    a. Drift is greatest in large populations.
    b. Drift is greatest in small populations.
    c. Drift is greatest in migrating populations.
    d. Drift is greatest in fixed populations.

17. A genetic bottleneck is a good explanation for all but one of the following patterns observed in natural populations. Which pattern is *not* caused by a genetic bottleneck?
    a. *achromatopsia* on Pingelap Atoll
    b. fixed loci in endangered plant populations
    c. long, thin necks in giraffes
    d. extremely low genetic diversity in cheetahs

18. Which of the following is *not* associated with inbreeding?
    a. Allele frequencies change in a population.
    b. The frequency of homozygote genotypes increases in a population.
    c. Individuals in a population experience reduced fitness.
    d. Individual plants self-fertilize.

19. Hominid brain size has increased over the last 3 million years. This increase is probably due to:
    a. directional selection
    b. stabilizing selection
    c. dispersive selection
    d. sexual selection
    e. mutation

20. Under extreme inbreeding like that shown in **Figure 24.13**, if genotype frequencies in one generation were $0.20 A_1A_1$, $0.60 A_1A_2$, and $0.20 A_2A_2$, what would be the genotype frequencies in the next generation?
    a. $0.5 A_1A_1$, $0.0 A_1A_2$, and $0.5 A_2A_2$
    b. $0.20 A_1A_1$, $0.60 A_1A_2$, and $0.20 A_2A_2$
    c. $0.45 A_1A_1$, $0.10 A_1A_2$, and $0.45 A_2A_2$
    d. $0.35 A_1A_1$, $0.30 A_1A_2$, and $0.35 A_2A_2$
    e. none of the above

(2)   Integrating Your Knowledge

(a) What five conditions must occur in order for the Hardy-Weinberg principle to accurately determine population allele frequencies?

(b) Can you think of a situation in which the Hardy-Weinberg expected genotypes could be found even though the Hardy-Weinberg assumptions were not strictly met?

(c) Explain how natural selection can lead to allele fixation.

(d) Why does genetic drift affect small populations more than it does large populations?

(e) Why did Kerr and Wright choose only four female and four male fruit flies from each population to breed in each generation?

(f) Could you do an experiment like Lenski's (**Figure 24.12**) on mice? In your answer, explain why or why not.

(g) In Lenski's experiment on the role of mutation in evolutionary change, why did new beneficial mutations increase in frequency in each succeeding generation?

(h) Sometimes mate choice is nonrandom, such that genetically *different* individuals are more likely to mate. This is called outbreeding. What would happen to allele frequencies and genotype frequencies in a population experiencing an outbreeding mating pattern?

CHAPTER 24—ANSWER KEY

D. Assessing What You've Learned

(1)   Testing Your Knowledge

1. b; 2. b; 3. a; 4. c; 5. a; 6. d; 7. b; 8. b; 9. d; 10. c; 11. d; 12. a; 13. b; 14. c; 15. d; 16. b; 17. c; 18. a; 19. a; 20. d

(2)   Integrating Your Knowledge

(a) no mutation, no gene flow, no genetic drift (or a large population so that chance events do not affect overall population allele frequencies), random mating, and no selection

(b) Theoretically this situation could occur if exactly the same allele proportions were migrating into a population as were migrating out. It could also occur if natural selection against some allele were exactly balanced by mutations and/or gene flow bringing that allele back into the population.

(c) If some individuals have an advantageous allele that leads to better survival and reproduction, that allele will increase in frequency in the population in each succeeding generation until all individuals have that allele (the allele is fixed). This occurs unless something else happens, such as an environmental change, that prevents allele fixation.

(d) Genetic drift does not usually affect allele frequencies in large populations, because random changes in one direction are balanced by other random changes in the other direction.

(e) They chose only four males and females because they wanted to observe the effects of genetic drift, and genetic drift has a greater impact on allele frequencies in small populations.

(f) Lenski's experiment could not be performed on mice, because you cannot freeze mice and then use them to compete against later generations. Also, observing 2000 mice generations would take way too much time.

(g) The bacteria with the beneficial mutation presumably were able to grow and reproduce faster than could the other bacteria.

(h) Outbreeding would not change allele frequencies either, but it would increase the frequency of heterozygotes relative to homozygotes.

25

Speciation

The **biological species concept** evaluates difference by determining if two populations are able to successfully interbreed. The **morphospecies concept** relies on difference in appearance; the **phylogenetic species concept** uses phylogenetic trees to evaluate evolutionary independence.

One key to the evolution of different species is isolation—the restriction of gene flow. Isolation is critical because gene flow tends to reduce differences among populations; but for divergence to occur, differences among populations must increase. The most obvious way for populations to become isolated is when some physical barrier prevents contact; these populations are then said to be **allopatric**.

But genetic isolation can also occur in **sympatric** populations (that live in the same geographic area) if something prevents the populations from successfully interbreeding. Prezygotic mechanisms prevent individuals in different populations from being able to produce a zygote (a fertilized ovum), and postzygotic mechanisms prevent a zygote from maturing and/or producing offspring.

A. KEY BIOLOGICAL CONCEPTS

- Populations that experience reduced gene flow may diverge genetically due to mutation, genetic drift, and/or natural selection. This genetic divergence may eventually lead to **speciation**, the creation of new species. It is most common for two or more distinct species to derive from one ancestral group (**Figure 25.1**).

25.1 Defining and Identifying Species

- A **species** is a distinct type of organism, formally defined as an evolutionarily independent population or group of populations. How different do two populations have to be before being considered different species? How is evolutionary independence determined?
- Gene flow reduces genetic differences among populations, so evolutionary independence begins with lack of gene flow. Once gene flow stops, mutation, selection, and drift can act on populations independently, and genetic divergence can occur. Genetic divergence, in turn, may lead to speciation.
- The biological species, morphospecies, and phylogenetic species concepts are three different methods used by biologists to evaluate whether populations show enough genetic divergence (are evolutionarily independent enough) to be considered separate species.

The Biological Species Concept

- The **biological species concept** considers populations to be evolutionarily independent if they are reproductively isolated from each other (no gene flow).
- **Prezygotic isolation** occurs when individuals from different populations are unable to mate (e.g., breed at different times or have incompatible genitalia). **Postzygotic isolation** occurs when individuals from different populations can breed, but the offspring produced have low fitness (**Table 25.1**).

- Applying the biological species concept of reproductive isolation may, however, be difficult for populations that are geographically separated (difficult to know if the geographic separation will continue to prevent interbreeding in the future). This concept cannot be applied to asexual or fossil species.

The Morphospecies Concept

- Evolutionary independence can also be estimated by looking at population morphology, because different appearances often evolve when populations experience different natural selection, mutations, and genetic drift. These different forms will persist only if gene flow is restricted.
- This morphospecies concept can easily be applied to most populations and species, but it is rather subjective (**Figure 25.2**). How much do populations need to differ to be considered different species? Which traits should be evaluated for difference?

The Phylogenetic Species Concept

- Data on genes, shape, physiology, behavior, and other traits of organisms can be used separately or together to estimate evolutionary relationships among organisms and to create phylogenetic trees depicting the relationships.
- On phylogenetic trees, an ancestral population plus all its descendants (but no other groups) is called a **monophyletic group**. A phylogenetic species is the smallest monophyletic group on a tree created from data on related populations (**Figure 25.3**).
- The phylogenetic species concept has the advantages of (1) being applicable to all populations; and (2) being logical—because monophyletic groups, by definition, differ in some traits and are isolated from gene flow with other groups (they have their own independent evolutionary history).
- Unfortunately, good phylogenetic trees have been created for only a few related populations. Use of the phylogenetic species concept may lead to recognition of more species.
- In practice, biologists use all three species concepts (**Table 25.2**). Sometimes the species they recognize differ, depending on the species concept used.

Species Definitions in Action: The Case of the Dusky Seaside Sparrow

- Several seaside sparrow (*Ammodramus maritimus*) subspecies with slightly different appearances have been named using the morphospecies concept. Populations with their own identifying traits, but not distinct enough to be a separate species, are called **subspecies**.
- The seaside sparrow subspecies were believed to be genetically isolated because the populations are geographically isolated and young birds breed near their hatching ground (**Figure 25.4**).
- Development of salt marshes where these birds live eliminates their natural habitat, causing population decreases. In 1980 the dusky seaside sparrow subspecies (*A. m. nigrescens*) had only six males left. In an effort to preserve this population, biologists bred these males with females from a nearby population (*A. m. peninsulae*) in hopes of reintroducing the hybrid offspring.
- Other biologists compared gene sequences from different seaside sparrow populations and determined the phylogeny shown in **Figure 25.4b**. They found that seaside sparrows belong to two monophyletic groups: one on the Atlantic Coast and one on the Gulf Coast.
- Thus, *A. m. nigrescens* is not distinct from other Atlantic seaside sparrows and does not need to be individually preserved. However, the Atlantic Coast and Gulf Coast groups should be separately preserved. Because *A. m. peninsulae* belongs to the Gulf group, the hybrid offspring should not be reintroduced to the wild.

25.2 Isolation and Divergence in Allopatry

- Physical separation of populations occurs either when a group colonizes a new habitat (dispersal) or when a new physical barrier divides a population (**vicariance**). Reduction or elimination of gene flow due to physical isolation can lead to **allopatric speciation**

(**Figure 25.5**) because populations living in different areas are said to be in **allopatry**.

Dispersal and Colonization Isolate Populations

- Peter and Rosemary Grant observed as five large ground finches from a nearby island colonized Daphne Major (the island where they were studying medium ground finches). Because finches normally stay on the same island, the colonists were allopatric to their original or source population.
- Colonization events often cause speciation because the physical separation reduces gene flow, and colonists are likely to experience genetic drift (both during the founder event and afterward, if the colonizing group is small). If the new habitat differs from the original habitat, natural selection may also cause divergence of the new population.
- The Grants caught, weighed, and measured large ground finches on Daphne Major over the next 12 years and found the average bill size of the new Daphne Major population was larger than that of other populations. The colonists were a nonrandom sample of the original population.
- Genetic drift and natural selection during and following colonization events appear to have caused repeated speciation in many groups of island organisms.

Vicariance Isolates Populations

- Sometimes an existing population is separated by a new physical barrier. For example, uplift of a new mountain range or a change in a river course may separate populations. This separation is known as a vicariance event.
- For example, vicariance events occurred during the last ice age as glaciers fragmented forest and grassland habitats. Physical isolation of populations in these fragmented habitats is thought to have led to the evolution of many modern species.
- Vicariance also occurred when a huge supercontinent called Gondwana broke up due to **continental drift**—the movement of continental plates (**theory of plate tectonics**, see **Figure 25.6**).

- As Gondwana broke up into the continents we know today, large flightless ratite birds and other species were split into isolated populations. Smaller populations might have experienced genetic drift, and natural selection would have acted differently on populations as the continents moved and environments changed. Today, five different ratite species exist, each on a different continent or island.
- A series of vicariance events can lead to a series of speciation events because genetic isolation followed by genetic divergence due to mutation, selection, and genetic drift often causes speciation.

25.3   Isolation and Divergence in Sympatry

- Populations or species that live in the same geographic region (close enough to mate) live in **sympatry**. Researchers used to think that speciation could not occur among sympatric populations, because if populations are close enough to mate, gene flow would tend to eliminate any evolutionary differences.
- Unbanded water snakes survive better on Lake Erie islands than do banded snakes, presumably since predators more easily spot banded snakes basking on the island's limestone rocks. But the banded pattern is more camouflaged in the mud and shrubs of mainland habitats.
- Unbanded water snakes predominate on islands, and banded snakes are more common on the mainland (**Figure 25.7**). However, banded mainland snakes regularly swim out and join island populations, reintroducing the genes for banded coloration. Gene flow prevents these populations from diverging.

Can Natural Selection Cause Speciation Even When Gene Flow Is Possible?

- In some situations, natural selection *can* overcome limited gene flow to cause **sympatric speciation**. Often this happens when populations become isolated by habitat preference.
- For example, the soapberry bug (found in the southern United States) feeds on plants in the Sapindaceae family by piercing fruits with their beaks and eating the seeds.

- In the mid-twentieth century, three Asian sapindaceous species were brought to North America. These introduced species spread, and soapberry bugs began using them as hosts even though their fruits are smaller than those of the American species are.
- Soapberry bug beak length is correlated with host fruit size in native plants; bugs on species with big fruit have longer beaks than do those on species with small fruit. Presumably this beak/fruit size correlation allows bugs to feed more efficiently.
- Researchers led by Scott Carroll measured beak length of bugs found on native and nonnative hosts to determine whether beak lengths had evolved shorter lengths to match the smaller introduced fruit.
- Bugs on large-fruited native plants have longer beaks than those found on small fruited nonnative plants (**Figure 25.8**). It seems that disruptive selection has occurred on beak length since the appearance of the new Asian host plants.
- Because soapbugs mate near their host plants, populations on native plants may continue to diverge from the populations on nonnative plants to eventually form a new species. Gene flow between the two populations is decreased; natural selection causes divergence.
- Many insects and parasites are associated with specific hosts. When some individuals in a population switch hosts, they often experience different selection pressures and reduced gene flow with the original population. Thus, host switching is an important cause of speciation in sympatric populations.

How Can Polyploidy Lead to Speciation?

- Mutation *alone* does not usually cause evolutionary change, because it is an inefficient evolutionary mechanism. But one type of mutation does cause speciation in sympatric plant populations. This type of mutation causes **polyploidy**—the presence of more than two sets of chromosomes.
- Polyploid mutations can cause sympatric speciation by reducing or eliminating gene flow between mutant and normal (wild-type) plants.

- Gene flow is reduced because if a normal diploid ($2n$) parent with haploid (n) gametes mates with a mutant **tetraploid** ($4n$) parent with diploid gametes, the offspring are **triploid** ($3n$). Even if this triploid survives, it cannot reproduce, because the three chromosome copies will be unable to undergo meiosis correctly (**Figure 25.9**).
- Mutations that result in a doubling of chromosome number produce **autopolyploid** individuals, whereas **allopolyploid** individuals are produced by a mating between parents of two different species.

Autopolyploidy

- Biologists found several polyploid maidenhair ferns. In the normal fern life cycle, individuals alternate between haploid (n) and diploid ($2n$) life stages (**Figure 25.10**). But the polyploid fern instead alternated between diploid and tetraploid ($4n$) life stages.
- The parent of the polyploid plants had a defect in meiosis, causing production of diploid instead of haploid gametes. Maidenhair ferns can self-fertilize, so these gametes combined to form tetraploid offspring that could in turn self-fertilize or mate with the parent plant.
- Because the polyploids are genetically isolated, they may experience different evolutionary forces (e.g., there are fewer polyploids than wild types, so the polyploids are more likely to experience genetic drift). If so, the two populations may diverge, creating two species where once there was just one.

Allopolyploidy

- Hybridization between two diploid species may occasionally create polyploidy. Many hybrid offspring are sterile because their chromosomes do not pair normally during meiosis. However, if a mutation doubles the chromosome number, then homologs can synapse normally and gametes can be produced. These gametes may be able to self-fertilize, producing a tetraploid offspring.
- The sequence of events just described seems to have caused speciation from three parent species of *Trogopogon*, a weedy European plant introduced to North America in the early 1900s. Two new tetraploid species ap-

pear to have multiple independent originations following allopolyploid events.

- Many diploid plant species have closely related polyploid species, indicating that this type of instantaneous sympatric speciation by autopolyploid and allopolyploid mutation has been relatively common in plants.

- Plant reproductive cells don't separate from somatic cells until late in development. If sister chromatids don't separate during an early mitotic division, a tetraploid cell is produced that may later undergo meiosis to form diploid gametes. Many plants can self-fertilize, so these gametes can fuse, creating a new, genetically isolated tetraploid population.

25.4   What Happens When Isolated Populations Come Into Contact?

- Populations that have been isolated may evolve differences in mating mechanics, time, place, or behavior (prezygotic isolation) that continue to prevent mating and reduce gene flow even if the populations regain contact. These populations may continue to diverge after contact is regained.

- When prezygotic isolation does not occur, populations may successfully interbreed and cause gene flow that might eventually eliminate differences that evolved during their separation. Other possible outcomes are reinforcement, hybrid zones, and speciation by hybridization.

Reinforcement

- Sometimes individuals from populations that had been separated can mate with each other, but the hybrid offspring are aborted, die young, or survive but are sterile. This is known as postzygotic isolation.

- Postzygotic isolation causes natural selection against interbreeding, because individuals mating within their own population will be more successful at passing their genes on to future generations. Selection for traits that reproductively isolate populations is called **reinforcement**.

- Researchers placed flies from different *Drosophila* species together in the lab and found that fruit flies from sympatric species typi-

cally will not mate with one another, whereas flies from allopatric species often will.

- This pattern suggests reinforcement selection occurred among sympatric *Drosophila* species. Reinforcement occurs *only* when species live in the same geographic area; if two species never actually interbreed in the wild, natural selection will not act to reduce interbreeding.

- Researchers continue to debate whether or not reinforcement selection has been important in other groups of organisms.

Hybrid Zones

- Sometimes hybrid offspring are healthy with traits that are intermediate between the two parental populations. A geographic area where interbreeding between the two populations is common and there are lots of hybrid offspring is called a **hybrid zone**.

- Researchers studied a hybrid zone between Townsend's warblers and hermit warblers in North America's Pacific Northwest. The species' ranges overlap in Washington state, and hybrid offspring with intermediate characteristics are common (**Figure 25.12**).

- Gene sequencing of the maternally inherited mitochondrial DNA (mtDNA) showed that most hybrids are the offspring of Townsend's males and hermit females.

- Townsend's males attack hermit males, but hermit males do not challenge Townsend's males. Thus Townsend's males may invade hermit territories, chase off the males, and mate with the females.

- Further investigation showed that many other northern Townsend's warblers have hermit mtDNA. On some islands all the warblers had hermit mtDNA, even though they looked like typical Townsend's warblers.

- The Townsend's warblers may have gradually expanded their range into hermit territory, hybridizing with female hermit warblers along the way. As each new generation mates with Townsend's males, the offspring look more and more like Townsend's warblers, but retain the original hermit mtDNA

- Continued extension of the Townsend's warbler range could lead to the extinction of hermit warblers; but hybridization can sometimes lead to the creation of a new species.

New Species through Hybridization

- The sunflower, *Helianthus anomalus*, resembles hybrids of *H. annus* and *H. petiolaris*. *H. anomalus* also contains some gene sequences similar to those in *H. annus* and others similar to those in *H. petiolaris*. Did *H. anomalus*, which has a distinct appearance and grows in drier habitats, originate from hybridization of *H. annus* with *H. petiolaris*?

- Biologists experimentally hybridized *H. annus* with *H. petiolaris* for several generations. These hybrids looked like *H. anomalus*, and genetic maps using genetic markers showed that the experimental hybrids had the same combination of *H. annus* and *H. petiolaris* genes found in *H. anamolus* (**Figure 25.14**).

B. CROSS-CUTTING THEMES

Looking Back—
Concepts from Earlier Chapters

Evolution—Chapters 1, 4, 23, and 24

Earlier chapters have focused on processes that cause evolution or change within a population. Chapter 25 focuses on how different evolution in different populations can cause speciation.

Mitochondrial DNA—Chapter 20

Maternally inherited mitochondrial DNA, studied to investigate the nature of warbler hybrid zones, was first mentioned in **Chapter 20**.

Meiosis—Chapter 12

If the description about why triploid offspring would be unable to make healthy gametes is not clear to you, consider reviewing the process of meiosis. What would happen to triploid chromosomes during synapsis?

Looking Forward—
Concepts from Later Chapters

Speciation Patterns—Chapter 26

Now that you understand how speciation occurs, you are ready to consider large-scale speciation patterns (e.g., adaptive radiations), discussed in the next chapter.

Conservation—Chapter 55

Understanding of speciation is critical to conservation efforts, as demonstrated by the dusky seaside sparrow example from this chapter. Biodiversity and conservation are further discussed in the final chapter of the text.

C. DIFFICULT TOPICS

In describing the process of speciation, this chapter shows that speciation is usually a gradual process. Because it is gradual, there is no obvious signpost announcing to researchers the point at which two populations are sufficiently diverged to be considered different species. Furthermore, questions that are relevant to these decisions are often unanswerable; for example, if two species are allopatric now, how long will they remain isolated?

This ambiguity can be frustrating for beginning biology students, but it is an important reality in many areas of research. Sometimes there are no easy answers or hard-and-fast rules for solving scientific questions. Researchers must then just do the best they can with the information they have.

D. ASSESSING WHAT YOU'VE LEARNED

(1) Testing Your Knowledge

1. The two key factors responsible for speciation among populations are:
   a. lack of gene flow and mutation
   b. mutation and genetic drift
   c. genetic isolation and genetic divergence
   d. postzygotic isolation and morphological change

2. The biological species concept defines species in terms of:
   a. reproductive isolation
   b. morphological differences
   c. genetic relatedness
   d. differences in behavior

3. Which of the following would *not* be a good example of prezygotic reproductive isolation?

a. two sympatric bird species that, while similar in plumage, engage in dramatically different courtship dances

b. two beetle species that are superficially similar in appearance; but the structure of the male penis and the female genitalia prevent males from one species copulating with females of the other

c. two plant species with wind-dispersed pollen that lands on the styles and grows a pollen tube through the ovary of either species; however, in hybrid matings, the sperm cannot fertilize the ovum

d. two frog species that meet and can mate with each other, but the hybrid offspring are infertile

4. A monophyletic group would be best described as:

a. a grouping of all species descended from a common ancestor, including that ancestor

b. a grouping of all species descended from a common ancestor, excluding that ancestor

c. a grouping of all species that share a similar set of traits

d. a grouping of species descended from two or more closely related species

5. Paleontologists studying fossilized therapsids (a group of mammal-like reptiles that are now extinct) would probably be using which of the following species concepts?

a. the biological species concept

b. the morphospecies concept

c. the phylogenetic species concept

d. None of the above; fossil species cannot be classified.

6. Which of the following situations would represent a problem in applying the biological species concept?

a. two populations of morphologically similar organisms that differ in breeding habits and physiology

b. two populations of asexual clonal organisms that are morphologically similar

c. two morphologically similar populations, one in Africa and one in South

America, that despite their geographic separation exhibit no apparent prezygotic or postzygotic isolation

d. lions and tigers do not interbreed in nature, but can hybridize in zoos

7. What is meant by a subspecies?

a. a species that is descended from another species

b. each population of a biological species

c. a population that is somewhat distinctive in behavior or appearance from other similar populations

d. a population that is in the process of diverging from another population of the same species but is not quite reproductively isolated

8. Why shouldn't the hybrid offspring of the male *A. m. nigrescens* and female *A. m. peninsulae* be released to replenish the dusky seaside sparrow's genetic diversity?

a. There is already enough genetic diversity in the *A. m. nigrescens* population.

b. Releasing these offspring would artificially cause gene flow between two genetically distinct populations.

c. Due to hybrid sterility, these hybrids cannot reproduce.

d. The morphospecies concept does not recognize subspecies.

9. Vicariance refers to:

a. A pattern of speciation in which a population is subdivided by a geographic barrier.

b. A kind of speciation where a small propagule founds a population on a habitat island.

c. A pattern of speciation whereby new species evolve reproductive isolation in sympatry.

d. A technique for constructing phylogenetic trees based on vicariant traits shared by all members of a taxon.

10. Peter and Rosemary Grant observed as a small group of the large ground finch colonized Daphne Major in the Galápagos Islands. In a few years, the descendents of the colonists had evolved beaks that were

much larger than those in the original source population. What factors did the Grants feel influenced this evolution?

a. The original small size of the colonist pool probably resulted in divergence due to genetic drift.

b. Natural selection was imposed by larger seeds found on Daphne Major than on the ancestral island.

c. Reinforcement was due to postzygotic isolation on secondary contact.

d. Both A and B were probably important.

11. Which of the following is the best example of isolation and divergence in allopatry?

a. Two similar ground-squirrel species are separated by the Grand Canyon.

b. Different species of tapeworms infect humans and cats.

c. Horticulturalists have created tetraploid flower species.

d. Soapberry bug populations appear to be diverging into a long-beaked and a short-beaked subspecies.

12. Which of the following would best be described as a case of speciation in sympatry?

a. A population of lizards is subdivided by a natural barrier and subsequently diverges to form two species that cannot interbreed.

b. A new, isolated population of fruit flies is founded by a small group of colonists, which then diverges from the ancestral source population.

c. An individual hermaphroditic plant undergoes meiotic failure, producing diploid pollen and ovules; these self-fertilize, germinate, and grow into several fully fertile tetraploid plants.

d. Speciation cannot take place in sympatry, only in allopatry where geography poses a barrier to gene flow.

13. Different ratite species are found on different continents. This is a good example of:

a. isolation and divergence in sympatry

b. postzygotic isolation

c. vicariance due to continental drift

d. phylogenic speciation

14. The Lake Erie water snakes do not appear to be diverging into banded and unbanded species. Why?

a. Gene flow keeps the populations from diverging any further.

b. The populations are sympatric.

c. The populations are allopatric.

d. No natural selection is acting on snake color.

15. Why have soapberry bugs feeding on non-native hosts evolved shorter beaks than those feeding on their native host plant?

a. The nonnative host species have smaller fruits; natural selection favors bugs with a shorter beak length on these hosts.

b. A small population of soapberry bugs with short beaks colonized the non-native hosts; the differentiation is due to genetic drift.

c. This difference is due to a change in ploidy, resulting in a genetically differentiated population on nonnative hosts.

d. Beak size is phenotypically plastic; the changes seen are not genetic and do not reflect genetic differentiation.

16. Parents from two different species mate and produce a fertile offspring with a polyploid chromosome number. This is called:

a. triploidy

b. allopolyploidy

c. autopolyploidy

d. mutation selection

17. Reinforcement of speciation refers to:

a. natural selection to increase prezygotic isolation imposed by reduced fitness in hybrid offspring

b. natural selection to increase postzygotic isolation imposed by differences in courtship rituals or pollination strategies

c. genetic drift making two subsets of a population become more differentiated

d. the evolution of compatibility between hybrids that leads to the fusion of two diverging populations

18. Researchers used mitochondrial DNA to study the hybrid zone between Townsend's warblers and hermit warblers. Why was

mitochondrial DNA so useful in this study?

a. The mtDNA is maternally inherited, so researchers were able to discover that most hybrids resulted from Townsend's males mating with hermit females.

b. The mtDNA is easy to extract, so the researchers could obtain lots of genetic information about the hybrids.

c. The mtDNA shows that reinforcement has been occurring in this hybrid zone.

d. The mtDNA is biparentally inherited, like most of the nuclear genetic material. Any gene could have been used in this study.

19. Biologists recently tested the hypothesis that the sunflower *Helianthus anomalus* originated as a hybrid between *H. annuus* and *H. petiolaris*. Which of the following findings would *not* support this hypothesis?

a *H. anomalus* phenotypically resembles hybrids between *H. annuus* and *H. petiolaris*.

b. Most of the gene sequences in *H. anomalus* are different from either *H. annuus* or *H. petiolaris*.

c. Experimental crosses between *H. annuus* and *H. petiolaris* produced hybrids with similar species-specific patterns of DNA sequences as found in *H. anomalus*.

d. *H. annuus* and *H. petiolaris* hybridize in the geographic region where *H. annuus* is found.

20. Imagine a lake containing a single population of snails. During a period of drought, the shallow middle of the lake dries up, creating two separate populations. The snails, being totally aquatic, cannot cross from one lake to the other. At some point in the future, the climate becomes wetter and the lake refills. Assuming an annual generation, how many years must the populations have remained divided for reproductive isolation between them to evolve?

a. 100 years

b. 5000 years

c. 500,000 years

d. In theory, speciation could occur within a few generations, or it could take thousands of generations. There is no general answer to this question.

(2)   Integrating Your Knowledge

(a) Which species definition (i.e., biological, morphospecies, or phylogenetic) would you use for (1) a new fossil species, (2) a well-studied bacterial species, and (3) a well-studied mammalian species? Why?

(b) Colonization followed by genetic drift and natural selection has also led to speciation in aquatic organisms and organisms that live at the top of some mountains. Explain how this could happen.

(c) Two ground squirrel populations that live on different sides of the Grand Canyon were separated by a vicariance event. Explain what happened.

(d) Why does the fact that the seaside sparrow phylogeny shows only two distinct groups suggest there may be more gene flow within those two groups than previously thought?

(e) Polyploid mutations are an important cause of speciation in plants but not in animals. Why?

(f) Why is genetic drift more likely to be an important evolutionary factor in colonization events than in vicariance events?

(g) Which of the following isolating mechanisms are prezygotic and which are postzygotic?

1. Land iguana eggs can't be fertilized by marine iguana sperm.

2. Mules—which are hybrids of the horse and donkey—are sterile.

3. In a forest, one species of beetle lives on oaks; another beetle species lives on pines.

4. In closely related bird species, males sing different courtship songs.

5. Hybrid seeds from pollination of one yucca by another yucca species are aborted.

(h) Why isn't there any Townsend's warbler mtDNA on some of the northern islands within the current Townsend's warbler range?

CHAPTER 25—ANSWER KEY

D. Assessing What You've Learned

(1) Testing Your Knowledge

1. c; 2. a; 3. d; 4. a; 5. b; 6. b; 7. c; 8. b; 9. a; 10. d; 11. a; 12. c; 13. c; 14. a; 15. a; 16. b; 17. a; 18. a; 19. b; 20. d

(2) Integrating Your Knowledge

(a) You would probably have to use the morphospecies concept for a new fossil species because phylogenetic trees with the species would be unavailable, and there is no way to determine with which other populations a fossil was able to mate. You would probably use the phylogenetic species concept for a well-studied bacterial group, although you might be able to use the morphospecies concept. Bacteria are asexual, so the biological species concept could not be used in this case. Finally, you could really use any of the species definitions for the well-studied mammal species; ideally, you would use all and compare your conclusions using each method.

(b) Movement of an aquatic organism from one lake to another (for example, during an unusual wet period that temporarily joins two lakes) also causes physical isolation and may involve a small colonist population. Similarly, mountaintops in the Southwest United States are separated by desert areas that are difficult for many mountaintop organisms to cross (in fact, these mountains are sometimes referred to as the sky islands). Colonization by a small population that then experiences genetic drift and natural selection can also occur in these habitats.

(c) The Colorado River has gradually cut down into the surrounding rock to form the Grand Canyon. Initially, there was just a river that could have been crossed by ground squirrels.

(d) Greater gene flow within the two monophyletic groups (perhaps by occasional dispersal of the young to a different population) would tend to eliminate any genetic differences.

(e) Many plants can self-fertilize; thus, polyploid individuals can often mate with themselves to produce healthy offspring. However, animals typically cannot self-fertilize, so that any surviving polyploid mutants would be unable to reproduce.

(f) Colonization events are more likely to produce small population sizes (at least initially).

(g) 1, 3, and 4 are prezygotic; 2 and 5 are postzygotic.

(h) Presumably there is no Townsend's warbler mtDNA on those islands because no female Townsend's have made it to the islands (perhaps the male Townsend's warblers disperse farther than do the females). Thus all the island birds are descended from the original female hermit warblers that first mated with the Townsend's males.

26

Phylogenies and the History of Life

A. KEY BIOLOGICAL CONCEPTS

- This chapter examines large-scale patterns in the family tree of life: What has been the history of life since life evolved on Earth? How do we study this history? How do evolutionary innovations occur, leading to adaptive radiation? Why do mass extinctions occur?

26.1 Tools for Studying History: Phylogenies and the Fossil Record

- The history of life is primarily studied using phylogenetic trees and the fossil record.
- The evolutionary history of an organismal group is called a **phylogeny**. A **phylogenetic tree** shows ancestor-descendant relationships among evolutionary groups (usually species or populations).
- **Fossils** are physical evidence left by organisms from the past. They provide the only direct evidence of what ancient organisms looked like and when they existed. The **fossil record** includes all fossils that have been found and recorded.

Using Phylogenies

A Field Guide to Reading Phylogenetic Trees

- **Figure 26.1** shows the parts of a **phylogenetic** tree. Populations are represented by **branches**, and **nodes** show where ancestral groups split into descendant groups. Adjacent branches are **sister taxa** (a **taxon** is any named group of organisms), and a **polytomy**

is a node where more than two descendant groups branch off. **Tips** are branch endpoints that represent living groups or a group's end in extinction.
- All phylogenies shown in this text are **rooted**, meaning that the most ancient node of the tree is shown at the bottom. The location of this node is determined using an **outgroup**, a taxonomic group that diverged before the rest of the taxa being studied.
- An ancestor and all its descendants form a **monophyletic group** (also called a **clade** or **lineage**).
- **Figure 26.3** shows how branches can be rotated around nodes without changing a tree's phylogenetic information.

How Do Researchers Estimate Phylogenies?

- Both morphological and genetic characteristics are used to estimate phylogenetic relationships among species. Closely related species should share many traits, while distantly related species should share fewer traits.
- The **phenetic approach** uses all traits to evaluate the related ness of species. For example, a researcher using this approach might assess similarity using *genetic distance*, a measure of the average percentage of difference in DNA bases between two populations or species.
- The **cladistic approach**, on the other hand, claims that some traits are more informative than other traits in determining relatedness.

Researchers using this approach focus on **synapomorphies**, the shared derived characters found in the species under study (see **Figure 26.4**).

- The cladistic approach can run into difficulties in cases of convergent evolution, where similar traits evolve independently in distantly related groups.
- In cases of convergent evolution, biologists use **parsimony** in trying to identify the phylogenetic tree that minimizes the overall number of convergent evolution events. This approach assumes that convergent evolution should be much rarer than similarity due to shared descent.

Whale Evolution: A Case History

- Traditionally, phylogenetic trees based on morphological data show whales outside of the artiodactyls (mammals with hooves and an even number of toes) and identify the evolution of a pulley-shaped astralagus as a synapomophy, a shared derived trait, for the artiodactyls group (**Figure 26.5a**).
- On the other hand, DNA sequence data suggests a close relationship between whales and hippos. A phylogenetic tree showing closely related whales and hippos is less parsimonious for morphological data because it requires evolution and then loss of the pulley-shaped astralagus in whales (**Figure 26.5b**).
- Recent data on short interspersed nuclear elements (SINEs) shows that whales and hippos share several SINE genes that are absent in other artiodactyl groups (**Figure 26.5c**). These SINEs are shared derived traits (synapomorphies) that support the hypothesis that whales and hippos are closely related.
- Recent fossil finds of whale-like artiodactyls with pulley-shaped astralagus bones lend further support to the hypothesis that whales and hippos are closely related and that whales are, in fact, artiodactyls.
- Other phylogenetic relationships are being reevaluated as new data sources are available.

Using the Fossil Record

- The fossil record is the only source of direct evidence about what prehistoric organisms looked like, where they lived, and when they existed.

How Do Fossils Form?

- Most fossils form when an organism is buried in sediment before decomposition occurs (see **Figure 26.6**).
- Four types of fossils are shown in **Figure 26.7**: (a) organic remains may resist decomposition and be preserved intact; (b) sediments above the organism may cement into rock and compress the organic material into a thin film; (c) remains may decay after burial, leaving a hole that fills with dissolved minerals, creating a **cast**; (d) if dissolved minerals enter cells, a *permineralized* fossil forms.
- Centuries later, fossils may be exposed due to erosion or human action. Once found, fossils must be carefully picked from the surrounding rock (**Figure 26.8**). Fossils are then named, described in a scientific publication, dated, and added to a collection so they are available for study as part of the fossil record.
- Most organisms decompose rapidly and are buried slowly, if at all. Thus, fossilization is an extremely rare event.

Limitations of the Fossil Record

- Habitat bias: Organisms living where sedimentation occurs (beaches, swamps, etc.) are more likely to become buried in sediments and be fossilized than are organisms living in other habitats. Burrowing organisms are more likely to fossilize because they are pre-buried at death.
- Taxonomic bias: Organisms with hard parts (bones, shells) are more likely to resist decay long enough to be fossilized than are organisms with only soft parts. Similarly, within an organism the hard parts (pollen, teeth) are more likely to fossilize than are the soft parts.
- Temporal bias: Rocks in Earth's crust may be worn down by erosion or melted as one tectonic plate is pushed underneath another one. Because these processes destroy rocks and the fossils in them, many ancient fossils have been destroyed; recent fossils are thus more common than ancient ones.
- Abundance bias: Because fossilization of any organism is a rare event, organisms that are

common, widespread, and extant as a species for long periods are more likely to become fossilized and be found by researchers.

- In sum, the fossil record is highly biased; but it gives the *only* direct evidence about the morphology and ecology of extinct organisms. Analysis of the fossil record (by **paleontologists**) is thus central to our understanding of the history of life.

Life's Timeline

- Major events in the history of life are marked on the timeline shown in **Figure 26.9**. Dates are primarily calculated using radiometric dating (see **Box 26.1** for another technique). Figure 26.9 has been broken into four segments (the **Precambrian**, the **Paleozoic**, the **Mesozoic**, and the **Cenozoid**).
- The **Precambrian era** (of almost 2 billion years!) includes the **Hadean, Archaean**, and **Proterozoic eons**. During most of this time period, almost all life was unicellular; hardly any oxygen was present in Earth's oceans or atmosphere.
- The appearance of many animal groups marks the beginning of the **Paleozoic era**. Land animals, land plants, and fungi appear and diversify during this period. This era ends with a mass extinction at the end of the Permian.
- The end-Permian extinction is followed by the **Mesozoic era**, also known as the Age of Reptiles. The Mesozoic ends with the extinction event that killed the dinosaurs.
- The Mesozoic is followed by the **Cenozoic era**, also known as the Age of Mammals because mammals diversified after the dinosaurs disappeared.

26.2   The Cambrian Explosion

- Animals first originated around 565 million years ago (Ma). Soon after that, at the start of the Cambrian period, animals diversified into almost all the major groups extant today. This is known as the **Cambrian explosion**.

Cambrian Fossils: An Overview

- Three major fossil beds record this explosion of animal life. In China, the Doushantuo fossils record the first animal life about 570 Ma;

the Ediacaran fossils in Australia date from 565 to 544 Ma; and the Burgess Shale fossils in Canada show animal life from 525 to 515 Ma (**Figure 26.12**).

- Each of these fossil beds is extraordinary because each includes more than one habitat and the fossils of soft-bodied as well as hard-shelled animals, greatly increasing number of **fauna** (animal species) represented.

The Doushantuo Microfossils

- Researchers identified several submillimeter-sized sponges in samples dated to 580 million years ago. Different cell types were present in these sponges, as were spicules like those found in sponges today. These are the first known animals on Earth!
- Other biologists found what seem to be animal embryos in the one-, two-, four-, and eight-cell stages. As in modern embryos, during these early divisions the cell number increases and the volume of the embryo stays the same (**Figure 26.12b**)
- Cyanobacteria and multicellular algae were also present in the samples, suggesting a shallow-water marine habitat.

The Ediacaran Faunas

- Sponges, jellyfish, comb jellies, and traces of other animals are found in these Australian deposits (**Figure 26.12c**). However, no animals with shells, heads, mouths, or feeding appendages are present either at this site or at other fossil sites from this time period (565–544 Ma). This shallow-water marine habitat had lots of filter feeders, but no animals that actively hunted for or captured food.

The Burgess Shale Faunas

- Just about every major animal group can be found among the Burgess Shale fossils: arthropods, molluscs, echinoderms, worms, even a chordate (**Figure 26.12d**). These animals show incredible morphological and ecological diversity. Besides filter feeders, there are predators, scavengers, and grazers. How did this incredible diversification happen?

26.3 The Genetic Mechanisms of Change

- Two important types of mutation make major morphological innovations possible: gene duplications and changes in gene expression. Research that combines evolutionary and developmental studies is called **evo-devo**.

Gene Duplications and the Cambrian Explosion

- *Homeotic genes* code for proteins that, in early development, determine the three-dimensional layout of multicellular organisms. Because these genes have such a major effect on morphology, researchers have examined correlations between the evolutionary history, morphology, and number of homeotic genes in different animal groups.
- One group of homeotic genes, the *Hox* genes (or *Hox* loci), is found clustered together. These genes probably derive from gene duplication events. Each gene has its own function in pattern formation (cell organization) within developing embryos.
- The number and type of *Hox* loci in different animal groups are shown in **Figure 26.13**, where each color represents a different type of *Hox* gene. Generally speaking, these data support the hypothesis that homeotic genes are responsible for the evolution of animal morphological diversity (the "new genes, new bodies" hypothesis).
- The number and type of *Hox* genes in animals is variable (supports new genes, new bodies hypothesis).
- In general, animal groups that branched off early in evolution have smaller, simpler bodies and fewer *Hox* genes than do groups that branched off later (supports new genes, new bodies hypothesis).
- Similar gene structure and base sequence supports the idea that gene duplications were responsible for the origin of many *Hox* loci. Animal phylogeny can be used to determine in which order different genes appeared. Gene duplication provides a mechanism for the new genes, new bodies hypothesis.
- Vertebrates, some of the largest and most complex animals, have several copies of the entire *Hox* cluster. This suggests the complete set of 13 genes was duplicated several times (though some clusters later lost some of their genes, these data generally support the new genes, new bodies hypothesis).
- Although greater complexity is associated with more *Hox* genes among the groups shown in **Figure 26.13**, this correlation does not always hold up within these groups. For example, zebrafish have more *Hox* genes than mice do.
- Duplication of *Hox* genes is an important factor in the diversification of animal body plans. However, changes in gene expression and function have also been important in creating animal morphological diversity.

Changes in Gene Expression: The Origin of the Foot

- The origin of **tetrapods** (four-footed animals) was a major innovation in vertebrate evolution. Limbs evolved from fins, but tetrapods have fewer *Hox* genes than fish do. Perhaps changes in homeotic gene expression created limbs.
- Biologists compared the gene expression of *hoxd-11* and *Sonic hedgehog* (*Shh*) in the zebra fish and the mouse. These genes affect spatial organization of embryo cells. *Hoxd-11* is expressed along the long axis of tetrapod embryo limb buds, whereas *Shh* is expressed along the front-to-back axis of limb buds.
- Zebra-fish and mouse limb buds were treated with molecules that bind to the mRNA transcripts of the *hoxd-11* and *Shh* genes in order to identify the pattern of protein production (in situ hybridization).
- Early in development, there is no difference between fish and mammal gene expression. Later in development, mouse *hoxd-11* gene expression is found in the part of the limb bud facing the head, where no *hoxd-11* product is found in fish. *Shh* gene expression also continues in mice after it ceases in fish (**Figure 26.14**).
- These changes in the timing and location of homeotic gene expression may have produced the transition from fin to limb 400 million years ago. Presumably the mutations causing these changes in gene expression were

selected for in some animals because they made movement on land easier.

26.4   Adaptive Radiations

- One broad pattern that can be observed in the tree of life is that dense groups of branches are scattered throughout the tree. These dense branching patterns, known as star phylogenies (**Figure 26.15**), represent major diversification of organism groups over a relatively short period of time.
- When a single lineage produces many descendant groups in a short period of time, an **adaptive radiation** has occurred. Adaptive radiations are an important pattern in the history of life.
- Hawaiian honeycreeper birds and Hawaiian silverswords both show adaptive radiations following colonization of the Hawaiian Islands; all major groups of mammals diversified from a common ancestor between 65 and 60 Ma.
- In all these examples, the descendant groups inhabit myriad habitats and have many different forms and lifestyles. What causes these adaptive radiations?

Colonization Events as a Trigger

- Adaptive radiations usually occur when habitats are unoccupied by competitors. For example, mammals evolved to occupy habitats left vacant after extinction of the dinosaurs.
- Biologists have documented adaptive radiations that occurred following the colonization of unoccupied island habitats.
- One study focused on *Anolis* lizards of the Caribbean islands. This group includes 150 species, and the size and shape of a species closely correlates with its habitat (e.g., species living on the ground or on tree trunks have longer legs than do those that spend their time clinging to twigs (**Figure 26.16a**).
- Researchers estimated *Anolis* phylogeny using DNA sequence data. They then compared habitat (and associated morphology) to the evolutionary relationships among species.
- For the islands of Hispaniola and Jamaica, the founding lizard colonists belonged to different ecological types; but over time, the lizards evolved to occupy all island habitats, forming four different species with four different morphologies (**Figure 26.16b**). Each island had its own adaptive radiation.
- Several of these miniature adaptive radiations, each on a different island, were possible following the colonization of each new available habitat that lacked competitors.

The Role of Morphological Innovation

- Morphological innovation can provide opportunities for exploiting new habitats. Multicellularity, shells, exoskeletons, and limbs were new traits that produced the adaptive radiation known as the Cambrian explosion.
- Other examples of traits that allowed exploitation of new habitats, causing adaptive radiations, are (1) the evolution of wings, three pairs of legs, and a protective external skeleton in insects; (2) the evolution of flowers in angiosperms; (3) the evolution of feathers and wings in birds; and (4) the evolution of a second set of jaws in cichlid fish (**Figure 26.17**).

26.5   Mass Extinctions

- **Mass extinctions** are the rapid extinction of many groups throughout the tree of life (loss of at least 60% of all species within 1 million years). These extinction events are caused by catastrophic episodes.
- Traditionally, five mass extinction events are distinguished from a lower **background extinction** rate that represents the normal loss of some species that always occurs (**Figure 26.18**).

How Do Background and Mass Extinctions Differ?

- Background extinctions typically occur when normal environmental change or competition reduces a population to the point that it dies out. Mass extinctions occur when unusual large-scale environmental change causes the extinction of many normally well-adapted species.
- Natural selection causes most background extinctions, whereas random chance plays a large role in mass extinctions.

What Killed the Dinosaurs?

- At the end of the Cretaceous, 65 Ma, an asteroid about 10 km across hit Earth and splashed material over much of North America. Destruction caused by the asteroid impact led to dinosaur extinction, paving the way for the mammalian radiation. This **impact hypothesis** is supported by several lines of evidence.
- (1) High levels of iridium, common in asteroids but rare on Earth, are found worldwide in sedimentary rocks from the Cretaceous-Tertiary (K-T) boundary; (2) minerals found only at meteorite impact sites (e.g., shocked quartz and microtektites) are common in Haitian and other rocks dated to 65 Ma; and (3) there is a huge crater off the northwest coast of the Yucatan peninsula (see **Figure 26.19**).
- Computer models and geologic data suggest that the asteroid impact caused a fireball of hot gas that spread out from the impact site, causing worldwide wildfires (evidenced by soot and ash deposits dated to 65 Ma). The dust, soot, and ash then blocked the Sun, causing rapid global cooling and decreased plant productivity. Additionally, SO_4^{2-} that was released from rock at the impact site formed sulfuric acid, causing acid rain.
- The large-scale environmental change triggered by the asteroid impact caused the extinction of 60 to 80% of all species.

Selectivity

- Some evolutionary lineages withstood the environmental change better than others did. Thus dinosaurs, pterosaurs, and large marine reptiles went extinct; but most mammals, crocodiles, amphibians, and turtles survived.
- One hypothesis is that extended darkness and cold affected large organisms more than small ones because they require more food. However, size selectivity is not found among marine clams and snails. Other hypotheses are still being investigated.

Recovery

- Ferns appear to have replaced diverse woody and flowering plants in many habitats following the K-T extinction. Mammals diversified to fill the niches left empty following the dinosaur extinction. Succeeding flora and fauna of the Tertiary were very different from those of the Cretaceous.

B. CROSS-CUTTING THEMES

Looking Back—
Concepts from Earlier Chapters

Geologic Time Scale and Radiometric Dating—Chapter 2

Radiometric dating is regularly used to date fossils. **Chapter 2** describes this technique, and provides information about the age of Earth and the appearance of first life.

Gene Duplication—Chapter 20

Chapter 20 describes how gene duplications may have contributed to the evolution of morphological diversity in animals.

Early Embryonic Development in Animals—Chapter 21

You learned about embryo development in **Chapter 21**. Here you learn additional details about how gene expression of homeotic genes controls limb development. Changes in expression of these genes may have been responsible for the evolution of tetrapod limbs.

Homeotic Genes—Chapter 22

Clearly, homeotic genes (e.g., *Hox* loci) have played a major role in evolution because they affect the overall morphology of organisms.

Evolution—Chapters 1, 4, 23, 24, and 25

Because evolution is one of the unifying themes of biology, many chapters have addressed this topic. Chapter 26, however, is the last to concentrate specifically on evolution. Nonetheless, watch out for continued references to evolution in future chapters.

Phylogenies—Chapters 1 and 25

Chapter 1 depicted a SSU-RNA phylogeny for the tree of life, and **Chapter 25** went into more detail on phylogenies and monophyletic groups. Now you are learning about large-scale patterns in the tree of life, and how phylogenetic information can be compared with information on specific genes to infer how different groups and morphologies evolved.

Looking Forward—
Concepts from Later Chapters
Diversity of Life—Chapters 27–34
Now that you know a bit about how evolution occurs and about evolutionary patterns in the tree of life, you are ready to learn specifics about the major groups of organisms comprising the tree of life.

Competition—Chapter 53
Chapter 26 mentioned how adaptive radiations tend to occur when niches are unoccupied by competitors. **Chapter 53** discusses what happens when two or more species do compete within the same habitat.

C. DIFFICULT TOPICS

To thoroughly understand evolution, you must integrate the concepts of how mutations and natural selection at the level of the individual are related to large-scale evolutionary patterns.

For example, consider a mutation in a fish that causes altered expression of *Hoxd-11*. Suppose that the altered expression causes the fish to have limb-like fins. If the fish can still swim effectively but can also more easily draw itself onto land to escape predators, it may survive better and leave more offspring (the altered *Hoxd-11* expression has been selected for).

Over time, the altered expression may increase in frequency throughout the fish species' population. Over time, other traits that favor the ability to exist on land may also be selected for. If eventually some of the population takes up living partly on land, they have colonized a new habitat and may experience sympatric speciation (different habitats in same location leading to decreased gene flow—just like the soapberry bugs in **Chapter 25**). Now the first vertebrate land dweller has lots of new habitats available without much competition. This situation causes an adaptive radiation of tetrapods on land.

In this way, individual and population-level evolutionary processes can lead to the large-scale patterns discussed in this chapter.

D. ASSESSING WHAT YOU'VE LEARNED

(1) Testing Your Knowledge

1. A monophyletic group is:
    a. a taxonomic group known to have diverged prior to the rest of the taxa in the study
    b. an ancestor and all its descendants
    c. the branch tips of a phylogenetic tree
    d. any two taxa that show convergent evolution

2. Which of the following is *not* true of the **cladistic** approach for inferring phylogenetic trees?
    a. Synapomorphies are used to infer evolutionary relationships.
    b. When convergent evolution occurs, biologists use parsimony to choose the best tree.
    c. Computer programs are used to identify traits that are unique to each monophyletic group.
    d. Genetic distance can be used to create phylogenetic trees.

3. Researchers conclude that whales are most closely related to hippos because (choose all that apply):
    a. Whales and hippos share several SINEs that are not present in other artiodactyls.
    b. Modern whales have tiny pulley-shaped astralagus bones.
    c. Non-SINE whale and hippo DNA sequences are also similar.
    d. Whales look more like hippos than they look like cows.

4. What is the first step in the creation of a fossil from the body of a plant or animal?
    a. The organism must fall into a body of water.
    b. The body must be buried in sediments.
    c. The body must absorb minerals and change into stone.
    d. Slow decomposition must begin, so that the rock forms around it.

5. The fossil record is an extremely biased sample of organisms that are now extinct.

Which of the following is *not* a reason for that bias?

a Organisms with hard parts, such as teeth, shells, and skeletons, are more likely to be preserved.

b. Organisms living in certain habitats were more likely to be buried in sediments and preserved than were organisms in other habitats.

c. Organisms with broad geographic distributions and large populations are more likely to be preserved than are rare endemic species.

d. Exposed fossils are often subject to weathering, whereas those beneath the surface are protected.

6. Which of the following correctly lists the geological eras from most ancient to most recent?

a. Paleozoic, Mesozoic, Precambrian, Cenozoic

b. Paleozoic, Mesozoic, Cenozoic, Precambrian

c. Precambrian, Paleozoic, Mesozoic, Cenozoic

d. Precambrian, Cenozoic, Paleozoic, Mesozoic

7. Which of the following did *not* occur during the Paleozoic?

a. appearance of the first multicellular organisms

b. diversification of animal lineages

c. a mass extinction at the end of the Permian

d. appearance and diversification of land plants and fungi

8. The Cambrian explosion refers to:

a. the rapid diversification of animals at the start of the Cambrian period

b. the explosion that followed the asteroid impact at the end of the Cambrian period

c. the origin and diversification of multicellular organisms during the Cambrian period

d. the high level of volcanic activity as Gondwana formed during the Cambrian period

9. Why are the Doushantuo, Ediacaran, and Burgess Shale fossil deposits so unusual?

a. because they preserve a diverse array of microorganisms that we believe gave rise to multicellular life at the start of the Cambrian period

b. because they capture a wide range of soft-bodied organisms that are rarely preserved in the fossil record at a time when the diversity of metazoans was greatly increasing

c. because these assemblages show precisely how animals began as single-celled organisms, then became two-celled, then four-celled, and so on to become the multicellular forms we know today

d. because these fossils record the origin of animals from seaweeds and other photosynthetic organisms

10. What was life like before the Cambrian explosion?

a. There was no life prior to the Cambrian explosion.

b. Several species of small, multicellular animals appear in the fossil record about 1 billion years ago, but none were terrestrial.

c. Only unicellular organisms existed; no multicellular forms appear until 565 Ma in the fossil record.

d. Bacteria, single-celled eukaryotes, and multicellular algae were all present 1 billion years ago.

11. Fossils of organisms from the pre-Cambrian are extraordinarily rare, but many species are found in the Burgess Shale deposits. Which of the following statements best describes the Burgess Shale organisms?

a. mostly bacteria and simple one-celled plants and animals

b. primarily sponges, algae, and seaweeds

c. arthropods, mollusks, chordates, and a variety of soft-bodied organisms

d. jellyfish and sponges, along with some tracks of various other organisms

12. What is one genetic hypothesis proposed for the rapid diversification of body plans during the Cambrian period?
    a. Many mutations accumulated due to high exposure to UV radiation, creating many new kinds of organisms.
    b. Reproductive isolation was nearly nonexistent; many "species" interbred to form hybrids that went on to become new species.
    c. New genes appeared in many different organisms through a mechanism that has not been identified; these are associated with new body plans.
    d. Several duplications of the homeotic genes that control major body plans apparently arose during the Cambrian, creating new body plans and appendage configurations.

13. Which of the following observations is *not* consistent with the hypothesis that increasing metazoan complexity is associated with an increase in number of *Hox* loci?
    a. Within phyla, there is no correspondence between the number of *Hox* genes and complexity.
    b. Phyla that branch off early, such as sponges and cnidarians, have simple body plans with relatively few *Hox* loci.
    c. *Hox* loci are similar in structure and DNA sequence and are grouped in clusters, hence are assumed to have arisen through duplication; when a new *Hox* gene appears within a lineage, most of the descendent taxa have a homologous *Hox* locus.
    d. In vertebrates, the *Hox* clusters themselves appear to be duplicated several times.

14. Tetrapod limbs are hypothesized to have evolved from the fins of fishes. Researchers studied the expression of two genes related to limb development in zebra fish and mice. They hypothesized that a change in timing of expression might be responsible for triggering the changes between fins and limbs. What pattern of expression for *hoxd-11* and Sonic hedgehog (*Shh*) did they observe in these two taxa?

    a. One gene, *hoxd-11*, was expressed only in zebra fish, whereas *Shh* was expressed only in mice, suggesting that limbs are not homologous with fins.
    b. Both genes are expressed early in development in fish and mice; the lack of differences suggests that these genes are not important in the differing development of limbs and fins.
    c. Both genes are expressed early in development in both taxa, but later in development there are changes in both the location and timing of expression that are correlated with development of the mouse limbs.
    d. Both genes were expressed early in development in fish, but neither was expressed early in development in mice. Instead, both were expressed later and in different patterns in the developing tetrapod limb.

15. Which of the following cases best illustrates an adaptive radiation?
    a. A fishless lake is colonized by a single fish species, which over a few thousand years gives rise to several species, each with a series of unique feeding adaptations.
    b. A small group of tree-dwelling lizards of a single species migrates to an uninhabited, treeless island and adapts to use the open grassland habitat.
    c. A population of a cricket species takes up residence in a cave. Over many generations, the cave crickets eventually lose their eyes, like many other cave-dwelling animals.
    d. A colonizing species of fruit fly takes up residence on a new continent and displaces two closely related native species.

16. Why are morphological innovations often associated with adaptive radiations?
    a. Morphological innovations often allow a species to replace competitors.
    b. Morphological innovations can open up new adaptive options that may allow a species to colonize an underutilized resource.

c. Morphological innovations disrupt natural selection, forcing species to find new adaptations.

d. Morphological innovations prevent adaptive radiations most of the time.

17. In the study of *Anolis* lizards colonizing various Caribbean Islands, researchers found similar habitats and similar ecological types on Hispaniola and Jamaica. They conducted a phylogenetic analysis of the species on each island. What observation suggested that adaptive radiations occur in response to habitat availability and the absence of competitors?

a. Different islands were initially colonized by different species that differed in habitat preference, but in both cases subsequent speciation produced a range of ecological specialists occupying similar habitat niches. Phylogeny of each island was unique, but ended up with species having similar sets of adaptations for each habitat.

b. The first colonists on every island were always the same species, a twig-dwelling specialist. But in every case, that species gave rise to the same set of other species that specialized on different habitats. The phylogeny for each island had the same root and the same pattern of speciation.

c. Each island was initially colonized by a different species that occupied its preferred habitat. Subsequent colonizing species were successful only if they could use an unoccupied habitat type. The phylogeny for each island was the same throughout the region because no evolution occurred, only colonization.

d. Each island was colonized by a different species initially, and that species underwent a radiation, giving rise to new species that occupied different habitat types. But there was no similarity in the adaptations from one island to another. The phylogeny for each island was completely unique and totally unlike that of any other island.

18. Which of the following statements represents an important contrast between background extinctions and mass extinctions?

a. Background extinctions occur only sporadically; mass extinctions happen on a regular, periodic basis.

b. Background extinctions are caused by rapid changes in the environment; mass extinctions occur primarily in response to biological competition.

c. During mass extinctions, adaptations for survival and competition make little difference in the likelihood of extinction; the opposite is true of background extinctions.

d. Mass extinctions, such as the K-T event, tend to take out large-bodied organisms; smaller-bodied organisms tend to be removed by background extinction.

19. Which of these events has been associated with the dinosaur extinction at the end of the Cretaceous 65 million years ago?

a. a worldwide change in climate, caused by increased carbon dioxide in the atmosphere, led to melting of the ice caps and dramatic changes in sea level

b. an outbreak of a highly pathogenic virus against which reptiles have no defense, but mammals are apparently immune

c. an asteroid impact that caused fires, acid rain, and a massive dust cloud that blocked sunlight for some time, causing rapid global cooling and low plant productivity

d. an outbreak of global volcanic activity that both directly, through eruption of magma, and indirectly, through ash clouds blocking the Sun, caused massive mortality in plants and animals

20. Which of the following statements best represents current thinking about the rise of the mammals after the K-T extinction?

a. Mammalian diversity increased dramatically prior to the impact, but the population sizes of every species remained small until after the dinosaurs were extinct.

b. Mammals were inherently better competitors than the dinosaurs; the K-T impact simply speeded up the replacement, which was already well under way.

c. By chance, the K-T impact caused extinction of the dinosaurs but did not have as great an effect on small, nocturnal, scavenging mammals. The release from competition allowed the mammals to diversify.

d. Both mammals and dinosaurs survived the asteroid, but the mammals were much better adapted to the new environments created by the impact and thus underwent adaptive radiation.

(2) Integrating Your Knowledge

(a) **Chapter 2** explained that there are no known fossils of the very first living organisms. Give at least two reasons why this lack of fossil evidence for these first organisms is not surprising.

(b) For the three periods 570 Ma, 565–544 Ma, and 525–515 Ma, give the name of the fossil bed representing that time period and summarize the types of fossils found in that fossil fauna.

(c) Why isn't the "new genes, new bodies" hypothesis sufficient to explain all animal morphological diversity?

(d) Explain why crustaceans and centipedes don't have the orange or red *Hox* loci while fruit flies and velvet worms do.

(e) Name three situations that can promote adaptive radiations.

(f) What does a star phylogeny show?

(g) What might happen if Earth were to be hit by another asteroid the size of the one that struck in 65 Ma?

CHAPTER 26—ANSWER KEY

D. Assessing What You've Learned

(1) Testing Your Knowledge

1. b; 2. d; 3. a, c; 4. b; 5. d; 6. c; 7. a; 8. a; 9. b; 10. d; 11. c; 12. d; 13. a; 14. c; 15. a; 16. b; 17. a; 18. c; 19. c; 20. c

(2) Integrating Your Knowledge

(a) It is no surprise that there are no known fossils from the very first living organisms, because (1) most of the sedimentary rocks that were formed then have since been eroded or otherwise destroyed; and (2) those organisms would have had no hard parts, and would thus have been less likely to be fossilized in the first place.

(b) 570 Ma = Doushantuo fossils—sponges with different cell types, and animal embryos at 1- to 8-cell stages. 565–544 Ma = Ediacaran fossils—large sponges, jellyfish, comb jellies, and traces of other animals; no animals with shells and no heads or feeding appendages. 525–515 Ma = Burgess shale—incredible diversity of animals including a chordate, arthropods, mollusks, echinoderms, and worms.

(c) The "new genes, new bodies" hypothesis cannot explain *all* animal diversity in morphology, because in the arthropod group and in the vertebrates group, there is no correlation between the number of *Hox* loci and organismal complexity.

(d) Gene deletions must have occurred in the Crustacean and Centipede groups.

(e) colonization events, mass extinctions, and morphological innovations

(f) A star phylogeny shows *rapid* evolution of many different groups (species, genera, or families) from a single common ancestor.

(g) The impact could cause worldwide wildfires and global cooling from all the dust and ash in the air. However, acid rain would occur again only if the impact site contained the sulfate-rich anhydrite.

27

Bacteria and Archaea

A. KEY BIOLOGICAL CONCEPTS

27.1 Why Do Biologists Study Bacteria and Archaea?

- Species in the domain **Bacteria** are prokaryotic and have cell walls made of peptidoglycan, plasma membranes similar to eukaryotes, and distinct ribosomes and RNA polymerase.
- Species in the domain **Archaea** are prokaryotic, unicellular, and have cell walls made of polysaccharides, unique plasma membranes, and ribosomes and RNA polymerase similar to eukaryotes.

Bacterial Diseases

- Bacteria that cause disease are said to be pathogenic.
- Robert Koch (1880) established that bacteria can cause disease. He proved his hypothesis using **Anthrax**. *disease in cattle*
- Koch's postulates: *microorganisms*
  1. The microbe must be present in individuals suffering from the disease and absent from healthy individuals.
  2. The organism must be isolated and grown in a pure culture away from the host organism.
  3. If organisms from the pure culture are injected into a healthy experimental animal, the disease symptoms should appear.
  4. The organism should be isolated from the diseased experimental animal, again grown in pure culture, and demonstrated by its

size, shape, and color to be the same as the original organism.

- Koch's experiments also became the basis for the **germ theory of disease**. *that infectious caused disease are by bacteria & virus*
- **Antibiotics** are molecules that kill bacteria. Widespread use of antibiotics has led to drug-resistant strains of bacteria.

Bioremediation

- Bioremediation is the practice of using bacteria and archaea to degrade pollutants.
  1. Fertilize contaminated sites to encourage the growth of existing bacteria that degrade toxic compounds.
  2. "Seed" or add specific species of bacteria to contaminated sites.

Extremophiles

- Bacteria or archaea living in high-salt, high- or low-temperature, or high-pressure habitats are **extremophiles**.
- Astrobiologists are using extremophiles as the model organisms in the search for extraterrestrial life.

Global Change

The Oxygen Revolution

- No free molecular oxygen existed for the first 2.5 billion years because:
  1. There was no plausible source of oxygen at the time the planet cooled to a solid.

2. The oldest Earth rocks indicate any oxygen that formed immediately reacted with iron atoms to produce iron oxides.
- Even in the ocean, oxygen is produced by **cyanobacteria** through photosynthesis.
- Quantities of oxygen did not begin to build up until 2.0 billion years ago. At 1.8 billion years ago, iron oxides in terrestrial environments appear.
- Once oxygen was common in the oceans, it could be used as an electron acceptor; aerobic respiration became a possibility.
- The rate of energy production and metabolism could now rise dramatically. Coincidentally, 2.1 billion years ago, the first macroscopic algae appeared in the fossil record.

The Nitrogen Cycle

- Plant growth and overall productivity are often limited by the availability of nitrogen.
- Redox reactions that result in the production of ammonia (NH_3) from nitrogen (N_2) are referred to as **nitrogen fixation**.
- Certain species of cyanobacteria are capable of fixing nitrogen.
- On land, nitrogen-fixing bacteria live in close association with plants—often taking up residence in **nodules**.
- Plants need this nitrogen to make up large quantities of proteins.

Nitrate Pollution

- Widespread use of ammonia fertilizers (NH_3) had an unforeseen consequence.
- Nitrate (NO_3^-) produced as a by-product of bacterial ammonia metabolism readily dissolves in water; eventually, it enters aquatic ecosystems.
- Nitrate in the body can cause cancer.
- Nitrate in an aquatic ecosystem can decrease oxygen content, thus causing anaerobic "dead zones" to develop.

27.2   How Do Biologists Study Bacteria and Archaea?

Using Enrichment Cultures

- Biologists rely heavily on the ability to culture organisms in the lab.

- **Enrichment cultures** are based on establishing a specific set of growing conditions.

Using Direct Sequencing

- **Direct sequencing** is a strategy for documenting the presence of bacteria and archaea that cannot be grown in culture and studied in the laboratory.
- This technique is enormously important in understanding diversity because only a tiny fraction of the bacteria and archaea that exist can be grown in the lab.
- Direct sequencing has been used to discover a new lineage of Archaea, called the **Korarchaeota**.

Evaluating Molecular Phylogenies

- **Ribosome complexes** are found in all organisms. Ribosomes are the site of protein synthesis and contain **small subunit** (SSU) RNA.
- In 1960 Carl Woese described base sequences from SSU in many bacterial species. He drew a diagram now known as the **universal tree**, or **tree of life** (**Figure 27.1**).
- Later, follow-up work documented that Bacteria were the first lineage to diverge from the common ancestor of all living organisms.

27.3   What Themes Occur in the Diversification of Bacteria and Archaea?

Morphological Diversity

- Bacteria and Archaea range from the smallest of all free-living cells—mycoplasmas have a volume of 0.03 vm³—to the largest known bacterium, *Thiomargarita namibiensis*, with volumes as large as $200 \propto 10^6$ vm³ (**Figure 27.11a**).
- Bacterial shapes include **filaments**, **spheres**, **rods**, **chains**, and **spirals** (**Figure 27.11b**). Some bacteria are **sedentary** and some are **motile**—using **flagella** or gliding.
- All bacteria are haploid and reproduce by **fission**—the splitting into two daughter cells. But **plasmids** (loops of DNA) can be transferred between bacteria by a **conjugation tube**.

Metabolic Diversity

- Bacteria produce ATP in three ways:
  1. **Phototrophs**—use light energy to promote electrons to the top of electron transport chains. ATP is produced by cellular respiration.
  2. **Organotrophs**—oxidize reduced organic molecules with high potential energy. ATP is produced by cellular respiration or fermentation.
  3. **Lithotrophs**—oxidize inorganic molecules with high potential energy. ATP is produced by cellular respiration with the inorganic compound serving as the electron donor.
- **Autotrophs** manufacture their own carbon-containing compounds, which **heterotrophs** must acquire.

Cellular Respiration: Variation in Electron Donors and Electron Acceptors

- **Cellular respiration**—A molecule with high potential energy serves as an electron donor and is oxidized, whereas a molecule with low potential energy serves as a final electron acceptor and becomes reduced. The potential energy difference is converted into ATP.
- Glucose is a common **electron donor**; oxygen is a **common electron acceptor**.
- When oxygen acts as the final electron acceptor, the by-product is water.
- Electron donors range from hydrogen molecules and hydrogen sulfide to ammonia and methane.
- Other electron acceptors are sulfate, nitrate, carbon dioxide, or ferric ions.

Fermentation

- **Fermentation** is a strategy for making ATP without using electron transport chains.
- No electron acceptor is used; redox reactions are internally balanced.
- Fermentation is less efficient compared to respiration and is often used as an alternative.
- Glucose is fermented into lactic acid or ethanol.
- *Clostridium aceticum* can ferment ethanol, fatty acids, and acetate.

- Species that ferment amino acids produce end-products like cadaverine and putrescine, responsible for the odors of rotting flesh.

Photosynthesis

- Bacteria can use kinetic energy in light to raise electrons to higher energy states.
- As electrons are stepped down to lower energy states, energy released is used to make ATP.
- This requires a source of electrons—cyanobacteria and plants use water. Splitting water to obtain electrons produces oxygen.
- Other bacteria split hydrogen sulfide or ferrous iron.
- It is important to recall the light-absorbing properties of chlorophylls *a* and *b*. Cyanobacteria have the same two pigments.

Pathways for Fixing Carbon

- Organisms must also obtain building-block molecules that contain carbon-carbon bonds.
- In cyanobacteria and plants, enzymes from the Calvin cycle transform CO_2 to organic molecules. Other bacteria, animals, and fungi must obtain carbon by eating plants or animals or by absorbing organic compounds released in dead tissues.
- **Methanotrophs** use methane as their primary electron donor and carbon source.

27.4   Key Lineages of Bacteria and Archaea

Bacteria

- **Spirochaeles**:
  1. They are distinguished by their corkscrew shape and unusual flagella.
  2. Syphilis is caused by a spirochete, and so is Lyme disease.
- **Chlamydiales**:
  1. They are spherical and very tiny.
  2. All known species live as parasites inside host cells (endosymbionts).
  3. *Chlamydia trachomatis* is a very common sexually transmitted disease.
- **High-GC Gram Positives**:
  1. Cell shapes in this lineage vary from rods to filaments. Many soil-dwelling species form branched filaments called **mycelia**.

2. Over 500 distinct antibiotics have been isolated from species in the genus *Streptomyces*. Tuberculosis and leprosy are caused by members of this group.

- **Cyanobacteria**: These organisms produce much of the oxygen and nitrogen and many organic compounds that feed other organisms in freshwater and marine environments.
- **Low-GC Gram Positives**: The species in this group cause anthrax, botulism, tetanus, gangrene and strep throat. *Lactobacillus* is used to make yogurt.
- **Proteobacteria**: Species in this group cause Legionnaire's disease, cholera, dysentery, and gonorrhea. Certain species can produce vinegars. *Rhizobium* can fix nitrogen.

Archaea

- **Crenarchaeota**: None of these species have yet caused disease in humans. These organisms can inhabit extreme environments.
- **Euryarchaeota**: None of these species have yet caused disease in humans. *Ferroplasma* live in piles of waste rock near abandoned mines. Methanogens live in the soils of swamps and guts of animals.

B. CROSS-CUTTING THEMES

Looking Back—
Concepts from Earlier Chapters
Cellular Respiration—Chapter 9
The diverse world of bacteria and archaea depends on the presence of electron donors and acceptors. There are many variations on the basic theme in bacteria and archaea, employing not only glucose, oxygen, and standard fermentation but also an entire suite of new substrates and reactions.

Photosynthesis—Chapter 10
Plants use water as their source of electrons, and bacteria can use many other molecules. Chlorophylls are also used in cyanobacteria in the same manner as in plants.

Carbon Fixing—Chapters 9 and 10
Plants use the Calvin cycle to get carbon from CO_2; and while some bacteria do this, others may fix carbon from other inorganic sources of carbon.

Looking Forward—
Concepts in Later Chapters
Biogeochemical Cycling—Chapter 54
Chapter 27 just touches on the importance of the nitrogen cycle in the global ecosystem, which is discussed in greater detail in **Chapter 54**. There are many other interrelated cycles that exist and function to maintain Earth's balance.

Behavior—Chapter 51
Animal behavior is intimately related to the process of disease transmission by bacteria. Many diseases travel to several animal species before ultimately settling on a host.

C. DIFFICULT TOPICS

In your first-year biology course, you are being exposed to many new terms in a short time. These terms are often introduced with a new branch of biology. In this chapter, terms describing the metabolic diversity of Bacteria can be quite confusing (phototroph, organotroph, lithotroph, autotroph, and heterotroph).

Take special care in learning what each term means, and be able to verbally explain the differences between them to a classmate, lab partner, or friend. "Compare and contrast" questions are often used on tests, so be able to create a list of similarities and differences between these terms. In the simplest approach, categorize the groups into terms explaining how bacteria make ATP (phototroph, organotroph, or lithotroph) and how bacteria obtain carbon (autotroph or heterotroph). After that, the prefixes will help you a lot in defining the terms.

D. ASSESSING WHAT YOU'VE LEARNED

(1) Testing Your Knowledge

1. The key substance making up bacterial cell walls:

   a. polysaccharides
   b. peptidoglycan
   c. ribosomes
   d. polymerases

2. Any bacteria that can cause disease is said to be
   a. pathogenic
   b. contagious
   c. resistant
   d. prolific

3. The use of bacteria and archaea to degrade pollutants is called
   a. biocontamination
   b. seeding
   c. biodetoxification
   d. bioremediation

4. The oxygen revolution was so important because available oxygen in the oceans could then be used as
   a. an electron donor
   b. an electron acceptor
   c. for oxygen fixation
   d. a reducer

5. Plant growth and overall productivity are often limited by the availability of
   a. oxygen
   b. nitrogen
   c. carbon
   d. xenon

6. Direct sequencing is used to document the presence of bacteria and archaea that
   a. cannot be found in animals
   b. cannot be detected
   c. cannot be grown in culture
   d. cannot infect humans

7. Researchers generally evaluate bacterial molecular phylogenies by using sequences from
   a. DNA
   b. cDNA
   c. membrane proteins
   d. ribosomes

8. Bacteria that use light to promote electrons to the top of electron transport chains and produce ATP using cellular respiration are called
   a. phototrophs
   b. autotrophs
   c. heterotrophs
   d. organotrophs

9. The strategy for making ATP without using electron transport chains is called
   a. aerobic respiration
   b. fermentation
   c. electron transport
   d. carbon fixation

10. The lineage of bacteria characterized by causing diseases such as tuberculosis and leprosy are the
    a. high-GC gram positives
    b. low-GC gram positives
    c. cyanobacteria
    d. spirochaeles

(2) Integrating Your Knowledge

(a) How does high nitrate affect aquatic ecosystems?

(b) Describe a few ways in which bacteria obtain carbon-carbon compounds.

(c) What is direct sequencing?

(d) How is fermentation relevant to obtaining energy?

(e) How do bacteria obtain energy from light?

CHAPTER 27—ANSWER KEY

D. Assessing What You've Learned

(1) Testing Your Knowledge
1. b; 2. a; 3 d; 4 a; 5. b; 6. c; 7. d; 8. a; 9. b; 10. a

(2) Integrating Your Knowledge
(a) Cyanobacteria and algae use nitrates and grow and die in huge numbers. Decomposers like heterotrophic bacteria and archaea then grow in huge numbers. These organisms then use up all available oxygen and produce "dead zones."

(b) Bacteria either make bonds themselves or get compounds from other organisms. Methanotrophs use methane (CH_4) as their primary carbon source. Several groups of bacteria can fix CO_2 using pathways other than the Calvin cycle.

(c) Direct sequencing is a strategy for documenting the presence of bacteria and archaea that cannot be grown in culture and studied in the laboratory. This technique is enormously important in understanding diversity because only a tiny fraction of the bacteria and archaea that exist can be grown in the lab.

(d) Fermentation is a strategy for making ATP without using electron transport chains. No electron acceptor is used; redox reactions are internally balanced. Fermentation is less efficient compared to respiration and is often used as an alternative. Glucose is fermented into lactic acid or ethanol.

(e) Bacteria can use kinetic energy in light to raise electrons to higher energy states. As electrons are stepped down to lower energy states, energy released is used to make ATP. Phototrophy requires a source of electrons—cyanobacteria and plants use water. Splitting water to obtain electrons produces oxygen.

28

Protists

A. KEY BIOLOGICAL CONCEPTS

28.1 Why Do Biologists Study Protists?

Impacts on Human Health and Welfare

- A protist caused the Irish potato famine, which was the most spectacular crop failure in history.
- Plasmodium is the causative agent of malaria and is transmitted to humans by mosquitoes. At least 300 million people worldwide are sickened by it each year, and over 1 million people die from the disease annually.
- Red tides occur when toxin-producing protists called dinoflagellates reach high densities in a particular area. Shellfish eat these protists, and people can become sick from eating the contaminated shellfish.

Ecological Importance of Protists

- Although protists represent 10% of the total number of named species, one teaspoon of pond water can contain over 1000 flagellated protists. Dinoflagellates can reach concentrations of 60 million cells per liter of seawater.
- Photosynthetic species are major players in the global carbon cycle. Primary productivity by the world's oceans, in turn, is responsible for almost half of the total carbon that is fixed on the planet.
- The photosynthetic protists (plankton) are the basis of marine and freshwater food chains.
- Protists could help reduce global warming by becoming carbon sinks.

28.2 How Do Biologists Study Protists?

Microscopy: Studying Cell Structure

- Detailed studies of cell structure show that many protists have a characteristic overall form.
- Researchers interpret these distinct morphological features as **synapomorphies**—shared, derived traits that distinguish major monophyletic groups (see **Table 28.2**).

Evaluating Molecular Phylogenies

- Using the gene that codes for RNA molecule in the small subunit of ribosomes, eight monophyletic groups from the Eukarya were found.
- The current tree (**Figure 28.7**) confirms that lineages grouped under the name *protist* are **paraphyletic**—meaning they do not comprise all of the descendants of a single common ancestor (see **Box 28.1**).

Combining Data from Microscopy and Phylogenies

- The eukaryotes share a common ancestor with the archaea. Early eukaryotes were probably single-celled organisms with a cytoskeleton and a nucleus, but no cell wall.

The Endosymbiosis Theory

- Lynn Margulis expanded on the original hypothesis of the origin of mitochondria. **Endosymbiosis theory** proposes that mitochondria originated when a bacterial cell took up residence inside a eukaryote about 2 billion years ago. The **host** supplied the bacterial **symbiont** with protection and reduced carbon compounds from its other prey; the symbiont supplied the host with ATP (**Fig. 28.9**).

- The endosymbiosis theory also contends that chloroplasts originated in an analogous way.

Do the Data Support the Endosymbiosis Theory?

- Observations are consistent with the theory:
  - Mitochondria and chloroplasts are about the size of an average bacterium.
  - Both organelles replicate by fission, and this process is independent of division by the host cell.
  - Mitochondria and chloroplasts have their own ribosomes and manufacture their own proteins. The ribosomes are similar to bacterial ribosomes.
  - The chloroplast glaucophyte alga has an outer layer containing the same molecule (peptidoglycan) found in the cell walls of cyanobacteria.
  - Some cyanobacteria look like chloroplasts—*Prochloron* even has thylakoids.
  - Both mitochondria and chloroplasts have double membranes, consistent with the engulfing mechanism.
  - Mitochondria and chloroplasts have their own genomes, which are organized as circular molecules, like bacteria.
- In the 1970s researchers compared gene sequences isolated from the nuclear DNA of eukaryotes, mitochondrial DNA from the same eukaryote, and DNA from several species of bacteria. Mitochondrial gene sequences were found to be much more closely related to the sequences from the bacteria.
- Researchers also showed that chloroplast genes are most closely related to sequences from cyanobacteria, not plants.
- Direct sequencing has led to the discovery of several new lineages of eukaryotes.

28.3 What Themes Occur in the Diversification of Protists?

Morphological Diversity

- The average eukaryotic cell is 10 times larger than an averaged-sized bacterial cell. Larger

volume is a problem, because volume increases by a power of 3 while surface area increases by a power of 2. Metabolic machinery in the volume of the cell will soon overcome the transporters of food and waste on the membrane.

Organelles Divide a Cell into Compartments

- Eukaryotes solve the problem of size by dividing their cell volume into compartments; consider the *Paramecium* (**Figure 28.12**).
- After ingesting a bacterium, a *Paramecium* surrounds it with an internal membrane, forming a compartment called a **food vacuole**. Food vacuoles combine with membrane-bound structures called **lysosomes** that hold digestive enzymes and circulate around the cell. Once the food is digested and absorbed, the vacuole merges with the external cell membrane and expels waste molecules.
- Recall the many membrane-bound organelles that may be present in eukaryotes.
- The cytoskeleton may be the most important feature of eukaryotes in organizing the cell volume. This system of filaments and tubules provides scaffolding that supports and organizes the cell.

The Evolution of Multicellularity

- **Differentiation** of cell types is a crucial criterion for defining multicellularity. In contrast, **colonial growth** defines cells that all perform the same function.
- *Volvox* cells cannot survive on their own, and each cell has a distinct function (**Figure 28.13**). Some cells are vegetative, some are photosynthetic, and some are sexual.
- Multicellularity has evolved several times, independently, in protists (**Box 28.2**, **Figure 28.10**).

Structures for Support and Protection

- Many protists have a rigid internal skeleton or a hard external structure, called a **test** or a **shell**. Tests vary in structure from intricate chambers of calcium carbonate found in foraminifera to glass-like, silicon-containing structures enclosing diatoms (**Figure 28.14**).

- Green algae have a tough outer wall made of cellulose.

How Do Protists Find Food?

Ingestive, Absorptive, and Photosynthetic Lifestyles

- **Predation and Scavenging**: Protists can eat bacteria, archaea, and other protists—dead or alive.
- The engulfing process is possible because protists lack a cell wall, and because their flexible membrane and cytoskeleton move as shown in **Figure 28.15a**.
- **Parasitism**: Some absorptive groups are decomposers, like the oomycetes. Many others live inside organisms as parasites, although many are beneficial or **mutualistic** relationships (**Figure 28.16**). Malaria is one of the most damaging of all human parasites.
- **Photosynthesis**: The red algae, brown algae, green algae, and other groups are distinguished by the pigments they contain, which absorb unique wavelengths of light (**Figure 28.17**). Harvesting different wavelengths of light may avoid competition.
- **Diversity**: All three lifestyles occur in many different eukaryotic lineages, and all three can occur in the same group of species.

How Do Protists Move?

- **Amoeboid movement** is a gliding motion. Fingerlike projections called **pseudopodia** stream forward over a substrate. This motion involves interactions between actin, myosin, and ATP (**Figure 28.20a**).
- The other major mode of locomotion involves **flagella** or **cilia**. Flagella and cilia have identical structures and are actually extensions of the cell. Both consist of nine sets of doublet microtubules around two central, single microtubules; and dynein is the major motor protein (**Figure 28.20b, c**).

How Do Protists Reproduce?

Sexual versus Asexual Reproduction

- **Asexual reproduction** produces offspring that are genetically identical to the parent.

- **Sexual reproduction** produces offspring that are genetically different from their parents.
- **Meiosis** is the reduction division that produces a set of haploid daughter cells from a diploid parent cell. Genetic differences between parents and offspring are increased when **crossing over** occurs during meiosis.
- Importance:
  1. Parents produce better-quality offspring.
  2. Genetically variable offspring have a better chance of surviving during environmental changes.

Variation in Life Cycles

- Every aspect of a life cycle is variable among protists (**Figure 28.21**).
- **Alternation of generations** is a phenomenon in which the haploid and diploid phases of the life cycle are multicellular (**Figure 28.22**).
- Multicellular haploid and diploid forms of the same individual can be identical in morphology, or radically different.

28.4 Key Lineages of Protists

Excavates

- **Diplomonads**: Each cell has two nuclei and all lack a cell wall (e.g., *Giardia*).
- **Parabasalids**: Cells lack a cell wall and feed by engulfing (e.g., *Trichomonas*).

Discicristates

- **Euglenids**: Cells lack a cell wall but have a system of interlocking protein strips. One-third of species perform photosynthesis. Euglenids are important components of freshwater plankton and food chains.

Alveolates

- **Ciliates**: Cells have a micronucleus and a macronucleus. These organisms can reproduce by conjugation, and they use cilia for locomotion.
- **Dinoflagellates**: Cells from sexual reproduction may form cysts to remain dormant. Dinoflagellates are second only to diatoms in amount of carbon fixed.

- **Apicomplexa**: All known species are parasitic. Malaria is caused by a *Plasmodium*.

Stramenopiles

- **Oomycetes**: These organisms are fungus-like and very important decomposers in aquatic ecosystems. An oomycete caused the Irish potato famine.
- **Diatoms**: Cells are supported by silicon-rich, glassy shells. These organisms are the most important producer of carbon compounds in the water.
- **Brown Algae**: They are photosynthetic and sessile, though their reproductive cells may be motile. Brown algae form forests that are important habitats.

Cercozoa

- **Foraminifera**: Cells have multiple nuclei, and their tests are usually made of organic material. Dead forams often form extensive deposits.

Plants

- **Red Algae**: Cell walls are composed of cellulose and other polymers and have no flagella. Species contribute to reef building.
- **Plasmodial Slime Molds**: Individuals form a huge supercell containing many diploid nuclei. Slime molds are important decomposers in forest ecosystems.

B. CROSS-CUTTING THEMES

Looking Back—
Concepts from Earlier Chapters
Cell Organelles—Chapter 7
All of the organelles of the cell, as well as their structure and function, are described in detail in **Chapter 7**. Chapter 28 explains some of the diversity in function of the same organelles.

Diversity—Chapter 25
The variation in photosynthetic pigments observed in bacteria echoes the theme in diversification of protists. Photosynthetic species harvest different wavelengths of light to avoid competition.

Genetic Recombination—Chapters 17 and 18
Chapter 17 introduces the mechanisms that bacteria use to transfer genes and undergo genetic recombination. These cells exchange short stretches of DNA between individuals, usually via plasmids. In protists, the entire genome is involved.

Meiosis—Chapter 12
The reduction division called meiosis is explored in **Chapter 12**. The process is now applied and explained in relation to the first organisms to adopt this process.

Looking Forward—
Concepts in Later Chapters
Ecosystems—Chapter 54
This chapter illustrates how all life is co-dependent upon other organisms through their key positioning in ecosystems.

C. DIFFICULT TOPICS

In this chapter, the theory of endosymbiosis is used to explain the origins of the first eukaryotes. Although the theory itself may be easy to understand, explaining it and the techniques used to support it is a different matter.

Try starting with the basics. Outline the theory and what it tries to explain. Then outline each experiment in the chapter that supports the theory, and consider why it supports the theory. If you do not understand the techniques used in a particular experiment, go back in the text to previous chapters that outline those techniques in more detail. Theories are modified constantly as new evidence (for or against) is published.

D. ASSESSING WHAT YOU'VE LEARNED

(1) Testing Your Knowledge

1. Which of the following, ranked as one of the most devastating diseases, is caused by a protist?
   a. red tides
   b. AIDS
   c. malaria
   d. plasmodium

2. The theory of endosymbiosis proposes that mitochondria originated
   a. when a virus invaded and stayed in a bacterial cell
   b. when a bacterial cell took up residence inside a eukaryote
   c. when a virus took up residence inside a eukaryote
   d. when a bacterial cell took up residence inside a prokaryote

3. Which of the following statements *does not* support endosymbiosis?
   a. Cyanobacteria look nothing like chloroplasts.
   b. Mitochondria and chloroplasts are about the same size as an average bacterium.
   c. Mitochondria and chloroplasts both have a double membrane.
   d. Mitochondria and chloroplasts both reproduce by fission.

4. The crucial criterion for defining multicellularity is
   a. differentiation
   b. colonial growth
   c. lysosomes
   d. cell walls

5. Which of the following is *not* a mode of locomotion in protists?
   a. amoeboid movement
   b. using flagella
   c. using cilia
   d. using mitochondria

6. The phenomenon in which the haploid and diploid phases of a life cycle are multicellular is called
   a. double diploid
   b. alternation of generations
   c. haplodiplo generation
   d. multigenerational

7. Which type of reproduction produces offspring that are genetically different from their parents?
   a. sexual reproduction
   b. asexual reproduction
   c. multisexual reproduction
   d. unisexual reproduction

8. This group of protists includes the species responsible for malaria.
   a. diatoms
   b. red algae
   c. apicomplexa
   d. dinoflagellates

9. This group caused the Irish potato famine.
   a. diatoms
   b. apicomplexa
   c. ciliates
   d. oomycetes

10. The key characteristic of the plasmodial slime molds is
    a. their single nuclei per cell
    b. their ability to parasitize
    c. their huge supercell with many nuclei
    d. their carbon fixing in aquatic habitats

(2)   Integrating Your Knowledge

(a) How is stable carbon deposited in the ocean?

(b) How does a *Paramecium* digest food?

(c) What are the key differences between cilia and flagella?

(d) Compare and contrast asexual and sexual reproduction.

(e) What is alternation of generations?

CHAPTER 28—ANSWER KEY

D. Assessing What You've Learned

(1)   Testing Your Knowledge

1. d; 2. b; 3. a; 4. a; 5. d; 6. b; 7. a; 8. c; 9. d; 10. c

(2)   Integrating Your Knowledge

(a) Carbon turnover in the ocean is significantly more rapid than in the terrestrial environment. However, limestone is a stable repository for carbon in the oceans. The sugars that protists produce are the basis for food chains in both freshwater and marine environments.

(b) After ingesting a bacterium, a *Paramecium* surrounds it with an internal membrane, forming a compartment called a food vacuole. Food vacuoles combine with membrane-bound structures called lysosomes that hold digestive enzymes and circulate around the cell. When the food has been digested and absorbed, the vacuole merges with the external cell membrane and expels waste molecules.

(c) Cilia are short and usually occur in large numbers, whereas flagella are long and are usually found singly or in pairs. Both are markedly different from those found in Bacteria and Archaea.

(d) Asexual reproduction produces offspring that are genetically identical to the parent Sexual reproduction produces off-spring that are genetically different from their parents. Meiosis is the reduction division that produces a set of haploid daughter cells from a diploid parent cell. Genetic differences between parents and offspring are increased when crossing over occurs during meiosis.

(e) Alternation of generations is a phenomenon in which the haploid and diploid phases of the life cycle are multicellular. Multicellular haploid and diploid forms of the same individual can be identical in morphology or radically different.

29

Green Plants

A. KEY BIOLOGICAL CONCEPTS

29.1   Why Do Biologists Study the Green Plants?

Plants Provide Ecosystem Services

- Plants produce oxygen via oxygenic photosynthesis.
- Plants build soil by providing food for decomposers.
- Plants hold soil and prevent nutrients from being washed or blown away (**Figure 29.1**).
- Plants hold water and increase water-holding capacity of an area.
- Plants moderate the local climate

Agricultural Research: Domestication and Selective Breeding

- The grains that form the basis of our current food supply were derived from wild species about 10,000 to 2000 years ago. Humans were actively selecting seeds to plant the next generation of crops—a process called **artificial selection**. Recently, increases in the oil content of corn and the 178 distinct varieties of potato being cultivated show the power of artificial selection (**Figure 29.3**).

Plant-Based Fuels and Fibers

- Sugars synthesized during the Carboniferous period laid the groundwork for the industrial revolution. There will likely not be another generation of plant-based fuels.
- The primary interest in woody plants is for building materials and fibers used in papermaking. Relative to its density, wood is a stiffer and stronger building material than concrete, cast iron, aluminum alloys, or steel. Refined cellulose fibers are stronger under tension than are nylon, silk, chitin, collagen, tendon, or bone—even though they are 25% less dense.
- Current research in forestry focuses on maintaining the productivity and diversity of forests that are managed for wood and pulp production. Most of the old-growth forests have now been cut. The current challenge to foresters is to reduce the demand for wood products through recycling and more efficient use of existing timber, and to sustain productivity of second- and third-growth forests.

Bioprospecting

- The effort to find naturally occurring compounds that can be used as drugs, fragrances, insecticides, herbicides, or fungicides is called **bioprospecting**.
- **Hydroponics** can be used to harvest large quantities of plant chemicals.
- **Ethnobiologists** (**Figure 29.5**) study how humans use plants. They are collecting plants in areas threatened by destruction.

29.2   How Do Biologists Study Green Plants?

Analyzing Morphological Traits

- The 12 most important phyla are grouped into three categories:
  1. **Nonvascular plants**: The liverworts, hornworts, and mosses (**Figure 29.7b**)

2. **Seedless vascular plants**: The horsetails, ferns, lycopods, and whisk ferns (**Figure 29.7c**)
3. **Seed plants**: The gymnosperms and angiosperms (**Figure 29.7d**)

Using the Fossil Record

- The fossil record for land plants is massive. It began 476 million years ago (**Figure 29.8**) and is broken up into five segments:
  1. **Origin of land plants**—476 million years ago; duration, 60 million years. Three types of microscopic fossils are present: (a) spores surrounded by a tough membrane, (b) spores that have a waxy cuticle, and (c) spores shaped like small tubes.
  2. **Silurian-Devonian explosion**—from 410 to 360 million years ago. Macroscopic fossils from most major plant lineages are found. Virtually all adaptations that allow plants to occupy dry, terrestrial habitats are present, including water-conducting cells, roots, and wood.
  3. **Carboniferous period**—from 350 to 290 million years ago. Extensive deposits of coal are found, which is a carbon-rich rock packed with fossil spores, branches, leaves, and tree trunks. Usually derived from lycopods and in the presence of water, this indicates the presence of extensive forested swamps.
  4. **Gymnosperms**—from 250 to 120 million years ago. These lineages include the cycads, conifers, and gingkoes. Because gymnosperms grow readily in dry habitats, biologists infer that both wet and dry environments on the continents became blanketed with green plants for the first time during this interval.
  5. **Angiosperms**—from 125 million years ago to the present. The woody plants that produced the first flowers are the ancestors of today's grasses, orchids, daisies, oaks, maples, and roses.

Evaluating Molecular Phylogenies

- The phylogenetic tree shown in **Figure 29.9** is a recent version from laboratories studying plant phylogenies. Note that:

  – Land plants most likely evolved from green algae.
  – Charales is the sister group to land plants.
  – The "Green algae" group is paraphyletic.
  – The land plants are monophyletic.
  – The nonvascular plants are the most basal groups among land plants.
  – The morphological simplicity of the whisk ferns is likely a derived trait.
  – Seeds and flowers evolved only once.

29.3   What Themes Occur in the Diversification of Green Plants?

The Transition to Land, I: How Did Plants Adapt to Dry Conditions?

- Plants had to adapt to situations where only a portion of their tissues were bathed in fluid.
- Solving the water problem was a breakthrough in evolution by natural selection. It was achieved in two steps:
  1. Preventing water loss from cells
  2. Transporting water from tissues with access to water to tissues without access to water

Preventing Water Loss: Cuticle and Stomata

- **Cuticle** is a waxy, watertight sealant that enables plants to survive in dry environments (**Figure 29.10a**). Gas exchange is accomplished on wax-covered plants by the presence of **stomata**, which consist of an opening covered by **guard cells**.
- The **pore** opens or closes as the guard cells change shape (**Figure 29.10b**). The presence of guard cells in later groups implies that many of the early land plants had the ability to regulate gas exchange and control water loss by opening and closing their pores.

Transporting Water: Vascular Tissue and Upright Growth

- Extraordinary fossils have been found in Scotland in a formation called the Rhynie Chert. This formation contains numerous plants fossilized in an upright position. Cells in these plants were able to pump cells up with water and make the early land plants rigid due to **turgor pressure**. Some cells contained thickened rings containing **lignin**, a polymer built from six-carbon rings and extraordinarily strong for its weight.

- At about 380 million years ago, the fossil record contains **tracheids** (**Figure 29.11a**). Tracheids are elongated cells that die after maturing, do not contain cytoplasm, and are the first true vascular tissues that made possible the efficient transport of water to aboveground tissues.
- About 250–270 million years ago, plants have true vessel elements (**Figure 29.11b**). Vessels are the most advanced type of water-conducting tissue known in plants and are present only in gnetophytes and angiosperms.

Transition to Land, II: How Do Plants Reproduce in Dry Conditions?

- All of the land plants have multicellular haploid phases (**gametophyte**) as well as multicellular diploid phases (**sporophyte**). This is known as **alternation of generations** (**Figure 29.14**).
- The relationship between gametophyte and sporophyte is variable. In bryophytes, the sporophyte is small; in ferns, it is larger; and in gymnosperms and angiosperms, the sporophyte is dominant. Mutation and selection resulted in male gametophytes that were dramatically reduced in size and surrounded by a coat of sporopollenin which resulted in **pollen grains**.

Retaining and Nourishing Offspring

- In the early history of land plants, two evolutionary changes occurred:
  1. Gametes were produced in complex, multicellular structures.
  2. The embryo was retained on the parent plant and nourished.

The Evolution of Pollen

- Pollen evolved, and seed plants lost their dependence on water for fertilization.

The Evolution of the Seed

- A **seed** is a structure that encloses and protects a developing embryo. Seeds are often attached to a structure that aids in dispersal by wind, water, or animals (**Figure 29.18**).

The Evolution of the Flower

- Flowers are diverse in size, shape, and color (**Figure 29.20**); they produce a wide range of scents. Scientists hypothesize that flowers are adaptations to increase the probability that pollination will occur. Flowers provide a food reward to the pollinator in the form of sugar-rich nectar or protein-rich pollen grains—a mutually beneficial relationship.
- About 100 Mya, the fossil record contains the first evidence of correlations between the size and shape of flowers and the size and shape of the mouthparts in their insect pollinators.

The Angiosperm Radiation

- Angiosperms represent one of the great adaptive radiations in the history of life.
- The diversification of angiosperms is associated with three key adaptations: (1) vessels, (2) flowers, and (3) fruits.
- **Monocots** are monophyletic, but **dicots** are paraphyletic. A new term for the monophyletic group of dicots is **eudicots** (**Fig. 29.25**).

29.4   Key Lineages of Green Plants

Green Algae

- **Ulvobionta**: Most green algae you have seen belongs to this group (e.g., *Volvox*). They are important primary producers in aquatic areas.
- **Coleochaetales**: Most grow as flat sheets of cells (e.g., water lilies), and the multicellular individuals are haploid.
- **Charales (Stoneworts)**: These organisms commonly accumulate crusts of calcium carbonate over their surfaces. They can form extensive beds on lake bottoms.

Nonvascular Plants ("Bryophytes")

- **Hepaticophyta (Liverworts)**: Liverworts have liver-shaped leaves and can grow on bare rock or tree bark, which helps in soil formation.
- **Anthocerophyta (Hornworts)**: Sporophytes have a hornlike appearance, and they have stomata.
- **Bryophyta (Mosses)**: Mosses can be abundant in extreme environments, and they can

become dormant. *Sphagnum* species are among the most abundant.

Seedless Vascular Plants

- **Lycophyta (Lycopods or Club Mosses)**: The lycopods are the most ancient plant lineage with roots. Tree-sized lycophytes dominated the coal-forming forests of the Carboniferous.
- **Psilotophyta (Whisk Ferns)**: These plants are restricted to tropical regions and have no fossil record.
- **Sphenophyta (Horsetails)**: Horsetails can flourish in waterlogged soils by allowing oxygen to diffuse down their hollow stem.
- **Pteridophyta (Ferns)**: Ferns are the only seedless vascular plants to have large, well-developed leaves.

Seed Plants

- **Gnetophyta**: This group has about 70 species in three living genera. The drug ephedrine has been isolated from this group.
- **Cycadophyta**: Cycads are mostly found in the tropics. They harbor many symbiotic, nitrogen-fixing cyanobacteria, which are important sources of nutrients.
- **Ginkgophyta**: One species of ginkgo is alive today. It is deciduous, and individual trees are either male or female.
- **Coniferophyta**: This group of land plants is named for its reproductive structure, the cone. Conifers dominate all high-latitude and high-altitude forests.
- **Anthophyta (Angiosperms)**: The defining adaptation of this group is the flower. They supply the food that supports virtually every other species.

B. CROSS-CUTTING THEMES

Looking Back—
Concepts from Earlier Chapters

Evolutionary Relationships—Chapter 23
Chapter 23 introduces us to how molecular and morphological data are used to infer evolutionary relationships. This investigation of evolutionary relationships is continued in Chapter 29 with the green plants.

Alternation of Generations—Chapter 28
The phenomenon of alternation of generations was introduced in **Chapter 28** on the protists. This phenomenon persists with further variation in the land plants in Chapter 29.

Genetic Engineering—Chapter 19
Mechanisms and procedures for genetic engineering are explained in **Chapter 19** and applied to the production of green plants in Chapter 29.

Looking Forward—
Concepts in Later Chapters

Turgor Pressure—Chapter 36
Turgor pressure is mentioned only as an explanation for the rigidity in early land plants. The mechanisms for producing turgor pressure are explored in detail in **Chapter 36**.

The Seed—Chapter 40
The structure of the seed is introduced in Chapter 29. How it protects the developing embryo, and the changes that occur during seed growth and development, are discussed in detail in **Chapter 40**.

C. DIFFICULT TOPICS

By far, understanding how alternation of generations works, and then changes during the evolution of land plants, is the toughest concept to master in this chapter.

First, think about the definition of alternation of generations, and then look at the differences between sporophytes and gametophytes. Next, slowly go through **Figure 29.14**. Really take in the general details, like how much time the groups spend in haploid versus diploid stages. Lastly, try to describe to your classmates these major changes in the different plant groups.

D. ASSESSING WHAT YOU'VE LEARNED

(1) Testing Your Knowledge

1. The process of actively choosing seeds to plant the next generation of crops is called:
   a. natural selection
   b. artificial selection

c. general selection

d. hydroponics

2. The effort to find naturally occurring compounds that can be used as drugs, fragrances, insecticides, herbicides, or fungicides is called:

a. bioprospecting

b. hydroponics

c. natural selection

d. plant hunting

3. Which of the following is *not* a major grouping of land plants?

a. seedless vascular plants

b. nonvascular plants

c. cone-bearing plants

d. seed plants

4. The latest and most impressive plant radiation was that of the

a. algae

b. angiosperms

c. gymnosperms

d. club mosses

5. Which of the following statements about plant molecular phylogenies is *true*?

a. The group "green algae" is paraphyletic.

b. The Charales is not the sister group to green algae.

c. The land plants are paraphyletic.

d. The nonvascular plants are the most recent group.

6. Plants can prevent water loss from cells by

a. secreting a waxy cuticle

b. exchanging respiratory gases through a pore using guard cells

c. regulating the opening and closing of these pores

d. all of the above

7. Upright growth in non-woody plants is attained by

a. extensive root systems for support

b. production of thicker leaves

c. absence of leaves

d. pumping water into cells to increase their rigidity

8. The structure that encloses and protects the developing plant embryo is

a. the dicot

b. the seed

c. the pollen grain

d. the ovule

9. The diversification of the angiosperms is associated with the following adaptation:

a. vessels

b. flowers

c. fruits

d. all of the above

10. The only major group of seed plants with only one species alive today is

a. Ginkophyta

b. Cycadophyta

c. Coniferophyta

d. Gnetophyta

(2)   Integrating Your Knowledge

(a) What key innovations allowed land plants to tolerate dry conditions?

(b) What are the two main problems in transporting water?

(c) What is the difference between sporophyte plants and gametophyte plants?

(d) Why is the flower a key innovation in plants?

(e) What is coal, and how was it formed?

CHAPTER 29—ANSWER KEY

D. Assessing What You've Learned

(1) Testing Your Knowledge

1. b; 2. a; 3. c; 4. b; 5. a; 6. d; 7. d; 8. b; 9. d; 10. a

(2) Integrating Your Knowledge

(a) Solving the water problem was a breakthrough in evolution by natural selection, achieved in two steps:

1. Preventing water loss from cells

2. Transporting water from tissues with access to water to tissues without access to water

(b) 1. Transporting water from tissues that are in contact with wet soil to tissues that are in contact with dry air, against the force of gravity

2. Becoming rigid enough to avoid falling over in response to gravity and wind

(c) All land plants have multicellular haploid phases (gametophyte) as well as multicellular diploid phases (sporophyte).

(d) Flowers are diverse in size, shape, and coloration. They also produce a wide range of scents. Scientists have hypothesized that flowers are adaptations that increase the probability that pollination will occur. Flowers provide a food reward to the pollinator in the form of sugar-rich nectar or protein-rich pollen grains—a mutually beneficial relationship.

(e) Coal deposits are dead plants that were laid down in the Carboniferous period. The plants did not have a chance to decay; therefore, they are a huge source of carbon.

Fungi

30

A. KEY BIOLOGICAL CONCEPTS

30.1 Why Do Biologists Study Fungi?

Fungi Feed Land Plants

- **Mycorrhizal** ("fungal-root") associations between fungi and the roots of land plants are extremely common (**Figure 30.1**).
- Tree species can grow three to four times faster with these fungal associations.

Fungi Speed the Carbon Cycle on Land

- **Saprophytes** are fungi that make their living by digesting dead plant material. They did not grow in the Carboniferous period, because the pH was too low.
- At the end of the great Permian extinction, researchers documented a huge fungal spike (**Figure 30.2**).
- Fungi make the carbon cycle turn faster.

Fungi Have Important Economic Impacts

- Incidence of fungal infections in humans is low. The major destructive impact is on crops (**Figure 30.4a**).
- In nature, fungi have killed 4 billion chestnut trees and tens of millions of American elm trees. The fungi were accidentally imported (**Figure 30.4b**).

30.2 How Do Biologists Study Fungi?

Analyzing Morphological Traits

- Only two growth forms exist: (1) single-celled forms called **yeasts**; and (2) multicellular, filamentous forms called **mycelia** (**Figure 30.5**).
- The filaments that make up a mycelium are called **hyphae**. Each filament is separated by cross-walls, called **septa** (**Figure 30.6**).
- Mycelia are adapted to an absorptive lifestyle. The only thick, fleshy structures are the reproductive structures.

Reproductive Structures

- By analyzing reproductive structures, biologists have identified four main groups of fungi (**Figure 30.7**):
  1. **Chytridiomycota (chytrids)**: The chytrids are the only group of fungi that live in water. Chytrids reproduce sexually and asexually. They are the only fungi that have motile cells.
  2. **Zygomycota**: Most Zygomycota live in soil, and many are key partners in the mutualistic interactions with plants. The hyphae of Zygomycota are haploid and come in several different mating types or "sexes." If **pheromones** released by two hyphae indicate that they are from different mating types, the individuals become yoked together.
  3. **Basidiomycota (club fungi)**: Mushrooms, bracket fungi, and puffballs are reproductive structures produced by members of the Basidiomycota. Their size, shape, and color vary enormously; but they all origiate from hyphae of mated individuals. They form specialized structures called **basidia**, which produce spores.

4. **Ascomycota (sac fungi)**: This group of fungi produces complex, cup-shaped reproductive structures. The tips of these hyphae have distinctive saclike cells called **asci**.

Evaluating Molecular Phylogenies

- Researchers compared results of ribosomal DNA sequence analyses with data from studies of cell-wall composition, life-cycle patterns, and reproductive structures. The phylogeny in **Figure 30.8a** is the current best estimate. The comparisons support 3 important conclusions:
  1. It is likely that a lineage of chytrids forms the most basal group of fungi, but it is not clear whether fungi moved from water to land just once or many times.
  2. The groupings called Chytridiomycota and Zygomycota appear to be paraphyletic. Instead, they have either evolved more than once or were present in a common ancestor and were retained in several lineages.
  3. Basidiomycota and Ascomycota are monophyletic and appeared late during the diversification of fungi.

Where Do Fungi Fit on the Tree of Life?

- Looking at **Figure 30.9**, the tree supports two conclusions:
  1. Closest living relatives of fungi are single-celled parasites called **microsporidians**.
  2. Fungi are more closely related to animals than they are to land plants.

Experimental Studies of Mutualism

- Fungi and land plants often have a **symbiotic** relationship.
- **Mutualism** is a relationship that benefits the host in return for the food the fungi absorbs.
- Researchers have begun to categorize these relationships as mutualistic, parasitic, or commensal (only the fungi benefit).
- Using radioactive isotopes of CO_2, phosphorus, and nitrogen, researchers have documented that up to 20% of sugars produced by a plant are transferred to fungi, which facilitate transfer of phosphorus and nitrogen to the plant.

30.3   What Themes Occur in the Diversification of Fungi?

Fungi Participate in Several Types of Mutualisms

Ectomycorrhizal Fungi (EMF)

- EMF form a dense network of hyphae around roots. They are found virtually on all of the tree species growing in temperate and boreal forests (**Figure 30.11a**).
- Where the growing season is so short, the decomposition of needles, leaves, twigs, and trunks is often sluggish. As a result, nitrogen tends to remain tied up in dead tissues.
- EMF, however, release peptidases that cleave the peptide bonds between amino acids. The nitrogen released by this cleavage is absorbed by the hyphae and transported close to the tree roots, where it can be absorbed by the plant. In return, the fungi receive sugars and other reduced carbon compounds from the plant.

Arbuscular Mycorrhizal Fungi (AMF)

- AMF grow *into* the cells of root tissue and are members of the Zygomycota. Found in 80% of all land plant species, they are especially common in grasslands and in the forests of warm, tropical habitats (**Figure 30.11b**).
- In these habitats, phosphorus is the limiting nutrient, so fungi supply plants with phosphorus in exchange for carbon.

Are Endophytes Mutualists?

- Endophytes do not cause disease in the tissues they invade. Most are mutualistic, and some are commensal.

Adaptations That Make Fungi Effective Decomposers

Extracellular Digestion

- Fungi secrete enzymes outside their hyphae into their food. Digestion takes place outside the fungus, or **extracellularly**.
- Only a few organisms on Earth can digest cellulose, and members of the basidiomycetes are the only organisms that can degrade lignin completely to CO_2 (**Figure 30.12**).

Lignin Degradation

- **Lignin peroxidase** is an enzyme that catalyzes the removal of a single electron from an atom in the aromatic rings of lignin. This oxidation step creates a free radical and leads to a series of uncontrolled, unpredictable reactions that end up splitting the polymer into smaller units.
- The uncontrolled oxidation reactions triggered by lignin peroxidase are analogous to the uncontrolled oxidation reactions that occur when wood burns. The uncontrolled nature of the reactions means that the oxidation of lignin cannot be harnessed to drive the production of ATP. However, by oxidizing lignin, the hyphae gain access to huge supplies of energy-rich cellulose within.

Cellulose Digestion

- Fungi secrete **cellulases** into the extracellular environment. Some cleave long strands of cellulose into a disaccharide called **cellobiose**. Other cellulases are equally specific; together, they eventually convert cellulose into glucose.
- Biologists have purified seven different cellulases from the fungus *Trichoderma reesei*. These could be used to make paper production more efficient.

Variation in Life Cycles

Unique Aspects of Fungal Life Cycles

- The process of sexual reproduction begins when hyphae from two individuals fuse to form a hybrid hypha (**Figure 30.13a**). When cytoplasm fuses, **plasmogamy** is said to occur. When the nuclei of the two individuals stay independent, the mycelium is **heterokaryotic**. When the nuclei fuse, it is called **karyogamy**.
- Fungi have four major types of life cycles:
  1. The Chytridiomycota are the only fungi to exhibit alternation of generations. Swimming gametes fuse and form a diploid zygote (**Figure 30.13b**).
  2. The hyphae of Zygomycota are haploid, with several different mating types or "sexes" (**Figure 30.13c**). The haploid nuclei from each hypha fuse, developing a tough, resistant coat. When conditions are favorable, meiosis occurs, and the products grow into a structure that produces haploid spores. When spores are released and germinate, they grow into new mycelia.
  3. In Basidiomycota, hyphae of mated individuals fuse. Then the cytoplasms fuse, but the nuclei remain independent. These are called **heterokaryotic hyphae**, and they eventually grow into fruiting bodies that we call mushrooms (**Figure 30.13d**). When they do, specialized cells called **basidia** form at the ends of the hyphae; and inside, the nuclei finally fuse. The diploid nucleus undergoes meiosis, and haploid spores mature from the meiotic products. Sexual reproduction concludes when the spores are expelled and dispersed by the wind.
  4. In Ascomycota, a short heterokaryotic hypha—containing one nucleus from each parent—emerges and eventually grows into a complex reproductive structure with a distinctive cell called the **ascus** (sac) at its tip (**Figure 30.13e**). The nuclei fuse inside the sac and then produce spores. The key difference is that ascomycetes perform meiosis within an **ascus**, and basidiomycetes go through meiosis in a **basidium**. Ascomycetes produce their spores inside a sac; basidiomycetes produce their spores at the end of a little pedestal.

30.4   Key Lineages of Fungi

Chytridiomycota

- Largely aquatic, members of this group are the only fungi that can produce motile cells.
- An epidemic of chytrid infections may be responsible for catastrophic declines that recently occurred in frog populations all over the world.

Zygomycota (Mucorales and Other Basal Lineages)

- Many members of this group are saprophytic, but some are important parasites of insects and spiders.

- The black bread mold, *Rhizus stolonifer*, is a common household pest. Some species of *Mucor* are used in the production of steroids.

Glomeromycota (AMF)

- All AMF are members of the Glomeromycota.
- Because grasslands and tropical forests are among the most productive habitats on Earth, the AMF are enormously important to both human and natural economies.

Basidiomycota (Club Fungi)

- Basidiomycetes are important saprophytes and are the only lineage capable of digesting lignin.
- The EMF are enormously important in forest management. Mushrooms are also cultivated for food. Some of the toxins found in poisonous mushrooms are used in medicine.

Ascomycota (Sac Fungi)

Lichen-Formers

- The fungal hyphae form a dense protective layer that shields the photosynthetic species. In return, the cyanobacterium or alga provides carbohydrates to the fungus.
- Lichens dominate the Arctic and Antarctic tundras. They are a major source of food; and they also break down minerals from rocks, the first step in soil formation.

Non-Lichen-Formers

- Most ascomycetes that do not form lichens are saprophytic, but parasitic forms exist.
- Some species are capable of growing on jet fuel or paint. They can be used with bacteria in efforts to clean up contaminated sites.
- *Penicillium* is an important source of antibiotics.

B. CROSS-CUTTING THEMES

Looking Back—
Concepts from Earlier Chapters
Monophyletic Groupings—Chapter 28
The term *monophyletic* was first used in describing protists, because they are not monophyletic.

In the same sense, fungi are not monophyletic; their phylogenetic groupings do not include all descendants of the same common ancestor.

Direct Sequencing—Chapter 27
The direct sequencing approach to analyzing genes was introduced in **Chapter 27**. This chapter analyzes the gene that codes for the RNA molecule in the small subunit of fungal ribosomes.

Cell Structure—Chapter 7
Fungi synthesize a tough structural material called chitin. Chitin is a prominent component of the cell walls of fungi.

Looking Forward—
Concepts in Later Chapters
Emergence—Chapter 34
Like the emerging viruses in **Chapter 34**, fungi that recently switched host organisms show the same outbreak trend.

C. DIFFICULT TOPICS

At first, it is usually difficult to understand the different associations between fungi and other organisms. Ectomycorrhizal fungi (EMF) and arbuscular mycorrhizal fungi (AMF) are often confused in the student's mind. It becomes even harder to follow the examples once the terms have been confused.

If you focus on the word *ecto*, meaning "outside," this should at least help you to get the two terms organized properly. So the EMF are associated in a network around the roots of plants, and AMF are actually within the root tissue itself. Once you have learned this, select a detailed example of each type of fungi and work through it. By understanding an example of each organism and being able to explain it to a classmate, lab partner, or friend, you will begin to master the material.

D. ASSESSING WHAT YOU'VE LEARNED

(1) Testing Your Knowledge

1. The associations between fungi and the roots of land plants are called
   a. saprophytic associations

b. parasitic associations
c. mycorrhizal associations
d. mycelial associations

2. Saprophytes are fungi that make their living by
a. digesting dead plant material
b. digesting live plant material
c. living inside the roots of land plants
d. living on the leaves of vascular plants

3. The filaments that make up the multicellular fungi are called
a. yeasts
b. hyphae
c. septa
d. gametes

4. Which of the following is *not* one of the four major groups of fungi?
a. Zygomycota
b. Chitinomycota
c. Basidiomycota
d. Ascomycota

5. The closest living relatives to the fungi are the single-celled parasites called
a. microsporidians
b. microsporangia
c. microsporogonia
d. microspermia

6. Which of the following symbiotic relationships offers benefits to *both* the host and the invader?
a. parasitism
b. commensalism
c. mutualism
d. all of the above

7. Which of the following best describes ectomycorrhizal fungi?
a. a mutualistic relationship in which the fungi grow into the cells of plant roots
b. a mutualistic relationship in which the fungi grow around the plant roots
c. a commensal relationship in which the fungi grow into the cells of plant roots
d. a commensal relationship in which the fungi grow around the plant roots

8. How do fungi digest their food?
a. They ingest particles and break them down.

b. They absorb complex molecules directly.
c. They have a rudimentary gut.
d. They secrete enzymes onto their food outside the animal.

9. Fungi have the ability to break down:
a. carbohydrates
b. cellulose
c. lignin
d. all of the above

10. When nuclei of two different individual fungi fuse, this is called
a. karyogamy
b. plasmogamy
c. nuclear fusion
d. homokaryotism

(2) Integrating Your Knowledge

(a) Name the four lineages of fungi; give the distinguishing feature and an example for each.
(b) What do current molecular data say about the lineages of fungi?
(c) Describe the morphology of hyphae.
(d) How is lignin degraded by fungi?
(e) How is cellulose digested by fungi?
(f) Give an example of parasitism in fungi, describing its effects.

CHAPTER 30—ANSWER KEY

D. Assessing What You've Learned

(1) Testing Your Knowledge
1. c; 2. a; 3. b; 4. b; 5. a; 6. c; 7. b; 8. d; 9. d; 10. a

(2) Integrating Your Knowledge
(a) **Chytridiomycota**: Chytrids are the only group of fungi that live in water. They have recently been the focus of research on an epidemic of chytrid infections responsible for amphibian die-offs. The chytrids reproduce sexually and asexually. They are the only fungi that have motile cells.
Zygomycota: Most Zygomycota live in soil, and many are key partners in

mutualistic interactions with plants. The black bread mold, *Rhizopus stolonifer*, is a common household zygomycete. The hyphae of Zygomycota are haploid, and they have several different mating types or "sexes."

Basidiomycota (club fungi): Mushrooms, bracket fungi, and puffballs are reproductive structures produced by members of the Basidiomycota. Their size, shape, and color vary enormously, but they all originate from hyphae of mated individuals. After hyphae of mated individuals fuse, the cytoplasms fuse; but the nuclei remain independent.

Ascomycota (sac fungi): Ascomycota resemble basidiomycetes in that they have a heterokaryotic stage. A short heterokaryotic hyphae, containing one nucleus from each parent, emerges and eventually grows into a complex reproductive structure with a distinctive cell called the ascus (sac) at its tip. After fusing inside the sac, the nuclei produce spores. The key difference is that ascomycetes perform meiosis within an ascus, and basidiomycetes go through meiosis in a basidium.

(b) Recently, a group of researchers compared the results of ribosomal DNA sequence analyses with data from studies of cell-wall composition, life-cycle patterns, and reproductive structures. The relationships support three important conclusions:

1. It is likely that a lineage of chytrids forms the most basal group of fungi, but it is not clear whether fungi moved from water to land just once or many times.

2. The groupings called Chytridiomycota and Zygomycota appear to be paraphyletic. Instead, they have either evolved more than once or were present in a common ancestor and were retained in several lineages.

3. The Basidiomycota and Ascomycota are monophyletic and appeared late during the diversification of fungi.

(c) Fungi produce spores that outnumber most pollen grains. If a spore falls on a food source and is able to germinate, a mycelium begins to form. As the fungus expands, hyphae grow in the direction in which food is most abundant. When food becomes limited, hyphae respond by making spores. The hyphae that make up the body of a fungus are long, narrow filaments; and most are broken into cell-like compartments by cross-walls called septa. Septa do not close off segments of hyphae completely; gaps exist.

(d) Lignin peroxidase is an enzyme that catalyzes the removal of a single electron from an atom in the aromatic rings of lignin. This oxidation step creates a free radical and leads to a series of uncontrolled, unpredictable reactions that end up splitting the polymer into smaller units. The uncontrolled oxidation reactions triggered by lignin peroxidase are analogous to the uncontrolled oxidation reactions that occur when wood burns. The uncontrolled nature of the reactions means that the oxidation of lignin cannot be harnessed to drive the production of ATP. However, by oxidizing lignin, the hyphae gain access to huge supplies of energy rich cellulose within.

(e) Cellulases are secreted into the extracellular environment by fungi. Some cleave long strands of cellulose into the disaccharide called **cellobiose**. Other cellulases are equally specific; and together, they eventually convert cellulose into glucose.

(f) During the twentieth century, epidemics caused by fungi killed 4 billion chestnut trees and tens of millions of American elm trees. These outbreaks were triggered by emerging fungi that were accidentally transported on species of chestnut and elm native to other regions of the world. Fungi can cause serious diseases in corn, wheat, barley, and other crops.

31

An Introduction to Animals

A. KEY BIOLOGICAL CONCEPTS

31.1 Why Do Biologists Study Animals?

- Animals are heterotrophs; they are the dominant consumers and predators in both aquatic and terrestrial habitats.
- Animals represent the most species-rich and morphologically diverse lineage of multicellular organisms on the tree of life.
- Humans depend on wild and domesticated animals for food.
- Humans study our closest relatives in order to understand ourselves.

31.2 How Do Biologists Study Animals?

Analyzing Comparative Morphology

The Evolution of Tissues

- The number of tissue layers that exist in an embryo helps to identify body plans. A **tissue** is a highly organized and functionally integrated group of cells.
- **Diploblasts** are animals whose embryos have two types of tissues. Only two groups of diploblastic animals are alive today: cnidarians and ctenophores.
- **Triploblasts** are animals whose embryos have three types of tissues. In triploblasts, the **ectoderm** gives rise to the skin and nervous system, the **endoderm** gives rise to the digestive tract, and the **mesoderm** gives rise to the circulatory system, muscle, and internal structures.

Symmetry and Cephalization

- A basic feature of a multicellular body is whether it has a plane of symmetry. All animals exhibit either **radial** or **bilateral** symmetry. Animals with radial symmetry have at least two planes of symmetry. Bilaterally symmetric organisms face their environment in one direction.
- This allowed **cephalization**: the evolution of a head, where structures for feeding, sensing the environment, and processing information are concentrated.

Evolution of a Body Cavity

- A body cavity creates a medium for circulation, along with space for internal organs. Fluid-filled chambers are central to the operation of a **hydrostatic skeleton** (**Figure 31.7**). This gave bilaterally symmetrical organisms the ability to move efficiently.
- Diploblasts do not have a body cavity; neither does the triploblast phylum of flatworms.
- Roundworms and rotifers have a **pseudocoelom** that forms between the endoderm and mesoderm.
- In all other triploblasts, the body cavity forms from within the mesoderm itself and is lined with cells from the mesoderm. Muscle and blood vessels can then form on either side of the body cavity or **coelom**.

The Protostome and Deuterostome Patterns of Development

- Except for echinoderms, all true coelomates are bilaterally symmetric and have three embryonic tissue layers. But this huge group can be divided into **protostomes** (arthropods, mollusks, and segmented worms) and **deuterostomes** (the vertebrates and echinoderms).
- **Gastrulation** is a series of cell movements that results in the formation of the three embryonic layers. This invagination of cells creates a pore that opens to the outside (**Figure 31.8b**). In **protostomes** this pore becomes the mouth, and the anus forms later in development. In **deuterostomes** this initial pore becomes the anus, and the mouth forms later in development.

The Tube-within-a-Tube Design

- The outer tube forms the body wall; the inner tube is the individual's gut (**Figure 31.9a**).

The Phylogeny of Animals Based on Morphology

- Sponges are shown as the earliest-branching lineage; this is because they lack tissues and they are asymmetrical. Radially symmetric phyla are placed next on the tree because their tubelike body plans are relatively simple. Among bilaterally symmetric phyla, the acoelomates and pseudocoelomates first appeared, followed by the coelomates.
- After this, two major events occurred:
  1. Radial symmetry evolved in some echinoderms.
  2. A type of body architecture called **segmentation** evolved independently in both protostomes and deuterostomes.

Using the Fossil Record

- Animals appeared starting about 580 million years ago, beginning with the **Doushantuo microfossils**, **Ediacaran faunas**, and fossils in the **Burgess Shale** deposits.
- Sponges are the first animals to appear in the fossil record. Biologists infer that animals diversified from an ancestor that looked like a simple contemporary sponge.

- In general, the fossil record of animal origins is consistent with the overall pattern of evolution described in **Figure 31.10**.

Evaluating Molecular Phylogenies

- The phylogenetic tree is based on genes for ribosomal RNA and several proteins (**Figure 31.13**). Several important observations have emerged from the data:
  1. The most ancient groups of triploblasts entirely lacked a coelom.
  2. The major event in the evolution of the Bilateria was the split between protostomes and deuterostomes. Also, within the protostomes, another fundamental split occurred between (1) the Ecdysozoa and (2) the Lophotrochozoa.
  3. Segmentation evolved independently in the annelids and the arthropods.
  4. Flatworms are acoelomate; but this condition is derived.
  5. Pseudocoeloms arose from coeloms twice in evolutionary history, giving rise to (1) nematodes and (2) rotifers.

31.3 What Themes Occur in Diversification of Animals?

Feeding

Suspension Feeding

- Suspension feeding is found in a wide variety of animal groups, from clams to whales, and it has evolved independently many times.
- A wide variety of animals feed by filtering out food particles suspended in water. Cilia on gills pump water in one siphon and out the other. Food particles are trapped by the gills and swept toward the mouth by cilia.
- Krill can suspension-feed as they swim. Projections on their legs trap food particles as they flow past (**Figure 31.14b**).

Deposit Feeding

- Deposit feeders, found in a wide variety of taxonomic groups, eat their way through a substrate. For example, earthworms swallow the soil as they move through it and leave mineral material behind as feces.

- Food for deposit feeders consists of soil-dwelling bacteria, protists, fungi, and archaea, along with **detritus**—the dead and partially decomposed remains of organisms.
- Deposit feeders are similar in appearance. They usually have simple mouthparts and a wormlike body shape (**Figure 31.15**).

Herbivory

- **Herbivores** are animals that digest algae or plant tissues. They have complex mouths with structures that make biting and chewing or sucking possible.
- A **radula** is found in snails and other organisms; it functions like a rasp or file to scrape away at leaf material.
- The **proboscises** of insects and the long beaks of hummingbirds are used to harvest nectar, while the chewing mouthparts of grasshoppers and grinding **molars** of horses process leafy tissues.

Predation

- **Sit-and-wait** predators, like web-spinning spiders, rarely move until prey is captured.
- **Stalkers**, like wolves and dogs, can run their prey down during an extended, long-distance chase. Mountain lions, in contrast, stalk their prey slowly and then run it down in a short sprint.

Parasitism

- Parasites, which are much smaller than their hosts are, often harvest nutrients without causing death.
- **Endoparasites** live inside their hosts. They are often wormlike in shape and can be extremely simple morphologically. The tapeworms found inside humans have no digestive system; however, most endoparasites ingest their food and have a digestive tract.
- **Ectoparasites** live outside their hosts. They usually have grasping mouthparts that allow them to pierce their host's exterior and suck the nutrient-rich fluids inside. A louse is an example of an insect ectoparasite (**Figure 31.18b**).

Movement

Types of Limbs: Unjointed and Jointed

- Unjointed limbs (i.e., velvet worms) are sac-like.
- Jointed limbs move when muscles that are attached to a skeleton contract or relax.
- Ecdysozoans have an exoskeleton, and vertebrates have an endoskeleton.

Are All Animal Appendages Homologous?

- Biologists have argued that at least a few of the same genes are involved in the development of all appendages observed in animals.
- The current controversial hypothesis is that all animal appendages are homologous.

Reproduction and Life Cycles

- Animal reproduction is extremely diverse.
- Animals can reproduce sexually or asexually.
- Sexual reproduction can be with internal or external fertilization.
- Eggs or embryos can be retained in the female's body during development, or the eggs may be laid outside the body.
- **Metamorphosis** is the change from juvenile to adult body type. This change can be either subtle or spectacular.
- **Hemimetabolous** development is a subtle change from metamorphosis; the term refers to the limited morphological difference between juvenile and adult. It is a one-step process of sexual maturation. Grasshoppers undergo this as a series of molts that gradually change it from a wingless, immature juvenile to a sexually mature adult capable of flight.
- **Homometabolous** development is a two-step process, from larvae to pupae to adult, involving dramatic changes in morphology and habitat use. This is found in 10 times as many species as the hemimetabolous process is.

31.4 Key Lineages of Animals

Choanoflagellates (Collar Flagellates)

- These are aquatic, unicellular animals that occur in colonies. The adults are sessile and reproduce by simple fission.

Porifera (Sponges)

- All are suspension feeders; the adults are sessile, larvae can swim with the aid of flagella.
- Reproduction can be asexual or sexual.
- Sponge cells are **totipotent**—an isolated cell can develop into a complete adult.

Cnidaria (Jellyfish, Corals, Anemones, Hydroids, Sea Fans)

- Cnidarians are radially symmetric diploblasts with ectoderm and endoderm layers that sandwich gelatinous material known as mesoglea.
- The gut has only one opening. A specialized cell, a **cnidocyte**, is used for prey capture.
- Most life cycles have a sessile polyp form and a mobile medusa.
- Polyps can reproduce asexually by budding, fission, or fragmentation. During sexual reproduction, fertilization occurs in open water.

Ctenophora (Comb Jellies)

- Ctenophores are transparent, ciliated gelatinous diploblasts that live in marine habitats.
- They are predators; adults move by beating of cilia.
- Most species have male and female organs and routinely self-fertilize, but fertilization is external.

Acoelomorpha

- These organisms lack a coelom. They are bilaterally symmetric worms with distinct anterior and posterior ends and are triploblastic.
- They feed on detritus and prey on small animals.
- Reproduce asexually by fission or budding. Sexual reproduction is also possible; fertilization is internal, and fertilized eggs are laid outside the body.

B. CROSS-CUTTING THEMES

Looking Back—
Concepts from Earlier Chapters
Fossil Record of Animals—Chapter 23
The Doushantuo microfossils, Ediacaran faunas, and Burgess Shale deposits are introduced in Chapter 23 and are then used in Chapter 31 to explain some of the origins and early diversification of animals.

Embryonic Tissue Layers—Chapter 21
The basic layers of embryonic tissues and their different paths of development are explained in **Chapter 21** and used in Chapter 31 to characterize the evolution of Chordates.

Homologous Structures—Chapter 23
These structures trace back to a common ancestor and are used to identify points of origin on an evolutionary tree.

Natural Selection—Chapter 23
Natural selection is explained in detail in **Chapter 23** and is applied to the many evolutionary processes of the animal kingdom in Chapter 31. The prevalence of this term in this unit should indicate its importance.

Looking Forward—
Concepts in Later Chapters
Mechanisms of Movement—Chapter 41
Movement is used to explain the diversity of feeding strategies used in animals. **Chapter 41** details the mechanisms of animal movement.

Mouthparts—Chapter 43
Mouthparts are used in Chapter 31 to classify animals according to feeding strategy. In **Chapter 43**, the structure-function relationship is explored in more detail.

Steroid Hormones—Chapter 47
Ecdysone affects gene regulation directly; this is explained in detail along with other steroid hormones in **Chapter 47**.

Reproduction—Chapter 48
Chapter 31 begins describing many reproductive strategies of animals. **Chapter 48** goes into further detail, focusing on vertebrate reproduction.

C. DIFFICULT TOPICS

In this chapter, the terminology referring to body tissue layers and body cavity organizations can be overwhelming and confusing, especially when you are trying to match these classifications with different animal groups. If these terms are not mastered during the study of this chapter, it will make the next two chapters on animals even more confusing.

Try to study the tissue layers first (endoderm, ectoderm, and mesoderm), and then list the animal groups that are diploblastic and triploblastic. Then define coelom, acoelomate, and pseudocoelome and try to list animal groups that match these terms. Finally, try to combine both sets and groups to get an overall summary. Explaining the trend to a classmate will help you organize your thoughts and prepare you for incorporating these terms into the next two chapters.

D. ASSESSING WHAT YOU'VE LEARNED

(1)   Testing Your Knowledge

1. Diploblasts and triploblasts are terms that describe
   a. the number of invaginations during development
   b. the number of head regions during development
   c. the number of tissue layers during development
   d. the number of cell types during development

2. In triploblasts, the body cavity or coelom forms
   a. from within the mesoderm
   b. between the endoderm and mesoderm
   c. between the ectoderm and mesoderm
   d. from within the endoderm

3. During gastrulation, when the initial pore eventually forms the mouth, this defines the lineage called:
   a. coelomates
   b. deuterostomes
   c. protostomes
   d. gastrosomes

4. What is the earliest-branching lineage of animals?
   a. insects
   b. sponges
   c. echinoderms
   d. chordates

5. Animals that eat and digest algae or plant materials are called
   a. herbivores
   b. carnivores
   c. detritivores
   d. parasites

6. Which parasites live inside their hosts?
   a. endoparasites
   b. ectoparasites
   c. mesoparasites
   d. none of the above

7. The current hypothesis regarding animal limbs is as follows:
   a. Animal limbs are not homologous structures.
   b. All animal limbs are homologous.
   c. Animal limbs all perform the same function.
   d. Animal limbs all arose at the same time.

8. Which specific term describes the change from larvae to pupa to adult?
   a. hemimetabolous
   b. metabolism
   c. heterometabolous
   d. homometabolous

9. A cell such as a sponge cell can be isolated and form a complete adult. What term is used to describe this cell?
   a. omnipotent
   b. totipotent
   c. heteropotent
   d. allipotent

10. The motile form of Cnidaria is called the
    a. medusa
    b. polyp
    c. cnidocyte
    d. mesoglea

(2)   Integrating Your Knowledge

(a) Describe the two main organizations of embryonic tissue.
(b) What is the difference between radial and bilateral symmetry?
(c) What is gastrulation?
(d) Describe the types of metamorphosis in insects.

CHAPTER 31—ANSWER KEY

D. Assessing What You've Learned

(1) Testing Your Knowledge

1. c; 2. a; 3. c; 4. b; 5. a; 6. a; 7. b; 8. d; 9. b; 10. a

(2) Integrating Your Knowledge

(a) Diploblasts are animals whose embryos have two types of tissues. Only two groups of diploblastic animals are alive today: cnidarians and ctenophores. Triploblasts are animals whose embryos have three types of tissues. In triploblasts, the ectoderm gives rise to the skin and nervous system, the endoderm gives rise to the digestive tract, and the mesoderm gives rise to the circulatory system, muscle, and internal structures.

(b) A basic feature of a multicellular body is whether it has a plane of symmetry. All animals exhibit either radial or bilateral symmetry. Animals with radial symmetry have at least two planes of symmetry. Bilaterally symmetric organisms face their environment in one direction.

(c) Gastrulation is a series of cell movements that results in formation of the three embryonic layers. This invagination of cells creates a pore that opens to the outside.

(d) Metamorphosis is the change from juvenile to adult body type. This change can be either subtle or spectacular. Hemimetabolous development is a subtle change from metamorphosis; it refers to the limited morphological difference between juvenile and adult. It is a one-step process of sexual maturation. Grasshoppers undergo this process as a series of molts gradually change it from a wingless, immature juvenile to a sexually mature adult capable of flight. Homometabolous development is a two-step process, from larvae to pupae to adult, involving dramatic changes in morphology and habitat use. This is found in 10 times as many species as the hemimetabolous process.

32

Protostome Animals

A. KEY BIOLOGICAL CONCEPTS

32.1 Why Do Biologists Study Protostomes?

Crustaceans and Mollusks Are Important Animals in Marine Ecosystems

- Of all seafood, they are the highest priced and most sought after.
- They are predators and scavengers in many marine food chains.

Insects, Spiders, and Mites Are Important Animals in Terrestrial Ecosystems

- Insects eat about one-third of crops planted by farmers.
- They are the dominant consumers, scavengers, and predators in all terrestrial habitats.

32.2 How Do Biologists Study Protostomes?

Analyzing Morphology Traits

- There is little variation in several aspects of the design and construction of the protostome body.
- All protostomes have three embryonic tissue layers and are bilaterally symmetrical.
- Radical changes occurred in coelom formation as protostomes diversified. Flukes and tapeworms are **acoelomate**, even though their ancestors weren't. Rotifers and nematodes are **pseudocoelomates**.
- Arthropods are segmented and have an exoskeleton. Mollusks typically have a shell and a hydrostatic skeleton.

Using the Fossil Record

- All of the protostome phyla appear very early in the history of animal evolution.
- After this, two major events occurred:
  1. Extinction of the trilobites
  2. The first appearance of insects

Evaluating Molecular Phylogenies

- Molecular phylogenies support the hypothesis that protostomes are a monophyletic group (**Figure 32.9**).
- Using ribosomal RNA supports the existence of the monophyletic groups Lophotrochozoa and Ecdysozoa.

32.3 What Themes Occur in the Diversification of Protostomes?

Feeding

- A wide diversity of feeding strategies, reflected in the wide diversity of mouthparts.
- Some have jaws and mouth, others have a proboscis. Arthropods show the most diversity; they can pierce, suck, grind, bite, mop, chew, engulf, cut, and mash (**Figure 32.10**).

Movement

- Variation in movement depends on (1) the presence or absence of limbs and (2) the type of skeleton that is present (**Figure 32.11**).
- Evolution of the jointed limb allowed movement in these unique ways, among others:

- The insect wing is one of the most important adaptations in the history of life.
- In mollusks, waves of muscle contractions sweep down the large, muscular foot, allowing the individuals to glide.
- Squid fill a cavity surrounded by their mantle and expel water through a siphon to propel themselves through the water.

Reproduction and Life Cycles

- Regarding variation of protostome life cycles, the most basic issue is whether a juvenile's body is rearranged during metamorphosis to form an adult.
- Reproduction is highly variable, from sexual to asexual. Many crustacean species can reproduce asexually via parthenogenesis.

32.4   Key Lineages: Lophotrochozoans

- This group is named for (1) a feeding structure called a lophophore and (2) a type of larvae called a trochophore (**Figure 32.12**).

Rotifera (Rotifers)

- Rotifers have a cluster of cilia at their anterior called a **corona**, which is used for suspension feeding (**Figure 32.13**).
- Females produce unfertilized eggs by mitosis; the eggs then hatch into new, asexually produced individuals (parthenogenesis).

Platyhelminthes (Flatworms)

- They include (1) free-living flatworms, (2) endoparasitic tapeworms, and (3) endo- or ectoparasitic flukes.
- They are unsegmented, do not have a coelom, and have no circulatory system. Their high surface-to-volume ratio allows for efficient absorption of nutrients from the host.
- In many cases, life cycles of tremadotes and cestodes involve 2 or 3 distinct host species.

Annelida (Segmented Worms)

- They have a segmented body and a coelom that functions as a hydrostatic skeleton.
- They are divided into two major subgroups:

1. Polychaeta are named for numerous, bristlelike extensions called chaetae, extending from appendages called parapodia.
2. Clitellata comprise the oligochaetes (including earthworms) and leeches.

- They can crawl, burrow, or swim using their hydrostatic skeletons.

Mollusca (Mollusks)

- By far the most species-rich and morphologically diverse group in the Lophotrochozoa.
- They have a specialized body plan based on a muscular foot, a visceral mass, and a mantle that may or may not secrete a calcium carbonate shell.

Bivalvia (Clams, Mussels, Scallops, Oysters)

- They have two separate shells, which are hinged.
- They are suspension feeders and use gills for respiration.
- Most are sessile, but some can move using their muscular foot.
- Only sexual reproduction occurs. Externally fertilized eggs develop into trochophore larvae. They metamorphose into a **veliger** that continues to feed before metamorphosing into the adult form.

Gastropoda (Snails, Slugs, Nudibranchs)

- They are named for the large, muscular foot on their ventral side.
- In many species the **radula** functions like a rasp to scrape away algae.
- Some species can reproduce via parthenogenesis, but most reproduction is asexual.

Polyplacophora (Chitons)

- They are named for the eight calcium carbonate plates along their dorsal side.
- They are marine and feed using a radula. They move like gastropods, using a muscular foot.
- Sexes are separate; fertilization is external.

Cephalopoda (Squids, Nautilus, Octopuses)

- Have a well-developed head and a foot modified to form long, muscular tentacles.

- Except for the nautilus, most have highly reduced shells or none at all.
- Highly intelligent predators that hunt by sight; have a radula and a beak for biting.
- They have separate sexes and highly elaborate courtship rituals. Males deposit sperm encased in a spermatophore.

32.5   Key Lineages: Ecdysozoans

- This lineage is so named because all its members grow by molting—shedding the external skeleton.

Nemotoda (Roundworms)

- They are unsegmented worms with a pseudocoelom.
- Several species are common parasites of humans. Pinworms infect about 40 million people in the United States.
- Sexes are separate; asexual reproduction is rare or unknown. Fertilization is internal, with egg laying and direct development.

Arthropoda (Arthropods)

- Arthropods are easily the most successful lineage of eukaryotes.
- Their body is organized into a distinct head and trunk. A compound eye contains many light-sensing structures, each with its own lens.
- Most arthropods have a pair of antennae on their head.

Myriapods (Millipedes and Centipedes)

- They have relatively simple bodies with a series of short segments.
- They have mouthparts that can bite and chew. Millipedes are detritivores and centipedes are predators.
- Sexes are separate; fertilization is internal. Males deposit sperm packets that the female picks up.

Chelicerata (Spiders, Ticks, Mites, Horseshoe Crabs, Daddy Longlegs, Scorpions)

- This group is named for appendages called **chelicerae** found near the mouth.

- The most prominent group is the arachnids (spiders, scorpions, mites, and ticks).
- Spiders, scorpions, and daddy longlegs capture and sting their prey. Mites and ticks are ectoparasitic.
- Sexual reproduction via internal fertilization and direct development is the most common in this group.

Insecta (Insects)

- They are distinguished by having a head, thorax, and abdomen. In most species, two pairs of wings are mounted on the dorsal side of the thorax.
- Mating usually takes place through direct copulation.

Crustaceans (Shrimp, Lobster, Crabs, Barnacles, Isopods, Copepods)

- Their segmented body is divided into the cephalothorax (combining head and thorax) and the abdomen.
- Many crustaceans have a carapace—a plate-like section of their exoskeleton that covers and protects the cephalothorax.
- Crabs and lobsters have mandibles that can bite or chew.
- Sexual reproduction via internal fertilization is the norm.

B. CROSS-CUTTING THEMES

Looking Back—
Concepts from Earlier Chapters
Protostome Development—Chapter 31
The protostomes develop in a dramatically different way than the other major lineage of bilaterians, the deuterostomes, do.

Genetic Defects during Early Development—Chapter 22
The use of flies in developmental studies has shown that flies with different genetic defects in early development could show how various gene products influence the development of eukaryotes.

Genomics—Chapter 20
The entire genome of *Caenorhabditis elegans*, as well as that of *Drosophila melanogaster*, has now been sequenced.

Phylogenies—Chapter 31

Recall from **Chapter 31** that careful descriptions of events in embryonic development of juvenile (often larvae) anatomy and adults, analyses of the fossil record, and phylogenies based on comparisons of DNA sequence data allowed researchers to make conclusions about the evolutionary history of protostomes.

Looking Forward—
Concepts in Later Chapters

Mechanisms of Movement—Chapter 41

Movement is used to explain the diversity of feeding strategies used in animals. **Chapter 41** details mechanisms of animal movement.

Gas Exchange—Chapter 44

In moving to land, animals had to solve the problem of exchanging gas and prevent drying out. As **Chapter 44** details, land animals exchange gases with the atmosphere readily, as long as a moist body surface is exposed to air.

C. DIFFICULT TOPICS

The hardest topic in this chapter is organization of the key lineages of protostomes. The sheer number of phyla and their characteristics present a large body of information that can be incredibly confusing to remember and recall on a test. Strict memorization can succeed, but it often yields undesirable results on some types of tests. The preferred way to learn this material is to outline the major divisions of lineages first, starting with what makes a protostome a protostome. Then, distinguish the Lophotrochozoans from the Ecdysozoans, even if you can only name a few things different. Last, fill in the key phyla under each group, starting with the easiest phyla to remember. Looking at pictures and explaining characteristics to classmates can also help you to learn all of these key lineages.

D. ASSESSING WHAT YOU'VE LEARNED

(1) Testing Your Knowledge

1. Rotifers have a cluster of cilia at their anterior, called:
   a. a pore
   b. a radula
   c. a corona
   d. a lophophore

2. The group Lophotrochozoans is so named because:
   a. They all have a feeding structure called a lophophore.
   b. They all a type of larvae called a trochophore.
   c. They are all wormlike.
   d. Answers a and b are both correct.

3. Platyhelminthes include
   a. free-living flatworms
   b. endoparasitic tapeworms
   c. endoparasitic and ectoparasitic flukes
   d. all of the above

4. The Annelida are divided into two major subgroups, called:
   a. Polyplacophora and monoplacophora
   b. Polychaeta and polyplacophora
   c. Polychaeta and Clitellata
   d. Polychaeta and mollusca

5. The most species-rich and morphologically diverse group of Lophotrochozoa are
   a. Mollusca
   b. Annelida
   c. Nemotoda
   d. Polyplacophora

6. Which of the following is not a Bivalvia?
   a. clam
   b. mussel
   c. snail
   d. scallop

7. The most intelligent of the mollusks are:
   a. Bivalvia
   b. Gastropoda
   c. Cephalopoda
   d. Polyplacophora

8. Which of the following are unsegmented worms with a pseudocoelom?
   a. nematodes
   b. annelids
   c. cephalopoda
   d. platyhelminthes

9. Spiders belong to the following major group:

a. Myriapods
b. Chelicerata
c. Insecta
d. Crustaceans

10. The anterior region of the Crustaceans is called:
a. thorax
b. head
c. cephalum
d. cephalothorax

(2) Integrating Your Knowledge

(a) Describe the changes in the coelom as protostomes diversified.

(b) After the protostome phyla appeared, what two major events occurred in their lineage?

(c) Describe a few unique ways that protostomes move.

(d) What is the most basic issue categorizing a protostome's life cycle?

CHAPTER 32—ANSWER KEY

D. Assessing What You've Learned

(1) Testing Your Knowledge

1. c; 2. d; 3. d; 4. c; 5. a; 6. c; 7. c; 8. a; 9. b; 10. d

(2) Integrating Your Knowledge

(a) Radical changes occurred in coelom formation as protostomes diversified. Flukes and tapeworms are **acoelomate**, even though their ancestors were not. Rotifers and nematodes are **pseudo-coelomates**.

(b) All of the protostome phyla appear very early in the history of animal evolution. After this, two major events occurred: (1) extinction of the trilobites, and (2) the first appearance of insects.

(c) Variation in movement depends on (1) the presence or absence of limbs, and (2) the type of skeleton that is present. The evolution of the joined limb allowed movement in unique ways, including the following:
  – The insect wing is one of the most important adaptations in the history of life.
  – In mollusks, waves of muscle contractions sweep down the length of the large, muscular foot, allowing individuals to glide.
  – Squid fill a cavity surrounded by their mantle and expel water through a siphon to propel themselves through the water.

(d) Regarding variation of protostome life cycles, the most basic issue is whether a juvenile's body is rearranged during metamorphosis to form an adult.

Deuterostome Animals

A. KEY BIOLOGICAL CONCEPTS

33.1 Why Do Biologists Study Deuterostome Animals?

- Humans are deuterostomes, and other species in our lineage are of interest to us.
- Understanding deuterostomes is critical to learning how energy and nutrients flow through marine and terrestrial ecosystems.
- Humans depend on vertebrates for food, and in preindustrial economies, for power.

33.2 How Do Biologists Study Deuterostome Animals?

Analyzing Morphological Traits

- Three central issues: (1) understanding diversity of body plans, (2) exploring how vertebrates evolved from invertebrates, and (3) exploring how vertebrates came onto land.

The Water Vascular System of Echinoderms

- The echinoderm body contains a series of fluid-filled tubes and chambers—the **water vascular system** (**Figure 33.3a**).
- **Tube feet** are elongated fluid-filled structures; **podia** are sections of the tube feet that project outside the body.
- Echinoderms have an **endoskeleton**, a supportive structure inside the body (**Fig. 33.4**).

The Origin of Chordates

- Phylum Chordata is defined by (1) pharyngeal gill slits, (2) a notochord, (3) a dorsal hollow nerve cord, and (4) a muscular tail extending past the anus.

- Hemichordates are not members of Chordata because they lack a notochord and a tail.

Using the Fossil Record

- In the Ordivician period (480 Mya), the first fossils to contain bone appear. They looked like plates, presumably used as armor.
- First vertebrates with jaws appear about 430 Mya. Jaws gave vertebrates the ability to bite and chew.
- The transition to land is dated at about 357 Mya. These were the first **tetrapods**—animals with four limbs.
- At about 20 Mya, the first amniotes were present.

Evaluating Molecular Phylogenies

- **Figure 33.7** is a phylogenetic tree based on morphology and DNA.
- It is clear that the three groups of deuterostomes are indeed monophyletic, and hemichordates and echinoderms are more closely related to each other than to chordates.
- The closest living relative of the vertebrates are the cephalochordates.

33.3 What Themes Occur in the Diversification of Deuterostomes?

Feeding

Feeding Strategies in Echinoderms

- Echinoderms suspension feed, deposit feed, or harvest algae or other animals. In most cases, the podia play a key role in obtaining food.

The Vertebrate Jaw

- The leading hypothesis is that natural selection acted on the gill arches by increasing its size and modifying its orientation.
- Three lines of evidence:
  1. Both gill arches and jaws consist of flattened bars of bony or cartilaginous tissue that hinges and beds forward
  2. Both jaws and gill arches derive from specialized cells—neural crest cells.
  3. The muscles that move jaws and gill arches are derived from the same population of embryonic cells.

Movement

- The hypothesis that tetrapod limbs evolved from fish fins is supported by fossil evidence as well as molecular genetic evidence.
- Once the limb evolved, natural selection elaborated it into structures used for running, gliding, crawling, burrowing, or swimming.

Reproduction

The Amniotic Egg

- These eggs have shells that minimize water loss as the embryo develops inside.
- It contains a membrane-bound supply of water in a protein-rich solution called **albumin** (**Figure 33.17**). The embryo is enveloped in a membrane called the **amnion**. The **yolk sac** is the membranous pouch containing nutrients, and the **allantois** is the pouch holding waste material. The **chorion** is the membrane that provides the surface for gas exchange.

The Placenta

- The mother retains the egg in her body. The placenta, an organ that is rich in blood vessels, facilitates the flow of oxygen and nutrients from mother to offspring (**Figure 33.18**).

Parental Care

- Defined as any action by a parent that improves the ability of its offspring to survive—feeding, keeping young warm, protection.
- The most extensive parental care is seen in mammals and birds.

33.4   Key Lineages: Echinodermata

Asteroidea (Sea Stars)

- They have star-shaped bodies with five or more long arms radiating from a central region containing the mouth, stomach, and anus (**Figure 33.21**).
- Sea stars are predators or scavengers. Some species can pull apart bivalves.

Echinoidea (Sea Urchins and Sand Dollars)

- Sea urchins have globular bodies and long spines, and they crawl along substrates. Most are herbivores of algae (**Figure 33.21a**).
- Sand dollars are flattened and disk-shaped, have short spines, and burrow in soft sediments (**Figure 33.22b**). They use mucus-covered podia to collect food particles in the sand.

33.5   Key Lineages: Chordata

Myxinoidea (Hagfish) and Petromyzontoidea (Lampreys)

- Both hagfish and lampreys belong to independent lineages, although both groups lack jaws. Hagfish lack any sort of vertebral column, but lampreys have small pieces of cartilage along their length.
- Hagfish are scavengers and predators (**Figure 33.23a**) that deposit-feed on the carcasses of dead fish and whales.
- Lampreys are **ectoparasites**—they attach to fish and suck blood and other body fluids. They are also **anadromous**—they spend their adult life in the ocean and swim up streams to breed.

Condrichthyes (Sharks, Rays, Skates)

- These animals are distinguished by their cartilaginous skeleton. Most species are marine, and sharks are top predators of marine ecosystems.

Actinopterygii (Ray-Finned Fishes)

- These fishes have fins supported by long, bony rods arranged in a ray pattern. They are the most ancient group of vertebrates that have a skeleton made of bone.

- They avoid sinking in the water with the aid of a gas-filled **swim bladder**.
- The most important subgroup of ray-finned fishes is the Teleosti, with 20,000 species.

Dipnoi (Coelocanths) and Actinistia (Lungfish)

- Although these are independent lineages, they are often grouped together and called the **lobe-finned fishes**.
- They represent a crucial evolutionary link between the ray-finned fishes and the tetrapods.

Amphibia (Frogs, Salamanders, Caecilians)

- Adults of most species feed on land; but many species have to lay their eggs in water.
- The larvae undergo metamorphosis into the adult form, which is mostly carnivorous.

Chordata: Mammalia (Mammals)

- Mammals have hair or fur. They have mammary glands for lactation.

Monotremata (Platypuses and Echidnas)

- The most ancient group of mammals living; they lay eggs and have lower metabolic rates than do other mammals (**Figure 33.28**).

Marsupiala (Marsupials)

- Although females have a placenta, the young are born poorly developed. They crawl to the female's nipples, where they suck milk. They stay there until they become independent (**Figure 33.29**).

Eutheria (Placental Mammals)

- Six (of the 18) most species-rich orders are rodents, bats, insectivores, artiodactyls, carnivores, and primates.
- At birth, the young are much better developed than marsupials are, although there is still a prolonged period of parental care.

Reptilia (Turtles, Snakes and Lizards, Crocodiles, Birds)

Testudinia (Turtles and Tortoises)

- These animals are distinguished by a shell comprised of bony plates that fuse to the vertebrae and ribs (**Figure 33.31**).

- Their skulls are more similar to amphibians than the skulls of other reptiles are. Although they lack teeth, their jawbone and lower skull form a bony beak.

Lepidosauria (Lizards and Snakes)

- Most lizards have well-developed jointed legs; but snakes are limbless, evolving from limbed ancestors (**Figure 33.32**).
- Most lay eggs, although a few give birth to live young that have hatched from the egg and begun development.

Crocodilia (Crocodiles and Alligators)

- In this group, only 21 species are known. They have eyes and nostrils located on the top of their skulls and can sit underwater for extended periods (**Figure 33.33**).
- They are top predators, and though they are oviparous, parental care is extensive.

Aves (Birds)

- Birds are the only **endotherms** within the Chordata. This means their metabolic and insulation rates are so high that they retain a high constant body temperature.
- Most birds are omnivores, and almost all species can fly (**Figure 33.34**).
- Birds are oviparous, though parental care is extensive.

33.6 Key Lineages: The Hominin Radiation

The Primates

- The **prosimians** are composed of lemurs, tarsiers, pottos, and lorises. Most of these live in trees and are active at night (**Figure 33.35a**).
- The anthropoids are the New World monkeys found in Central and South America, the Old World monkeys living in Africa and tropical regions of Asia, and the great apes—orangutans, gorillas, chimpanzees, and humans.
- Primates are distinguished by the placement of their eyes on the front of their face; anthropoids are distinguished by their opposable thumb.
- The lineage of the great apes is called the hominids. Humans are the only great ape that is bipedal (**Figure 33.35c**).

Fossil Humans

- The fossil record of human-like hominids is not nearly complete but is rapidly improving. They can be organized into four general groups (**Figure 33.36**):

  1. Three species of small apes called **gracile australopithecines** lived from 4.12 to 2.4 million years ago and were bipedal.

  2. Three distinct species of robust australopithecines lived from 2.7 to 1.0 million years ago and were bipedal. Robust forms had more massive cheek teeth and jaws.

  3. The earliest species in the genus *Homo* date from 2.4 to 1.5 million years ago. *Homo* species have flatter, narrower faces and smaller jaws. Their braincases are as much as three times larger than those of the australopithecines.

  4. More recent species of *Homo* date from 1.2 million years ago to the present.

- Humans did not evolve through a simple, steady progression from a chimpanzee-like ancestor. Instead, a complex radiation of bipedal hominids occurred in Africa during the past 4 to 5 million years

The Out-of-Africa Hypothesis

- Because the first lineage to branch off leads to descendant populations that live in Africa today, it is logical to infer that the ancestral population lived in Africa (**Figures 33.38 and 33.39**).

- **Assimilation hypothesis**: *H. sapiens* interbred with Neanderthals and *Homo erectus*. This hypothesis implies that the genetic composition and morphological features of *H. sapiens* are an amalgam of ancient traits of Neanderthals and *H. erectus*.

- **Out-of-Africa hypothesis**: *H. sapiens* evolved independently of the European and Asian species of *Homo*, and there was no interbreeding between *H. sapiens* and Neanderthals or *H. erectus*.

- **Key Research**: Svante Paablo extracted DNA from fossilized bone of a Neanderthal and compared a 379-base-pair section of the mitochondrial genome to humans living today. Researchers found the DNA to be extremely different; it supported the hypothesis that *H. sapiens* and *H. neanderthalensis* did not interbreed.

- Unfortunately, it has been impossible thus far to extract DNA *from H. erectus* fossils to perform the same test.

B. CROSS-CUTTING THEMES

Looking Back—
Concepts from Earlier Chapters

Deuterostome Development—Chapter 30
The deuterostomes developed in a dramatically different way than did the other major lineage of Bilaterians, the protostomes.

Genetic Defects during
Early Development—Chapter 24
The use of zebra fish in developmental studies has shown that zebra fish with different genetic defects in early development could show how various gene products influence the development of eukaryotes.

Genomics—Chapter 20
The entire genome of zebra fish, as well as that of humans, has now been sequenced.

Phylogenies—Chapter 31
Recall from **Chapter 31** that careful descriptions of events in embryonic development of the anatomy of juveniles (often larvae) and adults, analyses of the fossil record, and phylogenies based on comparisons of DNA sequence data allowed researchers to make conclusions about the evolutionary history of deuterostomes.

Looking Forward—
Concepts in Later Chapters

Mechanisms of Movement—Chapter 41
Movement is used to explain the diversity of feeding strategies used in animals. **Chapter 41** focuses on the mechanisms of animal movement in detail.

Gas Exchange—Chapter 44
In the transition to land, animals had to solve the problem of exchanging gas and preventing drying out. As **Chapter 44** details, land animals exchange gases with the atmosphere readily as long as a moist body surface is exposed to air.

C. DIFFICULT TOPICS

The particulars of hominin evolution are very confusing because they consist of hypotheses that are constantly being tested and updated. This is the problem inherent in studying the particulars of a branch of science that is unfolding as textbooks are being printed and classes are being taught.

The best place to start is **Figure 30.37**. Draw the lineages shown in this figure, and try to work out the timeline of presence and extinction of the different hominin species. Next, work through the hypotheses presented in this chapter and the experimental evidence for each.

D. ASSESSING WHAT YOU'VE LEARNED

(1) Testing Your Knowledge

1. The sections of echinoderm tube feet that project outside the body are called
    a. podia
    b. pseudopods
    c. postiums
    d. endoskeleton

2. The following statement regarding the Chordates is *false*:
    a. Chordates have pharyngeal gill slits.
    b. Chordates do not have a notochord.
    c. Chordates have a dorsal hollow nerve cord.
    d. Chordates have a muscular tail.

3. The leading hypothesis regarding the evolution of the vertebrate jaw is that it arose from natural selection acting on the
    a. gills
    b. gill arches
    c. opercula
    d. fins

4. The hypothesis that tetrapod limbs evolved from fish fins is
    a. rubbish
    b. only slightly supported
    c. supported only by fossil evidence
    d. supported by fossil and molecular genetic evidence

5. The organ rich in blood vessels facilitating the flow of oxygen and nutrients from the mother to offspring is called:
    a. the allantois
    b. the yolk
    c. the placebo
    d. the placenta

6. What is the major difference between lampreys and sharks?
    a. Lampreys lack jaws.
    b. Lampreys lack cartilage.
    c. Lampreys do not have tails.
    d. Lampreys can't swim.

7. What is the key difference between Condrichthyes and Actinoptergii?
    a. Actinoptergii have fins.
    b. Actinoptergii have jaws.
    c. Actinoptergii have true bone.
    d. Actinoptergii have no tail.

8. Endothermy is found among the Reptilia only within the following group:
    a. Testudinia
    b. Lepidosauria
    c. Crocodilia
    d. Aves

9. What is the only mammalian group to lay eggs?
    a. Monotremata
    b. Marsupiala
    c. Eutheria
    d. Primates

10. The prosimians include all of the following *except*:
    a. emurs
    b. gibbons
    c. tarsiers
    d. lorises

(2) Integrating Your Knowledge

(a) Describe the characteristics of the amniotic egg.

(b) Describe the water vascular system of the echinoderms.

(c) What lines of evidence show that gill arches and jaws are similar?

(d) Describe the two groups of primates.

(e) What is the Out-of-Africa hypothesis?

CHAPTER 33—ANSWER KEY

D. Assessing What You've Learned

(1) Testing Your Knowledge

1. a; 2. b; 3. b; 4. d; 5. d; 6. a; 7. c; 8. d; 9. a; 10. b

(2) Integrating Your Knowledge

(a) These eggs have shells that minimize water loss as the embryo develops inside. It contains a membrane-bound supply of water in a protein-rich solution called **albumin** (**Figure 33.17**). The embryo is enveloped in a membrane called the **amnion**. The **yolk sac** is the membranous pouch containing the nutrients; the **allantois** is the pouch holding waste material. The **chorion** is the membrane that provides the surface for gas exchange.

(b) The echinoderm body contains a series of fluid-filled tubes and chambers called the **water vascular system** (**Figure 33.3a**). **Tube feet** are elongated fluid-filled structures, and **podia** are sections of the tube feet that project outside the body. Echinoderms have an **endoskeleton**—a hard, supportive structure inside the body (**Figure 33.4**).

(c) The leading hypothesis holds that natural selection acted on gill arches by increasing the size and modifying the orientation. There are three lines of evidence:
1. Both gill arches and jaws consist of flattened bars of bony or cartilaginous tissue that hinges and bends forward.
2. Both jaws and gill arches derive from specialized cells—neural crest cells.
3. The muscles that move jaws and gill arches are derived from the same population of embryonic cells.

(d) **Prosimians** include lemurs, tarsiers, pottos, and lorises. Most of these live in trees and are active at night. The **anthropoids** are the New World monkeys found in Central and South America, the Old World monkeys that live in Africa and tropical regions of Asia, and the great apes—orangutans, gorillas, chimpanzees, and humans.

(e) The Out-of-Africa hypothesis states that *H. sapiens* evolved independently of the European and Asian species of *Homo*; there was no interbreeding between *H. sapiens* and Neanderthals or *H. erectus*.

34

Viruses

A. KEY BIOLOGICAL CONCEPTS

34.1 Why Do Biologists Study Viruses?

Recent Viral Epidemics in Humans

- Viruses have caused the worst **epidemics** in recent human history. The Spanish flu of 1918–1919 was the most devastating epidemic—over 50 million people died.

Current Viral Epidemics in Humans: HIV

- Human immunodeficiency virus will surpass influenza as the most deadly of diseases. This virus is the cause of acquired immune deficiency syndrome (AIDS).

How Does HIV Cause Disease?

- HIV parasitizes immune cells called T cells. The number of T cells produced by the body does not keep pace with the number of T cells destroyed by HIV (**Figure 34.3**).
- When the T cell drops, the body becomes less able to fight incoming pathogens.

What Is the Scope of the AIDS Epidemic?

- AIDS has already killed 25 million people. It is estimated that 42 million people worldwide are infected with HIV.

34.2 How Do Biologists Study Viruses?

Analyzing Variation in Replication Cycles: Lytic and Lysogenic Growth

- Though the hosts, morphology, and genomes of viruses vary widely, they all share the same basic replication cycle.

- **Lytic growth** refers to this basic replication cycle, which usually results in the death of the host cell (**Figure 34.9a**).
- Some viruses have variations of their life cycle; these include phages that parasitize bacteria and insert their DNA into the host's chromosome.
- The viral genome is then replicated with the host's, and this occurs without causing serious damage to the host cell (**Figure 34.9b**).
- **Lysogeny** refers to this integrated state in which the virus is often latent.

Analyzing the Phases of the Lytic Cycle
How Do Viruses Enter a Cell?

- Many are transmitted through mouthparts of sucking insects.
- Some gain entry by binding to specific proteins on cell membranes.

How Do Viruses Copy Their Genomes?

- Most DNA viruses copy their genomes by using their own DNA polymerase enzyme, a protein that synthesizes copies of the viral genome using nucleotides provided by the host.
- Most RNA viruses use a viral enzyme called **RNA replicase**. It is an RNA polymerase that synthesizes RNA from an RNA template, by using ribonucleotides provided by the host cell.
- In other RNA viruses, the genome is transcribed from RNA to DNA by a viral enzyme called **reverse transcriptase**. Reverse transcriptase is a DNA polymerase that makes a double-stranded DNA from a single-stranded RNA template.

- **Retroviruses** reverse-transcribe their genome. Enzymes then insert this copied DNA (cDNA) into a stretch of host-cell chromosome. Viral genes are then transcribed to mRNA, which is translated into proteins by the host cell's ribosomes.

Producing Viral Proteins

- Viruses must exploit the host cell's biosynthetic machinery to make viral proteins.
- During the third phase of the replication cycle, viral mRNAs and proteins are produced and processed.
- RNAs coding for a virus's envelope proteins go through the cell via a route identical to the RNAs of the cell's transmembrane proteins.
- Viral mRNAs are translated by ribosomes attached to the endoplasmic reticulum, are transported to the Golgi complex, where carbohydrates are attached, and are then inserted into the plasma membrane to be assembled (**Figure 34.11**).

How Are Viruses Transmitted to New Hosts?

- Viruses leave a host cell by **budding** from the cell membrane or by **bursting** out of the cell (**Figure 34.13**).
- If the host cell is part of a multicellular organism, the new generation of particles travels through the blood or lymph. Antibodies or macrophages can then attempt to destroy them before they infect another host cell.
- Natural selection favors alleles that allow viruses to replicate within a host and be transmitted to a new host.
- Transmission of HIV has taken place through direct exchange of blood or semen between an uninfected person and an infected person.
- HIV is quickly destroyed if heated or dried; therefore, these viruses do not survive long outside the host tissues.

34.3 What Themes Occur in the Diversification of Viruses?

The Nature of the Viral Genetic Material

- Researchers were able to separate the protein and nucleic acid components of the tobacco mosaic virus (TMV). They showed in this virus that RNA, *not* DNA, functions as the genetic material.
- In some viral groups, like measles and flu, the genome consists of RNA. In others, like herpes and smallpox, the genome is composed of DNA. RNA and DNA genomes of viruses can be either **positive sense** (single stranded) or **negative sense** (double stranded). In **positive-sense** viruses, the genome contains the same sequences as the mRNA required to produce viral proteins. In **negative-sense** viruses, the base sequences in the genome are complementary to those in viral mRNAs. **Ambisense** viruses contain both positive- and negative-sense sections.
- Viruses can have as few as three loci (tymoviruses), or they can have hundreds, as in the genome of smallpox (**Table 34.2**).

Where Did Viruses Come From?

- Many biologists think viruses are closely related to plasmids and transposable elements. Viruses are distinguished from plasmids by a protein coat or membrane-like envelope.
- The **escaped-genes hypothesis** states that these elements descended from clusters of genes that physically escaped from bacterial or eukaryotic chromosomes long ago.
- The **degeneration hypothesis** states that organisms gradually degenerated into viruses by gradually losing the genes required to synthesize ATP and other compounds.

Emerging Viruses, Emerging Diseases

- Hantavirus and Ebola are examples of emerging diseases—illnesses that suddenly affect significant numbers of individuals in a host population.
- Hantavirus and Ebola were considered emerging viruses because they had switched from their traditional host species to a new host—humans.
- In an outbreak, physicians need to (1) identify the agent causing the new illness, and (2) determine how it is being transmitted.

34.4   Key Lineages of Viruses

Double-Stranded DNA (dsDNA) Viruses

- This is a large group of 21 families and 65 genera, of which smallpox and herpes are the most familiar.
- These viruses parasitize hosts throughout the tree of life, with the exception of land plants (**Figure 34.15**).
- Because the viral genes have to enter the nucleus to be replicated, they can infect only cells that are actively dividing, such as those that line the respiratory tract or urogenital tract.

RNA Reverse-Transcribing Viruses

- There is only one family called the retroviruses, which includes HIV.
- Species are known to parasitize only vertebrates, specifically birds, fish, or mammals (**Figure 34.16**).
- When the virus's RNA genome and reverse transcriptase enters a host cell's cytoplasm, the enzyme catalyzes the synthesis of a single-stranded cDNA from the original RNA. Reverse transcriptase makes this cDNA double-stranded. The DNA then integrates with the host chromosome.

Double-Stranded RNA (dsRNA) Viruses

- This group includes 7 families and a total of 22 genera.
- Species in this group infect a wide variety of hosts, including fungi, land plants, insects, vertebrates, and bacteria.
- For humans, the most important viruses in this group cause disease in major crops (**Figure 34.17**).
- Once in the host cell, the double-stranded DNA synthesizes RNA, which are then translated into viral proteins.

Negative-Sense Single-Stranded RNA ([–]ssRNA) Viruses

- This group includes 7 families and 30 genera.
- The single-stranded virus genome is complementary to the viral mRNA.
- A wide variety of plants and animals are parasitized by these viruses. The flu, mumps, measles, Ebola, Hantaan, and rabies viruses all belong to this group.
- Once in a host cell, a viral RNA polymerase uses the negative-sense template to make viral mRNA. The viral mRNAs are then translated to form viral proteins and new negative-sense, single-stranded RNA copies.

Positive-Sense Single-Stranded RNA ([+]ssRNA) Viruses

- This is the largest group known, with 21 families and 81 genera.
- Because the sequence of bases is the same as for mRNA, it does not need to be transcribed before proteins are produced.
- Most of the commercially important plant viruses belong to this group (**Figure 34.19**). This group also includes colds, polio, and hepatitis A, C, and E.
- Single-stranded RNA is immediately translated into viral proteins.

B. CROSS-CUTTING THEMES

Looking Back—
Concepts from Earlier Chapters

Translation—Chapter 16
The translation process of creating proteins from codons attaching to ribosomes is applied in Chapter 34 to virus replication. An extensive review of DNA replication and protein synthesis would be beneficial.

Natural Selection—Chapter 23
In the same way that HIV has become resistant to drugs that were once effective, natural selection has recently favored strains of *Mycobacterium tuberculosis* that are resistant to the drug rifampin, and these strains are responsible for the resurgence of tuberculosis in the industrialized world.

DNA Polymerase—Chapter 14
Normal DNA polymerase is analyzed in **Chapter 14,** but the DNA polymerase of HIV in Chapter 34 has no 3≤ exonuclease activity. Comparing the two will help you to understand the overall functions of DNA polymerase.

Looking Forward—
Concepts in Later Chapters

The Immune System—Chapter 49

In Chapter 34, we just touch on the role of T cells and HIV. **Chapter 49** explains just how crucial these cells are to the immune system's response to invading bacteria and viruses. Chapter 49 also explores epidemic control through vaccination, and discusses why researchers have been unable to design an effective vaccine against HIV.

C. DIFFICULT TOPICS

In your first-year biology course, you are being exposed to many new terms in a short period of time. Often, these terms are introduced with each new branch of biology.

In this chapter, the HIV virus is used as an example in describing a viral replication cycle and the process of researchers in combating a deadly virus. To make matters more complex, HIV is only one virus and thus only one type of life cycle is explained.

To better understand the life cycle of HIV and how it differs from the cell cycle, it would be beneficial to review first the processes of DNA replication and then the processes of transcription and translation. Then, by consulting the diagrams in Chapter 34, make a list of differences and similarities between the viral life cycle as compared to the cell cycle. Finally, try to apply this to the HIV virus so you can easily explain to a classmate, lab partner, or friend why we haven't yet found a successful vaccine for HIV. Finally, consult the "Additional Reading" references for general background information on viral life cycles and some specific primary literature about HIV.

D. ASSESSING WHAT YOU'VE LEARNED

(1) Testing Your Knowledge

1. The basic viral replication cycle, which usually results in the death of the host cell, is called:
    a. lysogeny
    b. the lysogenic cycle
    c. the lytic cycle
    d. the translation cycle

2. Viruses can enter a cell by
    a. transmission through mouthparts of sucking insects
    b. binding to specific proteins on cell membranes
    c. diffusion
    d. Answers a and b are correct.

3. The RNA polymerase that synthesizes RNA from an RNA template using ribonucleotides provided by the host cell is called:
    a. RNA replicase
    b. reverse transcriptase
    c. retrovirus transcriptase
    d. RNA transcriptase

4. Viruses have to do which of the following in order to make viral proteins?
    a. ramp up their metabolism
    b. exploit the host cell's biosynthetic machinery
    c. enter a dormant phase
    d. create their own biosynthetic machinery

5. Viruses leave a host cell by
    a. budding from the cell membrane
    b. bursting out of the cell
    c. forming spores
    d. Answers a and b are correct.

6. The escaped-gene hypothesis states that viral elements descended from clusters of genes that
    a. physically escaped from bacterial or eukaryotic organisms
    b. degenerated over evolutionary time
    c. merged with other viral particles to become active
    d. none of the above

7. The main difference between emerging viruses and emerging diseases is
    a. emerging diseases involve only one isolated case
    b. emerging diseases involve a significant number of infected individuals
    c. emerging viruses involve a significant number of infected individuals
    d. none of the above

8. The group of viruses to which smallpox belongs is
   a. double-stranded DNA (dsDNA) viruses
   b. RNA reverse-transcribing viruses
   c. double-stranded RNA (dsRNA) viruses
   d. negative-sense single-stranded RNA ([−]ssRNA) viruses

9. The group of viruses to which HIV belongs is
   a. double-stranded DNA (dsDNA) viruses
   b. RNA reverse-transcribing viruses
   c. double-stranded RNA (dsRNA) viruses
   d. negative-sense single-stranded RNA ([−]ssRNA) viruses

10. Negative-sense single-stranded RNA must do the following in order to replicate successfully:
    a. Use the negative-sense template to make viral proteins.
    b. Use the negative-sense template to make viral RNA.
    c. Use the negative-sense template to make viral DNA.
    d. Use the negative-sense template to make replicase.

(2) Integrating Your Knowledge

(a) Compare and contrast negative-sense and positive-sense DNA. How does this relate to the types of viruses?

(b) What are retroviruses, and what specific viral enzyme do they use?

(c) Compare and contrast the lytic phase and the lysogenic phase of a viral life cycle.

(d) Compare the AIDS epidemic to another devastating epidemic.

(e) What is integrase?

CHAPTER 34—ANSWER KEY

D. Assessing What You've Learned

(1) Testing Your Knowledge

1. c; 2. d; 3. a; 4. b; 5. d; 6. a; 7. b; 8. a; 9. b; 10. b

(2) Integrating Your Knowledge

(a) RNA and DNA genomes of viruses can either be positive sense (single stranded) or negative sense (double stranded). In positive-sense viruses, the genome contains the same sequences as the mRNA required to produce viral proteins. In negative-sense viruses, the base sequences in the genome are complementary to those in viral mRNAs.

(b) Retroviruses are viruses that reverse-transcribe their genome. Enzymes then insert this copied DNA (cDNA) into a stretch of host-cell chromosome. The viral genes are then transcribed to mRNA, which is translated into proteins by the host-cell's ribosomes. In other RNA viruses, the genome is transcribed from RNA to DNA by a viral enzyme called reverse transcriptase. Reverse transcriptase is a DNA polymerase that makes a double-stranded DNA from a single-stranded RNA template.

(c) *Lytic growth* refers to this basic replication cycle and usually results in the death of the host cell. Some viruses have variations of their life cycle, which include phages that parasitize bacteria and insert their DNA into the host's chromosome. The viral genome is then replicated with the host's and occurs without serious damage to the host cell. *Lysogeny* refers to this integrated state in which the virus is often latent. A lysogenic phase in an infection cycle is possible if the virus's genome codes for an enzyme called integrase.

(d) The Spanish influenza of 1918–1919 qualifies as the most devastating epidemic on record, killing over 20 million people worldwide. Yet, AIDS is surpassing the Spanish epidemic; it has already killed 14 million people, and an additional 35 million people are currently infected with HIV. With no vaccine,

　　　researchers project that the virus could
　　　kill 100 million.

(e) Integrase catalyzes the cutting of host
　　　DNA, the insertion of the viral genome
　　　into the host DNA sequence, and the re-
　　　annealing of the DNA strand. A copy of
　　　the HIV genome is then integrated into
　　　the host chromosome.

Plant Form and Function

35

A. KEY BIOLOGICAL CONCEPTS

35.1 The Diversity of Plant Form

The Diversity of Roots: North American Prairie Plants

- J. E. Weaver conducted a series of important studies on the prairie plants of North America to understand the dynamics of prairie ecosystems. Weaver and colleagues dug deep trenches and examined exposed root systems. They found that the root systems of prairie plants that live side by side can be very different from one another.
- If all root systems were similar in size and shape, then every plant in the ecosystem would be in direct competition for water and nutrients. If a small patch of land has a diversity of root systems, it should be able to sustain a relatively large number of species.
- Even though root systems have similar functions—support, storage, and nutrient absorption—they are highly diverse in structure.

The Diversity of Shoots: Hawaiian Silverswords

- The **silverswords** are a group of closely related plant species found on the Hawaiian Islands and nowhere else. They are incredibly diverse in size, shape, and growth habit (**Figure 35.7**).
- Researchers have recently studied the evolutionary history, or phylogeny, of silverswords. These studies were inspired by the observation that the organization of vascular tissue in silverswords was strikingly similar to the vascular tissue of plants called tarweeds, which are native to California.
- Researchers sequenced the same stretch of DNA in a large sample of silverswords, tarweeds, and other plants. By comparing similarities and differences in the sequences, they were able to infer which species are most closely related and which are more distantly related (**Figure 35.6**).
- Tarweeds and silverswords appear to have a shared, fairly recent common ancestor. Some tarweeds have sticky seeds and can form viable offspring through **self-fertilization**. Possibly a seed made the trip on a seabird from California to Hawaii and founded a new population.

Modified Shoots

- Cactus stems are often modified into water-storage organs and also contain the plant's photosynthetic tissue (**Figure 35.8a**).
- **Stolons** are modified stems that run over the soil surface, producing roots and leaves at each node (**Figure 35.8b**).
- **Rhizomes** are similar to stolons—they are stems that grow horizontally producing new plants at nodes. However, these stems grow *below* the ground (**Figure 35.8c**).
- **Tubers** are rhizomes that are modified as carbohydrate storage organs (**Figure 35.8d**).
- **Thorns** are modified stems that help protect the plant from attacks by large herbivores.

Modified Leaves

- In most plant species, the vast majority of photosynthesis occurs in the leaves.

- Plants in deserts and in northern habitats tend to have leaves that are needle-like. These leaves are interpreted as adaptations that minimize transpiration in water-short habitats.
- Cactus spines are modified leaves that protect the stem.
- The leaves of the flowerpot plant are used as homes by ant colonies.

35.2   Plant Cells and Tissues

The Diversity of Plant Cells

- Review the generalized version of the plant cell (**Figure 35.11a**), including its most important organelles and structures.

Meristematic Cells

- The undifferentiated cells responsible for plant growth are called **meristematic cells**.

Parenchyma Cells

- **Parenchyma cells** serve as storage cells for starch deposits in roots, stems, and leaves (**Figure 35.12**). They are filled with chloroplasts and are the primary site of photosynthesis.
- These cells also make up phloem—the type of vascular tissue that transports nutrients throughout the plant body.

Collenchyma Cells

- **Collenchyma cells** have thickened primary walls and serve to stiffen leaves and stems (**Figure 35.15**).

Sclerenchyma Cells

- **Sclerenchyma cells** also stiffen stems and other structures, but they are distinguished by thickened secondary cell walls strengthened by tough lignin molecules (**Figure 35.16**).
- **Tracheids** and **vessel elements** are water-conducting cells that are dead at maturity.
- **Sieve-tube members** are food-conducting cells and remain alive.
- **Fibers** are usually associated with vascular tissue and are extremely elongated.
- **Sclereids** are relatively short, have variable shapes, and often function in protection.

The Diversity of Plant Tissues

- A **tissue** is defined as a group of cells that function as a unit.

Meristematic Tissue

- The rapidly dividing, undifferentiated parenchyma cells found in meristematic tissues are responsible for growth.
- Some daughter cells produced by mitosis remain as meristematic cells. Others differentiate into parenchyma, collenchyma, or sclerenchyma cells, which function as one of the following three mature plant tissues.

Dermal Tissue

- **Dermal tissue** or **epidermis** consists of a single layer of cells that covers the plant body.
- The epidermis also secretes the **cuticle**—a waxy layer that protects the leaves and reduces water loss.
- To allow CO_2 to enter photosynthetically active tissues, most land plant leaves have **stomata**—a hole produced by two bean-shaped **guard cells** (**Figure 35.17**).
- **Trichomes** are appendages from epidermal cells that function to minimize water loss and attacks by herbivores.

Ground Tissue

- **Ground tissue** is made up of cells below the epidermis and is made up of parenchyma stiffened by collenchyma or sclerenchyma.
- This tissue makes up the bulk of the plant body and is the primary location of photosynthesis and carbohydrate storage.

Vascular Tissue

- **Vascular tissue** is made up of conducting cells.
- **Xylem** conducts water and dissolved ions from the root system to the shoot system.
- Xylem contains two types of conducting cells (**Figure 35.19**):
1. **Tracheids** are dead cells that are long and slender with tapered ends. The sides and ends of the tracheids have pits that reduce resistance to water flow.

2. **Vessel elements** are shorter and wider and have perforations that produce little resistance to water flow.
- **Phloem** conducts sugar, amino acids, chemical agents, and other substances throughout the plant body.
- The cells that make up phloem are alive at maturity.
  1. **Sieve-tube members** are long, thin cells that have perforated ends called sieve plates. They lack major organelles.
  2. **Companion cells** are attached to sieve-tube members by plamodesmata. They provide materials to the sieve-tube members to maintain the cytoplasm and membrane.

Distribution of Tissue Types

- The meristematic, dermal, ground, and vascular tissues of plants are distributed throughout the root and shoot systems, as shown in **Figure 35.20**.

35.3   The Anatomy of Plant Growth

- **Growth** occurs in meristematic tissues and is the result of two processes: (1) production of new cells by mitosis and cytokinesis and (2) cell enlargement.
- **Apical meristems** are found at the tips of roots and shoots and increase their length—this process is termed **primary growth**.
- Species that form wood have a ring of meristematic cells called a **lateral meristem** that increases width—this is **secondary growth**.
- **Basal meristems** form in grass species that lack apical meristems. They form on the leaf-stem interface.

Primary Growth: The Root System

- The growing region of the root is covered by protective cells—the **root cap**. These cells also secrete **mucigel** for lubrication.
- Three distinct cell groups exist behind the root cap (**Figure 35.22**):
  1. The **zone of cellular division** contains the apical meristem.
  2. The **zone of cellular elongation** is where cells increase in length.

3. The **zone of cellular maturation** is where older cells complete their differentiation. In this region, **root hairs** increase in surface area and actively absorb water and nutrients.

Primary Growth: The Shoot System

- The top of the stem consists of a rounded dome of actively dividing cells called the **shoot apical meristem** (**Figure 35.23**).
- On both sides of this region, newly developing leaves emerge as **leaf primordia**.
- The center of the stem, or **pith**, consists of ground tissue that functions in carbohydrate storage.
- Vascular tissues are found around the pith in groups called **vascular bundles**. These bundles also branch and extend into leaf primordia, forming a continuous system for transport of water and nutrients through both stems and leaves.
- In the leaf, elongated **palisade mesophyll** cells (a type of parenchyma cell) are packed with chloroplasts and are the major site of photosynthesis.
- Rounded **spongy mesophyll** cells are loosely packed; they are often surrounded by air spaces near stomata.

Secondary Growth

- **Secondary growth** increases root and shoot width and provides the structural support necessary for extensive primary growth. Secondary growth produces wood (**Figure 35.26**).
- **Lateral meristems** are also called **secondary meristems** or **cambium**. Lateral meristems differ from apical meristems in two ways:
  1. Apical meristems are localized at root and shoot tips. Lateral meristems form cylinders that run the length of a root or stem.
  2. Cells in apical meristems divide in a plane parallel to the long axis of the root or stem. In the lateral meristem, the cells divide perpendicular to the long axis of the root or stem, which increases width.
- **Cork cambium** is a ring of meristematic cells located near the perimeter of the stem.

- **Vascular cambium** is a ring of meristematic cells just inside the cork cambium.

*How Does Cork Cambium
Contribute to Bark?*

- The cork cambium produces **cork cells**. These cells eventually die, forming a protective layer for the mature root or shoot.
- This is an important component for **bark**. Bark actually refers to all cells outside the vascular cambium.

*How Does Vascular Cambium Produce Wood
and Contribute to Bark?*

- The vascular cambium produces both phloem and xylem, which, as new layers form, develop into **secondary phloem** and **secondary xylem**.
- Secondary phloem contributes to bark, and secondary xylem forms the material we call **wood**.
- Much more secondary xylem is produced than secondary phloem. As a result, mature woody stems are dominated by wood.

The Structure of a Tree Trunk

- Trees are perennial—they can live for many years.
- A period of dormancy occurs in winter or dry seasons. When growth is rapid during the spring or wet season, large thin-walled cells are produced in the secondary xylem. When dormancy is approached, secondary xylem cells are smaller and have thick walls.
- When growth is seasonal, annual growth rings can be observed in cross sections of wood (**Figure 35.28**).
- The inner xylem region is called **heartwood**; the outer xylem is called **sapwood**.

B. CROSS-CUTTING THEMES

*Looking Back—
Concepts from Earlier Chapters*
Photosynthesis—Chapter 10
This chapter details how plants, along with algae, cyanobacteria, and a variety of protists, obtain the energy and carbon they need to grow and reproduce.

Phylogenetics—Chapter 1
Recall from **Chapter 1** that the branches on a phylogenetic tree represent populations through time. This information is used in Chapter 35 to explain the evolutionary relationships of some species of angiosperms.

The Plant Cell—Chapter 7
Chapter 7 introduced a generalized version of the plant cell. This should be reviewed before tackling the specialized cells seen in Chapter 35.

*Looking Forward—
Concepts in Later Chapters*
Water—Chapter 36
Plants must have water as a source of electrons to run photosynthesis. In **Chapter 36**, you'll see that plants also need water to keep their cells in proper working order.

Plant Nutrition—Chapter 37
Plants must obtain nitrogen, phosphorus, potassium, magnesium, and a host of other nutrients in order to synthesize nucleic acids, enzymes, phospholipids, and the other macromolecules needed to build and run cells. **Chapter 37** explores how plants acquire these key elements.

Sensory Response—Chapter 38
A shoot's growth is controlled by sophisticated sensory and response systems. **Chapter 38** explores how plants sense certain wavelengths of light and respond by bending and growing in that direction.

Reproductive Structures—Chapter 40
Twigs and stems are structures that arrange leaves in space. **Chapter 40** explores the reproductive structures housed in twigs and stems.

C. DIFFICULT TOPICS

The main aspect of this chapter that creates problems is the overwhelming amount of anatomical terms. Before you can understand the function of anatomy, you need to understand the anatomy itself. Drawing diagrams and labeling them is the best way to learn anatomical terms. Adding the functions on top of the anatomy will help to solidify all the terms and organize them.

To completely prove that you have mastered the anatomy and function of angiosperms, go out and collect various samples of plants and test yourself. This is a great way to see firsthand the variation of form and function.

D. ASSESSING WHAT YOU'VE LEARNED

(1) Testing your Knowledge

1. Which of the following statements about silverswords and tarweeds is *false*?
    a. Silverswords are a group of closely related plant species.
    b. Tarweeds and silverswords appear to have shared a fairly recent common ancestor.
    c. Silverswords are mainly found in Hawaii and other Pacific Islands.
    d. Some tarweeds can self-fertilize.

2. Which of the following is a modified stem that runs over the soil surface?
    a. stolon
    b. rhizome
    c. tuber
    d. thorn

3. Which of these terms refers to a modified stem used for carbohydrate storage?
    a. stolon
    b. rhizome
    c. tuber
    d. thorn

4. Plants in desert or northern habitats tend to have leaves that are needle-like because:
    a. They reduce herbivory.
    b. They reduce sun damage.
    c. They increase water absorption.
    d. They reduce water loss.

5. The undifferentiated cells responsible for plant growth are called
    a. meristematic cells
    b. parenchyma cells
    c. mesophyll cells
    d. apical cells

6. Which of the following cells serve as storage cells for starch deposits in roots, stems, and leaves?
    a. parenchyma cells
    b. meristematic cells
    c. collenchyma cells
    d. sclerenchyma cells

7. Which of the following are water-conducting cells that are dead at maturity?
    a. sclerenchyma cells
    b. tracheids
    c. sieve-tube members
    d. sclereids

8. The waxy layer that is secreted by the epidermis is called
    a. the dermis
    b. the stoma
    c. the cuticle
    d. the sterol

9. Which of the following zones is *not* a group of cells behind the root cap?
    a. zone of cellular division
    b. zone of cellular elongation
    c. zone of cellular communication
    d. zone of cellular maturation

10. Wood is
    a. produced by the cork cambium
    b. secondary xylem
    c. secondary phloem
    d. produced by the apical meristem

(2) Integrating Your Knowledge

(a) What are silverswords? Why are they significant?

(b) What are parenchyma cells?

(c) What is the difference between collenchyma cells and scleremchyma cells?

(d) Why is epidermal tissue called a tissue?

(e) What is the importance of secondary growth?

CHAPTER 35—ANSWER KEY

D. Assessing What You've Learned

(1) Testing Your Knowledge

1. c; 2. a; 3. c; 4. d; 5. a; 6. a; 7. b; 8. c; 9. c; 10. b

(2) Integrating Your Knowledge

(a) Silverswords are a group of closely related plant species found in Hawaii and nowhere else in the world. For being so closely related, they are incredibly diverse in size and structure.

(b) Parenchyma cells serve as storage cells for starch deposits in roots, stems, and leaves. They are jammed with chloroplasts and are the primary site of photosynthesis.

(c) Collenchyma cells have thickened primary walls and serve to stiffen leaves and stems. Sclerenchyma cells also stiffen stems and other structures, but are distinguished by thickened secondary cell walls strengthened by tough lignin molecules.

(d) A *tissue* is a group of cells that function as a unit. Epidermal cells function as guardians of the interior of the plant, bringing in nutrients and discarding waste.

(e) Secondary growth occurs when certain cell populations in the root and shoot never fully differentiate into specialized cell types. Instead, these cells continue to grow and produce new cells that, in turn, differentiate and cause the stem to increase in diameter.

Water and Sugar Transport in Plants

A. KEY BIOLOGICAL CONCEPTS

36.1 Water Potential and Cell-to-Cell Movement

- The **solute potential** is the difference in solute concentrations between two cells. This explains water's tendency to move via osmosis between two cells.
- **Turgor pressure** is the force of the cell membrane swelling and pushing against the cell wall. The cell wall exerts an equal and opposite force called **wall pressure**.
- **Pressure potential** is the sum of all the types of pressure on water.

What Factors Affect Water Potential?

- When the solute potential and the pressure potential of a cell are added together, the resulting quantity is called the **cell's water potential**. It can be thought of the tendency of water to move from one location to another (**Figure 36.1**).
- Water potentials measured in plants usually have a negative value. Solute potentials are always negative because they are measured relative to the solute potential of pure water.
- Pressure potential in a turgid cell is expressed as a positive value, because it increases the potential energy of the water inside.

Calculating Water Potential

- Ignoring the effects of gravity, water potential is defined algebraically as

$$\zeta = \zeta_p + \zeta_s$$

- The water potential is the pressure potential plus the solute potential (**Figure 36.2**).
- The unit of measurement is a pressure unit called the megapascal (MPa).

Water Potentials in Soils, Plants, and the Atmosphere

- The water contained within a leaf or root system or plant has a pressure potential and a solute potential, just as the water inside a cell does. The water potentials of plant tissues, air and soil are changing all the time in response to heat, cold, rain, and so on.
- **Water potential gradient** is the overall movement of water when a series of water potential differences are contrasted. Plants tend to gain water from the soil and lose it to the atmosphere (**Figure 36.3**).
- If the water potential in the space that surrounds a cell drops, water moves out of a cell in response, causing it to shrink. If the cells do not quickly regain turgor pressure, dehydration and death result (**Figure 36.6**).

36.2 Root Pressure and Short-Distance Transport

- Look at **Figure 36.7**, and note the different tissues. Starting at the outside of the root and working toward the inside, they are the **epidermis** and **root hairs**, the **cortex**, the **endodermis**, the **pericycle**, and the **vascular tissues**.

- When water enters a root along a water potential gradient, it does so through the root hairs. Before reaching the vascular tissue, water can be transported through two pathways (**Figure 36.8a**):
  1. The **apoplast** lies within cell walls, which are porous.
  2. The **symplast** consists of the continuous connections through cells that exist via plasmodesmata.
- **Endodermal** cells are tightly packed and secrete a waxy layer called the **Casparian strip**. It prevents water from creeping through the walls of the endodermal cells—it blocks the apoplastic pathway (**Figure 36.8b**).
- By forcing water to travel through the plasma membrane of endodermal cells, the water can be filtered. These cells can also prevent water from flowing out of the vascular tissue.
- The endodermis is also responsible for **root pressure**. The active transport of ions brings more water into the root that is lost via transpiration.
- Water droplets can even be forced out of leaves, via **guttation** (**Figure 36.9**).

36.3 Transpiration and Long-Distance Water Transport

- Water loss from the aerial parts of a plant is called **transpiration**. To avoid dehydration, plants must absorb huge quantities of water from soil and transport it to their leaves.
- **Cohesion** is the mutual attraction among like molecules, for example, the hydrogen bonding among water molecules.
- **Adhesion** is the attraction of unlike molecules.

The Cohesion-Tension Theory

- The leaf area just below the stoma is filled with moist air, such that when the stoma opens, the humid air is exposed to the atmosphere. Water exits through the stomatal pore as a result (**Figure 36.11**).
- The water potential of the atmosphere is extremely low compared to the water potential of the space inside the leaf, meaning that a steep water potential gradient exists between the leaf interior and the air.

- The water inside the cell walls, at the interface with the air, forms a concave boundary layer called a **meniscus**. If the water potential gradient is especially steep, then water molecules leave the surface rapidly and the menisci in the cell walls become more concave.

The Role of Surface Tension in Water Transport

- Water molecules are polar; they interact with one another through hydrogen bonding. Surface water molecules can form bonds in only one direction. As a result, the topmost layer of water molecules is pulled inward and the surface forms a meniscus.
- **Surface tension** is the pull that occurs on water molecules at the air-water interface.
- Dixon and Joly hypothesized that these menisci produced a force capable of pulling water up from roots, dozens of meters into the air. They proposed that the force generated at menisci is transmitted through the water present in leaf cells to the water in xylem tissue, on to the water in the vascular tissue of roots, and finally on to the water in the soil.
- The theory claims that because of the hydrogen bonding that occurs between water molecules, the water inside a plant acts as a continuous chain that transmits the pulling force generated by surface tension in the leaf cells.

What Evidence Do Biologists Have for the Cohesion-Tension Theory?

- Researchers also confirmed that water tension in trees is great enough to actually make tree trunks shrink. The **dendrograph** is the instrument used to measure trunk diameter.
- If you take a leaf and cut its petiole, the watery fluid in the xylem withdraws from the edge. This confirms that the xylem sap is under tension.
- Researchers placed a leaf or branch in an airtight container, applied a steadily increasing external pressure, and recorded the pressure required to push the xylem sap back to the cut surface. This pressure is equal to the tension experienced by the xylem sap.
- Researchers collected data on the water potential of Sitka spruce leaves. They harvested

branches from the tops of 10m-tall spruce trees several times a day for several days, used a pressure bomb to measure the water potential of the tissues, and produced a data set.
- Water potential of these tissues dropped during the day and rose at night. This finding correlates with stomata opening during the day and closing at night (**Figure 36.13**).
- Chunfang Wei, Melvin Tyree, and Ernst Steudle set out to study the cohesion-tension theory using a root bomb and a xylem pressure probe. This allowed the investigators to record changes in xylem pressure instantly and directly (**Figure 36.14**).
- First, they added pressure to the root systems of their experimental plants using a root bomb. Then they released pressure on the root system and began to alter light levels.
- Pulses of pressure applied through the root bomb produced sharp decreases in the tension experienced by xylem sap. The tension experienced by the xylem sap increased each time that light intensity was increased.

Water Absorption and Water Loss

- One of the most important features of the cohesion-tension theory is that it does *not require the expenditure of energy* by plants. The Sun furnishes the energy that breaks hydrogen bonds between water molecules at the air-water interface inside leaves and causes transpiration.
- Water simply moves along a potential water gradient by **bulk flow**—a movement in response to pressure.
- The balance between conserving water and maximizing photosynthesis is termed the **photosynthesis-transpiration compromise**.

Limiting Water Loss

- To cope with water loss, the oleander has the following adaptations (**Figure 36.15**):
  1. a thick cuticle on the upper surface of the leaves
  2. several layers of epidermis
  3. stomata located in deep pits on the undersurface of leaves

Obtaining Carbon Dioxide under Water Stress

- **Crassulation acid metabolism** (CAM) plants are able to continue photosynthesizing even though their stomata are closed during the day. CAM plants open their stomata at night and store the CO_2 that diffuses into their tissues by adding organic acids.
- When photosynthesis begins during the day, CO_2 is released from organic acids to **rubisco**—the enzyme that initiates the Calvin cycle.
- The C_4 plants use CO_2 so efficiently that they are able to keep their stomata closed more than competing plants can. The CO_2 is transferred to bundle-sheath cells, where rubisco is abundant.

36.4   Translocation

- **Translocation** is the movement of sugars through a plant. Sugars move from sources to sinks. In vascular plants, a **source** is defined as a tissue where sugar enters the phloem. A **sink** is a tissue where sugar exits the phloem.
- During the growing season, leaves and stems that are actively photosynthesizing and producing sugar in excess of their own needs act as sources. Early in the growing season, however, the situation is reversed.
- Experiments with ^{14}C have made two important generalizations possible (**Figure 36.17**):
  1. Sugars are translocated very rapidly.
  2. There is a strong correspondence between the physical location of certain sources and certain sinks.

The Anatomy of Phloem

- Phloem is made up of **sieve-tube elements** and **companion cells**. Both cells are alive at maturity. In most plants, sieve-tube elements lack nuclei and many major organelles; they are connected to one another, end to end, by open pores (**Figure 36.18**). These sieve-like pores create a direct connection between the cytoplasm of adjacent cells. Companion cells, in contrast, have nuclei and a large number of ribosomes, mitochondria, and chloroplasts.

The Pressure-Flow Hypothesis

- Ernst Münch proposed a mechanism for translocation in 1930. The pressure-flow hypothesis states that events at source tissues and at sink tissues create a steep pressure potential gradient in phloem. The force responsible for movement is generated by large differences in the turgor pressure of phloem sap between source and sink tissues.
- The net result of these events in **Figure 36.19** is high turgor pressure near the source and low turgor pressure near the sink. Because the pores of the sieve plate have no membranes, the difference in pressure potential drives phloem sap from source to sink. The pressure contrast is responsible for a one-way flow of sucrose molecules.

Phloem Loading

- At source cells, sugar has to be loaded into sieve-tube elements against a concentration gradient. Thus loading requires the use of ATP and a membrane transport mechanism. Unloading also requires ATP and a second membrane transport system.

How Are Sugars Concentrated in Sieve-Tube Members at Sources?

- The observation of strong pH differences between the interior and exterior of phloem cells led researchers to suspect that phloem loading depends on an H^+-ATPase.

Where Are H^+-ATPases Located?

- Proton pumps are found in a variety of organisms, including bacteria, fungi, and animals. Researchers found that the *Arabidopsis* (a widely studied plant model from the mustard family) genome actually codes for 10 different pump proteins. One of these genes, called *AHA3*, appears to be expressed primarily in vascular tissues.
- Natalie DeWitt and Michael Sussman used this information and hypothesized that *AHA3* encoded the proton pump responsible for phloem loading. Their goal was to treat *Arabidopsis* leaves with the *AHA3* antibody, examine treated leaves under the electron microscope, and determine exactly where the pump proteins are located.
- The proton pumps responsible for phloem loading are found almost exclusively in the membranes of companion cells.
- This result supported the following model for phloem loading (**Figure 36.23**):
  1. Proton pumps in the membranes of companion cells create a strong gradient that favors flow of protons into companion cells.
  2. A cotransporter protein in the membrane of companion cells uses the proton gradient to bring in sucrose.
  3. Once inside companion cells, sucrose travels into sieve-tube members via direct cytoplasmic connection.

Phloem Unloading

- Sugar transport is an energy-demanding process, but the membrane proteins involved and mechanism of movement vary among different types of sinks within the same plant, as well as among different species (**Figure 36.24**).

B. CROSS-CUTTING THEMES

Looking Back—
Concepts from Earlier Chapters

Leaf Structure—Chapter 35

Recall that the surfaces of leaves are dotted with structures called stomata. Chapter 36 begins to discuss the functions of the stomata and to apply the anatomy learned in **Chapter 35**.

Osmosis—Chapter 8

Water will begin moving into or out of a cell via osmosis—water moves from regions of low solute concentrations to regions of high solute concentrations. Chapter 36 discusses osmosis in relation to the plant and its transport of fluids.

Hydrogen Bonding—Chapter 2

Water molecules are polar, and they interact with one another through hydrogen bonding. Beneath an air-water interface, in the body of a solution, all of the water molecules present are surrounded by other water molecules and form hydrogen bonds in all directions. The water molecules on the surface, however, can form hydrogen bonds in only one direction—with the water molecules below them.

CAM and C_4 Photosynthesis—Chapter 10

Chapter 10 introduced two novel biochemical pathways that are found in species native to deserts and other hot, dry habitats. These pathways are called crassulacean acid metabolism (CAM) and C_4 photosynthesis. CAM plants are able to continue photosynthesizing even though their stomata close during the heat of the day.

Proton Pumps—Chapters 9 and 10

As Chapters 9 and 10 indicated, some proteins are proton pumps, or more formally, H^+-ATPases. Their activity establishes a large difference in the charge and hydrogen ion concentration on either side of the membrane.

Looking Forward—
Concepts in Later Chapters
Creating Antibodies—Chapter 49

Researchers raised antibodies to the AHA3 protein using the types of techniques introduced in **Chapter 49**. As that chapter points out, an antibody is a polypeptide that binds to a specific protein.

C. DIFFICULT TOPICS

Now that you have mastered the anatomy of vascular plants from Chapter 35, you are beginning to learn the physiology of fluid transport in Chapter 36. The terminology can be quite confusing, especially on a test. Try to explain, using plain language, how water and sugars move through a plant. Then apply the appropriate terms in your explanation. List the following terms, and make sure you are able to tell the difference between them:

> solute potential
> turgor pressure
> wall pressure
> pressure potential
> water potential
> water potential gradient

Try using each of these terms in a sentence that explains their meaning, and then compare and contrast the terms. If you can master the meaning of these terms, you are well on your way to understanding the physiology of fluid transport in vascular plants.

D. ASSESSING WHAT YOU'VE LEARNED

(1) Testing Your Knowledge

1. The force of the cell membrane swelling and pushing against the cell wall is called:
    a. turgor pressure
    b. wall pressure
    c. pressure gradient
    d. solute potential

2. When the solute potential and the pressure potential of a cell are added together, the resulting quantity is called the cell's
    a. turgor pressure
    b. wall pressure
    c. pressure gradient
    d. water potential

3. Water potential gradient is
    a. the overall movement of solutes
    b. the evaporation of water from the leaves
    c. the overall movement of water
    d. the pressure gradient minus the water potential

4. The barrier inside the root that prevents water from leaking out of the vascular tissue is the
    a. epidermis
    b. Casparian strip
    c. apoplast
    d. root hair

5. The pathway for water that lies within the cell walls is
    a. the apoplastic pathway
    b. the symplastic pathway
    c. the Casparian pathway
    d. the endodermal pathway

6. Cohesion is
    a. the mutual attraction among like molecules

b. the mutual attraction among unlike molecules

c. the repelling of like molecules

d. the repelling of unlike molecules

7. A movement in response to pressure can also be called
   a. solute flow
   b. bulk flow
   c. transpiration
   d. CAM metabolism

8. Which type of plants open their stomata at night to store CO_2?
   a. C_4 plants
   b. rubisco plants
   c. CAM plants
   d. transpiring plants

9. Translocation is
   a. the movement of water through a plant
   b. the movement of ions through a plant
   c. the movement of sugars through a plant
   d. the movement of xylem though the plant

10. Where are the proton pumps responsible for phloem loading located?
    a. on the membranes of companion cells
    b. on the membranes of sieve-tube members
    c. on the membranes of root cells
    d. on the membranes of root hairs

(2)   Integrating Your Knowledge

(a) What is the difference between turgor pressure and wall pressure?

(b) What is the difference between water potential and water potential gradient?

(c) Explain surface tension.

(d) Compare and contrast sieve-tube elements and companion cells.

(e) What is the difference between proton pumps and H^+-ATPases?

CHAPTER 36—ANSWER KEY

D. Assessing What You've Learned

(1)   Testing Your Knowledge

1. a; 2. d; 3. c; 4. b; 5. a; 6. a; 7. b; 8. c; 9. c; 10. a

(2)   Integrating Your Knowledge

(a) Turgor pressure is the force of a cell membrane swelling and pushing against the cell wall. The cell wall exerts an equal and opposite force called wall pressure.

(b) When the solute potential and the pressure potential of a cell are added together, the resulting quantity is called the cell's water potential. It can be thought of as the tendency of water to move from one location to another. Water potential gradient is the overall movement of water when a series of water potential differences are contrasted. Plants tend to gain water from the soil and lose it to the atmosphere.

(c) Surface tension is the pull that occurs on water molecules at the air-water interface.

(d) Phloem is made up of sieve-tube elements and companion cells. Both cells are alive at maturity. In most plants, sieve-tube elements lack nuclei and many major organelles; they are connected to one another, end to end, by open pores. These sieve-like pores create a direct connection between the cytoplasm of adjacent cells. Companion cells, in contrast, have nuclei and a large number of ribosomes, mitochondria, and chloroplasts.

(e) They are actually the same thing.

37

Plant Nutrition

A. KEY BIOLOGICAL CONCEPTS

37.1 Nutritional Requirements

Essential Nutrients

- An **essential nutrient** should fulfill the following criteria:
  1. It is required for growth or reproduction.
  2. No other element can substitute for it. Symptoms observed when the element is withheld are corrected only by supplying that element.
  3. It is required for a specific structure or metabolic function, not because it aids in the uptake of a different essential element.
- Some elements are essential for some plants and not others. For example, **silicon** is an essential nutrient for rice and corn but not for most vascular plants.
- **Micronutrients** (e.g., selenium) are required in relatively small quantities; they usually function as a cofactor for specific enzymes.

Nutritional Deficiencies

- Researchers who used hydroponic growth to explore the effect of copper deficiency on tomatoes (**Figure 37.3**) found that copper-deprived plants had stunted shoots and roots, dark foliage, curled leaves, and no flowers.
- Copper has since been revealed as a cofactor or component in several enzymes involved in redox reactions.

37.2 Soil

- The process of soil building begins with solid rock. Water, wind, and organisms continually break tiny pieces off large rocks—this is called **weathering** (**Figure 37.4**).
- As organisms occupy the substrate, they add their waste products and carcasses—this organic matter is **humus**.
- **Texture** refers to the proportions of different-sized particles present in soil and is important because:
  1. Texture affects the ability of roots to penetrate.
  2. Texture affects the soil's ability to hold water.
  3. Texture and water content dictate availability of oxygen.

Soil Conservation

- Soil erosion occurs when soil is carried away from a site by wind or water (**Figure 37.6**).
- When plant cover is removed for forestry, farming, or suburbanization, soil erosion is accelerated.
- Techniques that maintain long-term soil quality and productivity are collectively called **sustainable agriculture**.

Nutrient Availability

- The elements required for plant growth exist in the soil as **ions** (**Figure 37.7**). Ions with **negative charge** usually dissolve in water because they interact with water molecules via hydrogen bonding. Ions with **positive charges** often interact with negative charges found in organic matter and on the surfaces of the tiny sheetlike particles called clay—they are less readily available.

- As solutes, negatively charged ions are available to plants for absorption; however, they are also easily washed out of the soil by rain. The loss of nutrients via washing is called **leaching**.
- **Cation exchange** occurs when protons or other cations bind to negative charges on soil particles and cause bound cations such as magnesium or calcium to be released to nearby roots.

37.3   Nutrient Uptake

Mechanisms of Nutrient Uptake

- Some of the proteins found in certain cells span the lipid bilayer of a cell membrane and act as channels that allow the transit of specific ions.
- The large surface area of root hairs holds large numbers of membrane proteins, which contact the soil and selectively facilitate the passage of ions into the cell.

Establishing and Using a Proton Gradient

- The effect of concentration and electrical charge on an ion is called the **electrochemical gradient**. When an electrochemical gradient favors the movement of an ion, no energy expenditure is required and the movement is described as passive.
- **Proton pumps** use ATP and pump an excess of H^+ ions on the exterior of the plasma membrane of the root hair cells (**Figure 37.9**).
- This separation of charge is measured as a **voltage** across the membrane, and these pumps maintain a **membrane potential**.
- Researchers found that K^+ transport was passive. They found that the plant could import K^+ even when the outside concentrations of K^+ were extremely low.

Nutrient Transfer via Mycorrhizal Fungi

- Many plants living in northern forests receive large quantities of nitrogen from fungi that wrap themselves around the epidermal cells of roots and radiate out into the surrounding soil. Plants that live in grasslands and in tropical forests receive much of the phosphorus they need from species of fungi whose bodies actually penetrate into the plant root interior.

- Fungi that live in close association with roots are called **mycorrhizae**; they are symbiotic.
- Studies with the elements C, N, and P confirmed that mycorrhizal fungi receive sugar from plants in exchange for providing nitrogen or phosphorous.

Mechanisms of Ion Exclusion

- Many of the metal ions found in soils are poisonous to plants; even essential nutrients can become toxic if present in high levels.

Passive Exclusion

- The Casparian strip prevents some ions from entering the symplast and reaching the xylem.
- Passive exclusion also occurs in root hairs.
- Researchers explored the molecular mechanism responsible for variation in salt tolerance among corn varieties. They grew seedlings from salt-tolerant and salt-intolerant populations in hydroponic cultures containing a high NaCl concentration. They found that salt-intolerant plants had taken up almost twice as much Na^+ as did the salt-tolerant individuals.
- Researchers suggest that individuals from salt-tolerant populations have fewer sodium channels in their root hairs than do salt-intolerant individuals.

Active Exclusion

- Plants that grow on waste rock and soils from copper-mining operations experience large concentration gradients that favor an influx of this nutrient. How do plants neutralize excess nutrients?
- **Metallothioneins** are small proteins that bind to metal ions and prevent them from acting as a poison.
- The membrane surrounding the vacuole—the **tonoplast**—contains transport proteins that move sodium ions from the cytosol into the vacuoles, where they cannot poison enzymes.

37.4   Nitrogen Fixation

- Only selected bacteria are able to take up N_2, convert it to ammonia, and use it to fuel growth. This conversion process is called **nitrogen fixation**. Nitrogen fixation requires a series of specialized enzymes and cofactors, including an enzyme called **nitrogenase**.

- Bacteria in the genus *Rhizobium* are located in the roots of pea plants in structures called **nodules**. The bacteria provide the plant with ammonia; the legume provides the bacteria with sugar and protection.

How Do Nitrogen-Fixing Bacteria Colonize Plant Roots?

- Roots do not contain a population of *Rhizobia* from the moment of germination. The *Rhizobia* must colonize the plant.
- **Colonization** is a complex process involving a series of specific interactions between the *Rhizobia* and the legume.
- Recognition is possible because the root hairs of the pea family plants contain compounds called **flavonoids**. When rhizobia contact the flavonoids they produce **Nod factors**, which in turn bind to the proteins on the membrane surface of the root hairs.

How Do Host Plants Respond to Contact from a Symbiotic Bacterium?

- Nod factors bind to the root-hair surface, then set off a chain of events that leads to a dramatic morphological change in the legume.
- As in **Figure 37.12**, rhizobia proliferate at the tip of root hair, then enter epidermal cells through an invagination in the root hair membrane called an **infection thread**.
- The infected cortex cells begin to divide rapidly, forming root nodules.

37.5   Nutritional Adaptations of Plants

Epiphytic Plants

- **Epiphytes** are plants that never make contact with the soil and usually grow in the leaves or branches of trees. They absorb most nutrients they need from rainwater that collects in their tissues or in the crevices of bark (**Fig. 37.13**).

Parasitic Plants

- In most cases, parasitic plants use photosynthesis to make their own sugars and tap the root systems of their hosts for water and essential nutrients (**Figure 37.14**).

- Researchers investigated root parasitism versus competition in alfalfa plants. They found that the **biomass** was much smaller in plants with parasites than in pots with just host plants (**Figure 37.15**).
- Parasitism is much more damaging than competition in this host-parasite system, and parasitism lowers the total productivity of the individual involved.

Carnivorous Plants

- Carnivorous plants trap insects and other animals, kill them, and absorb the prey's nutrients (**Figure 37.17**).
- The Venus flytrap is an angiosperm native to bog habitats in the southeastern United States. It makes its carbohydrates via photosynthesis; but since bogs have very low nitrogen, it traps organisms for nitrogen.

B. CROSS-CUTTING THEMES

Looking Back—
Concepts from Earlier Chapters
Root Anatomy—Chapter 35
The general features of root anatomy were introduced in **Chapter 35**. This chapter investigates further how nutrition is obtained through the roots.

Root Function—Chapter 36
The specifics of root function in water uptake were analyzed in **Chapter 36**. Again, this topic is broadened in Chapter 37 to include roots and plant nutrition.

Phospholipid Bylayer—Chapter 6
The interior of the phospholipid bilayer is uncharged; therefore, it resists the passage of ions. Some of the proteins found in certain cells span the bilayer and act as channels that allow the transit of specific ions. The ions are discussed in relation to plant nutrition in Chapter 37.

Mycorrhizae—Chapter 30
The fungi that live in close association with roots are called mycorrhizae. **Chapter 30** introduced two major types of mycorrhizae and presented evidence that these fungi transfer

nitrogen and phosphorus, respectively, from soil to plant roots. This topic is further discussed in Chapter 37 with respect to nutrition.

Looking Forward—
Concepts in Later Chapters
Mutualism—Chapter 50
A mutually beneficial relationship was shown between mycorrhizae and plants in Chapter 37. This topic is further explored in **Chapter 50**.

Bog Habitats—Chapter 50
Bog habitats are notoriously poor in nutrients— particularly in nitrogen. **Chapter 50** explains why this is so, and Chapter 37 gives an example of a plant adaptation to this nutrient-poor environment.

Plants and Stimuli—Chapter 38
Chapter 38 focuses on how a signal travels from the hairs on a leaf to the rest of the plant, and on how plants respond to light.

Ammonia Fertilizers—Chapter 55
The use of ammonia fertilizers has been causing serious pollution problems, as discussed in **Chapter 55**. For this reason, there has been intense interest in the phenomenon of nitrogen fixation.

C. DIFFICULT TOPICS

You will probably find the concept of gradients is the most confusing topic in this chapter. Understanding the movement of ions based on concentration and based on charge is actually quite straightforward, if they are considered as separate entities. Once you incorporate both concentration and charge in one group of ions, predicting their movement begins to get quite complicated. Adding a selectively permeable membrane, like the plasma membrane of a cell, and then considering different ion types, can confuse even the well-versed physiologist.

Luckily, you only need to understand the basics, so make sure that you do. As with any other concept, start simple and gradually add complexity. Make sure you understand the concepts of concentration gradients and electrical gradients before you attempt to study the elec-

trochemical gradient. Stick with one type of ion and then try to set up gradients with other ion types and predict their movement. Save the study of the movement of mixed ions for a physiology class.

D. ASSESSING WHAT YOU'VE LEARNED

(1)   Testing Your Knowledge

1.   Which of the following is *not* a criterion for an essential nutrient?
   a.   It is required for growth and reproduction.
   b.   A cofactor is needed for absorption.
   c.   No other element can substitute for it.
   d.   It is required for a specific structure or metabolic function.

2.   Which of the following techniques can researchers use to explore plant nutrient deficiencies?
   a.   hydroponics
   b.   sun exposure
   c.   crop rotation
   d.   hyperbaric chambers

3.   The process of water and wind breaking off tiny pieces of rock is called
   a.   soil erosion
   b.   humus
   c.   blasting
   d.   weathering

4.   Soil texture is important because:
   a.   Texture affects the ability of roots to penetrate.
   b.   Texture affects the soil's ability to hold water.
   c.   Texture and water content dictate oxygen availability.
   d.   All of the above apply.

5.   Which of the following statements regarding soil erosion is *false*?
   a.   Soil erosion is a very slow natural process.
   b.   Forestry can accelerate soil erosion.
   c.   Suburbanization does not affect soil erosion.
   d.   Farming can accelerate soil erosion.

6. The loss of nutrients via washing by rain is called
   a. erosion
   b. filtering
   c. leaching
   d. binding

7. Fungi that live in close association with roots are called
   a. rhizomes
   b. mycorrhizae
   c. rhizobium
   d. rice

8. Passive exclusion of ions occurs mainly by
   a. the Casparian strip
   b. the root hairs
   c. both (a) and (b)
   d. sodium channels

9. Nitrogen fixation is
   a. the ability to convert N_2 to ammonia
   b. the ability to convert ammonia to N_2
   c. the ability to absorb N_2
   d. the ability to metabolize nitriles

10. What happens when rhizobia contact the flavonoids of pea root hairs?
    a. The rhizobia produce Nod factors.
    b. The rhizobia produce flavonoids.
    c. The pea root hairs produce Nod factors.
    d. Nothing happens.

(2) Integrating Your Knowledge

(a) What are the criteria of an essential nutrient?

(b) What is a membrane potential?

(c) Define *gradient*.

(d) Why would plants allow certain fungi to inhabit their roots?

(e) What are epiphytes?

CHAPTER 37—ANSWER KEY

D. Assessing What You've Learned

(1) Testing Your Knowledge

1. b; 2. a; 3. d; 4. d; 5. c; 6. c; 7. b; 8. c; 9. a; 10. a

(2) Integrating Your Knowledge

(a) An essential nutrient should fulfill the following criteria:
   1. It is required for growth or reproduction.
   2. No other element can substitute for it. Symptoms that are observed when the element is withheld are corrected only by supplying that element.
   3. It is required for a specific structure or metabolic function, not because it aids in the uptake of a different essential element.

(b) Proton pumps use ATP and pump an excess of H^+ ions on the exterior of the plasma membrane of the root hair cells. This separation of charge is measured as a voltage across the membrane, and these pumps maintain a membrane potential.

(c) A gradient is the potential or tendency of a substance to move.

(d) Certain plants allow fungi to inhabit their roots so that they have a mutualistic or symbiotic relationship. The plant can gain nitrogen from the fungi; in return, the fungi receive sugars and protection.

(e) An epiphyte is a plant that never makes contact with the soil.

38

Sensory Systems in Plants

A. KEY BIOLOGICAL CONCEPTS

38.1   Sensing Light

What Do Plants See?

- Charles Darwin and his son (1881) performed the first experiments showing that plants respond to light. They also concluded that the blue part of the spectrum was involved.
- **Phototrophism** is any type of directed movement in response to light. Plants are positively phototrophic (**Figure 38.1**).

Distinguishing Red and Far-Red Light

- Species from sunny habitats elongate their stems much more strongly in response to shade light (far-red wavelengths) than do species from forest-floor habitats.
- At least some plants can sense far-red wavelengths and respond by elongating their stems if they are adapted to open, sunny habitats.

Photoperiodism

- **Photoperiodism** is any response by an organism that is based on the photoperiod—the relative lengths of day and night.
- Plants fall into three categories (**Fig. 38.3**):
  1. **Long-day plants** bloom in midsummer when days are at their longest.
  2. **Short-day plants** bloom in spring, late summer, or fall.
  3. **Day-neutral plants** flower without regard to photoperiod.

The Red/Far-Red Switch

- In lettuce, seed sprouting is affected by light. Researchers discovered that germination rates peak at 660 nm (red spectrum) with a maximum inhibitory effect at 735 nm (far-red spectrum).
- A lettuce seed that germinated in shade with primarily far-red light available would have poor prospects because lettuce plants need full sun to thrive.
- Red and far-red light act like an on-off switch for seed germination in lettuce, just as they do for flowering in some plants.

How Do Plants See?

- Plants contain a wide variety of **pigments** that absorb light. These molecules include chlorophylls and carotenoids involved in photosynthesis. However, researchers hypothesized that the receptor molecules for light were different from these pigments.
- **Phytochromes** from young corn shoots were found to be **photoreversible**. When the phytochromes were placed in a solution and exposed to red or far-red light, solution color switched from blue to blue-green and black.

Phytochromes as Red and Far-Red Receptors

- The genome of *Arabidopsis thaliana* is a small, weedy mustard plant that serves as a model organism in the molecular studies of plants. It actually has five distinct loci that can encode phytochrome proteins—*PHYA*, *PHYB*, and so on.

- Each of these proteins holds a small pigment molecule that absorbs light in the red and far-red parts of the spectrum.

Phototropins as Blue-Light Receptors

- The chlorophylls and carotenoids involved in photosynthesis absorb strongly in the blue part of the spectrum.
- Researchers found a membrane protein that is abundant in the tips of emerging shoots and that becomes phosphorylated in response to blue light. But was this protein the membrane receptor itself?
- **Protein kinases** hydrolyze ATP and catalyze the addition of the phosphate group (phosphorylation) to another protein, which then activates another protein.
- Researchers began analyzing *A. thaliana* mutants that do not show a phototrophic response to blue light (**Figure 38.5**). The gene, which was named *phot1*, encodes the sunlight detector in plants. A phototropic response is triggered when this protein is phosphorylated (**Figure 38.6**).
- Most recent research indicates that multiple blue-light receptors—called **phototropins**—are related to PHOT1.

From Perception to Response: Signal Transduction

- **Signal transduction** is the process that translates the information that an individual receives about what is going on outside into a coordinated response. Make sure you study and are able to draw **Figure 38.7**, so that you understand the possible outcomes of signal transduction.
- In many cases, the response to a signal involves a phosphorylation event and a subsequent change in activity of a response protein or a series of proteins.

38.2   Gravity Perception

- **Gravitropism** is the ability of plants to move in response to gravity. Charles and Francis Darwin (1881) found that roots stop responding to gravity if their caps are removed.
- Researchers recently demonstrated that by killing tiny blocks of cells in *Arabidopsis* with a laser beam, they were able to determine that the cells directly under the epidermal cells at the tip of the root are the most important for initiating the gravitropic response (**Figure 38.9**).

The Statolith Hypothesis

- The **statolith hypothesis** holds that **amyloplasts** (starch-storing organelles) are the primary gravity sensors in plants (**Figure 38.10**). The idea is that gravity pulls the amyloplasts to the bottom of cells, where the force of the amyloplasts on the cell membrane activates pressure or stretch receptors that initiate the gravitropic response.
- Although recent experiments strongly support the statolith hypothesis, the receptor itself has yet to be found. In addition, *Arabidopsis* amyloplast mutants still show 25% of the normal response.

Is the Gravity Sensor a Transmembrane Protein?

- **Integrins** are membrane proteins that frequently bind to components of the extracellular matrix. As a result, they form a line of communication between the extracellular environment and the interior of the cell.
- Researchers recently made antibodies to the integrin proteins found in chickens and applied them to plant roots. They found that the membranes of amyloplasts also contain these proteins.
- If experiments confirm that integrins on the cell membrane *and* amyloplast membranes act as gravity receptors, it could turn out that plants sense gravity in both places; thus both hypotheses would be correct.

38.3   How Do Plants Respond to Wind and Touch?

An Introduction to Electrical Signaling

- The key to understanding electrical signaling in plants is to recognize that the interior of most plant cells has a negative charge relative to the exterior. When proton pumps move protons to the exterior of a plant cell, they create a charge separation across the membrane called a **membrane polarization**.

- Charge separation creates a **membrane voltage**, which is a form of potential energy. Because voltage represents a form of potential energy, voltage across a cell membrane is often called **a membrane potential** (**Figure 38.12**).

- The size of a membrane potential can be measured with a set of **electrodes** placed on the inside and outside of the cell. The units are expressed in **millivolts** (mV). By convention, membrane potentials are expressed as the state of the cell's interior relative to the exterior. As a result, the **resting potential** (normal state) of a plant cell is usually **negative**.

- **Voltage** is a form of **potential energy**, which can be thought of as the tendency for something to move. **Electrical potential energy** is the tendency of charged particles to move toward an area of opposite charge.

Action Potentials

- Impulses in sensory cells (nerve cells in animals) consist of a flow of charge in the form of ions across the cell membrane. Because of the charge flow, the voltage across the cell membrane changes drastically.

- The voltage change that occurs in the Venus flytrap has a characteristic pattern, called an **action potential** (**Figure 38.13**). This change is called a **depolarization** because the charges on either side of the membrane become more alike.

- An action potential is an extremely rapid change in membrane potential from negative to positive, and then back to negative (**repolarization**).

- Make sure you understand **Figure 38.13** in general terms. Chapter 45 will discuss details of this phenomenon. Explaining the action potential in general and applying it to the Venus flytrap will suffice for this chapter.

How Does the Venus Flytrap Close?

- The Venus flytrap closes through the following process:

1. The mechanical energy that moves the receptor hair on the trap surface needs to be transduced to an electrical signal.
2. The electrical signal must be propagated across the leaf in the form of one or more action potentials.
3. Effector (response) cells must respond to the action potential by undergoing a rapid change in turgor that moves the leaf and closes the trap (**Figure 38.14**).

- Recent research shows that H^+ rushes into the response cells when the electrical signal arrives.

B. CROSS-CUTTING THEMES

Looking Back—
Concepts from Earlier Chapters

Absorbing Light—Chapter 10

Recall from **Chapter 10** that land plants absorb many of the wavelengths in the red and blue portions of the spectrum, but they do not absorb wavelengths in the far-red.

Phosphorylation—Chapter 15

Many proteins switch from inactive to active states, or vice versa, when a phosphate group (PO_4^{2-}) from ATP is added to them.

Potential Energy—Chapters 2 and 36

As **Chapter 2** explained, the chemical potential electron describes its tendency to move closer to a nucleus. In **Chapter 36** it was pointed out that the water potential of a tissue or substance expresses the tendency of water to move toward it or away from it. In the same way, electrical potential is the tendency of charged particles to move toward an area of opposite charge.

Looking Forward—
Concepts in Later Chapters

Hormones—Chapter 39

Chapter 38 considers how plants receive information and how they respond to this information. Because many of these changes involve changes in growth patterns, **Chapter 39** follows up by introducing the chemical signals called hormones.

The Action Potential—Chapter 45

In **Chapter 45** we will more closely examine the molecular mechanisms responsible for the action potential.

C. DIFFICULT TOPICS

In this chapter you have explored the nature of physiological responses in plants to light, gravity, and touch. All of these responses involve signal transduction or the changing of environmental information into a form of information the plant can understand.

It is quite easy to memorize these processes of signal transduction without understanding them. However, this can get you into trouble down the road, because the upcoming chapters on sensory systems in animals are somewhat more complex. By taking the time to understand the sensory response systems at this rudimentary level in plants, you will set a good foundation for future chapters on animal sensory systems.

The most important link to the animal kingdom, and the hardest to understand, is the concept of the action potential. Draw an action potential, and make sure you understand all the terminology when tracing it. Try to visualize the movement of ions (just charges for now, not specific ions) during the action potential and after the action potential. By understanding the action potential, you will be in strong form for this chapter and the chapters to come.

D. ASSESSING WHAT YOU'VE LEARNED

(1) Testing Your Knowledge

1. Any directed movement in response to light is called
    a. photoperiodism
    b. tropism
    c. phototropism
    d. gravitropism

2. Plants that bloom in midsummer, when nights are shortest, are called
    a. short-day plants
    b. long-day plants
    c. day-neutral plants
    d. short-night plants

3. Why does lettuce germinating in the shade have a disadvantage over lettuce germinating in the sun?
    a. Lettuce needs far-red light (shade) to germinate.
    b. Lettuce needs full-spectrum light to germinate.
    c. Lettuce needs red-spectrum light (full sun) to germinate.
    d. None of the above apply.

4. Which molecules hydrolyze ATP and catalyze the addition of a phosphate group?
    a. pigments
    b. phototropins
    c. phosphate genes
    d. protein kinases

5. The technical name for blue-light receptors is
    a. phytochromes
    b. phototropins
    c. pigments
    d. transducins

6. The process that translates a signal from the environment into an energy form that the plant decodes is
    a. signal transduction
    b. signal deciding
    c. signal transmission
    d. neural transmission

7. The hypothesis stating that amyloplasts are the primary gravity sensors in plants is the
    a. statolith hypothesis
    b. gravitational pressure hypothesis
    c. amyloplast hypothesis
    d. none of the above

8. After all the research is weighed, which of the following hypotheses appear to be supported?
    a. the statolith hypothesis
    b. the gravitational pressure hypothesis
    c. both (a) and (b)
    d. No hypothesis has been supported.

9. What are integrins?
    a. membrane proteins that bind to components of the extracellular matrix

b. cytosolic proteins that bind to the components of the extracellular matrix

c. membrane proteins that bind to components of the intracellular matrix

d. nuclear proteins that bind to the interior of the cell membrane

10. An extremely rapid change in membrane potential from negative to positive, then back to negative again, is best termed
a. a resting potential
b. a depolarization
c. voltage
d. an action potential

(2)   Integrating Your Knowledge

(a) Plants are positively phototrophic— explain this statement.

(b) What are pigments, and why are they important to plants?

(c) What is a protein kinase, and what does it do?

(d) What are integrins, and why are they important?

(e) Define membrane potential.

CHAPTER 38—ANSWER KEY

D. Assessing What You've Learned

(1)   Testing Your Knowledge
1. c; 2. b; 3. c; 4. d; 5. b; 6. a; 7. a; 8. c; 9. a; 10. d

(2) Integrating Your Knowledge

(a) Phototropism is any type of directed movement in response to light. Plants are positively phototropic because they grow toward the light.

(b) Plants contain a wide variety of pigments that absorb light. These molecules include the chlorophylls and carotenoids involved in photosynthesis.

(c) Protein kinases hydrolyze ATP and catalyze the addition of the phosphate group to another protein, which then activates another protein. In this way, the receptor molecule and its adjacent protein kinase set off a chain of events that eventually leads to a response by the cell to the sensory stimulus.

(d) Integrins are membrane proteins that frequently bind to components of the extracellular matrix. As a result, they form a line of communication between the extracellular environment and the interior of the cell.

(e) Charge separation creates a membrane voltage, which is a form of potential energy. Because voltage represents a form of potential energy, voltage across a cell membrane is often called a membrane potential.

39

Communication: Chemical Signals

A. KEY BIOLOGICAL CONCEPTS

39.1 Plant Hormones: An Overview

- A **coleoptile** is a modified leaf that forms a sheath protecting the stems and leaves of young grasses.
- The Darwins confirmed that these coleoptiles were **phototropic**, meaning that they grow or bend toward light. The Darwins proposed that this bending might be due to a chemical substance.
- A **hormone** is an organic compound that is produced in small amounts in one part of the plant and transported to target cells in another region of the individual, where it causes a physiological response.
- Research on hormones is complex because:
  1. A single hormone may affect many target tissues.
  2. Several hormones may affect the same response.

39.2 Auxin and Phototropism

- Although the Darwins published their hypothesis about chemical signals in 1881, Peter Boysen-Jensen confirmed this in 1913. He cut the tips off young oat shoots and placed either a porous block or nonporous block between the shoot and tip. The stems with the porous block showed normal phototropism, so he concluded that the phototropic signal was indeed chemical (**Figure 39.2a**).

- Fritz Went (1925) collected the phototrophic hormone in gelatinous blocks of agar. Stems responded by bending away from the source of the hormone, without a source of light present (**Figure 39.2b**).
- The physical basis of the phototrophic response **is cell elongation**. Cells on the side of the shoot opposite to the source of light elongate in response to the phototrophic hormone. Because it promotes cell elongation, Went named the hormone **auxin** (from the Greek *auxein*, to increase).

The Cholodny-Went Hypothesis

- This hypothesis contends that auxin produced in the tips of coleoptiles is shunted from one side of the tip to the other, in response to light; then it is transported down the shoot.
- Others proposed an alternative hypothesis: that auxin is broken down by blue light, which produces an asymmetric distribution.
- Winslow Briggs grew corn seedlings in the dark, cut off their tips, and placed the tips on agar blocks. He either kept these in the dark or exposed them to light from one side. Later, he put agar blocks on one side of the decapitated shoots. In response, the shoots from each treatment bend the same amount (**Figure 39.3**).
- This result is inconsistent with the auxin destruction hypothesis. If light destroys auxin, then the block exposed to light should be much less effective in inducing bending.

- Briggs generated additional evidence in support of the Cholodny-Went hypothesis by dividing tips fully or partially with mica. He had shown that auxin had been transported from one side to the other (**Figure 39.4**).

Isolating and Characterizing Auxin

- Researchers succeeded in isolating and characterizing auxin. It is called IAA, or indole acetic acid.

How Does Auxin Produce the Phototropic Response?

The Auxin Receptor

- Researchers set out to attach a radioactive label to the hormone, treat cells with the labeled hormone, and purify the protein or proteins to which it binds.
- Researchers succeeded in isolating auxin-binding protein 1, or ABP1, from corn plants in 1985.
- Experiments have confirmed that ABP1 is an auxin receptor that is located in the plasma membrane of cells in the stem.
- When it binds auxin, ABP1 triggers cell elongation.

How Does Auxin Induce Cell Elongation?

- Researchers proposed that once ABP1 is bound to auxin, the receptor triggers a series of events that leads to an increase in the number of membrane H^+-ATPases.
- Because the pH of the cell wall goes down when H^+-ATPases are active, the **acid-growth hypothesis** resulted.
- Proton pumps could lead to water entry because potassium and other positively charged ions often enter a cell after protons are pumped out. Water follows K^+ via osmosis.
- Bringing water into the cell is only part of the process. The cell wall must also expand. Cell wall expansion occurs in an **acidic** environment. Researchers believe that proteins called **expansins** are somehow responsible.

39.3   Auxin and Apical Dominance

- **Apical dominance** is a growth in which most stem elongation occurs at the **apical meri-stem** of the main shoot. The presence of this topmost meristem inhibits growth by apical meristems that are present lower down on the plant, in nodes (**Figure 39.5**).
- Apical dominance occurs because a continuous flow of auxin from the tips of growing shoots to the tissues below signals the direction of growth. If the signal stops, it means that growth has been interrupted. In response, lateral branches sprout and begin to take over for the main shoot.

Polar Transport of Auxin

- Transport of auxin is **polar**, or **unidirectional**. It travels in a "fountain" pattern. At the root, auxin is redistributed before it travels back up the plant (**Figure 39.6**).

An Overview of Auxin Action

- Auxin clearly plays a key role in phototropism, gravitropism, and apical dominance. However, auxin has other important effects.
- Fruit development is influenced by auxin produced by seeds within the fruit.
- Falling auxin concentrations are involved in the abscission, or the shedding of leaves and fruits, associated with senescence, or aging. Auxin interacts with ethylene in these processes.
- The presence of auxin in growing roots and shoots is essential for proper differentiation of xylem and phloem cells in vascular tissue and the development of vascular cambium.
- Auxin stimulates the development of adventitious roots in tissue cultures and cuttings.

39.4   Cytokinins and Cell Division

- **Cytokinins** are a group of plant hormones that promote cell division.
- Cytokinins are synthesized in root tips, young fruits, seeds, growing buds, and other developing organs.
- The cytokinin that has been found in the most species is derived from adenine and is called **zeatin**.
- Cytokinins act as growth hormones by activating the genes that keep the cell cycle going. In the absence of cytokinins, cells

arrest at the G_1 checkpoint in the cell cycle and cease growth.

39.5   Gibberellins and ABA: Growth and Dormancy

- Growth responses and dormancy are mediated by two main hormones, **abscisic acid (ABA)** and the **gibberellins (GAs)**.

Gibberellins Stimulate Shoot Elongation

- In stems, gibberellins appear to promote both cell elongation and rates of cell division. In seeds, GAs activate the transcription of digestive enzymes that support seed germination and shoot growth.
- In these tissues, GA action must be coordinated with the effects of cytokinins and auxin.

Analyzing Stem-Length Mutants

- The strong association between shoot elongation and gibberellin dosage gave researchers an important tool for dissecting how GAs work.
- By using forward genetics, researchers began with mutant phenotypes and attempted to characterize the loci responsible for the defect (**Figure 39.8**).
- The locus responsible for the stem-length differences came to be known as *Le*. Early work on *Le* mutants showed that they attain normal height if they are treated with the gibberellin called GA_1.
- Researchers confirmed that the *Le* locus encodes an enzyme involved in GA synthesis by finding a locus in the pea genome that encodes an enzyme called **3χ-hydroxylase**. This enzyme adds a hydroxyl group to GA_{20} to produce GA_1. Mutant enzymes were unable to convert GA_{20} to GA_1.

Gibberellins and ABA Interact during Seed Dormancy and Germination

- By applying hormones to seeds, researchers learned that in many plants, ABA is the signal that inhibits seed germination, and gibberellins are the signal that triggers embryonic development.

- During the germination of a barley seedling, **β-amylase** is released from a tissue called the **aleurone layer** (**Figure 39.9a**). The β-amylase is significant because it acts as a digestive enzyme that breaks the bonds between sugar subunits of starch. The enzyme diffuses into the carbohydrate-rich endosperm tissue and releases sugars that can be transported and used by the growing embryo.
- Adding GA to the aleurone layer increases production of β-amylase; adding ABA decreases β-amylase levels.

How Does GA Activate the Production of α-Amylase?

- Researchers noticed that the promoter sequence near the β-amylase gene resembles the DNA sequences targeted by a class of transcription factors called the **Myb proteins**. Mybs are DNA-binding proteins that turn gene transcription on or off.
- It was proposed that a Myb might turn β-amylase on in response to a signal from GA or off in response to a signal from ABA (more specifically, see **Figure 39.9b**).
- Researchers began by isolating all of the mRNAs produced in the aleurone layer of a germinating barley seed. After they used reverse transcriptase to make DNA copies of the mRNAs, they tested each DNA sequence to see if it would react with the DNA-binding sequence that characterizes Mybs. One DNA sequence did, and they confirmed that a Myb protein exists in activated aleurone tissue.
- Was this transcription factor produced specifically in response to GA? Follow-up experiments confirmed that this protein, which they named GAMyb, binds to the β-amylase promoter and acts as a transcription activator that stimulates β-amylase production.

How Do GA and ABA Interact?

- Recent studies show that ABA also induces the production of Myb proteins. Preliminary data suggest that these ABA-dependent transcription factors bind to the β-amylase promoter and shut down transcription. The key observation is that the transcription activators and repressors compete for the same binding sites near genes.

- Summary of hormone action:
    1. A cell's response to a hormone often occurs because specific genes are turned on or off.
    2. Hormones rarely act on DNA directly. Instead, a receptor on the surface of a cell usually receives the message and responds by initiating a chain of events that leads to gene activation or repression.
    3. Different hormones interact at the molecular level because they induce different transcription activators and repressors.

ABA Closes Guard Cells in Stomata

- Early work on stomatal opening and closing suggested that ABA is involved in communicating from roots to leaves about water conditions. Applying ABA to the exterior of guard cells causes them to close.
- Researchers were able to confirm two important predictions (**Figure 39.11**):
    1. ABA concentrations in roots on the dry side of the pot were extraordinarily high.
    2. ABA concentrations in the leaves of experimental plants were much higher than in the controls.
- These results suggested that ABA from roots is transported to leaves, and that it actually does serve as an early warning system of drought stress.

How Do Stomata Open and Close?

- Changes in cell shape are related to changes in the activity of H^+-ATPases in the plasma membrane. See **Figure 39.12** and the preceding section on stem elongation and the acid growth hypothesis.

39.6   Ethylene and Senescence

- Ethylene is strongly associated in three aspects of senescence in plants:
    1. Fruit ripening, eventually leading to rotting
    2. The fading of flowers
    3. Abscission, or the detachment of leaves
- The abscission zone is a region of the leaf petiole that is more sensitive to ethylene in

the tissue. Thus it degrades first, and the leaf breaks off at that point (**Figure 39.13**).

B. CROSS-CUTTING THEMES

Looking Back—
Concepts from Earlier Chapters
Plant Response—Chapter 38
Plants constantly monitor their environment and respond to light, gravity, touch, and other stimuli in ways that increase their ability to survive and reproduce.

Signal Transduction—Chapter 38
Chapter 38 introduced the concept of signal transduction. When a sensory receptor perceives a change in stimulus such as light, gravity, or touch, the receptor molecule changes its conformation or composition or activity. The action by the receptor is the first step in a sequence of events that culminates in the cell's response.

Gene Hunting—Chapter 19
The gene-hunting strategy for finding the *ABP1* locus and other strategies were discussed in detail in **Chapter 19**.

Membrane Proteins—Chapter 36
Chapter 36 discussed proton pumps and how they drive protons out of the cell against an electrochemical gradient. Potassium and other positively charged ions often enter a cell after protons are pumped out. The method for visualizing membrane proteins was also discussed in **Chapter 36**.

Looking Forward—
Concepts in Later Chapters
Fruit Ripening—Chapter 40
As **Chapter 40** will show, fruit ripening is interpreted as an adaptation that enhances the attractiveness of fruits to birds, mammals, and other animals that disperse seeds to new locations.

C. DIFFICULT TOPICS

The world of plant hormones is a very complicated one, and this chapter just begins to scratch the surface of its complexity. The key concepts you must understand are that a hormone can have: (1) a multitude of actions, depending on the location in the organism, and (2) a multitude of interactions, depending on both the location and the nature of the interactive chemical. You will encounter more actions and interactions of ABA and the GAs, so it is important to learn their present functions as discussed in this chapter.

Organize both hormones in a chart and list their actions, the locations of those actions, and possible interactions between these two hormones. Continue to describe their actions in a broad sense, and then proceed to describe these actions at the molecular level, citing the research used to determine each. This process will be of great benefit as you proceed to the chapters dealing with hormones in animals.

D. ASSESSING WHAT YOU'VE LEARNED

(1) Testing Your Knowledge

1. A coleoptile is
    a. a modified root that forms a sheath protecting itself
    b. a modified leaf that forms a sheath protecting itself
    c. a modified stem that forms a sheath protecting itself
    d. none of the above

2. The Darwins confirmed that coleoptiles were
    a. phototropic
    b. gravitropic
    c. pigmented
    d. long-day plants

3. Name the organic substance that is produced in small quantities in one part of the plant and transported to target cells, causing a physiological response.
    a. sodium
    b. a protein
    c. a hormone
    d. nucleotides

4. What is the physical basis of the phototropic response?
    a. cell division
    b. cell elongation
    c. cell differentiation
    d. cytokinesis

5. What is the best explanation for how cell elongation occurs?
    a. active pumping of water
    b. active pumping of ions
    c. active pumping of ions, passive movement of water
    d. active pumping of ions, active movement of water

6. Growth in which most of the stem elongation occurs at the apical meristem is called
    a. apical dominance
    b. apical elongation
    c. top growth
    d. antilateral growth

7. Which of the following statements is *true* about auxin transport?
    a. Auxin transport is random.
    b. Auxin transport is bidirectional.
    c. Nothing is known about the specifics of auxin transport.
    d. Auxin transport is unidirectional.

8. Which of the following is *not* an additional function of auxin?
    a. Fruit development is influenced by auxin.
    b. Auxin stimulates the development of adventitious roots in tissue cultures and cuttings.
    c. Auxin is responsible for promoting cell division.
    d. Auxin interacts with ethylene to cause fruit ripening.

9. The two main hormones that mediate growth responses and dormancy are
    a. auxin and cytokinins
    b. gibberellins and cytokinins
    c. gibberellins and auxin
    d. gibberellins and abscisic acid

10. Which of the following is the main hormone responsible for plant senescence?
    a. auxin
    b. ethylene
    c. gibberellins
    d. cytokinins

(2) Integrating Your Knowledge

(a) Is there an auxin receptor? What proof can you provide?

(b) What does apical dominance mean, and why is it important for plant growth?

(c) Is auxin presence in roots important?

(d) What does the application of ABA do to seeds? What about the application of GA?

(e) What is a Myb protein, and how does it relate to plant growth?

CHAPTER 39—ANSWER KEY

D. Assessing What You've Learned

(1) Testing Your Knowledge

1. b; 2. a; 3. c; 4. b; 5. c; 6. a; 7. d; 8. c; 9. d; 10. b

(2) Integrating Your Knowledge

(a) Yes; Lobler and Klambt succeeded in isolating auxin-binding protein 1, or ABP1, from corn plants in 1985.

(b) Apical dominance is a growth in which most of the stem elongation occurs at the apical meristem of the main shoot. The presence of this topmost meristem inhibits growth by apical meristems that are present lower down on the plant, in nodes Apical dominance occurs because a continuous flow of auxin from the tips of growing shoots to the tissues below signals the direction of growth. If the signal stops, it means growth has been interrupted. In response, lateral branches sprout and begin taking over for the main shoot.

(c) The presence of auxin in growing roots and shoots is essential for the proper differentiation of xylem and phloem cells and their organization into vascular tissue.

(d) By applying hormones to seeds, researchers learned that in many plants, ABA is the signal that inhibits seed germination. They also learned that gibberellins are the signal that triggers embryonic development.

(e) Gubler and associates noticed that the promoter sequence near the β-amylase gene resembles the DNA sequences targeted by a class of transcription factors called the Myb proteins. Mybs are DNA-binding proteins that turn gene transcription on or off. Gubler's team proposed that a Myb might turn β-amylase on in response to a signal from GA or off in response to a signal from ABA.

40

Plant Reproduction

A. KEY BIOLOGICAL CONCEPTS

40.1   An Introduction to Plant Reproduction

Sexual Reproduction

- Most plants reproduce sexually. **Sexual reproduction** is based on the reduction division known as **meiosis** and on **fertilization**, the fusion of haploid cells called **gametes**.
- **Sperm** are small cells from the male that contribute genetic information in the form of DNA but few or no nutrients to the offspring. Female gametes are called **eggs**, which contribute a store of nutrients to the offspring.
- **Perfect flowers** contain both male and female structures. **Imperfect flowers** contain either male or female parts.
- **Outcrossing** occurs when male and female gametes are exchanged between individuals of the same species. When **self-fertilization** occurs, a sperm and an egg from the same individual unite to form a progeny.

Plant Life Cycles

- Plants are the only organisms that have both a multicellular form that is diploid and a multicellular form that is haploid. This life cycle is called **alternation of generations** (**Figure 40.3**).
- The diploid phase is called the **sporophyte**; the haploid phase is called the **gametophyte**.
- A **spore** is a reproductive cell that grows into a new individual directly. A **gamete** is a reproductive cell that must fuse with another gamete before growing into a new individual.

Asexual Reproduction

- **Asexual reproduction** does not involve meiosis or fertilization. It leads to offspring that are **genetically identical** to the parent plant. The major advantage of asexual reproduction is its high efficiency.
- However, if a fungus or other disease-causing agent infects an individual, it will probably succeed in infecting the plant's asexual offspring as well.

40.2   Reproductive Structures

The General Structure of the Flower

- **Sepals** are leaflike structures comprising the outermost part of the flower (**Figure 40.6**).
- **Petals** are arranged on a stem in a whorl. Often brightly colored, they advertise the flower to visually oriented animals such as bees, wasps, and hummingbirds.
- The **nectary** produces sugar-rich fluid called **nectar**, which is harvested by many of the animals that visit.
- An entire group of petals is called a **corolla**. The male reproductive structure of angiosperms is called a **stamen**, while the female reproductive structure is called the **carpel**.

Producing the Female Gametophyte

- The carpel consists of three parts: the **stigma**, the **style**, and the **ovary**. The **ovule**, inside the ovary, is where meiosis takes place and where the female gametophyte is produced (**Figure 40.8**). Four nuclei result from meiosis, but three of these degenerate. The remaining haploid nucleus is the **megaspore**. The megaspore divides by mitosis to produce **eight nuclei** (the number varies among species), which segregate to different positions and form seven cells, comprising the **embryo sac**. The most important elements of the embryo sac are the **egg** and the **polar nuclei**. The egg is located near the base of the structure by an opening called the **micropyle**.

Producing the Male Gametophyte

- The **stamen** consists of the **anther** and a **filament**. Cells inside the anther undergo meiosis. Each of the haploid cells that results is called a **microspore** (**Figure 40.9**).
- Each microspore becomes a **pollen grain**. When mature, pollen grains consist of a small **generative cell** enclosed within a large **vegetative cell**. The vegetative cell develops a **hard coat** that protects the cell contents and generative cell when pollen grains are shed.
- Pollen grains **do not** contain sperm; rather, they are tiny **gametophytes**. The haploid generative cell produces sperm cells via mitosis.

40.3 Pollination and Fertilization

Pollination

- **Pollination** is the transfer of pollen from an anther to a stigma.
- In many angiosperms, pollen is carried from the anther of one individual to the stigma of a different individual (**cross-pollination**) on the body of an insect, bird, or bat that moves from flower to flower. Animal pollination is an example of **mutualism**.

What Is the Adaptive Significance of Pollination?

- The more recently evolved plant groups do not need water in order for sexual reproduction to occur. As a result, the evolution of pollen made these species less dependent on wet habitats and allowed colonization of dryer habitats.
- Pollination evolved into a much more precise process when animals began to act as pollinators. Insect pollination is an important adaptation because it makes sexual reproduction much more efficient.

Does Pollination by Animals Encourage Speciation?

- Insect pollination appears to be closely associated with the formation of new flower and insect species.
- A biologist documented differences in alpine skypilot flowers (**Figure 40.12**). In the lower timberline regions, flowers are small and smell skunky; in the tundra regions, flowers are large and smell sweet.
- Large bumblebees pollinate the tundra flowers, while small flies pollinate the timberline flowers. Flower morphology is in tune to attract the type of insect that is abundant in that area. These two types of flowers are on their way to evolving into two distinct species.

Fertilization

- The male gametophyte produces a long projection called a **pollen tube** that grows down the length of the style. When the pollen tube reaches the micropyle, the sperm are discharged into the ovule (**Figure 40.13**).
- **Fertilization** occurs when a sperm and egg actually unite to form a diploid zygote.
- In angiosperms, an event called **double fertilization** takes place (**Figure 40.14**).
- One sperm unites with the egg nucleus to form the zygote. The other sperm nucleus moves through the embryo sac and fuses with the two polar nuclei to form a single triploid ($3n$) cell.
- The triploid cell resulting from the second fertilization begins a series of mitotic divisions that form tissue called **endosperm**, which stores nutrients.

40.3 The Seed

Embryogenesis

- When a zygote divides, two **daughter cells** are produced: The **basal cell** divides to form a row of single cells, and the **terminal cell** is the parent cell of all the cells in the embryo (**Figure 40.15**).

- The terminal cell and its progeny divide and sort into three groups:
  1. The exterior layer forms the **epidermis**.
  2. The cells just inside the exterior layer form the **ground tissue**.
  3. A group of cells in the core of the embryo become the **vascular tissue**.
- **Cotyledons** are the seed leaves that take up the nutrients in the endosperm and store them. The **hypocotyl** is the initial stem, and the **radicle** is the first root structure.
- By the time the embryo matures, the three tissue types have differentiated; the structures mentioned earlier have formed. The seed tissues then dry, and the embryo becomes **quiescent**.

The Role of Drying in Seed Maturation

- Water loss is an adaptation that prevents seeds from germinating on the parent plant.
- Researchers have shown that as water leaves the seed during drying, **sugars** begin to interact with the hydrophilic parts of cell components. These interactions stabilize the proteins and membranes before damage occurs.

Fruit Development and Seed Dispersal

- After fertilization occurs in angiosperms, the cells that make up the ovary develop a structure called the **pericarp** (**Figure 40.17**). This part protects the seeds. The mature structure is called a **fruit**.
- Most dry fruits are either dispersed by wind or simply fall to the ground. Animals are the most common dispersal agent of fleshy fruits (**Figure 40.19**).

Seed Dormancy

- Even though water and oxygen are available, seeds may not germinate. This process is called **seed dormancy**.
- Researchers have studied lotus plant seeds from an old dried-up lakebed in China. The oldest seed that germinated was 1300 years old.
- Dormancy is usually a feature of seeds from species that inhabit seasonal environments.

- Dormancy is usually interpreted as an adaptation that allows seeds to remain viable until conditions improve.

What Role Does ABA Play in Dormancy?

- ABA levels have not been universally correlated with dormancy. Researchers have concluded that there is no single universal mechanism for initiating and maintaining seed dormancy.

How Is Dormancy Broken?

- For some seeds, the seed coat must be disrupted, or **scarified**, for germination to occur. For others, exposure to water, light, oxygen, or fire will initiate germination.
- The cue that triggers germination is a reliable signal that conditions for seedling growth are favorable for a particular species in a particular environment.

Seed Germination

- Seeds do not germinate without water, because water uptake is the first event in germination.
- During the first phase of water uptake, oxygen consumption and protein synthesis in the seed increase dramatically, but no new messenger RNAs are transcribed (**Figure 40.20**). Based on these observations, biologists have concluded that the earliest events in germination are driven by mRNAs that are stored in the seed.
- During the second phase, when water uptake stops, newly transcribed mRNAs appear and are translated into protein products (**Figure 40.21**). Mitochondria also begin to multiply. In effect, seeds take up enough water to hydrate their existing proteins and membranes, and then begin to manufacture the proteins and mitochondria needed to support growth. Water uptake resumes as growth begins. This second bout of water uptake enables cells to enlarge and the embryo to burst from the seed coat.

B. Cross-Cutting Themes

Looking Back—
Concepts from Earlier Chapters
Inbreeding—Chapter 23
Self-fertilization is the most extreme form of inbreeding. Inbreeding is defined as reproduction among relatives. **Chapter 23** explored the consequences of inbreeding in detail and presented data showing that inbred offspring have poor fitness.

Diploid and Haploid Cells—Chapters 11 and 12
Except for certain red, brown, and green algae, plants are the only organisms that have both a multicellular form that is diploid and a multicellular form that is haploid. Recall from **Chapters 11 and 12** that diploid cells have two copies of each chromosome, while haploid cells have one.

Life Cycles of Plants—Chapter 29
In flowering plants, the gametophyte generation is tiny, short-lived, and dependent on the sporophyte for nutrition. **Chapter 29** explored the diversity of life cycles found in plants in more depth.

Tissue Layers—Chapter 35
The three tissue types formed during embryogenesis were discussed in detail in **Chapter 35**. Chapter 40 shows the relationship between these tissue types and the originating cells.

Looking Forward—
Concepts in Later Chapters
Disease—Chapter 50
If a fungus or other disease-causing agent infects a big bluestem individual, it will probably succeed in infecting the plant's asexual offspring as well. As **Chapter 50** will show, plants fight disease with a wide variety of molecules.

C. DIFFICULT TOPICS

The concepts in this chapter are fairly straightforward, but there are many new anatomical terms and physiological processes to learn. The best way to begin studying the anatomy is by drawing pictures and labeling them. Once you are comfortable with the anatomy, proceed to the processes of pollination, fertilization, and embryogenesis.

Using the figures in your textbook, map out each process while drawing panels and describing each step. Use the proper terms for each structure, as you have learned in the anatomy section. Lastly, try describing each process with only words, no pictures. This will help you to understand the concepts and not just simply memorize facts.

D. ASSESSING WHAT YOU'VE LEARNED

(1) Testing Your Knowledge

1. Which kind of flowers contain both male and female structures?
    a. perfect flowers
    b. imperfect flowers
    c. haploid flowers
    d. diploid flowers

2. Which kind of reproduction increases the genetic variation of the offspring?
    a. asexual reproduction
    b. alternation of generations
    c. sexual reproduction
    d. sporophyte production

3. What are the leaflike structures that comprise the outermost part of the flower?
    a. the petals
    b. the sepals
    c. the nectary
    d. the leaves

4. The sugar-rich fluid in the flower that is harvested by many animals is called
    a. juice
    b. nectar
    c. pollen
    d. fruit

5. Which of the following statements regarding reproductive structures is *false*?
    a. The stamen is the female reproductive structure of angiosperms.
    b. The carpel consists of the stigma, the style, and the ovary.

c. The stamen consists of the anther and filament.

d. The pollen grains are tiny gametophytes.

6. When pollen is carried from the anther of one individual to the stigma of a different individual, it is called
a. self-pollination
b. self-fertilization
c. cross-pollination
d. cross-fertilization

7. Animal pollination is an example of
a. parasitism
b. competition
c. commensalism
d. mutualism

8. "One sperm unites with an egg nucleus and another sperm unites with two polar nuclei." This scenario best describes
a. fertilization in gymnosperms
b. double fertilization in angiosperms
c. pollination in angiosperms
d. fertilization in angiosperms

9. Which of the following statements regarding seed dormancy is *false*?
a. ABA is the universal hormone initiating seed dormancy.
b. Seeds can lay dormant for hundreds of years.
c. Dormancy is usually a feature of seeds from species that inhabit seasonal environments.
d. Dormancy is usually interpreted as an adaptation that allows seeds to remain viable until conditions improve.

10. Which of the following is absolutely required for seed germination?
a. fire
b. light
c. water
d. soil

(2) Integrating Your Knowledge

(a) What is the difference between outcrossing and self-fertilization?

(b) Compare and contrast a spore and a gamete.

(c) What is the difference between the megaspore and the embryo sac?

(d) What is double fertilization and what makes it different from just fertilization?

(e) What is responsible for protecting the seeds when they dry up?

CHAPTER 40—ANSWER KEY

D. Assessing What You've Learned

(1) Testing Your Knowledge
1. a; 2. c; 3. b; 4. b; 5. a; 6. c; 7. d; 8. b; 9. a; 10. c

(2) Integrating Your Knowledge

(a) Outcrossing occurs when male and female gametes are exchanged between individuals of the same species. When self-fertilization occurs, a sperm and an egg from the same individual unite to form a progeny.

(b) A spore is a reproductive cell that grows into a new individual directly. A gamete is a reproductive cell that must fuse with another gamete before growing into a new individual.

(c) Four nuclei result from meiosis, but three of these degenerate. The haploid nucleus that remains is called the megaspore. The megaspore divides by mitosis to produce eight nuclei (the number varies among species), which segregate to different positions and form seven cells. These seven cells are called the embryo sac.

(d) Fertilization occurs when a sperm and an egg actually unite to form a diploid zygote. In angiosperms, an event called double fertilization takes place.

(e) Researchers have shown that as water leaves the seed during drying, sugars begin to interact with the hydrophilic parts of cell components. These interactions stabilize the proteins and membranes before damage occurs.

41

Animal Form and Function

A. KEY BIOLOGICAL CONCEPTS

41.1　Form, Function and Adaptation

- Adaptive evolution shapes animal populations.
- Adaptive evolution occurs when the frequency of alleles subject to natural selection increases from one generation to the next *and* the resulting changes in the characteristics of the population lead to higher average fitness in a particular environment.
- Natural selection is not the only process that leads to changes in allele frequencies over time. Evolution occurs via the random process called genetic drift, the movement of alleles into and out of populations by migration (gene flow), and the constant introduction of new alleles by mutation.
- Researchers studying medium ground finches on the Galápagos island of Daphne Major found that beak size and shape vary among individuals, and that this trait is heritable.
- During a major drought, 84 percent of the population disappeared. The researchers found that survivors tended to have much deeper beaks than did the birds that died. What was the likely cause of this enduring morphological trait?

Trade-Offs

- One of the most important constraints on adaptation involves **trade-offs**. This may involve expenditures of time or energy.
- Barry Sinervo and colleagues investigated the predicted trade-off in egg size and egg num-

ber by manipulating these parameters in side-blotched lizards (**Figure 41.1**). They were able to induce the production of large clutches of small eggs by catching females and surgically removing yolk from their eggs. By removing eggs early on, they found that the remaining eggs grew larger.

Adaptation and Acclimatization

- **Adaptation** occurs when a population changes in response to natural selection. **Acclimatization** occurs when an individual changes in response to a change in environmental conditions.

41.2　Tissues, Organs, and Systems: How Does Structure Correlate with Function?

Tissues

- A **tissue** is a group of cells with the same structure and function.
- There are four basic types of tissues (**Figure 41.3**):
  1. The **epithelial tissues** cover the outside of the body and line the surfaces of organs. This tissue consists of layers of tightly packed cells.
  2. **Connective tissue** is made up of cells that are loosely arranged in a liquid, jellylike, or solid extracellular matrix. **Loose connective tissue** serves as a packing material between organs. **Cartilage** and **bone** support the body, and **blood** transports material throughout the body.

3. **Muscle tissue** functions in movement. Muscle cells are packed with specialized proteins that move in response to **phosphorylation**.

4. Nerve cells or neurons make up **nervous tissue**. Although their shapes vary widely, all neurons have connections to other cells and deliver signals in the form of electrical impulses.

Organs and Systems

- An **organ** is a structure that serves a specialized function and consists of several tissues (**Figure 41.5a**).

- A **system** consists of tissues and organs that work in conjunction to perform a function (**Figure 41.5b**).

41.3 How Does Body Size Affect Animal Physiology?

Surface Area/Volume Relationships

- As a cell gets larger, its volume increases much faster than its surface area. Examine **Figure 41.7** and convince yourself that this is true.

- Given this relationship, the transport processes that occur across the cell membrane would have to support disproportionately more and more cell volume as the cell increased.

Comparing Mice and Elephants

- **Metabolic rate** is defined as the overall rate of energy consumption by an individual. Because it is usually based on aerobic respiration, it is measured in units of ml O_2 consumed per hour.

- Even though an elephant consumes a great deal more oxygen per hour than a mouse does, if you looked at 1 gram of mouse tissue, it would consume about 12 times more oxygen per hour than 1 gram of elephant tissue would (**Figure 41.8**).

- The leading hypothesis to explain this pattern is based on surface area/volume ratios. Many aspects of metabolism, digestion, the delivery of nutrients, and the removal of wastes depend on the exchange of materials across surfaces.

Adaptations That Increase Surface Area

- Because gills have flattened, sheetlike structures called lamellae, the organ has an extremely high surface area relative to its volume (**Figure 41.11**). Diffusion of gases from water into blood can take place rapidly enough to keep up with the growth in volume of the developing fish.

- If the function of a cell or tissue depends on diffusion, its structure most likely has a shape that increases its surface area relative to its volume.

- Can you explain how the structure of the digestive system and circulatory systems are consistent with these observations?

Do All Aspects of an Animal's Body Increase in Size Proportionately?

- **Allometry** occurs when changes in body size are accompanied by disproportionate changes in anatomical structures or physiological processes.

- In skeletal size and body size, the volume involved is the mass that must be supported by the skeleton. The area involved is the cross-sectional area of the bones supporting the mass.

- As mass increases, the amount of bone required for support has to increase disproportionately (**Figure 41.12**).

- Structures that display allometry can sometimes be interpreted as adaptations to a particular lifestyle or environment. For example, at any given body size, dogs have much larger hearts than cats do. Garland and Huey hypothesize that the large hearts of dogs are an adaptation that increases blood flow to muscles, making long-distance chases possible.

41.4 Homeostasis

- **Homeostasis** is the maintenance of relatively constant chemical and physical conditions in an animal's cells, tissues, and organs.

- Epithelium plays a vital role in creating an internal environment that is dramatically different from the external environment, and in maintaining physical and chemical conditions inside an animal that are relatively constant.

Its most basic function is to control the exchange of materials across surfaces.

Regulation and Feedback

- Achieving homeostasis requires a regulatory system that has a specific **set point**, or normal value. The temperature that you set on the dial of a room's thermostat is an example of a set point.
- Regulatory systems typically consist of three components: a sensor, an integrator, and an effector (**Figure 41.14a**).
- A **sensor** is a structure that senses some aspect of the external or internal environment.
- An **integrator** is a component of the nervous system that evaluates the incoming sensory information and "decides" if a response is necessary to achieve homeostasis.
- An **effector** is any structure that helps to restore the desired internal condition.
- Regulatory systems depend on **negative feedback** to maintain constant conditions. Negative feedback occurs when the regulatory system makes a change in the opposite direction to a change in internal conditions (equivalent to a heater turning on and raising temperature when a room has cooled off).

41.4   How Do Animals Regulate Body Temperature?

- Many animals can control their body temperature through the process of **thermoregulation** (**Figure 41.14b**). In mammals, this process is coordinated by neurons in the hypothalamus. Neural signals from temperature receptors in the skin (the "sensors") are interpreted by the hypothalamus (the "integrator"), which then causes varying responses, ranging from metabolic to behavioral (the "effectors").

Gaining and Losing Heat

- An **endotherm** produces heat in its own tissue, while an **ectotherm** relies on heat gained from the environment. Birds and mammals are endotherms; most other animals are ectotherms.

- **Homeotherms** keep their body temperature constant, whereas **heterotherms** experience changes in body temperature.
- As always, there are exceptions—ectothermic pythons can generate body heat when they need to warm their eggs; tuna, mackerel, and certain other species of ectothermic fish generate heat to warm certain sections of their bodies (**Figure 41.15**).
- Compared to endotherms, ectotherms have low metabolic rates and lower mitochondrial enzyme activity.

Sources of Body Heat

- Heat is produced by muscle activity during movement and sometimes by the involuntary muscle contractions known as shivering.
- **Brown adipose tissue** (BAT) is a collection of specialized heat-generating cells (**Figure 41.17**). BAT has a high density of mitochondria and stored fats. When fats are oxidized by the mitochondria in BAT, no ATP is produced. Instead, all of the stored energy is released as heat. Which animals do you think would particularly benefit from BAT, and why?

Exchanging Heat with the Environment

- **Conduction** is the direct transfer of heat between two physical bodies that are in contact with each other. The rate at which conduction occurs depends on the surface area, the steepness of temperature difference, and how well each body conducts heat (**Figure 41.18**).
- **Convection** occurs when air or water moves over the body surface. As the speed of water or air flow increases, so does the rate of heat transfer.
- **Radiation** is the transfer of heat between two bodies that are not in direct physical contact. The major source of radiant energy is the sun.
- **Evaporation** is the phase change that occurs when liquid water becomes a gas. It leads to heat loss only. As a result, water is an efficient coolant on a hot day.

Conserving Heat

- Air conducts heat poorly; thus, it is a good insulator. Endothermic animals have elaborate external structures that trap air, slow the rate of heat exchange, and conserve body heat.

- Water is a good conductor; thus, the body temperature of most aquatic invertebrates and fish is the same as the water they inhabit.
- To conserve heat in water, otters have dense, **water-repellent fur** that maintains a layer of trapped air against the fur. Thick layers of **fatty blubber** insulate seals and whales.
- Some aquatic mammals and birds have **countercurrent heat exchangers**, where arteries and veins lie beside each other and freely exchange heat (**Figure 41.19**). Explain how this anatomical arrangement conserves heat.

Endothermy vs. Ectothermy

- Endotherms are able to maintain enzymes at optimal temperatures at all times, so mammals and birds can remain active in winter and at night. Due to their high metabolic rates and insulation, endotherms are also able to sustain very high levels of aerobic activities like running or flying.
- To fuel their high metabolic rates, endotherms have to obtain large quantities of energy-rich food.
- Ectotherms are able to thrive with much lower intakes of food. They can also use a greater proportion of their energy intake to support reproduction.
- Chemical reactions are temperature-dependent, so muscular activity and digestion slow down dramatically as body temperature drops. As a result, ectotherms are more vulnerable to predation in cold weather.

B. CROSS-CUTTING THEMES

Looking Back—
Concepts from Earlier Chapters
Natural Selection and
Adaptation—Chapter 23
As **Chapter 23** explained, an adaptation is a trait that allows individuals to survive and reproduce better than individuals that lack this trait. This chapter also introduced some of the techniques that biologists use to establish that traits have a genetic basis.

Genetic Drift—Chapter 24
Natural selection is not the only process that leads to changes in allele frequencies over time. Evolution occurs via the random process called genetic drift—the movement of alleles into and out of populations by migration (gene flow) and the constant introduction of new alleles by mutation.

Form and Function—Chapter 7
The shape of proteins often relates to their role as enzymes or structural components of the cell. **Chapter 7** pointed out that strong correlations exist between the structure and function of the rough ER, Golgi apparatus, mitochondrion, chloroplast, and other organelles.

Looking Forward—
Concepts in Later Chapters
The Vertebrate Eye—Chapter 46
The importance of historical constraint will surface again in the discussion of the vertebrate eye in **Chapter 46**. Although nonbiologists sometimes use the vertebrate eye as an example of a "perfect" adaptation, optometrists can attest that the organ is often defective.

Membrane Transport and
Surface Area—Chapter 42
Nutrients such as glucose must diffuse into the cell, and waste products such as urea and carbon dioxide must diffuse out. As Chapter 41 explains, the rate at which these and other molecules and ions diffuse depends in part on the amount of surface area available. In contrast, the rate at which nutrients are used and waste products are produced depends on the volume of the cell. **Chapter 42** will also explore how the gill functions as an adaptation for increased surface area.

C. DIFFICULT TOPICS

When studying the section on body size and scaling, you may find that some of the concepts are rather abstract, and you will need concrete examples for each concept described. Surface area and volume relationships cannot simply be written down and studied. You should draw the examples listed in your text to learn what it means for the volume of a cell or organism to outgrow its surface area.

The study of allometry is very complex, and this chapter only touches on the basics. It is intuitive to understand that an elephant consumes more oxygen than a mouse does, but do you really understand why the mouse consumes more oxygen per gram of tissue than an elephant does? As you imagine these comparisons between animals, also place the animals in their natural environment and try to apply what you find about their anatomy and physiology to how they interact with their environment. Maintaining a "big picture" of these concepts in addition to learning everything in little chunks will also help you in understanding this chapter.

D. ASSESSING WHAT YOU'VE LEARNED

(1) Testing Your Knowledge

1. Which of the following descriptions is an example of *acclimatization*?
    a. Over many generations, the average length of giraffe necks has increased.
    b. The optimal temperature at which goldfish can swim at maximum speed will decrease if the goldfish is maintained in cold water.
    c. The average brain size of the ancestors of modern humans has increased dramatically over time.
    d. The emergence of the seed represents a very important event in the evolution of land plants.

2. Consider several objects that are heated and then allowed to cool. Which of the following would cool at the fastest rate?
    a. an object with a surface area of 20 and a volume of 10
    b. an object with a surface area of 40 and a volume of 30
    c. an object with a surface area of 8 and a volume of 2
    d. an object with a surface area of 10 and a volume of 30

3. What happens to the mass-specific metabolic rate of an organism as it gets bigger?
    a. Mass-specific metabolic rate would decrease.
    b. Mass-specific metabolic rate would remain constant.
    c. Mass-specific metabolic rate would increase.

4. Patrick Wells and Alan Pinder conducted experiments to explore how gas exchange occurs in Atlantic salmon at various stages of their life. They found that the percentage of oxygen uptake by the gills increases as the organism grows. Which of the following could explain this result?
    a. The gills provide a much lower surface-to-volume ratio, so that gas transfer becomes more efficient as the organism gets larger.
    b. The surface area of skin is much larger than the surface area of the gills.
    c. As an individual grows, the skin surface area decreases in relation to its volume (the surface-to-volume ratio drops). To keep the individual from suffocating, gills must take over most of the gas exchange activity.
    d. As an individual grows, the skin surface area increases in relation to its volume (the surface-to-volume ratio increases). To keep the individual from suffocating, gills must take over most of the gas exchange activity.

5. Which of the following describes an allometric relationship?
    a. a relationship in which two quantities change at the same rate
    b. a relationship in which one quantity increases while one quantity decreases
    c. a relationship in which two quantities both decrease
    d. a relationship in which two quantities change at different rates

6. The relationship between heart mass and body mass differs for dogs and cats. Which of the following statements explains these differences?
    a. Cat hearts are smaller because they have smaller bodies. The heart does not have to work as hard in a smaller animal.

b. There is no difference in the size of dog and cat hearts when you compare animals of similar size.

c. Cats are much more active than dogs. This increased activity is reflected in the data when comparing heart mass to body mass for dogs and cats.

d. The hunting style of dogs involves long-distance chases. Cats use short, quick sprints to ambush their prey. The increased activity in the hunting style of dogs requires a larger heart.

7. Which of the following terms implies the maintenance of relatively constant physical conditions?
a. homeostasis
b. acclimatization
c. adaptation
d. isometry

8. Which of the following terms describes an animal that maintains body temperature by producing heat in its own tissues?
a. isotherm
b. ectotherm
c. endotherm

9. Which of the following describes an ectotherm?
a. a person shivering in the cold
b. a turtle basking in the sun on a log
c. a small animal increasing its rate of oxidation of fat in brown adipose tissue to produce heat
d. a hormonally induced increase in basal metabolism for the purpose of generating heat

10. The direct transfer of heat between two physical bodies that are in contact with each other would be:
a. evaporation
b. radiation
c. convection
d. conduction

11. Which of the following would explain the observation that marine mammals maintain very thick layers of insulating fat?
a. The fat helps prevent heat loss through the process of evaporation.

b. The fat helps prevent heat loss through the process of conduction.
c. The fat prevents heat loss through the process of convection.

12. When researchers compare the cells in tissues collected from endotherms and ectotherms of similar size, what can they expect to find?
a. Endotherm cells will have a much greater density of mitochondria.
b. Ectotherm cells will have a much greater density of mitochondria.
c. Endotherm cells will have a much greater density of ribosomes.
d. Ectotherm cells will have a much greater density of ribosomes.

(2) Integrating Your Knowledge

(a) What processes lead to changes in allele frequency in a population?

(b) What is the difference between adaptation and acclimatization?

(c) Why aren't all traits adaptive? Give an example.

(d) What is the difference between a tissue and an organ?

(e) Why does surface area increase more slowly than volume in a growing organism?

(f) Define allometry, and give a biological example.

(g) Would you expect to find brown fat in a reptile? Why or why not?

CHAPTER 41—ANSWER KEY

D. Assessing What You've Learned

(1) Testing Your Knowledge
1. b; 2. c; 3. a; 4. c; 5. d; 6. d; 7. a; 8. c; 9. b; 10. d; 11. b; 12. a

(2) Integrating Your Knowledge

(a) Changes in allele frequency in a population can occur by natural selection, genetic drift, gene flow, and mutation.

(b) Adaptation occurs when a population changes in response to natural selection. Acclimatization occurs when an individual changes in response to a change in environmental conditions.

(c) Animals possess a wide variety of structures that were present in ancestral populations but are not currently adaptive. Vestigial traits like this are widespread. Some traits exist as holdovers from structures that appear early in development. Human males have rudimentary mammary glands only because nipples form in the early embryo before sex hormones begin influencing the development of organs.

(d) A tissue is a group of cells with the same structure and function. An organ is a structure that serves a specialized function and consists of several tissues.

(e) In a cell, the surface area grows as a power of 2; the volume increases as a power of 3. As a result, its volume increases much faster than its surface area.

Given this, the transport processes that occur across the cell membrane would have to support disproportionately more and more cell volume as the cell increased.

(f) Allometry occurs when changes in body size are accompanied by disproportionate changes in anatomical structures or physiological processes. In skeletal size and body size, the volume involved is the mass that must be supported by the skeleton. The area involved is the cross-sectional area of the bones supporting the mass. As mass increases, the amount of bone required for support has to increase disproportionately.

(g) Reptiles are ectothermic, so you would not expect to see brown fat in them because this would mean they would be producing heat from within, a requirement for endothermy.

42

Water and Electrolyte Balance in Animals

A. KEY BIOLOGICAL CONCEPTS

42.1. Osmoregulation and Osmotic Stress

- Substances move from regions of higher concentrations to lower concentrations by the process of **diffusion** (**Figure 42.1a**).
- The diffusion of water from areas of higher to lower concentration is called **osmosis** (**Figure 42.1b**). The concentration of dissolved substances in a solution, measured in moles per liter, is referred to as its **osmolarity**.
- Cells and tissues that are gaining too much water are under **osmotic stress**. Fish must rid these cells of incoming water, or the cells will burst and the individual will die.
- The ability to maintain homeostasis with respect to water and electrolyte balance in the face of osmotic stress is called **osmoregulation**.
- In freshwater fish, gill epithelium is **hypertonic** in relation to the surrounding water (**Figure 42.2**). Gill epithelial cells can gain water via osmosis. Marine fish are **hypotonic** to seawater, so water tends to flow out of the gill epithelium. Marine fish drink large quantities of seawater and excrete salt so that their cells do not shrivel and die.
- Freshwater animals must replace electrolytes that are lost by obtaining them in food or from the surrounding water. Marine animals must rid the body of excess ions they gain from drinking seawater (**Figure 42.3**).

- Depending on conditions, terrestrial animals may need to conserve or to excrete electrolytes to maintain homeostasis.

42.2 Water and Electrolyte Balance in Aquatic Environments

How Do Sharks Osmoregulate?

- The shark rectal gland secretes a concentrated salt solution. Ions can be concentrated in this way only if they are actively transported against a concentration gradient.
- Solutes move across membranes by passive or active transport. **Passive transport** often occurs via channels or by transmembrane proteins that act as carriers.
- **Active transport** requires membrane proteins that act as a pump, requiring energy in the form of **ATP**.

The Role of Na^+/K^+-ATPase

- Epithelial cells along the inner surface, or lumen, of the shark rectal gland contain Na^+/K^+ -ATPase. The location of the pumps was paradoxical because they were situated along the **basolateral membrane** of the cells and not on the **apical membrane**. This meant that sodium would be pumped opposite to the direction it is secreted.

A Molecular Model for Salt Excretion

- In 1977, Silva and colleagues developed the following molecular model for salt excretion:

1. Na$^+$/K$^+$-ATPase pumps sodium ions out of epithelial cells across the basolateral surface and into the surrounding extracellular fluid (**Figure 42.5**). The pump creates a large electrochemical gradient favoring the diffusion of Na$^+$ into the cell.

2. A Na$^+$/Cl$^-$ cotransporter, powered by the gradient favoring Na$^+$ diffusion, brings these two ions from the extracellular fluid into epithelial cells across their basolateral surfaces.

3. Although sodium ions are pumped back out by Na$^+$/K$^+$-ATPase, chloride ion concentrations build up inside the cell as a result of the cotransport process. A chloride channel located in the apical membrane of the epithelial cells allows Cl$^-$ to diffuse out down its concentration gradient.

4. Sodium ions also diffuse into the lumen of the gland, following their charge and concentration gradient. But instead of passing through the epithelial cells as Cl$^-$ does, Na$^+$ diffuses out through spaces between the cells.

- Using the steps provided here, can you draw the model for salt excretion?

A Common Molecular Mechanism Underlies Many Instances of Salt Excretion

- Marine birds and reptiles drink saltwater and must excrete NaCl. They have **salt-excreting glands** in their nostrils that function much like the shark rectal gland.

- Because marine teleost (bony) fish are hypertonic to seawater, salt constantly diffuses in through their **gills**. Their gills contain specialized cells that are configured precisely like the cells lining the rectal gland.

- Cells with the same configuration of **pumps**, **cotransporters**, and **channels** are responsible for transporting salt in the **kidneys** of mammals.

How Do Salmon Osmoregulate?

- Young chum salmon living in salt water have a large number of chloride cells at the base of their gill filaments. In contrast, most of the chloride cells observed in freshwater chum are located in the sheetlike lamellae that extend from the base of gill filaments.

- Chloride cells in the gill lamellae import electrolytes, while chloride cells at the base of the gills secrete salt.

42.3   Water and Electrolyte Balance in Terrestrial Insects

How Do Insects Minimize Water Loss from the Body?

- Water loss is an inevitable by-product of respiration. Insects have a relatively large surface area with which to lose water and a small volume in which to retain it.

- An insect's **trachea** connects with the atmosphere through openings called **spiracles**. Muscles just inside each spiracle close or open the pore. This is an important adaptation for minimizing water loss (**Figure 42.9**).

- The insect **exoskeleton** consists of **chitin** and layers of protein. A layer of **wax** covers the surface, which is highly impermeable to water.

Types of Nitrogenous Wastes: Impact on Water Balance

- **Ammonia** is a by-product of **catabolic** reactions. Those animals that excrete ammonia directly usually lose a lot of water. Fish detoxify ammonia by diluting to a low concentration in watery urine.

- Humans convert ammonia to nontoxic **urea** and excrete it. Urea requires water for excretion, however, but not as much as excreting ammonia directly.

- Birds, reptiles, and terrestrial arthropods convert ammonia to **uric acid**. Because uric acid is not soluble in water, it can be excreted as a dry paste.

- Does the type of nitrogenous waste produced by an animal correlate with the amount of osmotic stress it endures?

Maintaining Homeostasis: The Excretory System

- Insects must carefully regulate the composition of a blood-like fluid called **hemolymph**.

- Pre-urine formation occurs in the **Malpighian tubules,** which have a large surface area and are in direct contact with the hemolymph.
- Because the urine that is finally expelled from the anus is often strongly hyperosmotic to pre-urine, a large amount of water must be reabsorbed. Selective reabsorption of electrolytes and water occurs in the hindgut. The epithelial cells transport ions out of the hindgut and water follows by osmosis.
- The formation of pre-urine is not particularly selective. Reabsorption, in contrast, is highly selective. Waste products do not pass through the rectal membrane, instead remaining in the urine and feces.
- Water and electrolyte recovery in the hindgut is precisely controlled to maintain homeostasis.

42.4   Water and Electrolyte Balance in Terrestrial Vertebrates

The Structure of the Kidney

- The paired **kidneys** receive considerable blood flow via the **renal artery**. Urine formed in each kidney is carried to the **bladder** via a long, thin tube called the **ureter**. Urine leaves the bladder and exits the body via the **urethra** (**Figure 42.11**).

Filtration: The Renal Corpuscle

- **Nephrons** are the functional units of the kidney—they perform the work in maintaining water and electrolyte balance (**Figure 42.12**).
- Urine formation begins in the **renal corpuscle** (**Figure 42.13**). Water and small solutes are pushed out of the capillary's pores, through the slits in the surrounding cells, and into the fluid-filled space inside **Bowman's capsule**. Because proteins, cells, and other large components of blood would not fit through the pores, they would not enter the nephron.
- Pressure is much higher inside the capillaries than it is in the surrounding capsule. This pressure differential forces water and solutes out of the blood and into the capsule space.
- Can you predict how glomerular filtration would be affected by high blood pressure?

Reabsorption: The Proximal Tubule

- Fluid leaves the Bowman's capsule and enters a convoluted structure called the **proximal tubule**. The fluid entering this area has the composition of plasma minus the blood cells and large proteins.
- The epithelial cells of the tubule have a prominent series of small projections called **microvilli** facing the lumen (**Figure 42.14**).
- Na^+/K^+-ATPase in the basolateral membranes removes intracellular Na^+ and creates a gradient for Na^+ entry. In the apical membrane, Na+-dependent cotransporters simultaneously bind Na^+ and another solute such as glucose, an amino acid, or Cl^-.
- Water follows the movement of these solutes by osmosis. It leaves the proximal tubule via membrane proteins called **aquaporins**.

Creating an Osmotic Gradient: The Loop of Henle

- In 1942 Werner Kuhn hypothesized that the loop of Henle functions as a countercurrent multiplier, setting up an osmotic gradient. He proposed that an exchange of water and solutes occurs between each of the segments in the loop and cells outside.

Testing Kuhn's Hypothesis

- A strong gradient in osmolarity existed from the cortex to the medulla, as predicted in Kuhn's hypothesis (**Figure 42.15**).

How Is the Osmotic Gradient Established?

- In the ascending loop of Henle, Na^+ and Cl^- constitute at least 60% of the solutes; urea is about 10%. In contrast, in the collecting duct both ions were rare, and the major solute was urea (**Figure 42.16**).
- Na^+ and Cl^- are actively pumped out of the ascending limb.
- The descending limb is highly permeable to water but almost completely impermeable to solutes. The thick ascending limb, in contrast, is highly permeable to Na^+ and Cl^-, moderately permeable to urea, and almost completely impermeable to water.

Regulating Water and Electrolyte Balance: The Distal Tubule and the Collecting Duct

- The amount of Na^+, Cl^-, and water that is reabsorbed in the distal tubule and collecting duct varies with the animal's condition and is under **hormonal control**.
- If an individual is dehydrated, a molecule called **antidiuretic hormone (ADH)** is released from the brain. When ADH interacts with cells lining the distal tubule and collecting duct, aquaporin channels are inserted into the membrane. The cells thus become highly permeable to water and large amounts of water are reabsorbed (**Figure 42.18**).
- What happens to water reabsorption in the distal tubule and collecting duct in the absence of ADH?

B. CROSS-CUTTING THEMES

Looking Back—
Concepts from Earlier Chapters
Osmosis and Diffusion—Chapter 6
Chapter 6 introduced the processes called diffusion and osmosis. Diffusion describes the movement of substances from regions of higher concentration to regions of lower concentration. The movement of water from regions of higher to lower concentration is called osmosis.

Gas Exchange in Plants—Chapter 35
Water balance in land animals mirrors the situation in land plants. Land plants also lose water as an inevitable by-product of gas exchange. As **Chapter 35** pointed out, land plants lose huge quantities of water in the course of obtaining CO_2 through their stomata.

Surface-to-Volume Ratio—Chapter 41
Small organisms have a high surface area to volume ratio. Insects have a relatively large surface area with which to lose water, and a small volume in which to retain it.

Looking Forward—
Concepts in Later Chapters
Membrane Transport and Surface Area—Chapter 44
Nutrients such as glucose must diffuse into the cell, and waste products such as urea and carbon dioxide must diffuse out. As explained in **Chapter 44**, the rate at which these and other molecules and ions diffuse depends partly on the available surface area. In contrast, the rate at which nutrients are used and waste products are produced depends on the volume of the cell. This chapter will also explore how the gill functions as an adaptation for increased surface area, and how animals exchange gases in detail.

The Sodium-Potassium Pump—Chapter 45
The best-characterized membrane protein pump involving ions is the sodium-potassium pump, also known as Na+/K+-ATPase. **Chapter 45** discusses how this enzyme was discovered and explores its function in the transmission of electrical signals in animals. That discussion also introduces a plant defense compound called ouabain. Ouabain is toxic to animals because it binds to Na+/K+-ATPase, prevents it from functioning, and poisons transmission of nerve impulses.

C. DIFFICULT TOPICS

The entire chapter has a central theme of sodium-potassium pumps. They are always involved in some way with water and ion balance. Once you have understood this basic tenet, the difficult part is understanding the most complex process of water and ion balance—the vertebrate kidney. By working through this chapter and learning how more primitive vertebrates and invertebrates maintain water and ion homeostasis, you can focus on the vertebrate kidney. Just like the chapter and study guide organize the functions of the kidney, so should you. Study the filtration process first, then reabsorption, and then secretion, as you trace the path of urine formation beginning at the renal corpuscle and ending at the collecting duct. Since the formation of urine is such a dynamic process, make sure to note changes in urine formation under stressful conditions—dehydration, high blood pressure, and response to the various hormones discussed.

D. ASSESSING WHAT YOU'VE LEARNED

(1) Testing Your Knowledge

1. The major challenges to water and ion balance for most marine vertebrates are:

    a. gaining water from and losing ions to the environment

    b. gaining water and ions from the environment

    c. losing water to and gaining ions from the environment

    d. losing water and ions to the environment

2. Marine elasmobranchs, like the shark, avoid water loss to the environment by:

    a. drinking seawater and using their rectal gland to excrete excess ions

    b. maintaining the osmolarity of their body fluids hypotonic relative to seawater

    c. gaining water through their gills and rectal gland

    d. gaining water through their gills

3. Which of the following statements accurately describes the role of Na^+K^+-ATPase in the shark rectal gland?

    a. Na^+K^+-ATPase, localized in the apical membrane of the lumen cells, pumps Na^+ into the lumen of the rectal gland.

    b. Na^+K^+-ATPase, localized in the basolateral membrane of the lumen cells, pumps Na+ into the blood.

    c. Na^+K^+-ATPase, localized in the apical membrane, pumps Na+ into the blood.

    d. Na^+K^+-ATPase, localized in the basolateral membrane of the lumen cells, pumps Na+ out of blood and into the lumen cells.

4. The major site for osmoregulation in teleosts (bony fishes) is:

    a. the rectal gland

    b. the gill

    c. the kidney

    d. the integument (skin)

5. An animal that lives in a dry environment needs to minimize water loss and will thus most likely secrete nitrogenous waste in the form of:

    a. uric acid

    b. urea

    c. creatinine

    d. ammonia

6. Which of the following statements correctly summarizes the primary role of the Malpighian tubule system of insects in formation of pre-urine?

    a. K^+ ions are actively secreted into the lumen of the Malpighian tubules; other electrolytes, water, and waste products follow passively.

    b. Na^+ and Cl^- ions are actively transported into the lumen of the Malpighian tubules; water and waste products follow.

    c. Cl^- is actively transported into the lumen by the actions of chloride cells; Na^+, waste products, and water follow.

    d. Waste products such as uric acid are actively secreted into the lumen, and water follows.

7. Which of the following statements concerning the composition of the filtrate formed by the renal corpuscle of the human kidney is correct?

    a. Blood and filtrate are identical in composition.

    b. Filtrate has a higher concentration of the waste product urea than blood does.

    c. The filtrate is similar to blood but without proteins and blood cells.

    d. The filtrate has a lower concentration of glucose than blood does.

8. Which of the following represents the primary driving force for the filtration of blood within the renal corpuscle?

    a. Pressure within the glomerular capillaries is lower than pressure in Bowman's capsule.

    b. Pressure within the glomerular capillaries is greater than pressure in Bowman's capsule.

    c. The collecting duct creates a suction that draws fluid into Bowman's capsule.

    d. The filtration slits are present in the glomerular capillaries.

9. Which statement most accurately describes the primary role of the loop of Henle in urine formation?

    a. The loop of Henle deposits Na^+ and Cl^- in the medullary region of the kidney, increasing its osmolarity.

    b. The hormone ADH acts on the loop of Henle to increase water reabsorption.

c. The ascending limb of the loop of Henle contributes to the high osmolarity of the medullary region by depositing urea.

d. The loop of Henle is responsible for most of the Na^+ reabsorbed by the nephron tubule.

10. The amount of Na^+ excreted in the urine is primarily determined by:
    a. the amount of Na^+ secreted into the proximal convoluted tubule
    b. the amount of Na^+ reabsorbed by the proximal convoluted tubule
    c. the amount of ADH released by the posterior pituitary gland
    d. hormonal regulation of Na^+ reabsorption in the distal convoluted tubule

11. Which of the following would result in the production of hypotonic urine?
    a. dehydration in an individual
    b. an increase in the permeability of the collecting duct to water
    c. a decrease in ADH release by the brain
    d. a decrease in water reabsorption by the loop of Henle

12. From an evolutionary perspective, which of the following adaptations makes the most sense?
    a. long loops of Henle in the kidneys of a river otter
    b. long loops of Henle in the kidneys of a desert fox
    c. an extensive Malpighian tubule system in a freshwater beetle
    d. renal corpuscles that are few in number and small in size in the kidney of a freshwater fish

(2) Integrating Your Knowledge

(a) How do freshwater fish and saltwater fish differ in their osmotic stresses?

(b) How did the study of the shark rectal gland yield practical applications for human medicine?

(c) Why is a migrating salmon such an interesting animal with respect to osmotic stress?

(d) How do terrestrial insects minimize water loss from the body?

(e) Why is uric acid such an efficient waste product?

(f) What is a Malpighian tubule, and which animals have them?

(g) What are aquaporins, and how do they relate to water and ion balance?

(h) How does ADH affect water balance in the human kidney?

CHAPTER 42—ANSWER KEY

D. Assessing What You've Learned

(1) Testing Your Knowledge
1. c; 2. d; 3. b; 4. b; 5. a; 6. a; 7. c; 8. b; 9. a; 10. d; 11. c; 12. b

(2) Integrating Your Knowledge

(a) Freshwater animals must replace electrolytes that are lost by obtaining them in food or from the surrounding water. Marine animals must rid the body of excess ions they gain from drinking seawater.

(b) Investigators realized that the amino acid sequence of CFTR (cystic fibrosis transmembrane regulator) was 80% identical to the shark chloride channel. It was their first hint that CFTR involved chloride transport.

(c) Young chum salmon living in salt water have a large number of chloride cells at the base of their gill filaments. In contrast, most of the chloride cells observed in freshwater chum are located in the sheetlike lamellae that extend from the base of gill filaments. This allows researchers to study the distinctive challenges of osmoregulating in fresh and in salt water.

(d) Water loss is an inevitable by-product of respiration. Insects have a relatively large surface area with which to lose water, and a small volume in which to retain it. The tracheal system connects with the atmosphere at openings called spiracles. Muscles just inside each spiracle close or open the pore. This is an important adaptation for minimizing water loss. The insect exoskeleton consists of chitin and layers of protein. A layer of wax covers the surface, which is highly impermeable to water.

(e) Birds, reptiles, and terrestrial arthropods convert ammonia to uric acid. Because uric acid is not soluble in water, it can be excreted as a dry paste. In this way, the maximum amount of water can be conserved compared to excreting other types of nitrogenous wastes.

(f) Insects must carefully regulate the composition of a blood-like fluid called hemolymph. Malpighian tubules have a large surface area and are in direct contact with the hemolymph. They are the water and ion balance organs in insects. J. A. Ramsay and colleagues found that K^+ accumulated inside the tubules against a concentration gradient. They hypothesized that cells in the membranes of Malpighian tubules contain a pump that transports K^+ into the organ.

(g) Na^+/K^+-ATPase in the basolateral membranes removes intracellular Na^+ and creates a gradient for Na^+ entry. In the apical membrane, Na+-dependent cotransporters simultaneously bind Na^+ and another solute such as glucose, an amino acid, or Cl^-. Water follows the movement of these solutes via osmosis. It leaves the proximal tubule through membrane proteins called aquaporins.

(h) The amount of Na^+, Cl^-, and water that is reabsorbed in the distal tubule and collecting duct varies with the animal's condition and is under hormonal control. If an individual is dehydrated, a molecule called antidiuretic hormone (ADH) is released from the brain. When ADH interacts with the cells lining the distal tubule and collecting duct, aquaporin channels are inserted into the membrane. As a result, the cells become highly permeable to water and large amounts of water are reabsorbed.

43

Animal Nutrition

A. KEY BIOLOGICAL CONCEPTS

43.1   Nutritional Requirements

Meeting Basic Needs

- **Nutrients**, substances needed by an organism to remain alive, are contained within **food**. An organism needs chemical energy in the form of reduced carbon compounds and the elements and molecules needed to synthesize body components and sustain cells.
- Recommended Daily Allowances (RDAs) are placed on packaging in the USA to specify amounts of each essential nutrient that must be ingested to meet the needs of practically all healthy people. Specifically:
  1. **Essential amino acids** are the 8 (out of a total of 20) amino acids that are required to manufacture proteins, but that cannot be synthesized by humans. We can synthesize the other 12 amino acids.
  2. **Vitamins** are carbon-containing molecules that function as coenzymes in critical reactions (**Table 43.1**).
  3. **Essential elements** cannot be synthesized by the body, but they serve a wide variety of functions (**Table 43.2**).
  4. **Electrolytes** form ions in solution. They influence osmotic balance and are required for normal membrane function.

Nutrition and Athletic Performance

- RDAs identify nutrients that healthy people need to stay healthy, but target people with normal needs and activities.
- The dietary regime known as **carbohydrate loading** is now a routine part of race prepar-ations. For endurance athletes, a high-carbo-hydrate meal is the breakfast of champions (**Figure 43.2**).

43.2   Obtaining Food: The Structure and Function of Mouthparts

- **Heterotrophs** are organisms that obtain ener-gy and nutrients from other organisms. Ani-mals are heterotrophs.
- Given the diversity of food sources that animals exploit, they have a wide variety of structures in and around the mouth to obtain food (**Figure 43.3**).
- A mosquito's proboscis allows it to pierce the hide of an animal and suck up blood, and the mandibles of some ants are modified to cut leaves and other plant material.
- The mouthparts of mammals are adapted to different diets. Human teeth include sharp canines for tearing meat and flattened molars for crushing seeds, roots, and other sources of carbohydrates.

43.3   How Are Nutrients Digested and Absorbed?

- Digestion takes place in the **alimentary ca-nal**, also known as the **gastrointestinal tract** (**Figure 43.6**).
- Mechanical breakdown occurs primarily in the mouth and stomach, but chemical break-down begins with enzymes in the saliva.

The Mouth and Esophagus

- **Amylase**, contained in saliva, cleaves the bonds that link glucose monomers in starch, glycogen, and other glucose polymers.

- **Salivary glands** also release water and gly-coproteins called **mucins**. The combination of water and mucus makes food soft and slippery enough to be swallowed.
- Food then enters the esophagus and is propelled by a wave of contractions called **peristalsis** (**Figure 43.8**). Food then enters the stomach.

The Stomach

- When the stomach is filled by a meal, muscular contractions result in churning that mixes the contents (**Figure 43.9**).
- The predominant acid in the stomach is hydrochloric acid. There is also the enzyme pepsin, which breaks down proteins. Pepsinogen is converted to active pepsin by contact with the acidic environment of the stomach.

Which Cells Produce Stomach Acid?

- **Parietal cells** are the source of the HCl in gastric juice, whose pH can be as low as 1.5. **Goblet cells** secrete **mucus**, which lines the gastric epithelium and protects the stomach from damage by HCl (**Figure 43.10**).

How Do Parietal Cells Secrete HCl?

- The enzyme **carbonic anhydrase** is present in high concentration in parietal cells. In solution, the carbonic acid that is formed immediately dissociates to form the bicarbonate ion and a proton.
- Protons formed by dissociation of carbonic acid are actively pumped into the lumen of the stomach. Chloride ions enter parietal cells in exchange for bicarbonate ions and then move into the lumen via a chloride channel.

The Ruminant Stomach

- Cattle, sheep, deer, and other ruminants have complex, four-chambered stomachs (**Figure 43.11**). The largest chamber, the **rumen**, serves as a vat for fermenting food before food is regurgitated and re-chewed as **cud.** It then passes to the **reticulum** for bacterial breakdown of cellulose, to the **omasum** for water removal, and to the **abomasum** (true stomach) for further chemical digestion.

The Small Intestine

- In the small intestine, partially digested material mixes with secretions from the pancreas and liver and begins a journey of six meters.

Protein Processing by Pancreatic Enzymes

- The activating enzyme in the intestinal juice is **enterokinase,** which activates a pancreatic enzyme called **trypsinogen**, yielding **trypsin**. Trypsin triggers the activation of other protein-digesting enzymes secreted by the pancreas.
- **Secretin** is produced in the small intestine in response to the arrival of food to the stomach. The discovery of secretin was important because it confirmed that digestion is under both neural and hormonal control.

Digesting Lipids: Bile and Transport

- The pancreatic secretions include digestive enzymes that act on fats and carbohydrates as well as proteins. **Lipase** breaks certain bonds present in complex fats.
- Can you outline the steps in digestion of lipids in the small intestine? (See **Figure 43.13**.)
- **Bile salts** are synthesized in the liver and secreted in a complex solution called **bile** that is stored in the **gallbladder**. When bile enters the small intestine, it raises the pH and emulsifies fats so that they can be digested.

Digestion and Absorption of Carbohydrates

- Cells lining the small intestine absorb the sugars released by amylase digestion. The enormous surface area of the small intestine (because of **villi** and **microvilli**) increases the efficiency of nutrient absorption.
- Nutrient absorption depends on the presence of an electrochemical gradient favoring an influx of sodium ions into the epithelium. This process depends on a series of Na^+ cotransporters, like the **Na^+/glucose cotransporter** (**Figure 43.15**).
- Water follows the solutes into the epithelium passively, via osmosis. This movement of water is the mechanistic basis of an extremely important medical strategy called **oral rehydration therapy**.

The Large Intestine

- The primary function of the large intestine is to compact wastes that remain and absorb enough water to form **feces**.
- To date, four distinct transmembrane water channels called **aquaporins** have been found in the large intestine of rats, mice, or humans. AQP3 and AQP4 are located in the basolateral membrane of cells in the epithelium of the large intestine of the rat.
- Mechanisms of water absorption in the large intestine are not well understood, but aquaporins are hypothesized to have a major role.

43.4   Nutritional Homeostasis— Glucose as a Case Study

- People with **diabetes mellitus** experience abnormally high glucose levels in their blood.
- Banting and Best (1921) removed a pancreas from a dog, injected an extract made from it, and injected into a diabetic dog; its blood glucose stabilized. They had discovered **insulin**.

Insulin's Role in Homeostasis

- **Insulin** is a hormone produced in the pancreas when blood glucose levels are high. It travels through the bloodstream and binds to receptors on cells throughout the body. In response, the cells increase their rate of glucose uptake and respiration.
- If blood glucose levels fall, cells in the pancreas secrete a hormone called **glucagon** (**Figure 43.17**). Cells in the liver and skeletal muscle catabolize glycogen, and cells that store lipids catabolize fatty acids. The result is that glucose levels in the blood rise.
- **Diabetes mellitus** develops in people who do not synthesize insulin or who have defective versions of the insulin receptor. Type I diabetes mellitus is treated with insulin injections; type II diabetes is managed mainly via diet.

The Type II Diabetes Mellitus Epidemic

- In the United States, about 6.6% of people age 20–74 have type II diabetes. Among the Pima, almost 50% of the adults over 35 years old have type II diabetes (**Figure 43.18**).

- Researchers have concluded that some individuals are genetically predisposed to developing the disease, but evidence is strong that environmental conditions have an impact.
- Nutrition-related diseases like obesity are being linked to diabetes, especially in young people.

B. CROSS-CUTTING THEMES

Looking Back—
Concepts from Earlier Chapters

Osmosis and Diffusion—Chapter 6
Chapter 6 introduced the processes called diffusion and osmosis. Diffusion describes movement of substances from regions of higher concentration to regions of lower concentration. Movement of water from regions of higher to lower concentration is called osmosis.

Surface-to-Volume Ratio—Chapter 41
Organs like the small intestine increase surface area in order to increase absorption efficiency.

Glycogen—Chapter 5
Recall that glycogen is a polysaccharide made up of glucose molecules and that glucose is the preferred starting compound for the production of ATP via cellular respiration.

Adaptive Radiation—Chapter 25
Cichlids inhabiting the Rift Lakes of East Africa are a spectacular example of adaptive radiation. Recall from **Chapter 25** that adaptive radiation refers to diversification of a single ancestral population into many species, each of which lives in a different habitat or utilizes a distinct feeding method.

Embryonic Tissue Layers—Chapter 22
The interior tube of the GI tract communicates directly with the environment. Embryologically, it derives from the hollow tube that forms as sheets of cells invaginate during gastrulation.

Aquaporins—Chapter 42
In an attempt to identify the mechanism of water absorption in the large intestine, researchers focused on aquaporins. Recall that aquaporins are water channels and are common in the descending loop of Henle of the kidney.

Looking Forward—
Concepts in Later Chapters

**Membrane Transport and
Surface Area—Chapter 44**

Nutrients such as glucose must diffuse into the cell, and waste products such as urea and carbon dioxide must diffuse out. As **Chapter 44** explains, the rate at which these and other molecules and ions diffuse depends partly on the amount of surface area available. In contrast, the rate at which nutrients are used and waste products are produced depends on the volume of the cell. This chapter also explores how the gill functions as an adaptation for increased surface area, and how animals exchange gases in detail.

The Sodium-Potassium Pump—Chapter 45

The best-characterized membrane protein pump involving ions is the sodium-potassium pump, also known as Na^+/K^+-ATPase. **Chapter 45** discusses how this enzyme was discovered and explores its function in the transmission of electrical signals in animals. That discussion also introduces a plant defense compound called ouabain. Ouabain is toxic to animals since it binds to Na^+/K^+-ATPase, prevents it from functioning, and poisons transmission of nerve impulses.

Muscle Types—Chapter 46

The action of peristalsis and churning is unconscious. **Chapter 46** explains the differences between the structure and function of different types of muscle and related nerve actions.

C. DIFFICULT TOPICS

Following food through the digestive system is a fairly simple process, but the chemical reactions that occur during this process coupled with the neural and hormonal signals can get very confusing very quickly. Make sure you have mastered the anatomy of the digestive system before you tackle the physiology. It will be difficult to study the functions of the liver if you can't picture where it's located in relation to the digestive system and the body.

Start by following a piece of food through the digestive system, starting at the mouth. Note the chemical reactions occurring at each stage, and any neural or hormonal signals involved. The most confusing section will be the small

intestine because secretions from the liver and pancreas are involved with digesting food entering from the stomach. Draw arrows mapping the hormonal and neural signals so you are clear on which way the signal is traveling.

D. ASSESSING WHAT YOU'VE LEARNED

(1) Testing Your Knowledge

1. Which of the following represent examples of electrolytes?
    a. K^+ and Cl
    b. valine and glycine
    c. niacin and folate
    d. glucose and fructose

2. As you've likely noticed, recommendations for basic nutritional requirements for humans change fairly frequently. Which of the following statements provides the most likely explanation?
    a. As humans evolve, their nutritional requirements change.
    b. The nutritional information provided on packaged food labels contains many errors.
    c. It is difficult to perform controlled experiments on humans in order to make reliable nutritional recommendations.

3. Which of the following experimental results led Jonas Bergstrom and co-workers to challenge the prevailing hypothesis concerning the nutritional requirements of endurance athletes?
    a. Athletes eating a diet rich in complex carbohydrates had higher blood glucose levels while exercising than did those who ate a diet rich in fats and proteins.
    b. Athletes eating a diet rich in complex carbohydrates experienced a decrease in the length of time it took to reach exhaustion when compared to those who ate a diet rich in fats and protein.
    c. Athletes eating a diet rich in fat and protein experienced an increase in levels of stored glycogen in their muscles when compared to those who ate a diet rich in complex carbohydrates.

d. Athletes who ate a diet rich in simple carbohydrates (such as glucose) experienced an increase in performance level due to enhanced glycogen storage.

4. Which of these adaptations to the pharyngeal jaw system of cichlids does *not* represent an example of how mouthparts are adaptations to allow for efficient feeding?
   a. Because the pharyngeal jaws articulate with the braincase, they can be used for biting as well as for moving food to the back of the throat.
   b. Because the lower pharyngeal jaw exists as two separate pieces, the pharyngeal jaws are better able to grasp prey.
   c. The pharyngeal jaws of cichlids have more muscles attached, improving their use for biting prey.
   d. The toothlike projections of the cichlid pharyngeal jaws vary in shape and size depending on the primary food source of each species.

5. Chemical digestion begins in the mouth. Which of the following best describes the type of digestion that occurs in the mouth?
   a. Lipase begins to break down proteins.
   b. Amylase begins to break down starch.
   c. Pepsin begins to break down proteins.
   d. Amylase begins to break down fats.

6. Which of these statements concerning acid production by the stomach is *not* correct?
   a. The enzyme carbonic anhydrase, located in the parietal cells, catalyzes the formation of carbonic acid from carbon dioxide and water.
   b. Carbonic acid is actively secreted by parietal cells into the stomach lumen.
   c. Protons are actively pumped into the stomach lumen by the parietal cells.
   d. Chloride ions move into the lumen of the stomach through chloride channels located in the parietal cells.

7. Based on what you've learned about the role of the parietal cells in HCl production in the stomach, predict the relative acidities of these three compartments: A = the stomach lumen; B = blood supply arriving to the stomach; C = blood supply leaving the stomach.
   a. pH A > pH C > pH B
   b. pH B > pH C > pH A
   c. pH A > pH B > pH C
   d. pH C > pH B > pH A

8. Which of the following does *not* represent a function of HCl in the stomach?
   a. HCl cleaves peptide bonds.
   b. HCl activates digestive enzymes like pepsin.
   c. HCl acts as a barrier to pathogens present in food.
   d. HCl disrupts the tertiary and secondary structure of proteins, enhancing their enzymatic degradation.

9. Which of the following represents the correct sequence for protein digestion and absorption?
   a. In the stomach, pepsin degrades large polypeptides to smaller polypeptides. In the small intestine, the smaller polypeptides are further broken down into amino acids by the actions of pancreatic en-zymes, like trypsin, and brush border en-zymes. Amino acids are absorbed in the small intestine by a Na^+-amino acid cotransport mechanism.
   b. In the stomach, enterokinase degrades large polypeptides to smaller polypeptides. In the small intestine, the smaller polypeptides are degraded to amino acids by the actions of pancreatic proteases, like trypsin, and brush border enzymes. Amino acids are absorbed in the small intestine by a Na^+-amino acid cotransport mechanism.
   c. In the stomach, pepsin degrades large polypeptides to small polypeptides. In the small intestine, the smaller polypeptides are broken down into amino acids by the actions of pancreatic enzymes, like trypsin, and brush border enzymes. Amino acids are absorbed in the large intestine by a Na^+-amino acid cotransport mechanism.
   d. In the mouth, large polypeptides are degraded to smaller polypeptides by the actions of amylase. In the small

intestine, the smaller polypeptides are further broken down into amino acids by the actions of pancreatic enzymes, like trypsin, and brush border enzymes. Amino acids are absorbed in the small intestine by a Na^+-amino acid cotransport mechanism.

10. Which of the following represents the correct sequence for fat digestion?
    a. The bulk of fat digestion takes place in the small intestine with the release of lipases from the gallbladder. Lipases cleave fats into fatty acids and other small lipids that are absorbed by the small intestine.
    b. The bulk of fat digestion takes place in the small intestine. Bile released from the gallbladder helps to break up large fat globules to smaller ones, while lipases released from the pancreas cleave the fats to fatty acids and other small lipids. Absorption of these small lipids occurs via a Na^+-fatty acid cotransport mechanism in the small intestine.
    c. The bulk of fat digestion takes place in the small intestine. Bile released from the gallbladder helps to break up large fat globules to smaller ones, while amylase released from the pancreas cleaves the fats to fatty acids and other small lipids. Absorption of these small lipids occurs via a Na^+-fatty acid cotransport mechanism in the small intestine.
    d. The bulk of fat digestion takes place in the small intestine. Bile released from the gallbladder helps to break up large fat globules to smaller ones, while lipases released from the pancreas cleave the fats to fatty acids and other small lipids. Absorption of these small lipids occurs using fatty acid-binding proteins located in the small intestine.

11. Which of these statements concerning pancreatic secretion is *false*?
    a. The introduction of HCl into the upper regions of the small intestine stimulates pancreatic secretion.
    b. Nerves innervating the upper region of the small intestine sense the arrival of food from the stomach and signal the pancreas to secrete enzymes.
    c. The arrival of food into the small intestine from the stomach stimulates release of the hormone secretin, which triggers pancreatic secretion.
    d. Digestive enzymes secreted by the pancreas are activated by an enzyme, called enterokinase, secreted by the small intestine.

12. All of the following occur with the condition known as diabetes *except*:
    a. high blood levels of insulin
    b. high blood levels of glucose
    c. reduced rates of glucose uptake by cells
    d. presence of glucose in the urine

(2) Integrating Your Knowledge

(a) What were the four areas of emphasis of the Recommended Daily Allowances?

(b) Give an example of an animal with a specialized mouth for feeding.

(c) What is peristalsis, and where in the body does it occur?

(d) Which stomach cells secrete acid?

(e) How is the hormone secretin involved in the digestion process?

(f) Briefly, how and where is fat digested?

(g) What is the main function of the large intestine?

(h) What is the difference between insulin and glucagon?

CHAPTER 43—ANSWER KEY

D. Assessing What You've Learned

(1) Testing Your Knowledge
    1. a; 2. c; 3. a; 4. b; 5. b; 6. b; 7. d; 8. a; 9. a;
    10. d; 11. b; 12. a

(2) Integrating Your Knowledge
    (a) In 1943, the Recommended Daily Allowances (RDAs) were published to specify the amount of each essential nutrient that must be ingested to meet the needs of

practically all healthy people. The following were their focus:

1. Of the 20 amino acids required to manufacture proteins, humans can synthesize 12. The other 8 must be ob-tained from food and are called essen-tial amino acids.

2. Vitamins are carbon-containing molecules that function as coenzymes in critical reactions.

3. Essential elements serve a wide variety of functions.

4. Electrolytes form ions in solution; they influence osmotic balance and are required for normal membrane function.

(b) Human teeth include sharp canines for tearing meat and flattened molars for crushing seeds, roots, and other sources of carbohydrates.

(c) Peristalsis is a wave of contraction that propels food through the digestive system. It occurs primarily in the esophagus and the intestines.

(d) Parietal cells in the lining of the stomach produce hydrochloric acid.

(e) Bayliss and Starling discovered a hormone called secretin, which is produced in the small intestine in response to the arrival of food to the stomach. This dis-covery was important because it con-firmed that digestion is under both neural and hormonal control.

(f) The pancreatic secretions include digestive enzymes that act on fats and carbohydrates, as well as proteins. Lipase breaks certain bonds present in complex fats. Bile salts are synthesized in the liver and secreted in a complex solution called bile, which is stored in the gallbladder. When bile enters the small intestine, it raises the pH and emulsifies fats so that they can be digested.

(g) The primary function of the large intestine is to compact wastes that remain and absorb enough water to form feces.

(h) Insulin is a hormone produced in the pancreas when blood glucose levels are high. It travels through the bloodstream and binds to receptors on cells throughout the body. In response to insulin, the cells increase their rate of glucose uptake and respiration. If blood glucose levels fall, cells in the pancreas secrete a hormone called glucagon. Cells in the liver and skeletal muscle catabolize glycogen and cells that store lipids catabolize fatty acids. The result is that glucose levels in the blood rise.

44

Gas Exchange and Circulation

A. KEY BIOLOGICAL CONCEPTS

44.1 Air and Water as Respiratory Media

How Do Oxygen and Carbon Dioxide Behave in Air?

- **Partial pressure** is the pressure of a particular gas in a mixture. To calculate this, multiply the percent composition of that gas by the total pressure exerted by the entire mixture (**Figure 44.2**).

How Do Oxygen and Carbon Dioxide Behave in Water?

- The amount of gas that dissolves in a liquid depends on several factors:
  1. Solubility of the gas in that liquid
  2. Temperature of the liquid
  3. Presence of other solutes
- CO_2 is almost 30 times more soluble in water than O_2. As a result, fish rid themselves of CO_2 more easily than they can obtain oxygen.
- As the temperature of the water increases, the amount of gas that dissolves decreases. Cold-water habitats have much more oxygen available than warm-water habitats.
- Because seawater has a much higher concentration of solutes than freshwater, it can hold much less dissolved gas.
- Mixing improves the oxygenation of water. Rapids and waterfalls tend to be the most highly oxygenated of all aquatic environments (**Figure 44.3**).

- A liter of air can contain up to 209 ml of O_2 while a liter of water may contain up to 7 ml of O_2. To extract the same amount of oxygen, water-breathers must breathe 30 times more water than air.
- Water is also about 1000 times denser than air. As a result, water-breathers expend considerably more energy to ventilate their respiratory surfaces.

44.2 Organs of Gas Exchange

Design Parameters: The Law of Diffusion

- The rate of diffusion, as defined by Fick's law of diffusion (**Figure 44.4**), depends on five factors:
  1. Solubility of the gas
  2. Temperature
  3. Surface area available for diffusion
  4. Difference in partial pressures of the gas across the respiratory medium
  5. Thickness of the barrier to diffusion

How Do Gills Work?

- **Gills** are outgrowths of the body surface that are used for gas exchange in aquatic animals.
- Among invertebrates, the structure of gills is extremely diverse, whereas the gills of bony fish are similar in structure (**Figure 44.5**).
- To move water through these structures, fish open and close their mouths and the stiff flap of tissue that covers the gills. Fish can also force water through their gills by swimming with their mouths open; this process is called **ram ventilation**.

- Movement of water over the gills is **unidirectional**, which sets up a **countercurrent exchange system**. This ensures that the difference in the amount of O_2 and CO_2 in water versus blood is large over the entire respiratory surface (**Figure 44.6**).

How Do Tracheae Work?

- **The tracheal system** transports air close enough to cells for gas exchange to take place directly. As a result, insects do not require a circulatory system to transport gases from a respiratory structure to the tissues.
- Trachea connect to the exterior via an opening called a **spiracle** that can be closed. Spiracles are interpreted as adaptations to minimize water loss.

How Do Lungs Work?

- Mammalian lungs are divided into tiny sacs called **alveoli**. Each human lung contains about 150 million of these structures (**Figure 44.8**). It has about 100 m^2 of surface area, and the barrier of diffusion is only 0.2 vm thick.
- In the simple lungs of snails and spiders, air movement takes place by diffusion only. Vertebrates actively ventilate their lungs by pumping air via muscular contractions.

44.3   Blood

- Blood is a **connective tissue** consisting of cells in a fluid extracellular matrix called **plasma**. **Formed elements** in the blood include red blood cells, platelets, and several types of white blood cells.
- The functions of blood include:
  1. Carrying O_2 and CO_2 between mitochondria and lungs
  2. Transporting nutrients from the digestive tract to other tissues in the body
  3. Moving waste products to the kidney and liver for processing
  4. Conveying hormones from glands to target tissues
  5. Delivering immune system cells to sites of infection
  6. Distributing heat from deeper organs to the surface

- **White blood cells** are part of the immune system and fight infections. **Platelets** are involved in clot formation.
- **Red blood cells** contain an oxygen-carrying molecule called **hemoglobin**. It consists of four polypeptide chains, each of which binds to a nonprotein group called **heme**. Each heme contains an iron molecule that can bind to an oxygen molecule.

Oxygen Delivery: The Bohr Effect

- Blood leaving the lungs has a P_{O_2} of 100 mm Hg; muscle, and other tissues have P_{O_2} levels of about 40 mm Hg at rest. This partial pressure difference drives the unloading of O_2 from the hemoglobin to the tissues via diffusion (**Figure 44.12**).
- Because of the sigmoidal relationship, hemoglobin responds quickly and effectively to small changes in oxygen demand.
- Hemoglobin is also sensitive to changes in pH and temperature. During exercise, the temperature and the P_{CO_2} increase in the tissue. This results in a drop in pH.
- Decreases in pH and increases in temperature alter hemoglobin's conformation such that it is more likely to release O_2 at all values of P_{O_2}. This is known as the **Bohr shift**.
- Kicenuik and Jones wanted to determine how the oxygen transport system in trout responded to sustained exercise. They found that arterial O_2 levels remained fairly constant as swimming speed increased, but O_2 levels in the venous blood dropped steadily as swimming speed increased. In hard-working tissues, the combination of increased temperature, lower pH, and lower P_{O_2} caused hemoglobin to become almost completely deoxygenated.

CO_2 Transport and the Buffering of Blood pH

- **Carbonic anhydrase** in red blood cells catalyzes the formation of carbonic acid from carbon dioxide and water (**Figure 44.13**). As a result, CO_2 that diffuses into red blood cells is quickly converted into bicarbonate ions and protons. The same reaction occurs much more slowly in the plasma surrounding the red blood cells.

- When Hb is not carrying O_2, it has a high affinity for protons. As a result, Hb takes up much of the H+ that is produced by the dissociation of carbonic acid. In this way, Hb acts as a **buffer**—a compound that minimizes changes in pH.

- At the lungs, CO_2 diffuses from the blood into the alveoli and the Pco_2 in the blood declines. Bicarbonate is converted back to CO_2, which then diffuses into the alveoli and is exhaled from the lungs. Hb picks up O_2 during inhalation and the cycle begins again.

44.4   The Circulatory System

Open Circulatory Systems

- One or more hearts pump hemolymph into an artery that empties into many small open fluid-filled spaces. Hemolymph is returned to the heart when it relaxes and creates suction (**Figure 44.14**).

- Hemolymph is not tightly constrained inside vessels, so the overall pressure in the system is low.

Closed Circulatory Systems

- Blood is completely contained within blood vessels and flows in a continuous circuit through the body under pressure generated by the heart.

Blood, Interstitial Fluid, and the Lymphatic System

- **Interstitial fluid** is constantly augmented by water and proteins from blood that leak out between the cells that form the walls of the capillaries (**Figure 44.16**).

- The lymphatic system is a collection of vessels that branch throughout the body. The fluid inside the lymph vessels is called lymph and is eventually returned to the body.

The Heart

- The atrium receives blood returning from circulation and the ventricle generates force to propel the blood through the system.

- The number of distinct chambers in the heart has increased as vertebrates diversified. The circulatory system in fish forms a single loop, and in all other vertebrates there are two separate circuits, to the lungs and the body (**Figure 44.17**).

- The right ventricle powers the blood to the lungs and back to the heart in a section called the **pulmonary circulation**. Blood flow from the atrium to the ventricle is one way because of tissue flaps called **valves**. The left ventricle pumps blood to the entire body and back, or to the **systemic circulation**.

The Cardiac Cycle

- A cardiac cycle consists of one complete contraction phase, called **systole**, and one complete relaxation phase, called **diastole** (**Figure 44.20**).

Electrical Activation of the Heart

- The "pacemaker cells" that initiate contraction are located in a region of the right atrium called the **SA (sinoatrial) node** (**Figure 44.21**).

- The pace at which the SA node initiates electrical signals determines heart rate. The pace, in turn, is regulated by electrical signals from the brain.

- Heart rate may be altered by electrical signals from emotional centers in the brain or by signals from the chemical messengers called epinephrine and norepinephrine.

- Heart rate can also vary with exposure to drugs.

Regulation of Blood Pressure and Blood Flow

- **Blood pressure** is the force that blood exerts on the walls of arteries, capillaries, and veins.

- Looking at **Figure 44.23**, as the same volume of blood travels through a greater cross-sectional area, the pressure in the fluid drops. Overall, blood pressure is highest in the artery that leads away from the left ventricle (100 mm Hg) and lowest in the veins that return blood from the body to the right atrium (10 mm Hg).

- Peak pressure is due to the contraction of the ventricle and is called **systolic pressure**. The lowest pressure occurs at the end of ventricular relaxation and is called **diastolic pressure**.
- A typical blood pressure might be 120/80 (systolic/diastolic). People with pressures of 150/90 and higher are considered to have high blood pressure or **hypertension**.

Circulatory Homeostasis in Blood Pressure and Blood Chemistry

- Sensory cells are located next to an artery in the neck, an artery near the heart, and the wall of the heart. These cells act as **baroreceptors**, or pressure receptors. When they detect a change in blood pressure, they trigger electrical signals that change the heart's output and vessel diameter.
- Changes in blood P_{O_2}, P_{CO_2}, and pH are detected by specialized nerve cells in the neck, near the heart, and in the brain region just above the spinal cord. These cells act as **chemoreceptors** and send signals to the lungs and heart to maintain homeostasis.

B.   CROSS-CUTTING THEMES

Looking Back—
Concepts from Earlier Chapters

Osmosis and Diffusion—Chapter 6
Chapter 6 introduced the processes called diffusion and osmosis. Diffusion describes the movement of substances from regions of higher concentration to regions of lower concentration. The movement of water from regions of higher to lower concentration is called osmosis.

Surface to Volume Ratio—Chapter 41
Organs such as the small intestine have a large internal surface area to increase absorption efficiency.

Water Loss—Chapter 42
As explained in **Chapter 42**, many marine mammals lose water across their respiratory surface via osmosis; animals that live in freshwater lose ions and other solutes by diffusion. **Chapter 42** also introduced the trachea of insects. Biologists found that when spiracles

were kept open experimentally, insects were likely to die of dehydration.

Carbonic Anhydrase—Chapter 43
Recall from **Chapter 43** that carbonic anhydrase catalyzes the formation of carbonic acid from carbon dioxide and water. As a result, CO_2 that diffuses into red blood cells is quickly converted to bicarbonate ions and protons.

Looking Forward—
Concepts in Later Chapters

Electrical Signals—Chapter 45
Because intercalated discs of cardiac muscle contain numerous gaps, electrical signals pass directly from one cell to the next. **Chapter 45** explores how electrical signals in animals are initiated and propagated.

Fight or Flight—Chapter 47
As part of the fight-or-flight response analyzed in **Chapter 47**, blood is directed away from the skin and digestive system and toward the heart, brain, and muscles.

The Lymphatic System—Chapter 49
The lymphatic system acts as a type of circulatory system. Because its primary function is in defense against disease-causing agents, however, the lymphatic system is analyzed in more detail in **Chapter 49**.

C.   DIFFICULT TOPICS

The two most difficult topics in this chapter have to do with respiratory properties of blood. The first topic deals with the partial pressures of respiratory gases. This concept is central to respiratory physiology, so you must feel comfortable using partial pressures to describe elements of respiration and know the difference between partial pressures and percent composition in respiratory gases.

In air, think of percent composition as constant but partial pressures changing with altitude. Run through a few calculations and follow the partial pressures of O_2 and CO_2 in the blood going from the lungs to the tissues and back. Have classmates ask you questions, and try to stump each other. Mastering this concept will make future lessons in respiratory physiology more enjoyable.

The second topic that most students have difficulty with is the sigmoidal curve of hemoglobin and O_2 loading/unloading. Always think of the loading and unloading of O_2 as a dynamic process. Think of the binding of one oxygen molecule like a person getting into a car. Once this person is inside, he or she can open the doors of the car to allow three other people easier access to the car. In this way, when one oxygen molecule binds to Hb, it changes the conformation such that it is very easy for the next three O_2 molecules to bind. This is the primary reason for the sigmoidal curve in Hb-O_2 loading.

D. ASSESSING WHAT YOU'VE LEARNED

(1) Testing Your Knowledge

1. Which of the following most accurately describes the gas composition of the atmosphere?
    a. nitrogen > oxygen > carbon dioxide > inert gases (such as argon)
    b. oxygen > nitrogen > carbon dioxide > inert gases (such as argon)
    c. nitrogen > oxygen > inert gases (such as argon) > carbon dioxide
    d. oxygen > carbon dioxide > nitrogen > inert gases (such as argon)

2. Of the following gases found in the atmosphere—nitrogen, oxygen, argon, and carbon dioxide—which are considered to be physiologically active?
    a. carbon dioxide and oxygen
    b. nitrogen, carbon dioxide, and oxygen
    c. nitrogen and oxygen
    d. oxygen only

3. All of the following represent examples of how the design of respiratory organs conforms to the dictates of Fick's law of diffusion *except*:
    a. The respiratory surface of the human lung is 70 square meters.
    b. The lung epithelium and blood vessels are extremely thin.
    c. The epithelium of both internal and

external gills is in direct contact with oxygen-bearing water.
    d. The flow of blood through the gills is in the same direction as the flow of water over the gill surface.

4. Which of the following represents a disadvantage of breathing air (at lower elevations) versus breathing water?
    a. A liter of water holds less O_2 than a liter of air.
    b. Water is denser than air.
    c. Carbon dioxide is quite soluble in liquids like water.
    d. Air-breathing animals are better able to retain heat generated by their bodies.

5. The insect tracheal system differs from the mammalian respiratory system in all of the following ways *except*:
    a. O_2 is delivered by hemolymph, not blood, to cells.
    b. Spiracles are used to reduce evaporative water loss.
    c. There is an absence of respiratory pigment for gas transport.
    d. Muscles of locomotion help to ventilate the respiratory system.

6. Which of the following statements most accurately describes the structure of hemoglobin?
    a. Hemoglobin consists of a single polypeptide chain with a heme center containing a copper ion to which a molecule of O_2 binds.
    b. Hemoglobin consists of a single polypeptide chain with a heme center containing an iron ion to which a molecule of O_2 binds.
    c. Hemoglobin consists of four polypeptide chains, each of which have a heme group containing a copper ion. One hemoglobin molecule can bind a total of four O_2 molecules.
    d. Hemoglobin consists of four polypeptide chains, each of which have a heme group containing an iron ion. One hemoglobin molecule can bind a total of four O_2 molecules.

7. Which of the following best describes the significance of the sigmoidal relationship between Po_2 levels and hemoglobin saturation?
   a. As tissue Po_2 levels decline, the amount of O_2 released increases proportionately.
   b. As it circulates, hemoglobin releases the same amount of oxygen to all tissues.
   c. As tissue Po_2 levels decline with an increase in activity, the oxygen reserve increases.
   d. Once tissue Po_2 levels drop below 40 mm Hg, the amount of oxygen released by hemoglobin increases greatly.

8. Which of the following best describes the phenomenon known as the Bohr shift?
   a. As tissue CO_2 levels increase, pH decreases, resulting in an increase in O_2 delivery to the tissues.
   b. As tissue CO_2 levels decrease, pH decreases, resulting in an increase in O_2 delivery to the tissues.
   c. As tissue CO_2 levels increase, pH increases, resulting in a decrease in O_2 delivery to the tissues.
   d. As tissue CO_2 levels decrease, pH increases, resulting in an increase in O_2 delivery to the tissues.

9. Which of the following situations would result in the greatest degree of O_2 saturation for hemoglobin, assuming PO_2 remains constant?
   a. increased CO_2 levels, decreased temperature
   b. increased CO_2 levels, increased temperature
   c. decreased CO_2 levels, decreased temperature
   d. decreased CO_2 levels, increased temperature

10. The transport of CO_2 by the blood is primarily dependent on:
    a. the solubility of CO_2 in blood
    b. the presence of carbonic anhydrase in red blood cells
    c. the ability of hemoglobin to bind and transport CO_2

   d. the ability of other blood proteins to bind CO_2

11. Hemoglobin acts as a buffer by:
    a. binding O_2 and CO_2
    b. binding CO_2
    c. binding H^+
    d. binding bicarbonate ions

12. One of the main advantages of a closed over an open circulatory system is:
    a. the ability to direct the flow of blood to certain tissues
    b. the use of a heart to create pressure gradients necessary for blood flow
    c. the ability to return blood to the heart
    d. the use of a respiratory pigment to transport O_2

13. What prevents the atria and the ventricles from contracting at the same time?
    a. Pacemaker cells located in the atria fire before the pacemaker cells in the ventricles.
    b. It takes time for epinephrine to diffuse from the atria to the ventricles to trigger contraction.
    c. The electrical signal generated in the right atrium is delayed at the AV node before passing to the ventricles.
    d. The Na^+ channels responsible for initiating ventricular contraction are inactivated and need to return to activated configuration to be electrically stimulated.

14. Your blood pressure is reported to you as two values (e.g., 120/80). These represent:
    a. the pressure at the beginning of atrial contraction and the beginning of ventricular contraction
    b. peak pressure during ventricular contraction and lowest pressure at end of ventricular relaxation phase
    c. peak pressure at the end of atrial contraction and lowest pressure at the end of atrial relaxation
    d. the pressure at the end of atrial contraction and at the end of ventricular contraction

(2) Integrating Your Knowledge

(a) The composition of air at 760 mm Hg is 20.95% oxygen, 78.89% nitrogen, and 0.03% carbon dioxide. What are the partial pressures for each of these gases? As you climb Mt. Everest, what happens to these gas percentages and these gas partial pressures?

(b) What are the three factors that determine the amount of gas that will dissolve in water? Which factor contributes the most in the difference between O_2 and CO_2 dissolving in water?

(c) The rate of gas diffusion depends on what five factors?

(d) Do insects have a "respiratory system" by definition? If not, what do they have instead?

(e) List the functions of the blood.

(f) What is the Bohr shift?

(g) Describe the significance of carbonic anhydrase.

(h) What is the difference between systole and diastole? What does this have to do with blood pressure?

(i) What is the electrical event that regulates heart rate?

CHAPTER 44—ANSWER KEY

D. Assessing What You've Learned

(1) Testing Your Knowledge
1. c; 2. a; 3. d; 4. c; 5. a; 6. d; 7. d; 8. a; 9. c; 10. b; 11. c; 12. a; 13. c; 14. b

(2) Integrating Your Knowledge
(a) $760 \propto 0.2095 = 159.2$ mm Hg
$760 \propto 0.7889 = 599.6$ mm Hg
$760 \propto 0.0003 = 0.228$ mm Hg
The percentages of these gases would remain the same as you climbed up in altitude; however, the partial pressures of these gases would all decrease because the total atmospheric pressure decreases as your altitude increases.

(b) The amount of gas that dissolves in a liquid depends on several factors:

1. Solubility of the gas in that liquid
2. Temperature of the liquid
3. Presence of other solutes

CO_2 is almost 30 times more soluble in water than in O_2. As a result, fish rid themselves of CO_2 more easily than they can obtain oxygen.

(c) The rate of diffusion depends on these five factors:

1. Solubility of the gas
2. Temperature
3. Surface area available for diffusion
4. Difference in partial pressures of the gas across the respiratory medium
5. Thickness of the barrier to diffusion

(d) In insects, the tracheal system transports air close enough to cells for gas exchange to take place directly. As a result, insects do not require a circulatory system to transport gases from a respiratory structure to the tissues.

Trachea connect to the exterior via an opening called a spiracle, which can be closed. Spiracles are interpreted as adaptations to minimize water loss.

(e) The functions of blood include:

1. Carrying O_2 and CO_2 between mitochondria and lungs
2. Transporting nutrients from the digestive tract to other tissues in the body
3. Moving waste products to the kidney and liver for processing
4. Conveying hormones from glands to target tissues
5. Delivering immune system cells to the sites of infection
6. Distributing heat from deeper organs to the surface

(f) Decreases in pH and increases in temperature alter hemoglobin's conformation such that it is more likely to release O_2 at all values of P_{O_2}. This effect is known as the Bohr shift.

(g) **Carbonic anhydrase** in red blood cells catalyzes the formation of carbonic acid from carbon dioxide and water. As a result, CO_2 that diffuses into red blood cells is quickly converted into bicarbonate ions and protons. The same reaction occurs much more slowly in the plasma surrounding the red blood cells.

(h) A cardiac cycle consists of one complete contraction phase, called systole, and one complete relaxation phase, called diastole. Peak pressure is due to the contraction of the ventricle and is called systolic pressure. The lowest pressure occurs at the end of ventricular relaxation and is called diastolic pressure.

(i) "Pacemaker cells" that initiate contraction are located in a region called the sinoatrial (SA) node. The pace at which the SA node initiates electrical signals determines heart rate. The pace, in turn, is regulated by electrical signals coming from the brain.

45

Electrical Signals in Animals

A. KEY BIOLOGICAL CONCEPTS

45.1 Principles of Electrical Signaling

- A **sensory receptor cell** transmits the information it receives via a **sensory neuron**, and this travels to the brain or spinal cord (**Figure 45.1**). Together, the brain and spinal cord form the **central nervous system** (CNS).
- The CNS integrates information from many sensory neurons and sends signals to **effector cells** via **motor neurons**. All of the components outside the CNS are part of the **peripheral nervous system** (PNS).

The Anatomy of a Neuron

- Most neurons have a **cell body**, a highly branched group of short projections called **dendrites**, and one or more long projections called **axons** (**Figure 45.2**).
- A dendrite receives electrical signals from the axons and dendrites of adjacent cells. The axon then sends the signal to the dendrites of other neurons.

An Introduction to Membrane Potentials

- Due to the separation of ions across a membrane, and because ions carry charge, there is a separation of charge across the membrane called a **membrane potential**.
- If the difference in charges on either side of the membrane is large, so is the membrane voltage.

The Resting Potential

- When a neuron is not transmitting an electrical signal but is merely sitting in extracellular fluid at rest, its membrane has a voltage called the **resting potential**.
- Most proteins inside the cell are negatively charged. The major positively charged ion inside neurons is potassium. In the extracellular fluid, however, sodium and chloride predominate (**Figure 45.3**).
- In neurons, only K^+ can cross the membrane easily along its concentration gradient. It does so via potassium channels.
- As K^+ moves from the interior of the cell to the exterior via K^+ channels, the inside of the cell becomes more negatively charged relative to the outside. Eventually, the buildup of negative charge inside the cell begins to attract K^+ and counteract the concentration gradient that favors movement of K^+ out. As a result, the membrane reaches a voltage where there is equilibrium between the concentration gradient that moves K^+ out and the electrical gradient that moves K^+ in (**Figure 45.4**).
- The voltage where there is no net movement of K^+ is called the **equilibrium potential**.

Using the Nernst and Goldman Equations

- The **Nernst equation** converts energy stored in a concentration gradient into the energy stored as an electrical potential.

- The equilibrium potential given by the Nernst equation is calculated independently for each ion.
- The **Goldman equation** accounts for the fact that the nerve membrane is permeable to three different ions; thus it interrelates the potentials of Na^+, K^+, and Cl^- to generate the membrane potential, in volts.

Using Microelectrodes to Measure Membrane Potentials

- Hodgkin and Huxley pioneered the study of electrical signaling in animals. To record membrane voltages in the giant squid axon, they cut off an intact section of the axon, bathed it in seawater or other known solution, and inserted a microelectrode. Their recordings suggested that the giant squid axon had a resting potential of -45 mV, but later work confirmed that most neurons have a resting potential of around -70 mV.

What Is an Action Potential?

- The initial event is a rapid depolarization of the membrane. Current flow causes the inside of the membrane to become less negative and then positive with respect to the outside (**Figure 45.5**).
- The membrane then experiences a rapid repolarization. The membrane actually becomes more negative than the resting potential. This is called the undershoot.
- An action potential is all or none. There is no such thing as a partial action potential.
- Action potentials can be triggered by artificially depolarizing the membrane with an injection of electrical current through a microelectrode.
- Action potentials are propagated down the length of the axon.

45.2   Dissecting the Action Potential

Distinct Ion Currents are Responsible for Depolarization and Repolarization

- The action potential consists of a strong inward flow of Na^+ followed by a strong outward flow of K^+ (**Figure 45.6**).

Voltage-Gated Channels

- **Voltage-gated channels** are proteins that change conformation in response to charges present at the membrane surface. These changes will open or close the channel.
- **Key Research**: Kenneth Cole and colleagues came up with a technique—called **voltage clamping**—that holds the membrane potential of a neuron constant and measures the flows of ions (**Figure 45.7**).
- They found that action potentials result from staggered activity of voltage-gated Na^+ and K^+ channels.

Patch Clamping and Studies of Single Channels

- Neher and Sakmann perfected a technique called patch clamping, whereby they suctioned a microelectrode to a membrane (**Figure 45.8**). They could then document the currents flowing through individual transmembrane channels.
- The opening of sodium channels exhibits **positive feedback**, which takes place when the occurrence of an event makes the same event more likely to occur.

Using Neurotoxins to Identify Channels and Dissect Currents

- In giant axons from lobsters treated with tetrodotoxin (found in puffer fish), the outward-directed K^+ current was normal; however, the inward-directed flow of Na^+ was blocked.

The Role of the Sodium-Potassium Pump

- To maintain the concentration gradients of Na^+ and K^+, a protein that hydrolyzes ATP must pump Na^+ out and K^+ in.
- Skou isolated a protein from crab neurons that began hydrolyzing ATP if both Na^+ and K^+ were present (**Figure 45.10**). When he added ouabain to this reaction mix, the enzyme's activity stopped. This enzyme, called Na^+/K^+-ATPase, transports three sodium ions out of the cell for every two potassium ions transported inward.

How Is the Action Potential Propagated?

- Positive charges inside the cell are repulsed by the influx of Na^+ and negative charges are attracted. As a result, sections of the membrane close to the site of an action potential are depolarized. Nearby voltage-gated channels pop open in response, positive feedback occurs, and an action potential results (**Figure 45.11**).
- Na^+ channels are **refractory**. Once they have opened and closed, they are less likely to open again for a short period of time.
- **Myelination** acts as insulation. It prevents charge from leaking out as it spreads down an axon. An unmyelinated area of the neuron, called a **node of Ranvier**, has a dense concentration of voltage-gated channels (**Figure 45.12**). Electrical signals jump from node to node much faster than if the entire neuron were unmyelinated.

45.3   The Synapse

- Otto Loewi (1920) showed that the mechanism that transmits electrical signals from cell to cell involves **neurotransmitters**. He stimulated the vagus nerve in a frog heart and slowed the heart. He then took some of the solution this heart was bathed in and applied it to another heart, and it slowed as well. This result supplied conclusive evidence for the chemical transmission of electrical signals (**Figure 45.13**).
- The interface between two neurons is called a **synapse** (**Figure 45.14**). Just inside the synapse, the axon contains numerous saclike structures called synaptic vesicles. At this site, three steps occur (**Figure 45.15**):

  1. The action potential arrives and triggers entry of Ca^{2+}.

  2. In response to Ca^{2+}, synaptic vesicles fuse with the presynaptic membrane and release a neurotransmitter.

  3. Ion channels open when the neurotransmitter binds; ion flows cause change in postsynaptic cell potential.

What Do Neurotransmitters Do?

- Many receptors function as **ligand-gated ion channels**, meaning that a molecule binds to a specific site on the receptor to activate it. (See **Table 45.2** for categories of neurotransmitters.)
- Some other receptors activate enzymes that lead to the production of a **second messenger**, which may induce changes in enzyme activity, gene transcription, or membrane potential.

Postsynaptic Potentials and Summation

- If the receptors at the synapse admit an influx of sodium ions in response to the arrival of the neurotransmitter (NT), then the postsynaptic membrane depolarizes (**Figure 45.16**). These are called **excitatory postsynaptic potentials** (**EPSPs**).
- In other receptors, binding of the NT leads to influx of K^+ or Cl^-. These events hyperpolarize the membrane and make action potentials less likely to occur. These are called **inhibitory postsynaptic potentials** (**IPSPs**).
- ESPSs and ISPSs are not all-or-none events. They are graded and short-lived. Their size depends on the amount of NT released at the synapse.
- The sodium channels that trigger action potentials in the postsynaptic cell are located near the start of the axon at a site called the **axon hillock** (**Figure 45.17**).

What Happens when Ligand-Gated Channels Are Defective?

- **Schizophrenia** is a mental illness that is highly variable in nature. Experiments using **phencyclidine** induced significant changes in number or function of a ligand-gated channel called **N-methyl-D-aspartic acid (NMDA)-sensitive glutamate receptor**. Glutamate is an amino acid that functions as a neurotransmitter in certain neurons.
- This may suggest that the NMDA receptor is somehow involved in the onset of schizophrenia (**Figure 45.18**).

45.4   The Vertebrate Nervous System

A Closer Look at the Peripheral Nervous System

- The PNS consists of two distinct systems (**Figure 45.19**):
  1. An afferent division that transmits sensory information to the CNS
  2. An efferent division that carries commands from the CNS to the body
- The efferent division of the PNS is further divided into two systems:
  1. A somatic system that controls the skeletal muscles
  2. An autonomic system that controls internal processes like digestion and heart rate
- The autonomic system is composed of the **parasympathetic nerves**, which stimulate digestion and other processes, and the **sympathetic nerves**, which prepare organs for stressful situations (**Figure 45.20**).

Functional Anatomy of the CNS
Mapping Functional Areas: Lesion Studies

- Broca (1861) studied an individual who could understand language but could not speak. After this person's death, Broca found a damaged area in the left front lobe of the cerebrum (**Figure 45.21**).
- Broca's claim that functions are localized to specific brain areas has been verified through extensive efforts to map the cerebrum.

Electrical Stimulation of Conscious Patients

- Wilder Penfield worked with epileptics scheduled to have seizure-prone areas of their brains surgically removed. While the patients were awake, Penfield electrically stimulated portions of their cerebrum and recorded each patient's sensations (**Figure 45.22**).
- This technique was used to map the general areas of brain function.

How Does Memory Work?
Recording from Single Neurons during Memory Tasks

- The researchers George Ojemann and Julie Schoenfield-McNeill recorded individual neurons in the temporal lobes of humans (**Figure 45.23**).

Documenting Changes in Synapses

- Work on the molecular basis of memory is founded on two main ideas:
  1. Learning and memory must involve some type of short- or long-term change in the neurons responsible for these processes.
  2. It will be much easier to understand what these changes are if an extremely simple system of neurons can be studied.
- Eric Kandel and colleagues have focused on the sea slug *Aplysia californica*. These slugs have a simple reflex that can be modified by learning (**Figure 45.24**). Follow-up studies have shown that the neurons involved in learning release the NT **serotonin**, and that serotonin causes an EPSP in the motor neuron in the gill.
- More serotonin is released after learning takes place (**Figure 45.25**). Therefore, EPSPs are higher and the motor neuron is more likely to generate action potentials.

B.   CROSS-CUTTING THEMES

Looking Back—
Concepts from Earlier Chapters
Osmosis and Diffusion—Chapter 6

Chapter 6 introduced the processes called diffusion and osmosis. Diffusion describes the movement of substances from regions of higher concentration to regions of lower concentration. The movement of water from regions of higher to lower concentration is called osmosis. This chapter also discussed the electrochemical gradient and ion channels.

Looking Forward—
Concepts in Later Chapters
Senses—Chapter 46

Before we discuss how hormones work, the electrical signals involved in vision, hearing, taste, and movement must be discussed. Sensory systems and movement are the subject of **Chapter 46**.

Second Messengers—Chapter 47

Second messengers may induce changes in enzyme activity, gene transcription, or membrane potential. In **Chapter 47** you'll explore the role of second messengers and hormones in detail.

C. DIFFICULT TOPICS

The resting and action potentials are some of the first concepts you will encounter in any physiology or neurology course. They are therefore the foundation of this chapter on electrical signals.

   The movement of ions generating a resting potential across a membrane is sometimes difficult to visualize as occurring in the human body. Nevertheless, visualizing these ion movements is a great place to start.

   First, visualize the movement of ions across the membrane in a resting potential and then try to draw this from memory while explaining the process aloud. Once you are comfortable with how the resting potential is maintained, try your skills on the action potential. Concentrate on where it starts, how it is propagated, and where it ends. Visualize the ions moving through channels as the action potential is created, and graph the voltage across the membrane as the action potential proceeds. Last, as you consider how this process works in the entire body, integrate the concepts of EPSP and IPSP so that you are sure not to confuse them in a broader sense.

D. ASSESSING WHAT YOU'VE LEARNED

(1) Testing Your Knowledge

1. Which of the following describes the role of a dendrite on a nerve cell?
   a. receives electrical signals
   b. transmits electrical signals from the neuron
   c. maintains the DNA of the neuron in a nucleus

2. Which of the following is true about the distribution of ions when a cell is at rest?
   a. Potassium and sodium ion concentrations are higher inside the cell than outside.
   b. Potassium and chloride ion concentrations are higher inside the cell than outside.
   c. Sodium and chloride ion concentrations are higher inside the cell than outside.
   d. Potassium ion concentration is higher inside, and sodium ion concentration is higher outside.

3. What would happen if voltage-gated sodium channels on a resting neuron were to open?
   a. Sodium ions would move into the cell and the membrane potential would become more negative.
   b. Sodium ions would move into the cell and the membrane potential would become more positive.
   c. Potassium ions would move out of the cell and the membrane potential would become more negative.
   d. Sodium ions would move out of the cell and the membrane potential would become more negative.

4. What would happen if voltage-gated potassium channels on a resting neuron were to open?
   a. Potassium ions would move into the cell and the membrane potential would become more negative.
   b. Sodium ions would move into the cell and the membrane potential would become more positive.
   c. Potassium ions would move out of the cell and the membrane potential would become more positive.
   d. Potassium ions would move out of the cell and the membrane potential would become more negative.

5. What phase of the action potential is associated with the influx of sodium ions that occurs when voltage-gated sodium ion channels open?
   a. repolarization
   b. depolarization
   c. resting potential
   d. undershoot

6. The action potential is produced by a combination of an influx of sodium ions and an efflux of potassium ions. The ions carry similar charge and move in opposite directions, so why don't their movements

simply cancel each other out so there is no change in the membrane potential?

a. Voltage-gated potassium channels open at the same time as voltage-gated sodium channels.

b. Voltage-gated potassium channels open at the same voltage as voltage-gated sodium channels.

c. The opening of voltage-gated potassium channels is delayed.

d. The opening of voltage-gated sodium channels is delayed.

7. What would happen to the resting membrane potential of a neuron if the extracellular fluid were flooded with a high concentration of potassium ions?

a. The resting membrane potential would become more negative.

b. The resting membrane potential would become more positive.

c. There would be no change in the resting membrane potential.

8. In patients suffering from multiple sclerosis (MS), the myelin surrounding the axons degenerates. Which of these statements describes the effect of loss of myelination?

a. Action potentials would spread down the axon more quickly if the myelin is removed.

b. Action potentials would spread down the axon more slowly if the myelin is removed.

c. Action potentials would spread down the axon more slowly because voltage-gated ion channels would open more slowly.

d. Action potentials would spread down the axon more quickly because voltage-gated ion channels would open more quickly.

9. Otto Loewi determined that the vagus nerve of the frog releases a neurotransmitter that causes the heart to slow down. Which of the following explains the effect that this neurotransmitter is having?

a. It is causing voltage-gated sodium channels in the heart muscle to open more quickly.

b. It is causing voltage-gated sodium channels in the heart muscle to become more "sensitive" to changes in voltage.

c. It is causing an increase in the rate at which action potentials can be generated in the heart muscle.

d. It is causing a decrease in the rate at which action potentials can be generated in the heart muscle.

10. Eventually, it was discovered that the neurotransmitter being released from the vagus nerve was acetylcholine (ACh). Heart muscle contracts following the depolarization that occurs during an action potential. In these cells, the negative resting potential must become less negative and eventually reach what is called a threshold. Once reaching this threshold potential, the action potential will spontaneously generate, and the muscle can contract. Which of the following descriptions could explain the effect of acetylcholine on the heart muscle?

a. The cells become leaky to potassium ions. Potassium ions leak out of the cell, which causes the resting potential to become more negative.

b. The cells become leaky to sodium ions. Sodium ions leak out of the cell, which causes the resting potential to become more negative.

c. The cells become leaky to potassium ions. Potassium ions leak into the cell, which causes the resting potential to become more negative.

d. The cells become leaky to sodium ions. Sodium ions leak into the cell, which causes the resting potential to become more positive.

11. Under which of the following conditions would a ligand-gated ion channel open?

a. when the membrane potential becomes more positive

b. when a molecule binds to a specific site on the channel

c. when the membrane potential becomes more negative

12. The ability of a postsynaptic neuron to generate an action potential (a signal) is dependent on information that is being received from surrounding presynaptic neurons. These "connections" can be either inhibitory or excitatory. What action would be associated with an inhibitory presynaptic neuron?
    a. The signal from the presynaptic neuron causes an influx of sodium into the postsynaptic neuron.
    b. The signal from the presynaptic neuron causes an efflux of sodium ions from the postsynaptic neuron.
    c. The signal from the presynaptic neuron causes an influx of potassium ions from the postsynaptic neuron.
    d. The signal from the presynaptic neuron causes an efflux of potassium ions from the postsynaptic neuron.

13. Neurotransmitters released from the vagus nerve cause a decrease in heart rate. In which part of the vertebrate nervous system would the vagus nerve be?
    a. parasympathetic nervous system
    b. sympathetic nervous system
    c. somatic system

14. Which of the following two structures make up the central nervous system (CNS)?
    a. spinal cord and parasympathetic system
    b. brain and sympathetic nervous system
    c. brain and spinal cord
    d. spinal cord and sympathetic nervous system

15. A 42-year-old male patient comes into the emergency room with paralysis on the right side of his body. The attending physician also notices that the patient is having trouble with language. Assuming that there is a head injury, on which part of the head did the injury occur?
    a. the back
    b. the left side
    c. the right side

16. Which part of the brain is associated with conscious thought, memory, and personality?

a. cerebellum
b. cerebrum
c. medulla

(2) Integrating Your Knowledge

(a) Briefly describe the anatomy of a neuron.

(b) How is K^+ involved in the maintenance of the resting membrane potential?

(c) What are voltage-gated channels?

(d) Why doesn't an action potential go backward?

(e) How does myelination aid in the propagation of an action potential?

(f) What are ligand-gated ion channels?

(g) Compare and contrast the EPSP and the IPSP.

(h) How were the brain's functional areas mapped out?

CHAPTER 45—ANSWER KEY

D. Assessing What You've Learned

(1) Testing Your Knowledge
1. a; 2. c; 3. b; 4. d; 5. b; 6. c; 7. a; 8. b; 9. d; 10. a; 11. b; 12. d; 13. a; 14. c; 15. b; 16. b

(2) Integrating Your Knowledge

(a) Most neurons have a cell body, a highly branched group of short projections called dendrites, and one or more long projections called axons. A dendrite receives electrical signals from the axons and dendrites of adjacent cells. The axon then sends the signal to the dendrites of other neurons.

(b) In neurons, only K^+ can cross the membrane easily along its concentration gradient. It does so via potassium channels. As K^+ moves from the interior of the cell to the exterior via K^+ channels, the inside of the cell becomes more negatively charged relative to the outside. Eventually, the buildup of negative charge inside the cell begins to attract K^+ and counteract the concentration gradient that favors movement of K⁺ out. As a result, the membrane reaches a voltage where there is equilibrium between the concentration gradient that moves K⁺ out and the electrical gradient that moves K⁺ in. The voltage where there is no net movement of K⁺ is called the equilibrium potential.

(c) Voltage-gated channels are proteins that change conformation in response to charges present at the membrane surface. These changes open or close the channel.

(d) Na⁺ channels are refractory. Once they have opened and closed, they are less likely to open again for a short period of time.

(e) Myelination acts as insulation. It prevents charge from leaking out as it spreads down an axon. An unmyelinated area of the neuron, called a node of Ranvier, has a dense concentration of voltage-gated channels. Electrical signals jump from node to node much faster than if the entire neuron were unmyelinated.

(f) Many receptors function as ligand-gated ion channels, meaning that a molecule binds to a specific site on the receptor to activate it.

(g) If the receptors at the synapse admit an influx of sodium ions in response to the arrival of the neurotransmitter (NT), then the postsynaptic membrane depolarizes. These are called excitatory postsynaptic potentials (EPSPs). In other receptors, binding of the NT leads to the influx of K⁺ or Cl⁻. These events hyerpolarize the membrane and make action potentials less likely to occur. These are called inhibitory postsynaptic potentials (IPSPs). ESPSs and ISPSs are not all-or-none events. They are graded and short-lived. The size depends on the amount of NT released at the synapse.

(h) Wilder Penfield worked with several epileptics scheduled to have seizure-prone areas of their brains surgically removed. While the patients were awake, Penfield electrically stimulated portions of their cerebrum and recorded each patient's sensations. This technique was used to map the general areas of brain function.

46

Animal Sensory Systems and Movement

A. KEY BIOLOGICAL CONCEPTS

46.1 How Do Sensory Organs Convey Information to the Brain?

Sensory Transduction

- Animals have sensory receptors that detect a remarkable variety of stimuli, but they all **transduce** sensory input (light, sounds, touch, and odors) to a change in membrane potential. In this way, different types of information are transduced into a common type of signal (**Figure 46.2**).
- The amount of depolarization that occurs in a sound receptor cell is proportional to the loudness of the sound. Louder sounds induce a higher frequency of action potentials than do softer sounds (**Figure 46.3**).

Transmitting Information to the Brain

- *Receptor cells tend to be highly specific*. For example, each receptor in a human ear responds best to certain frequencies of sound. Therefore, the train of action potentials from a cell contains information about the frequency of the sound, its intensity, and how long the stimulus lasts.
- *Each type of sensory neuron sends its signal to a specific portion of the brain*. Axons from sensory neurons in the human ear project to a particular area in the side of the brain.

46.2 Hearing

- **Hearing** is the ability to sense the changes in pressure called **sound**. A sound consists of **waves** of air or water pressure. The number of pressure waves that occur in one second is called the **frequency**.

How Do Sensory Cells Respond to Sound Waves and Other Forms of Pressure?

- In **pressure receptors**, direct physical pressure on a cell membrane or distortion by bending causes ion channels on a cell membrane to open or close. In response to change in ion flows, the membrane depolarizes or hyperpolarizes. The result is a new pattern of action potentials from the sensory neuron.
- In many pressure-sensing organs, **hair cells** are named for stiff outgrowths, **stereocilia**, occurring at one end of the cell (**Fig. 46.4**). If they bend in one direction in response to pressure, channels open and depolarize the cell. If they bend in the other direction, channels close and the cell hyperpolarizes.

The Mammalian Ear

- After reaching the head, sound waves go through a canal and strike the **tympanic membrane** separating the outer ear from the middle ear (**Figure 46.5**). Membrane vibrations are passed to three tiny bones called **ear ossicles**. The last ossicle, the **stapes**, vibrates against a membrane called the **oval window** that separates the middle ear from the inner ear.
- The oval window vibrates, generating waves in the fluid inside the **cochlea**. These pressure waves are sensed by **hair cells** in the cochlea.

The Middle Ear Amplifies Sounds

- The **size difference** between the tympanic membrane and oval window is important. The amount of vibration induced by sound waves is increased by a factor of 15. The ossicles act as **levers** and further amplify the sound such that the overall amplification factor is 22.

The Cochlea Detects the Frequency of Sounds

- Georg von Békésy (1920s) performed experiments on cochleas painstakingly dissected from human cadavers. He was able to vibrate the oval window and record how the cochlea's internal membranes moved in response. His key finding was that sounds of different frequencies caused the basilar membrane to vibrate in specific spots along its length (**Figure 46.7**).
- Certain portions of the basilar membrane vibrate in response to specific pitches, and result in the bending of hair-cell stereocilia. In this way, hair cells in a specific place on the membrane respond to sounds of certain frequency.

Sensory Worlds: What Do Other Animals Hear?

- Elephants use **infrasounds** (sound frequencies too low for humans to hear) to communicate.
- Bats use ultrasonic sound to **echolocate** (navigate by sound). Bats with cotton in their ears or their mouths taped shut are unable to fly without running into objects. Blindfolded bats, in contrast, are fully able to navigate.

46.3 Vision

The Vertebrate Eye

- The outermost layer of the eye is a tough rind of white tissue called the **sclera** (**Figure 46.9**). The front of the sclera is transparent and forms the **cornea**. Just inside the cornea is a colored, round muscle called the **iris**. The hole in the center of the iris is the **pupil**.
- Together, the cornea and **lens** focus the light onto the retina in the back of the eye. The **retina** contains a layer of **photoreceptors** and several layers of neurons.

What Do Photoreceptor Cells Do?

- **Photoreceptors** in the vertebrate eye are small rod- or cone-shaped cells called rods and cones (**Figure 46.10a**). **Rods** are very sensitive to dim light but not to color. **Cones** are much less sensitive to faint light but are stimulated by different colors.

How Do Rods and Cones Detect Light?

- Rods and cones are packed with membrane-rich disks containing large quantities of a transmembrane protein called **opsin**. Each opsin molecule is associated with a much smaller molecule called **retinal**. This two-molecule complex is called **rhodopsin** (**Figure 46.10b**).
- Retinal changes shape when it absorbs a photon of light, leading to a change in opsin's conformation. This in turn leads to a series of events that culminates in a change in the cell's membrane potential (**Figure 46.11**).

Color Vision: The Puzzle of Dalton's Eye

- John Dalton and his brother (1794) could not differentiate between the colors red and green (this condition is called red-green color blindness). Thomas Young analyzed cones and found that there were three different cone types, each having a different type of opsin. These proteins are called the blue, green, and red (or S, M, and L for short, medium, and long wavelength) opsins (**Figure 46.12**).
- David Hunt (1990) showed that color-blind people lack either functional M or L cones, or both.
- Hunt analyzed Dalton's DNA and found that he had a perfectly normal L allele but lacked a functional M allele. As a result, Dalton did not have green-sensitive cones.

Sensory Worlds: Do Other Animals See Color?

- A marine fish called a coelocanth lives in water 200 meters deep and has two opsins that respond to the blue region of the spectrum. It is likely that they perceive several hues of blue that we humans would consider as one color. Birds and insects can also see ultraviolet light, which has shorter wavelengths than the human eye can see. Why would this ability to detect UV light be adaptive?

- Rattlesnakes and other pit vipers sense **infrared light** using specialized infrared-sensing organs on their snouts (**Figure 46.13**). Because endothermic animals radiate heat in the form of infrared radiation, pit vipers can detect and strike at prey in complete darkness.

46.4 Taste and Smell

Taste: Detecting Molecules in the Mouth

- Each **taste bud** contains about 100 spindle-shaped **taste cells** that synapse to taste neurons (**Figure 46.14**).
- The four basic tastes are salty, sweet, sour, and bitter.
- The sensation of **saltiness** is primarily due to sodium ions (Na^+) dissolved in food. These ions flow into certain taste cells through open channels and depolarize their membranes.
- **Sourness** is due to the presence of protons that flow directly into certain taste cells via channels. In general, the lower the pH of food, the more it depolarizes a taste-cell membrane and the more sour the food tastes.

Why Do Many Different Foods Taste Bitter?

- Arthur Fox (1931) blew some phenylthiocarbamide (PTC) into the air and found that a colleague could taste bitterness in the air while he could not. Follow-up research confirmed that the ability to taste PTC is inherited and polymorphic.
- Independently, two teams identified a family of 40–80 genes that encode transmembrane receptor proteins. Each protein in this family binds to just one particular type of bitter molecule. A taste cell can have many different receptor proteins from this family.

What Is the Molecular Basis of Sweetness and Other Tastes?

- Membrane receptors that respond to sugars have yet to be identified by the same techniques employed in the search for bitterness receptors.

Olfaction: Detecting Molecules in the Air

- **Smell** allows animals to monitor airborne molecules that convey information. Wolves and domestic dogs can distinguish millions of different odors at extremely small concentrations.
- Odor molecules reach the nose and diffuse into a **mucus layer** in the roof of the nose. There they activate **olfactory receptor neurons** via membrane-bound receptor proteins. Axons from these neurons project up to the **olfactory bulb** of the brain (**Figure 46.15**).
- A gene family containing hundreds of distinct coding regions encodes receptor proteins on the surface of olfactory receptor neurons.
- Each olfactory neuron has only one type of receptor, and neurons with the same type of receptor project to different regions in the olfactory region of the brain.

46.5 Movement

Skeletons

- Skeletons provide attachment sites for muscles and a support system for the body's soft tissues. **Exoskeletons** are hard, hollow structures enveloping the body. **Hydrostatic skeletons** use the pressure of internal body fluids to support the body. **Endoskeletons** are hard structures buried inside the body.
- Endoskeletons are composed of cartilage and bone (**Figure 46.16**). **Cartilage** is made up of cells scattered in a gelatinous matrix of polysaccharides and scattered protein fibers. **Bone** is made up of cells in a hard extracellular matrix of calcium phosphate with small amounts of calcium carbonate and protein fibers.
- Ends of skeletal muscle are often attached to two different bones by **tendons**, which are bands of tough, fibrous connective tissue (**Figure 46.18**).

How Do Muscles Contract?

- A **muscle fiber** is a long, thin muscle cell. Within each cell there are many small strands called **myofibrils** (**Figure 46.19**).

The Sliding-Filament Model

- The **sarcomere**, which is the functional unit of skeletal muscle, appears as light and dark bands on the myofibril (**Figure 46.20**).
- Andrew Huxley and Jean Hanson (1954) proposed that the banding patterns in the sarcomere are actually caused by two types of long filaments—thick and thin—and that the filaments slide past one another during contraction. This explanation became known as the **sliding-filament model** (**Figure 46.21**).
- Follow-up research showed that the thin filaments are composed of two coiled chains of a globular protein called **actin**. One end of each chain is bound to the sarcomere; the other interacts with the thick filament.
- Thick filaments are composed of multiple strands of a long protein called **myosin**; it is anchored to the middle of the sarcomere.

How Do Actin and Myosin Interact?

- The myosin head can bind actin and catalyze the hydrolysis of ATP into ADP and a phosphate ion (**Figure 46.22**). Myosin and actin are locked together when an animal dies and its muscles enter the stiff state known as **rigor mortis**. This suggested that ATP must be present for myosin to release from actin.
- Ivan Rayment and colleagues proposed the following model (**Figure 46.23**):
  1. When a molecule of ATP binds to the myosin head, the myosin releases from actin.
  2. When ATP is subsequently hydrolyzed, the conformation of the protein changes. Specifically, the myosin neck straightens out, the head pivots, and the myosin head then binds to action in a new location.
  3. When inorganic phosphate is released, the myosin neck bends back into its original position. This bending is called the **power stroke** because it moves the entire actin filament.
  4. A new ATP molecule then binds to myosin, and the cycle starts again.

How Does Relaxation Occur?

- Sarcomeres contain proteins called **tropomyosin** and **troponin**. They work together to block the myosin binding sites on actin. As a result, actin and myosin cannot slide past each other.

An Overview of Events at the Neuromuscular Junction

1. Action potentials trigger the release of **acetylcholine** (ACh) from the **motor neuron** onto the muscle cell (**Figure 46.24**).
2. Membrane depolarization occurs on the muscle fiber in response to ACh. Action potentials propagate throughout muscle cells via axon-like structures called **T tubules**.
3. T tubules intersect extensive sheets of smooth endoplasmic reticulum called the **sarcoplasmic reticulum**. As an action potential passes down a T tubule, a protein in the T tubule membrane changes conformation and opens a **calcium channel** in the sarcoplasmic reticulum (SR).
4. Calcium ions are released from the SR, causing a conformational change in troponin, which then moves tropomyosin away from the myosin binding sites on actin.

B. CROSS-CUTTING THEMES

Looking Back—
Concepts from Earlier Chapters

Osmosis and Diffusion—Chapter 6

Chapter 6 introduced the processes called diffusion and osmosis. Diffusion describes the movement of substances from regions of higher concentration to regions of lower concentration. The movement of water from regions of higher to lower concentration is called osmosis. This chapter also discussed the electrochemical gradient and ion channels.

Gene Hunting—Chapter 19

To find the gene responsible for tasting PTC, biologists compared the distribution of genetic markers observed in "tasters" and "nontasters." The effort narrowed its location down to several candidates. **Chapter 19** detailed how this type of gene hunt is done.

Amoebae and Slime Molds—Chapter 27

Actin and myosin are also responsible for the amoeboid movement observed in amoebae and slime molds and the streaming of cytoplasm observed in algae and land plants.

Pollination—Chapter 40
Certain flowers have patterns that are apparent only in ultraviolet wavelengths that serve as signals for pollinating insects.

Electrical Signals—Chapter 45
Because intercalated discs of cardiac muscle contain numerous gaps, electrical signals pass directly from one cell to the next. **Chapter 45** explored how electrical signals in animals are initiated and propagated.

Looking Forward—
Concepts in Later Chapters
Second Messengers—Chapter 47
The second messenger may induce changes in enzyme activity, gene transcription, or membrane potential. The role of second messengers and hormones is explored in **Chapter 47**.

C. DIFFICULT TOPICS

You may have had difficulty with the action potential in the last chapter; the difficulty you may have in this chapter relates to the action potential in muscle cells. The sliding-filament model and the electrical and mechanical events that occur in muscle contraction and relaxation are quite difficult to absorb in one sitting.

As you prepare for the test, break muscle contraction down into a few discrete sections. First, go over how the action potential gets to the motor neuron and depolarizes the muscle surface. Once you have mastered that part, focus on how the depolarization travels to the actin and myosin filaments and how the release of calcium is involved in initiating muscle contraction. Lastly, go over the molecular action of muscle contraction itself and how the myosin heads ratchet along the actin filaments.

Once you have mastered the process in smaller pieces, try to put it all together and write out all the steps as if you were answering an essay question. This should prepare you for an essay question you may see on your test.

D. ASSESSING WHAT YOU'VE LEARNED

(1) Testing Your Knowledge

1. When sound strikes a sound-receptor cell, the cell depolarizes. Which of the following could explain why this depolarization occurs?
    a. an influx of chloride ions
    b. an influx of positively charged ions
    c. an efflux of potassium ions
    d. an efflux of sodium ions

2. A light-receptor cell becomes hyperpolarized in response to a flash of light. Which of the following statements could explain why this hyperpolarization occurs?
    a. The light promotes an influx of sodium ions into the receptor cell.
    b. The light promotes an efflux of chloride ions.
    c. The light promotes the opening of sodium on channels.
    d. The light promotes the closing of sodium on channels.

3. Cells found in the ear contain small hairs called stereocilia. These hair cells respond to the pressure of a sound wave to either depolarize or hyperpolarize. Which of the following hypotheses could explain the mechanism of this effect?
    a. The pressure waves take away or deliver potassium ions to the cell.
    b. The pressure waves bend the stereocilia, which causes ion channels to open or close.
    c. The pressure waves take away or deliver sodium ions to the cell.
    d. The pressure waves carry electrical signals to the cells.

4. When sound waves enter the ear, they cause a membrane called the tympanic membrane to vibrate. These vibrations are then transmitted across the middle ear to another membrane called the oval window. Why is it beneficial to have the middle ear involved in the delivery of these vibrations to the oval window?

a. The lever action of the sound waves and the smaller surface area of the oval window help to boost sound energy in the ear, so that the ear is more sensitive to sound.

b. The ear ossicles produce electrical signals that help to amplify the sound waves as they are transmitted to the oval window.

c. The middle ear is not important, and our hearing would be the same if the ear canal led directly to the oval window.

d. The middle ear helps to dampen the sound energy to prevent the sound energy from damaging the inner ear.

5. In what part of the human ear are sound waves transmitted in a fluid?
   a. the outer ear
   b. the middle ear
   c. the inner ear

6. Which of the following organisms has an eye that most closely resembles a camera?
   a. flatworm
   b. housefly
   c. spider
   d. field mouse

7. Which of the following lists represents the path of light through the vertebrate eye?
   a. cornea, iris, pupil, retina, lens
   b. cornea, iris, pupil, lens, retina
   c. cornea, iris, lens, retina, pupil
   d. cornea, lens, pupil, iris, retina

8. What role does the iris play in the vertebrate eye?
   a. controls amount of light entering the eye
   b. focuses light on the retina
   c. contains light-sensitive cells that send signals to the brain to process the visual image
   d. is the opening through which light passes into the eye

9. What role does the cornea play in the vertebrate eye?
   a. controls amount of light entering the eye
   b. focuses light on the retina

c. contains light-sensitive cells that send signals to the brain to process the visual image
d. is the opening through which light passes into the eye

10. John Dalton, an 18th-century chemist, worked on the cause of the color blindness he and his brother both experienced. To them, red and green appeared to be the same color. Dalton incorrectly hypothesized that the problem was in the color of the fluid filling his eye. We now know he was missing cones that are important for telling the difference between red and green light. How do we know that he had a normal amount of S opsin cones?
    a. S opsin cones reflect red light, and Dalton could see red objects.
    b. M and L opsin cones absorb blue light, and Dalton could see blue objects.
    c. S opsin cones absorb blue light, and his problem was with the colors red and green.
    d. S opsin cones are not involved in sensing color.

11. In the 1990s, a team led by David Hunt was able to extract DNA from the preserved eyes of 18th-century chemist John Dalton. Dalton and his brother experienced a form of red-green color blindness. The DNA revealed that Dalton lacked the *M* opsin gene. What color would Dalton have the most difficulty discerning?
    a. red
    b. green
    c. blue
    d. gray

12. Taste research has focused on four basic tastes (salty, sour, sweet, and bitter). Which of these tastes is caused by elevating the sodium ion concentration and sensed by a depolarization of taste cells due to an influx of sodium ions?
    a. salty
    b. sour
    c. sweet
    d. bitter

13. Which of the following muscle types is associated with the intestinal tract and arteries?
    a. smooth muscle
    b. skeletal muscle
    c. cardiac muscle

14. Which of the following muscle types is considered voluntary?
    a. smooth muscle
    b. skeletal muscle
    c. cardiac muscle

15. Which of the following muscle types contains multinucleated fibers?
    a. smooth muscle
    b. skeletal muscle
    c. cardiac muscle

16. Which of the following two points will get closer to each other when the muscle contracts?
    a. A and C
    b. B and D
    c. B and C
    d. A and B

17. What component of the sarcomere contains a binding site for ATP?
    a. myosin heads
    b. actin
    c. troponin
    d. tropomyosin

18. Which of the following events produces the "pull" associated with myosin and actin sliding past one another?
    a. the binding of ATP to the myosin head
    b. the hydrolysis of ATP
    c. the release of a phosphate from the myosin head after ATP is hydrolyzed
    d. the myosin head with an empty ATP binding site

19. Research has shown that a contraction occurs in response to a release of calcium ions from the sarcoplasmic reticulum found in muscle fibers. What is the role of these calcium ions in the contraction?
    a. The calcium ions are incorporated into ATP, which fuels the contraction process.

    b. The calcium ions cause the thick filaments to shorten. This shortening of the thick filaments is what produces the contraction.
    c. The calcium ions bind to the thin filaments and pull on them to produce the contraction.
    d. The binding of calcium to troponin causes the tropomyosin to rotate out of the way of the binding sites for myosin on the actin molecules. This allows the myosin to bind to the actin to initiate a contraction.

20. In animal joints, at least two muscles are required to flex and extend at the joint. Which of the following explains why a single muscle cannot act as both a flexor and an extensor?
    a. A single muscle would not be strong enough to produce both actions.
    b. The muscle would have to move to a different bone in order to produce both actions.
    c. A muscle can "pull" at a joint only through a contraction. Because muscles cannot push, a different muscle would be required to move the joint in an opposite direction.
    d. The contraction of a muscle that is acting as a flexor is different from the contraction that produces extension.

(2)   Integrating Your Knowledge

(a) Define a pressure receptor.

(b) How does the ear amplify sound?

(c) Compare and contrast the rods and cones of the eye.

(d) How are rattlesnakes able to hunt prey at night?

(e) Describe and characterize the four basic tastes.

(f) Why does an animal that dies undergo rigor mortis?

(g) List the steps of myosin movement.

CHAPTER 46—ANSWER KEY

D. Assessing What You've Learned

(1) Testing Your Knowledge

1. b; 2. d; 3. b; 4. a; 5. c; 6. d; 7. b; 8. a; 9. b; 10. c; 11. b; 12. a; 13. a; 14. b; 15. b; 16. d; 17. a; 18. c; 19. d; 20. c

(2) Integrating Your Knowledge

(a) In pressure receptors, direct physical pressure on a cell membrane or distortion by bending causes ion channels on a cell membrane to open or close. In response to change in ion flows, the membrane depolarizes or hyperpolarizes, and the result is a new pattern of action potentials from the sensory neuron.

(b) The size difference between the tympanic membrane and oval window is important. The amount of vibration induced by sound waves is increased by a factor of 17. The ossicles act as levers and further amplify the sound such that the overall amplification factor is 22.

(c) Photoreceptors in the vertebrate eye are small rod- or cone-shaped cells called rods and cones. Rods are very sensitive to dim light but not to color. Cones are much less sensitive to faint light but are stimulated by different colors.

(d) Rattlesnakes and other pit vipers can sense infrared light using specialized infrared sensing organs on their snouts. Because endothermic animals radiate heat in the form of infrared radiation, pit vipers can detect and strike at prey in complete darkness.

(e) The four basic tastes are salty, sweet, sour, and bitter. The sensation of saltiness is primarily due to sodium dissolved in food. Sourness is due to the presence of protons that flow directly into certain taste cells through channels. For bitterness, researchers identified a family of 40–80 genes that encode transmembrane receptor proteins. Each protein in this family binds to just one particular type of bitter molecule. A taste cell can have many different receptor proteins from this family. Membrane receptors that respond to sugars have yet to be identified by the same techniques employed in the search for bitterness receptors.

(f) The myosin head can bind actin and catalyze the hydrolysis of ATP into ADP and a phosphate ion. Myosin and actin are locked together when an animal dies, and its muscles enter the stiff state known as rigor mortis. This suggests that ATP must be present for myosin to release from actin.

(g) 1. ATP binds.
2. ATP undergoes hydrolysis.
3. Phosphate is released.
4. ADP is released.

47

Chemical Signals in Animals

A. KEY BIOLOGICAL CONCEPTS

47.1 Cataloging Hormone Structure and Function

- A **hormone** is a chemical signal that circulates through the blood or other bodily fluids to affect target cells. The **endocrine system** is responsible for the production and secretion of hormones.
- Hormones can also act locally through **autocrine effects,** when hormones released from a cell influence that same cell, and **paracrine effects** where hormones affect neighboring cells (**Figure 47.1**).

How Do Researchers Identify a Hormone?

- Three steps are usually taken in identifying and purifying hormones (**Figure 47.2**):
  1. Remove suspected tissue and study it.
  2. Extract a solution from the tissue in intact animals.
  3. Purify the active ingredient in the raw tissue extract, inject the hormone into animals that cannot produce the molecule, and determine if their symptoms are cured.
- Studies on the adrenal glands of dogs identified **cortisol** and **corticosterone** as adrenal hormones that increase glucose concentrations in the blood and increase glycogen formation in the liver. Follow-up studies documented that these compounds, which together are referred to as **glucocorticoids**, are produced in the outer section of the adrenal glands—the **adrenal cortex.**

- The central part of the gland, called the **adrenal medulla,** produced hormones called **epinephrine** and **norepinephrine**. These molecules promote the breakdown of glycogen in the liver.

Chemical Characteristics of Hormones

- The three types of chemical messengers in animals are (**Figure 47.3**):
  1. polypeptides
  2. amino acid derivatives
  3. steroids
- The key **similarity** in these hormone types is that all are present in extremely small concentrations, yet have large effects. For example, 1000 grams of cow pituitary tissue yields just 0.04 grams of growth hormone.
- The key **difference** in these hormone types is that steroids are lipid soluble while polypeptides and amino acid derivatives are not. Steroids thus can cross cell membranes much more readily than do other types of hormones.

The Human Endocrine System— an Overview

- Hormone-secreting organs are called **endocrine glands**. They are diverse in size, shape, and location (**Figure 47.4**).
- Hormones are released into the bloodstream and act on **target cells** that are distant from the source gland. The role of hormones is to coordinate the activities of diverse groups of cells in response to changes in the internal or external environment.

47.2   What Do Hormones Do?

- Hormones coordinate the activities of cells in response to three general situations:
  1. Environmental challenges
  2. Growth, development, and reproduction
  3. Homeostasis

How Do Hormones Coordinate Responses to Environmental Challenges?

- When a person is put into a dangerous situation, hormones regulate both short-term and long-term responses. The short-term reaction, called the **fight-or-flight response**, occurs in conjunction with the activation of the **sympathetic nervous system**.
- If you were being chased by a grizzly bear, action potentials from sympathetic nerves would stimulate the adrenal medulla and lead to the release of epinephrine.
- Glucocorticoids prepare an individual for long-term stress by conserving glucose for use by the brain at the expense of other tissues and organs. They also conserve glucose by actively suppressing wound healing and other aspects of immune system function.

How Do Hormones Direct Developmental Processes?

- *Early embryonic development:* **Primary sex determination** dictates whether the sex organs of an embryo will become male or female. Once testes or ovaries develop, they begin producing male- or female-specific hormones.
- **Testosterone** leads to early development of the male reproductive tract and genitalia and inhibits the formation of breast primordia.
- **Estrogens** induce the initial formation of the female reproductive tract, the genitalia, and the breast tissue.
- *Juvenile-to-adult transition:* In boys, surges of sex hormones lead to changes that include enlargement of penis and testes and growth of facial and body hair.
- In girls, increased concentrations of estrogen called **estradiol** lead to the enlargement of breasts, the onset of menstruation, and other changes.

- *Seasonal or cyclical sexual activity:* Most long-lived animals reproduce seasonally. In many species, environmental cues trigger the release of sex hormones.

How Are Hormones Involved in Homeostasis?

- Messages from sensory receptors travel to the brain for integration, and the messages often travel from integrators to effectors in the form of hormones.
- **Erythropoietin** (EPO) is released from the kidneys and other tissues when blood oxygen levels are low. EPO acts to restore homeostasis by stimulating the production of red blood cells, which increase the oxygen-carrying capacity of the blood (**Figure 47.6**).
- **Calcitonin** and **parathyroid hormone** work in tandem to keep Ca^{2+} levels in the blood close to a set point, much as **insulin** and **glucagon** interact to maintain blood glucose concentrations at preferred levels.

47.3   How Is the Production of Hormones Regulated?

The Hypothalamus and the Pituitary Gland
Controlling the Release of Glucocorticoids

- Philip Smith (1930) showed that rats suffer from various debilitating symptoms when their pituitary gland is removed. In addition, their genitals, thyroid glands, and adrenal cortexes atrophy. Smith proposed the existence of a molecule that affects the adrenal gland.
- In 1943, Choh Hao Li and co-workers purified and characterized **adrenocorticotrophic hormone** (ACTH). They confirmed that glucocorticoids are secreted from the adrenal cortex when ACTH is released from the pituitary. ACTH is therefore called a **regulatory hormone**.
- The presence of cortisone inhibits the release of ACTH. When the presence of a molecule inhibits its production, **feedback inhibition** is occurring (**Figure 47.10**).
- These hormones all act as regulators, and all of them are involved in feedback inhibition. CRF triggers ACTH production, and ACTH triggers glucocorticoid release; but ACTH

also inhibits CRF function through a feedback loop, and the glucocorticoids provide feedback by inhibiting ACTH release.

Hypothalamic-Pituitary Axis—an Overview

- Posterior and anterior sections of the pituitary gland are each influenced by different populations of neurons in the **hypothalamus**. Populations of hypothalamic neurons are called **neurosecretory cells** because they synthesize the release of hormones (**Figure 47.12**).
- Release of hormones by the cells in the hypothalamus is under the control of brain regions responsible for integrating information about the external or internal environment.
- The posterior pituitary is actually an extension of the hypothalamus itself. Neurosecretory cells that project from the hypothalamus produce the hormones **ADH** and **oxytocin**, which are then stored in the posterior pituitary. From there, these hormones are released into the bloodstream.
- The hypothalamus and anterior pituitary are connected indirectly by blood vessels. In response to the arrival of releasing hormones from the hypothalamus, the anterior pituitary secretes hormones that enter the bloodstream and act on target tissues and glands.

Control of Epinephrine by Sympathetic Nerves

- Epinephrine functions not only as a neurotransmitter but also as a hormone. In some cases, the mode of actions of hormones and neurotransmitters is similar.
- By altering synapses during memory functions in the brain, neurotransmitters play a central role in learning and memory. Similarly, many hormones exert their effects by activating particular genes in target cells.

47.4 How Do Hormones Act on Target Cells?

Steroid Hormones and Intracellular Receptors
Identifying the Estrogen Receptor

- **Key Research**: David Toft and Jack Gorski (1964) succeeded in isolating the estradiol receptor in laboratory rats (**Figure 47.13**).

- Follow-up experiments confirmed that the estradiol receptor is located in the nucleus but not associated with the nuclear envelope. It is also found in all target tissues, including the uterus, hypothalamus, and mammary glands.
- Bert O'Malley found that the gene for the estradiol receptor has a sequence and structure similar to the receptors for the glucocorticoids, testosterone, and other steroids.

Documenting Changes in Gene Expression

- Bert O'Malley's laboratory found a distinctive DNA-binding region in the steroid hormone receptor. This region codes for the DNA-binding domain called a **zinc finger**. Zinc fingers are found in all proteins in the steroid hormone receptor family.
- Investigators also confirmed that steroid hormone-receptor complexes bind to specific DNA sites. These sites are called **hormone-response elements**. They function as the type of regulatory DNA sequence called an enhancer (**Figure 47.14**).

Hormones That Bind to Cell-Surface Receptors

- Epinephrine and the peptide hormones are not lipid soluble, so they have to bind to receptors on the cell surface. Because the messenger never enters the target cell, it must activate a receptor on the cell; this is called **signal transduction**.

Identifying the Epinephrine Receptor

- Raymond Ahlquist (1948) found that epinephrine and its **agonists** produce two distinct patterns of responses because two types of receptor exist. He called these receptors **alpha receptors** and **beta receptors**.
- Follow-up work has documented that there are two types of alpha receptor and two types of beta receptor.

Signal Transduction and the Role of Second Messengers

- To understand how epinephrine triggers the release of glucose from target cells, biologists focused on an enzyme called **phosphorylase**. It catalyzes the cleavage of glucose molecules off glycogen. Phosphorylase exists in active and inactive forms and the enzyme (**Figure 47.15a**).

- Earl Sutherland's laboratory found that when they added epinephrine to their cell-free system, large amounts of phosphorylase were activated (**Figure 47.15b**). By purifying components of the cell-free system, they found **cyclic adenosine monophosphate** (cAMP) as the molecule that activates phosphorylase.

- A **second messenger** is a signaling molecule that increases in concentration inside a cell in response to a molecule that binds at the surface.

- cAMP binds to an enzyme called **cAMP-dependent protein kinase**. This enzyme responds by phosphorylating an enzyme, which then phosphorylates phosphorylase. cAMP is produced from ATP in a reaction that is catalyzed by the enzyme **adenylyl cyclase** (**Figure 47.17**).

- The second messenger cAMP plays a primary role in transmitting the signal from the cell surface to the **signal transduction cascade**.

- The primary function of the other events in the sequence is to amplify the signal.

B. CROSS-CUTTING THEMES

Looking Back—
Concepts from Earlier Chapters

Animal Nerves—Chapter 43
Recall from **Chapter 43** that researchers investigating how secretion from the pancreas is regulated had cut all the nerves connected to the upper part of the small intestine and pancreas of a dog. After adding a small amount of dilute hydrochloric acid (HCl) to the small intestine, they found that the pancreas secretes compounds that neutralize the acid in the small intestine.

Lab Techniques—Chapters 17 and 18
As explained in **Chapter 17**, DNA-binding domains are protein regions that make physical contact with DNA. As shown in **Chapter 18**, transcription of a gene begins when a regulatory protein like a steroid-receptor complex binds to an enhancer element for that locus.

Homeostasis—Chapter 41
Chapter 41 introduced the concept of homeostasis, or the maintenance of relatively constant

physical and chemical conditions inside the body.

ADH—Chapter 42
Recall from **Chapter 42** that when an individual is dehydrated, ADH is released from the pituitary gland. ADH increases the permeability of the kidney's distal tubules and collecting ducts to water. As a result, water is reabsorbed from the urine and saved. In this way, ADH is instrumental in achieving homeostasis with respect to water balance. If ADH action is inhibited, homeostasis fails and illness may occur.

Metamorphosis—Chapter 31
Insect metamorphosis is controlled by hormones. If the steroid called juvenile hormone (JH) is present at a high concentration, surges of the steroid hormone ecdysone induce growth of a juvenile insect via molting. But if JH levels are low, ecdysone triggers metamorphosis and the transition to adulthood and sexual maturity.

Electrical Signals—Chapter 45
Recall from **Chapter 45** that neurotransmitters have two major effects on target cells. All neurotransmitters trigger a post-synaptic potential, which can make the postsynaptic neuron more or less likely to deliver an action potential. This chapter also illustrated the effect of neurotransmitters on transcription by reviewing experiments on the sea slug *Aplysia*.

Looking Forward—
Concepts in Later Chapters

Reproduction—Chapter 48
After discovering what is on the forefront of research in chemical signaling, the way has been paved for exploring how hormones regulate reproduction by humans and other animals.

Animal Mating—Chapter 48
Chapter 48 details how during development, a flush of testosterone or estrogen induces the development of seasonal traits like singing in male birds and sexual receptivity in female lizards.

C. DIFFICULT TOPICS

The most complicated system to learn in this chapter is the last item you studied—the second messenger system. As with other physiological cascades, break the system down bit by bit, learn each piece, and then integrate the puzzle. For chemical reactions in the cAMP second messenger system, drawing the reactions in the cascade is of great benefit. It is the best way to answer an essay question. Rather than memorize a lot of key points, draw a diagram; you'll find great amounts of information to extract.

D. ASSESSING WHAT YOU'VE LEARNED

(1) Testing Your Knowledge

1. All of the following accurately describe the functions of hormones *except*:
   a. Hormones are released into the bloodstream to act on distant targets.
   b. Hormones coordinate the activities of groups of cells.
   c. Hormones respond to changes in the internal, but not external, environment.
   d. Hormones act only on cells that possess specific receptors for that hormone.

2. All of these would help in the identification of a suspected hormone *except*:
   a. removal of hormone target tissues
   b. removal of suspected source (endocrine) gland
   c. injection of source gland extract to animal that lacks source gland
   d. purification of source gland extract

3. These statements concerning the chemical classes of hormones are all correct *except*:
   a. Steroid hormones are lipid soluble; polypeptide hormones are water soluble.
   b. Steroid hormones readily cross cell membranes; polypeptide hormones do not.
   c. Some amino acid derivative hormones are lipid soluble; most are water soluble.
   d. Steroid hormones are lipid soluble; polysaccharide hormones are water soluble.

4. Which of the following represents a consequence of the observation that hormones fall into different chemical classes?
   a. Polypeptide hormones bind to receptors located inside target cells.
   b. Steroid hormones bind to receptors located inside target cells.
   c. Steroid hormones bind to receptors located on the surface of target cells.
   d. Most amino acid derivative hormones bind to receptors located inside target cells.

5. Hormones coordinate the activities of cells in response to which of the following?
   a. maintaining homeostasis
   b. growth and development
   c. environmental challenges
   d. all of the above

6. Which of these situations do you predict might lead to epinephrine release by the adrenal glands, but *not* to cortisol release?
   a. witnessing a car accident
   b. preparing for final exams
   c. skipping meals for two days
   d. training for a marathon

7. Which of the following descriptions best illustrates the concept of feedback inhibition within the endocrine system?
   a. An increase in CRH release by the hypothalamus causes a decrease in ACTH release by the anterior pituitary gland.
   b. An increase in CRH release by the hypothalamus causes an increase in ACTH release by the anterior pituitary gland.
   c. An increase in cortisol release by the adrenal glands causes a decrease in CRH release by the hypothalamus.
   d. An increase in cortisol release by the adrenal glands causes a decrease in CRH release by the anterior pituitary gland.

8. Predict the effects of the administration of ACTH on a normal rat.
   a. decrease in cortisol release and an increase in CRH release
   b. increase in cortisol release and a decrease in CRH release
   c. increase in cortisol release and an increase in CRH release
   d. increase in growth hormone release and a decrease in CRH release

9. Predict the effects of the administration of ACTH to a rat that has had its pituitary gland removed.
   a. The adrenal glands will grow in size.
   b. The adrenal glands will shrink (or atrophy).
   c. There will be an increase in CRH levels.
   d. There will be a decrease in cortisol production.

10. All of these effects would be expected to occur if a normal rat were given cortisol injections *except*:
   a. a decrease in size of the adrenal glands
   b. a decrease in ACTH release
   c. a decrease in CRH release
   d. an increase in cortisol release from the adrenal glands

11. Studies by various researchers on the estradiol receptor determined that all of the following are characteristics of the receptor *except*:
   a. The estradiol receptor is found on the surface of target cells.
   b. The estradiol receptor is a protein.
   c. The estradiol receptor is similar in structure to other steroid receptors.
   d. The estradiol receptor was found in uterine tissue.

12. Which of the following best defines a hormone-response element?
   a. A hormone-response element is a receptor with hormone bound to it.
   b. A hormone-response element is a DNA-binding region located on the hormone receptor.
   c. A hormone-response element is a DNA sequence that binds a specific hormone-receptor complex.
   d. A hormone-response element is an RNA transcript.

13. What is the significance of the discovery of different hormone receptor types for the hormone epinephrine?
   a. Two types of epinephrine hormone response elements were identified.
   b. Different tissues responded differently to the presence of epinephrine.

c. It was discovered that slightly different forms of epinephrine (agonists) were secreted by the adrenal glands.
   d. All target tissues for epinephrine possess all four types of epinephrine receptors.

14. Which of these molecules serves as a second messenger for epinephrine actions on glucose production?
   a. phosphorylase
   b. cAMP
   c. protein kinase
   d. adenylyl cyclase

15. Which of these items represents the correct sequence of events following the binding of epinephrine to a liver cell?
   a. increase in cAMP levels; phosphorylation of protein kinase; activation of phosphorylase; decrease in glucose levels
   b. increase in cAMP levels; activation of protein kinase; inhibition of phosphorylase; decrease in glucose levels
   c. increase in cAMP levels; activation of protein kinase; phosphorylation of phosphorylase; increase in glucose levels
   d. increase in cAMP levels; phosphorylation of protein kinase; inhibition of phosphorylase; increase in glucose levels

16. The primary functions of a hormonally regulated second messenger system are:
   a. to relay a signal from the cell surface to the interior and to stimulate expression of certain genes
   b. to relay a signal from the cell surface to the interior and to initiate long-term metabolic effects in that cell
   c. to relay and amplify the signal from the cell surface to the interior
   d. to change levels of cAMP in target cells

17. Which of the following best explains the observation that certain hormones, like epinephrine, can have different effects on different cells?

a. Second messenger molecules, like cAMP, can bind to different proteins within cells.

b. Second messenger molecules, like cAMP, can result in the transcription of different genes.

c. Different receptor types can activate the same signal transduction pathway.

d. Different receptor types can activate different signal transduction pathways.

(2) Integrating Your Knowledge

(a) What three steps are usually taken to identify and purify a hormone?

(b) Of the three structural types of hormones, identify the key similarity and the key difference between them.

(c) Compare and contrast the short-term and the long-term response to stress.

(d) How do hormones affect the maturation of juveniles into adults?

(e) Compare and contrast the anterior and posterior regions of the pituitary gland.

(f) What are hormone-response elements, and what do they do?

CHAPTER 47—ANSWER KEY

D. Assessing What You've Learned

(1) Testing Your Knowledge
1. c; 2. a; 3. d; 4. b; 5. d; 6. a; 7. c; 8. b; 9. a; 10. d; 11. a; 12. c; 13. b; 14. b; 15. c; 16. c; 17. d

(2) Integrating Your Knowledge

(a) The following steps are usually taken in identifying and purifying hormones:
1. Remove suspected tissue and study it.
2. Extract a solution from the tissue in intact animals.
3. Purify the active ingredient in the raw tissue extract, inject the hormone into animals that cannot produce the molecule, and determine if their symptoms are cured.

(b) The key similarity in these hormone types is that all are present in extremely small concentrations, yet have large effects. For example, 1 kilogram of cow pituitary tissue yields 0.04 grams of growth hormone. The key difference in these hormone types is that steroids are lipid soluble, whereas polypeptides and amino acid derivatives are not. As a result, steroids cross cell membranes much more readily than do other types of hormones.

(c) When a person is put into a dangerous situation, hormones regulate the short-term and the long-term response. The short-term reaction, called the fight-or-flight response, occurs in conjunction with activation of the sympathetic nervous system. If you were being chased by a grizzly bear, action potentials from sympathetic nerves would stimulate the adrenal medulla and lead to the release of epinephrine. Glucocorticoids prepare an individual for long-term stress by conserving glucose for use by the brain at the expense of other tissues and organs. They also conserve glucose by actively suppressing wound healing and other aspects of immune system function.

(d) In boys, surges of sex hormones lead to changes that include enlargement of the penis and testes and growth of facial and body hair. In girls, increased concentrations of estrogen called estradiol lead to the enlargement of breasts, the onset of menstruation, and other changes.

(e) The posterior pituitary is actually an extension of the hypothalamus itself. Neurosecretory cells projecting from the hypothalamus produce the hormones ADH and oxytocin, which are then stored in the posterior pituitary. From there, these hormones are released into the bloodstream. The hypothalamus and anterior pituitary are connected indirectly by blood vessels. In response to the arrival of releasing hormones from the hypothalamus, the anterior pituitary secretes hormones that enter the bloodstream and act on target tissues and glands.

(f) Investigators also confirmed that steroid hormone-receptor complexes bind to specific sites in DNA. These sites, which are called hormone-response elements, function as the type of regulatory DNA sequence called an enhancer.

48

Animal Reproduction

A. KEY BIOLOGICAL CONCEPTS

48.1   Asexual and Sexual Reproduction

Mechanisms of Asexual Reproduction

- The sponge generates offspring by **budding**. Budding is completed when a miniature version of the parent breaks free and begins to grow on its own (**Figure 48.1**).
- Some groups, like the bdelloid rotifers, some species of fish, and lizards, reproduce exclusively by parthenogenesis.
- In ants, honeybees, and related insects, males are produced by parthenogenesis, but females in these groups develop from fertilized eggs produced through sexual reproduction.

Switching Reproductive Modes: A Case History

- *Daphnia* reproduce asexually throughout the spring and summer, with the females producing diploid eggs without sex (**Figure 48.2**). The production of offspring via unfertilized eggs is called **parthenogenesis**.
- In late summer or early fall, many females begin producing male offspring parthenogenetically. Sperm from the males fertilize haploid eggs that females produce via meiosis.
- High population densities and food availability are also factors (in addition to day length) in causing this switch.

Mechanisms of Sexual Reproduction: Gametogenesis

- The mitotic cell divisions, meiotic cell divisions, and developmental events that result in

the production of male and female gametes are collectively called **gametogenesis**. This process usually occurs in a sex organ, or **gonad**, in which the male gonads are called **testes** and the female gonads are called **ovaries** (**Figure 48.3**).
- In the male and female gonad, diploid cells called spermatogonia and oogonia divide by mitosis to generate cells that then undergo meiosis.

48.2   Fertilization and Egg Development

- **Fertilization** is the joining of a sperm and an egg to form a diploid zygote. The most basic aspect of diversity in fertilization is where the union of sperm and egg takes place.

External Fertilization

- Most animals that rely on external fertilization live in aquatic environments and tend to produce very large numbers of gametes.
- Gametogenesis occurs in response to environmental cues like lengthening days and warmer water temperatures.

Internal Fertilization and Sperm Competition

- **Internal fertilization** occurs in the vast majority of terrestrial animals as well as in a significant number of aquatic animals.
- Males deposit sperm directly into the female reproductive tract with an organ called the **penis**, or males may package their sperm into a structure called a **spermatophore**, which is

411

then placed into the female's reproductive tract by the male or female.

- Geoff Parker (1970) published results of an experiment that confirmed the existence of **sperm competition**. Whichever male was last to copulate fathered an average of 85% of the offspring produced.
- In species where multiple mating is common, males have extraordinarily large testes for their size and produce proportionately larger numbers of sperm. This was first documented in primates, but it has also been observed in many other animals, including fruit bats.

Oviparity and Viviparity

- In **oviparous** animals, the embryo develops in the external environment. In some animals there is no parental care, but a substantial number of species continue to care for their young after the eggs have emerged from the mother's body.
- In **viviparous** species, development occurs in the mother's body. An embryo attaches to the female's reproductive tract and receives nutrition directly from the mother's circulatory system.
- Miriam Benabib and colleagues studied *Sceloporus* to investigate how oviparity and viviparity evolved (**Figure 48.5**). Egg laying likely represents the ancestral condition, and viviparity evolved from this, probably due to cold or high-altitude habitats.

48.3   Reproductive Structures and Their Functions

The Male Reproductive System
Variation in External Anatomy Is Important in Sperm Competition

- The external anatomy of the male reproductive system consists of a saclike **scrotum** and the **penis**. The scrotum holds the testes; the penis functions as the organ of intromission prior to fertilization.

Internal Anatomy

- Following are the structural components of internal reproductive structures in the human male (**Figure 48.7**):

1. *Spermatogenesis and sperm storage*: The sperm are produced in the **testes** and then stored in the **epididymis**.
2. *Production of additional fluids*: Complex solutions that form in the **seminal vesicles, prostate gland**, and **bulbourethral gland** are added to sperm prior to **ejaculation**. The combination of sperm and these **accessory fluids** is called **semen**.
3. *Transport and delivery*: These functions are the domain of the **vas deferens, urethra**, and **penis**.

The Female Reproductive System

- The external anatomy of the female reproductive system features two folds of skin that cover the **clitoris**, the opening of the **urethra**, and the opening of the **vagina**. The **clitoris** develops from the same population of embryonic cells that give rise to the penis in males. The **vagina** is where semen is deposited during sexual intercourse and where the fetus is delivered during birth (**Figure 48.8**).
- Eggs are produced in the female's paired **ovaries**. When **ovulation** occurs, a mature egg is expelled from the ovary and enters the **oviduct**, where fertilization may take place. Fertilized eggs are transported to the muscular sac called the **uterus**, where embryonic development continues.

48.4   The Role of Sex Hormones in Mammalian Reproduction

Puberty

- Changes that occur during puberty are triggered by increased levels of **testosterone** in boys, and of **estradiol** in girls.
- Researchers isolated a hormone from the hypothalamus, which they called **gonadotropin-releasing hormone (GnRH)**. Investigators also noted that boys and girls who were entering puberty experienced distinctive pulses in the pituitary hormone called **luteinizing hormone (LH)** (**Figure 48.9**).
- R. Stanhope and colleagues administered the hormone GnRH to boys and girls who had deficits in their hypothalamus and were experiencing delays in the onset of puberty.

Female Sex Hormones and the Menstrual Cycle

- Although the length of the **menstrual cycle** varies among women, 28 days is about average. In conjunction with changes in the ovary, the lining of the uterus undergoes a dramatic thickening and regression. **Menstruation**, by definition, is the expulsion of the uterine lining, which designates day 0 in the menstrual cycle.
- **Follicular phase** (14 days)—a **follicle** matures and **ovulation** occurs when the follicle is mature; it releases its **oocyte** into the oviduct (**Figure 48.10**).
- **Luteal phase** (14 days)—the **corpus luteum** forms from the ruptured follicle and subsequently degenerates.

How Do Pituitary and Gonadal Hormones Change during a Menstrual Cycle?

- Dramatic changes in hormone concentrations (including estradiol, LH, FSH, and progesterone) occur during the cycle (**Figure 48.11**).
- LH stays constant except during a spike just prior to ovulation, whereas FSH levels are high during the follicular phase and low during the luteal phase.
- Progesterone is low during the follicular phase but high during the luteal phase.

How Do the Pituitary and Gonadal Hormones Interact?

- Changes in the concentration of estradiol and progesterone affect the release of the pituitary hormones LH and FSH.
- The following steps summarize the interaction of these hormones (**Figure 48.12**):
  1. As the uterus is shedding its lining, a follicle is being stimulated to develop under the influence of FSH.
  2. As the follicle grows, its production of estradiol begins to increase. While estradiol is low, it suppresses LH secretion via negative feedback inhibition.
  3. When the follicle produces large quantities of estradiol, it begins exerting positive feedback on LH, producing a spike in both hormones.
  4. The LH surge triggers ovulation and ends the follicular phase.
  5. As the corpus luteum develops from the ruptured follicle, it begins to secrete progesterone in response to LH.
  6. Increased progesterone converts the thickened uterine lining to an actively secreting tissue with a well-developed blood supply.
  7. If fertilization does *not* occur, corpus luteum degenerates and progesterone levels fall, causing the uterine lining to degenerate. Lower progesterone levels release LH and FSH from inhibitory control; their levels rise, and a new cycle begins.

48.5   Human Pregnancy and Birth

- Once an egg is released from the ovary, it is viable for less than 24 hours. Sperm can survive for 5 days, so sexual intercourse must occur less than 5 days before ovulation for pregnancy to result.
- Although an ejaculate may contain hundreds of millions of sperm, only about 100–300 actually reach the oviduct.

Major Events during Pregnancy

- As smooth-muscle contractions in the oviduct gradually move the zygote toward the uterus, the cell begins to divide by mitosis.
- Once the embryo has become implanted in the uterine lining, its cells begin synthesizing and secreting a hormone called **human chorionic gonadotropin (hCG)**, a chemical messenger that prevents the corpus luteum from degenerating.

The First Trimester

- The mass movement of cells results in formation of the embryonic tissues called ectoderm, endoderm, and mesoderm. By eight weeks, these tissues have differentiated into the various organs and systems of the body (**Figure 48.13a**).
- Another key event in the first trimester is formation of the **placenta**. This structure starts to form on the uterine wall a few weeks after implantation.

The Second and Third Trimesters

- During this time, the remainder of development focuses mainly on growth (**Figures

48.13b and c). The brain and lungs undergo dramatic growth and development.

How Does the Mother Nourish the Fetus?

- The developing embryo is completely dependent on the mother for oxygen; chemical energy in the form of sugars, amino acids, and other raw materials for growth; and waste removal.
- A mother's blood volume expands by as much as 50% during pregnancy, and dramatic increases occur in the stroke volume of the heart (**Figure 48.14**).
- A mother's breathing rate and volume also increase to accommodate the fetus's need for oxygen and its production of carbon dioxide.
- The fetus' blood has a higher affinity for oxygen than the mother's blood does.

Birth

- The rapid growth of the brain during the last trimester makes birth challenging. The pituitary hormone **oxytocin** is clearly important in stimulating smooth-muscle cells in the uterine wall to begin contractions.
- During the last stage of birth, the uterus contracts at a low frequency; then, as labor progresses, the cervix at the base of the uterus begins to dilate (**Figure 48.16**). Once the cervix is fully dilated, uterine contractions become more forceful, longer, and more frequent. Eventually, the fetus is expelled through the cervix and into the vagina.
- After the baby is delivered, the medical staff clamps the umbilical cord connecting the infant to the placenta. Gently tugging the cord, they help the mother deliver the placenta.

B. CROSS-CUTTING THEMES

Looking Back—
Concepts from Earlier Chapters
Fertilization—Chapter 21
Chapter 21 introduced this process by describing how sperm makes contact with an egg and penetrates its membrane. The discussion in that chapter focused on the molecular mechanisms of fertilization using the sea urchin as a model organism. Recall that contact between sperm and egg triggers the acrosomal reaction, in which enzymes released from the head of the sperm chew a path through the material surrounding the egg membrane.

Countercurrent Exchangers—Chapter 44
Maternal and fetal arteries in the placenta are arranged in a countercurrent fashion. As explained in **Chapter 44**, countercurrent flows maintain a concentration gradient that makes diffusion efficient.

Hormones—Chapter 47
Chapter 47 introduced the female sex hormone estradiol and the male sex hormone testosterone. Also recall that human sex hormones play a key role in three events—development of the male and female reproductive tract in embryos, maturation of the reproductive tract during the transition from childhood to adulthood, and regulation of spermatogenesis and oogenesis in adults.

Looking Back—
Concepts from Earlier Chapters
Hormonal Control of Behavior—Chapter 51
As will be evident in **Chapter 51** on behavior, especially the reproductive behaviors of courtship and mating are under strong influence of the endocrine system.

C. DIFFICULT TOPICS

You have studied mitosis and meiosis in previous chapters, so make sure you have reviewed these processes extensively. They were presented as difficult topics in the chapters in which they first appeared. With regard to new material, the menstrual cycle is by far the most complex portion of this chapter.

As with everything else you've studied, begin by learning how each major hormone changes during this cycle. Then study the physical and physiological changes that occur due to hormonal changes. Finally, try to put everything together by learning how all of the hormones interact by positive and negative feedback. Drawing diagrams and explaining them to classmates will help to reinforce these concepts.

D. ASSESSING WHAT YOU'VE LEARNED

(1) Testing Your Knowledge

1. The production of offspring via unfertilized eggs is called:
   a. gametogenesis
   b. parthenogenesis
   c. oogenesis
   d. spermatogenesis

2. Which of the following represents an example of asexual reproduction?
   a. budding
   b. parthenogenesis
   c. regeneration
   d. all of the above

3. All of the following conditions led to a switch from asexual to sexual reproduction in the aquatic crustacean *Daphnia except*:
   a. longer day lengths
   b. low food availability
   c. high population densities
   d. season

4. Many animals, including *Daphnia*, are able to reproduce both sexually and asexually. Which of the following represents the leading hypothesis explaining the advantage associated with the ability to switch to sexual reproduction?
   a. Sexual reproduction is advantageous when energy is readily available.
   b. Sexual reproduction does not require the process of meiosis.
   c. Sexual reproduction produces more genetically diverse offspring capable of surviving stressful environmental conditions.
   d. Sexual reproduction allows for more rapid production of offspring.

5. All of the following represent differences between the processes of spermatogenesis and oogenesis in humans *except*:
   a. Production of oogonia stops before birth; production of spermatogonia occurs throughout adult life.
   b. Sperm are haploid; eggs are diploid.
   c. Oogenesis produces polar bodies; spermatogenesis does not.

   d. Sperm cells are smaller in size than oocytes are.

6. All of the following are general characteristics of external fertilization *except*:
   a. production of large numbers of gametes
   b. use of terrestrial environments
   c. synchronization of release of gametes
   d. production of diploid offspring

7. In animals that have eyes and use external fertilization, synchronization of gamete release is most likely triggered by:
   a. courtship rituals
   b. pheromone release
   c. change in day length
   d. change in water temperature

8. Which of these experimental results support the hypothesis that, for some species, pheromones may coordinate spawning?
   a. Fish maintained in groups, but not individually, exhibited coordinated spawning activity.
   b. Only 10% of sea cucumbers raised in isolation under natural conditions released their gametes.
   c. Observations of sea stars in the wild have demonstrated the coordination of spawning activity.
   d. Male fruit flies mate with a female only when her eggs have achieved the appropriate level of maturity.

9. Studies of sperm competition with fruit flies found all of the following *except*:
   a. Females mate with more than one male prior to fertilization.
   b. Sperm from different males compete with each other to fertilize eggs.
   c. When two males mated with one female, the proportion of offspring fathered by each male was 50:50.
   d. The last male to mate with a female fathered a greater proportion of offspring.

10. Which of these mechanisms likely play a role in conferring "the second male advantage" during internal fertilization in fruit flies? Note: There may be more than one correct choice.

a. The fluid surrounding the sperm is poisonous to the sperm of other males.

b. The second male's sperm physically dislodge the first male's sperm from storage areas in the female reproductive tract.

c. The female reproductive tract produces toxins that eventually harm stored sperm.

d. The most recently deposited sperm are more vigorous in their swimming movements toward the egg.

11. What characteristics are common to animals for which multiple matings are common?
a. They tend to live in aquatic habitats.
b. This characteristic is observed only in insects (like the fruit fly).
c. The females tend to produce large numbers of offspring.
d. The males tend to have large testes.

12. All of the following observations challenge the "lottery model" of sperm competition *except*:
a. In some species, females actively choose the last male with whom they mate.
b. In some species, females will eject the sperm of undesirable males.
c. In some species, females exert control over which sperm are successful at fertilization.
d. In some species, the ejaculates of males are so large that they cannot be successfully retained by the female.

13. Which of the following statements best explains the difference between oviparity and viviparity?
a. Oviparous animals utilize external fertilization, while viviparous animals use internal fertilization.
b. Oviparous animals do not care for their young following fertilization, while viviparous animals do.
c. Oviparous animals are egg-bearing, and embryonic development takes place in the environment; viviparous animals retain their embryos for development

within the mother's body.
d. Oviparous animals receive nutrition directly from the mother's body during development, while the embryos of viviparous animals receive nourishment from yolk sacs.

14. How have studies of the lizard species *Sceloporus* helped scientists to understand how natural selection has favored changes between oviparity and viviparity?
a. Oviparity evolved independently in two different groups of lizards, so it is thought to represent the ancestral condition.
b. Viviparous lizard species appear to have evolved from oviparous species.
c. There seems to be a correlation between cold weather and the evolution of oviparity.
d. Oviparity is favored in high-altitude environments.

15. The seminal vesicles, bulbourethral glands, and prostate gland contribute all of the following components to the semen *except*:
a. an energy source in the form of sucrose
b. prostaglandins to stimulate smooth muscle contractions in the female reproductive tract
c. alkaline compounds to counteract the acidity of the female reproductive tract
d. antibiotic substances

16. Which is the correct route for sperm migration in the human, starting from its storage site and finishing at the site of fertilization? Note: All possible structures may not be included in these answers.
a. epididymis; vas deferens; seminal vesicle; uterus; oviduct
b. epididymis; vas deferens; ejaculatory duct; uterus
c. testes; vas deferens; urethra; uterus
d. epididymis; vas deferens; urethra; uterus; oviduct

17. Puberty in both males and females is triggered by:
a. testosterone in males and estrogen in females
b. LH release from the hypothalamus in

both males and females

c. GnRH release from the hypothalamus in both males and females

d. GnRH release from the anterior pituitary gland in both males and females

18. How does a new menstrual cycle become initiated when fertilization fails to occur?

a. Progesterone levels drop due to regression of the corpus luteum, releasing the hypothalamus and pituitary from feedback inhibition. FSH levels rise.

b. hCG produced by the uterine lining signals the corpus luteum to regress, releasing the hypothalamus and pituitary from feedback inhibition. FSH levels rise.

c. The nervous system detects the absence of implantation, signaling the release of GnRH from the hypothalamus, which triggers the release of FSH from the pituitary.

d. hCG, produced by the embryo following implantation in the uterus, signals the corpus luteum to regress. Progesterone levels drop and FSH levels rise.

19. Which statement is correct concerning how long the human egg and sperm are viable?

a. Both the egg and the sperm are viable for 1 day.

b. Both the egg and the sperm are viable for 5 days.

c. Sperm are viable for 1 day, while eggs are viable for 5 days.

d. Sperm are viable for 5 days, while eggs are viable for 1 day.

20. Which of the following best describes the function of the placenta?

a. The placenta produces the hormone FSH, which is needed for embryonic development.

b. The placenta provides the embryo with a protective cushion.

c. The placenta allows for gas and nutrient exchange between the maternal and fetal circulations.

d. The placenta is one of the three main types of embryonic tissues. Eventually, it develops into the fetal heart.

21. All of the following physiological adaptations occur within the mother during pregnancy *except*:

a. increased heart rate

b. increased blood pressure

c. increased blood volume

d. increased breathing rate

22. All of the following represent adaptations that help to ensure that oxygen transport always occurs from the mother to the fetus (and not in the reverse direction) *except*:

a. Fetal hemoglobin has a higher affinity for O_2 than does maternal hemoglobin.

b. There is a countercurrent arrangement between maternal and fetal arteries in the placenta.

c. The PO_2 of fetal blood is lower than that of maternal blood.

d. For a given PO_2 value, fetal hemoglobin has a lower percentage of O_2 saturation than does maternal hemoglobin.

23. Which of the following are symptoms of FAS (fetal alcohol syndrome)?

a. degeneration of neurons in fetal brains

b. hyperactivity in children born with FAS

c. depression in children born with FAS

d. all of the above

(2)   Integrating Your Knowledge

(a) What is parthenogenesis, and which animals use it?

(b) Compare and contrast external and internal fertilization.

(c) Compare and contrast oviparity and viviparity.

(d) Describe the three main functions of the internal anatomy of the male reproductive system.

(e) What hormonal changes occur during puberty, and what physiological changes do they induce?

(f) Briefly describe the two phases of the menstrual cycle.

CHAPTER 48—ANSWER KEY

D. Assessing What You've Learned

(1) Testing Your Knowledge

1. b; 2. d; 3. a; 4. c; 5. b; 6. b; 7. a; 8. b; 9. c;
10. a and b; 11. d; 12. d; 13. c; 14. b; 15. a;
16. d; 17. c; 18. a; 19. d; 20. c; 21. b; 22. c;
23. d

(2) Integrating Your Knowledge

(a) *Daphnia* reproduce asexually throughout the spring and summer, and the females produce diploid eggs without sex. The production of offspring via unfertilized eggs is called parthenogenesis. In late summer or early fall, many females begin producing male offspring parthenogenetically. Sperm from these males fertilize haploid eggs that females produce via meiosis.

(b) Most animals that rely on external fertilization live in aquatic environments and tend to produce exceptionally large numbers of gametes. Gametogenesis occurs in response to environmental cues like lengthening days and warmer water temperatures. Internal fertilization occurs in the vast majority of terrestrial animals as well as in a significant number of aquatic animals. Males deposit sperm directly into the female reproductive tract with an organ called the penis, or males may package their sperm in a structure called a spermatophore, which is then placed into the female's reproductive tract by the male or female.

(c) In oviparous animals, the embryo develops in the external environment. In some animals there is no parental care, but a substantial number of species continue to care for their young after the eggs have emerged from the mother's body. In viviparous species, development takes place within the mother's body. The embryo attaches to the female's reproductive tract and receives nutrition directly from the mother's circulatory system.

(d) Structural components of the internal reproductive structures in the human male:
   1. *Spermatogenesis and sperm storage*— sperm are produced in the testes and stored in the epididymis.
   2. *Production of additional fluids*—complex solutions that form in the seminal vesicles, prostate gland, and bulbourethral gland are added to sperm prior to ejaculation. The combination of sperm and these accessory fluids is called semen.
   3. *Transport and delivery*—these functions are the domain of the vas deferens, urethra, and penis.

(e) Changes that occur during puberty are triggered by increased levels of testosterone in boys and estradiol in girls. Researchers isolated a hormone from the hypothalamus called gonadotropin-releasing hormone (GnRH). Investigators also noted that boys and girls who were entering puberty experienced distinctive pulses in the pituitary hormone, called luteinizing hormone (LH).

(f) 1. *Follicular phase* (14 days)—a follicle matures, and ovulation occurs when the follicle is mature; it releases its oocyte into the oviduct.
   2. *Luteal phase* (14 days)—the corpus luteum forms from the ruptured follicle and subsequently degenerates.

The Immune System in Animals

49

A. KEY BIOLOGICAL CONCEPTS

49.1   Innate Immunity

- **Innate immunity** refers to immune system cells that are ready to respond to foreign invaders at all times.
- An **antigen** is any foreign molecule. Most antigens are proteins or glycoproteins from bacteria or viruses or other invaders, but foreign carbohydrates and lipids also function as antigens.
- The innate immune system responds the same way to all antigens.

Barriers to Entry

- In humans and other animals, the most important barrier to pathogen entry is the **skin**. In addition to a physical barrier, skin cells present a chemical barrier because they secrete **lactic acid** and **fatty acids** that lower pH of the surface and prevent bacterial growth.

The Innate Immune Response

- Cells involved in the innate response are alerted to the presence of foreign invaders by certain molecules. For example, all bacteria have proteins that begin with an N-formyl-methionyl molecule instead of methionine. When **pattern-recognition receptors** in the innate immune systems detect specific shapes or patterns present in foreign molecules, **leukocytes** (white blood cells) respond (**Figure 49.2**).

- The main steps in the **inflammatory response** are as follows (**Figure 49.3**):
  1. Bacteria and other pathogens enter wound.
  2. At the wound, clotting reduces bleeding.
  3. Mast cells secrete factors that mediate **vasodilation** and **vascular constriction**, thus bringing blood, plasma, and cells to the injured area.
  4. **Neutrophils** secrete factors that kill and degrade pathogens.
  5. Neutrophils and **macrophages** remove pathogens by **phagocytosis**.
  6. Macrophages secrete hormones called **cytokinins** that attract immune system cells to the site and activate cells involved in tissue repair.
  7. Inflammatory response continues until the foreign material is eliminated and the wound is repaired.

49.2   The Acquired Immune Response: Recognition

An Introduction to Lymphocytes and the Immune System

- An **antibody** is a protein that binds to a specific antigen. Binding of an antibody to an antigen leads to destruction of the foreign molecule or cell.
- Cells involved in the acquired immune response are called **lymphocytes**. Lymphocytes form and mature in the primary organs of the immune system: the **bone marrow** and **thymus** (**Figure 49.4**).

- The **lymph nodes** and **spleen** are important sites where lymphocytes encounter antigens. The mixture of fluid and lymphocytes in the lymph nodes and ducts is called **lymph**.

The Discovery of B Cells and T Cells

- Investigations of the immune system's response to *Salmonella typhimurium* show that antibodies are important in neutralizing the antigen and that the **bursa** is critical for antibody production.
- Mice lacking a thymus developed pronounced defects in their immune system. Follow-up experiments showed that lymphocytes from the bursa and thymus have two different functions.
- The bursa-dependent and thymus-dependent lymphocytes became known as **B cells**, which produce antibodies, and **T cells**, which are involved in an array of functions.

Antigen Recognition and Clonal Selection

- The clonal-selection theory consists of three central claims:
  1. Each lymphocyte formed in the bone marrow or thymus has a unique receptor on its surface that recognizes one antigen.
  2. When the receptor on a lymphocyte binds to an antigen, the lymphocyte is activated.
  3. Some of these cloned cells persist after the pathogen is eliminated. As a result, they are able to respond quickly and effectively if the infection recurs in the future.

The Discovery of B-Cell Receptors and T-Cell Receptors

- The **B-cell receptor** (BCR) and antibodies belong to a family of proteins called the gamma globulins and are more specifically called **immunoglobulins**.
- The BCR has three distinct components (**Figure 49.6a**):
  1. The **light chain** of about 25,000 daltons
  2. The **heavy chain** of about 50,000 daltons
  3. A **transmembrane domain** that anchors the protein in the B-cell membrane. Antibodies lack this domain and are secreted from the cell.

- The **T-cell receptor** (TCR) binds to antigens only after they have been processed by other immune system cells and presented on their cell membranes. It is composed of a single **alpha chain** and a single **beta chain**. The shape is similar to the arm of the BCR molecule (**Figure 49.6b**).
- Antibodies and lymphocyte receptors do not bind to the entire antibody molecule, but to a selected region called an **epitope**. An antigen may have many different epitopes, where the binding by antibodies and lymphocytes actually takes place (**Figure 49.7**).

What Is the Molecular Basis of Antigen Specificity?

- Tai Te Wu and Elvin Kabat counted how many of the 77 proteins on the light chain had a different amino acid at each position on the chain. Certain amino acids in the light chain are extremely variable among B cells.
- The presence of specific amino acids at certain positions in the light chain is responsible for the ability of the BCR and antibodies to bind to unique epitopes (**Figure 49.9**).
- The binding occurs at the tips of the antibody and BCR molecule's arms.

The Discovery of Gene Recombination

- Dryer and Bennett (1965) suggested that a single gene codes for the C region of the light chain and that a separate gene codes for the V region. They hypothesized that early in the life of a lymphocyte, sequences from the V gene are inserted into the C gene.
- Hozumi and Tonegawa (1976) reasoned that if light-chain diversity was produced by DNA recombination, then the variable and constant regions of mature B cells should be shorter than in the same region in immature cells.
- Note that two large DNA fragments from the immature cell hybridized with the entire L chain, and that the larger of these two fragments also hybridized with the **constant-region mRNA**.
- Biologists proposed that the smaller fragment contained the variable region, and the larger fragment included the constant region. This was strong evidence that DNA recombination

brings V and C genes closer together as B cells mature (**Figure 49.10**).

- The following steps summarize how the light-chain locus changes (**Figure 49.11**):

  1. Start with immature DNA containing full complement of V, J, and C segments.
  2. One V segment joins with one J segment and the C segment by recombination.
  3. Transcription occurs.
  4. RNA processing occurs [removal of introns, poly (A) tail, etc.].
  5. Translation results in a protein with a unique amino acid sequence. Depending on which V and J segments join during recombination, 629 other amino acid sequences are possible.

How Does the Immune System Distinguish Self from Non-Self?

- Researchers introduced B cells and T cells with anti-self receptors into mice. These experimental lymphocytes were eliminated.
- The great majority of all immature B cells and T cells produced in the bone marrow and thymus are anti-self, and are destroyed before they mature; this process is termed **negative selection**.
- The mechanism for recognition of self is being intensively studied, because it holds the key to **autoimmunity**, where immune system cells turn on the normal cells of the body and destroy them as if they were non-self.

49.3   The Acquired Immune Response: Activation

Antigen Presentation by MHC Proteins: Activating T Cells

- If bacteria migrate rapidly at a wound site, leukocytes called **dendritic cells** are recruited to the area. They take up some of the antigens present and migrate to the nearest lymph node (**Figure 49.12**).
- As CD4$^+$T cells move through the lymph node in search of an antigen, those with complementary receptors interact with their corresponding epitope on dendritic cells.
- Once a T cell receives signal 1 and 2, it begins to divide and produce a series of daughter cells. This process, called **clonal expan-**

sion, is a crucial step in the acquired immune response; it leads to a large population of lymphocytes capable of responding specifically to the antigen that has entered the body.

- If the original T cell is CD4$^+$, the daughters differentiate into one of two types of **helper T cells**: T_H1 and T_H2.
- If the original T cell is CD8$^+$, its daughter cells develop into **cytotoxic T lymphocytes (CTLs)**.
- Helper T cells and CTLs are often referred to as **effector T cells**.

B-Cell Activation and Antibody Secretion

- Receptors on B cells interact directly with the bacterial or viral antigens that are floating free in lymph or blood. Once a free antigen is bound, B cells **internalize** the molecule and process it via the same Class II pathway used by dendritic cells (**Figure 49.13**).
- The second part of the activation process occurs when an activated CD4$^+$ T_H2 cell with a **complementary receptor** arrives. Helper T cell binds to the **receptor-MHC complex** on the B cell.
- The interaction between the B cell and helper T cell supplies an initial activation signal, followed by a co-stimulatory signal 2 analogous to the one that occurs during T cell activation.

Antigen Presentation by Infected Cells: A Signal for Action by CD8$^+$ T Cells

- Once T cells and B cells have been activated, most major elements of the acquired immune response are in place.
- Infected cells respond to the arrival of a virus by processing antigens from the invader. Viral antigens are attached to **MHC Class I proteins** and presented on the surface of the infected cell. This sends a message to the immune system to kill the cell.

49.4   The Acquired Immune Response: Culmination

Killing Bacteria

- During the innate response to bacteria that enter a wound, macrophages at the site **phagocytose** some of the invaders.

- Macrophages also process and present antigens via the **Class II pathway**. As a result, macrophages at the site of infection display epitopes on their surfaces that can be recognized by helper T cells.
- If an activated T_H1 cell binds to these antigen-laden macrophages, two things happen:
  1. The phagocytic activity of the macrophages is enhanced.
  2. The T_H1 cells secrete cytokines that kill bacteria and viruses, recruit additional phagocytic cells to the site, and increase the inflammatory response.

Destroying Viruses

- The immune system has two major ways to eliminate viruses: **cell-mediated response** and **humoral response**.
- In the cell-mediated response, cytotoxic T lymphocytes (CTLs) contain granules that punch holes in cells. CTL makes contact with a virus-infected cell and releases granules, causing the cell to lyse (**Figure 49.14**).
- In the humoral response, antibodies coat free virus particles. The virus envelope cannot fuse with the host-cell membrane. The antibody-coated virus is recognized and phagocytosed by a macrophage.

Why Does the Immune System Reject Foreign Tissues and Organs?

- Organ transplants must be matched to the donor, because the MHC cells of the transplanted tissue will otherwise be recognized as foreign antigens
- Transplanted organs are carefully chosen to match the immune system characteristics of the recipients, who are also given immune suppressing drugs to ensure success of the transplant operation.

Responding to Future Infections: Immunological Memory

- **Memory cells** do not participate in the initial, or primary immune response. Instead, they provide a surveillance service after the original infection has been cleared.
- If the same antigen enters a body a second time, a **secondary acquired immune response** takes place (**Figure 49.16**). This response is faster and more efficient than the primary response.
- The process of **somatic hypermutation** leads to a fine-tuning of the immune response. The B cells that result from somatic hypermutation produce antibodies that bind to the antigen more tightly than their ancestor cells did during the primary response (**Figure 49.15**).
- A **vaccine** contains epitopes from a pathogen, but not the pathogen itself. After vaccination occurs, the body mounts a primary immune response that results in production of memory cells.

B. CROSS-CUTTING THEMES

Looking Back—
Concepts from Earlier Chapters
Viruses—Chapter 34
The only evidence that Dryer and Bennett could offer in support of their gene recombination idea was that viral genes insert themselves into the genomes of host cells.

DNA Recombination—Chapter 17
Children who are born with a genetic defect in the enzyme responsible for DNA recombination in maturing lymphocytes are unable to generate normal T cell and B-cell receptors. As detailed in **Chapter 17**, their acquired immune response is badly impaired as a result.

C. DIFFICULT TOPICS

Because the topics covered in this chapter are very visual and involve multistep processes, a lot of diagram drawing will be helpful in learning the immune responses. The most confusing aspect of the immune system is the different responses, including the differences in B-cell and T-cell activation. Compare-and-contrast questions will appear on your exam, so study similarities as well as differences in these systems. As with every chapter, learn the responses and cell types separately until you are comfortable with them. Then attempt to draw charts, first outlining similarities and then differences.

D. ASSESSING WHAT YOU'VE LEARNED

(1) Testing Your Knowledge

1. The barriers faced by invading pathogens are considered part of the innate immune system. Which of the following statements regarding innate immunity is true?
   a. Components of the innate immune system must first be activated to respond to an antigen.
   b. Components of the innate immune system respond the same way to all antigens.
   c. Components of the innate immune system respond in a specific way to each antigen.
   d. The response of the innate immune system depends on the type of proteins or glycoproteins found on the bacteria, viruses, or other invaders.

2. Which of the following would be associated with innate immunity?
   a. lymphocytes
   b. antibodies
   c. inflammation
   d. B cells

3. An important step in our understanding of acquired immunity came when scientists discovered the receptors on B cells and T cells that bind to antigens. What part of these protein receptor molecules shows the highest variation in amino acid sequence from cell to cell?
   a. the antigen-binding sites
   b. the transmembrane domains
   c. the heavy chain of the B-cell receptor and alpha chain of the T-cell receptor
   d. the light chain of the B-cell receptor and the beta chain of the T-cell receptor

4. In acquired immunity, the epitope is:
   a. the type of bacteria or pathogen that is invading
   b. the molecule produced by B cells for antigen recognition
   c. a molecule involved in promoting inflammation
   d. a selected region of an antigen where antigen recognition occurs

5. In 1976, Hozumi and Tonegawa published data supporting the idea that sequences for constant and variable regions of immunoglobulin genes recombine in mature B cells. They carried out the following restriction enzyme digestion of DNA from immature (embryonic) and mature (B-cell myeloma) cells. Hozumi and Tonegawa showed that two large DNA fragments from the embryonic cell hybridized with the entire light chain, and the larger of these two fragments also hybridized with the C region of the mRNA (see Figure 49.10). What do the results reveal about how the restriction enzymes digested the DNA from the two different cell types?
   a. The restriction enzyme cut the DNA in both cell types in an identical way.
   b. The restriction enzyme cut the DNA in two places in the embryonic DNA and one place in the mature B-cell DNA.
   c. The restriction enzyme cut out a fragment of the gene containing the variable region, but failed to cut out the fragment containing the constant region.
   d. The restriction enzyme cut in three places in the embryonic DNA and two places in the mature B-cell DNA.

6. In 1976, Hozumi and Tonegawa used experimental data to support their conclusion that in mature B cells, the segment of DNA between the variable (V) and constant (C) regions of the immunoglobulin gene is lost as the V and C genes recombine. Why is this result so important for understanding acquired immunity?
   a. It was important to show that the V and C genes get closer together on the chromosome where they are located.
   b. Recombination helps to explain the diversity of immunoglobulin molecules that are part of acquired immunity.
   c. Recombination is not important for understanding acquired immunity.
   d. Recombination is an important step in initiating the inflammatory response.

7. When T cells encounter an antigen, they begin the process of clonal expansion. What is meant by clonal expansion?
   a. The T cell divides rapidly to produce a large population of genetically identical cells that are all capable of recognizing the same antigen.
   b. The T cell divides rapidly to produce a large population of genetically diverse cells that are capable of recognizing a variety of antigens.
   c. The T cell secretes chemicals to destroy the antigen.
   d. The T cells bind to B cells to activate the B cells and promote the production of antibodies that are capable of recognizing the antigen.

8. What process leads to the activation of T cells and triggers the clonal expansion of these cells?
   a. T-cell clonal expansion is triggered when T cells encounter a free-floating antigen in blood or lymph.
   b. T cells are always active and producing clones.
   c. T-cell clonal expansion is triggered when T cells encounter an antigen that is presented by another cell.

9. During HIV infection, the number of cells drops as the virus infects these cells and destroys many of them. What makes HIV selective for this specific type of cell?
   a. HIV contains surface proteins that bind to the CD4 receptor.
   b. HIV contains surface proteins that bind to the antigen-binding portion of the heavy chain of an antibody receptor.
   c. HIV contains surface proteins that bind to the CD8 receptor.

10. Which of the following is associated with the humoral response to a viral infection?
    a. activated cells
    b. cytotoxic T cells
    c. MHC Class I proteins bound to an antigen found at the surface of an antigen-presenting cell
    d. plasma cell production of antibodies

11. Which of the following can explain why the secondary immune response is both stronger and faster than the primary immune response?
    a. Exposure to an antigen is always stronger the second time it is encountered.
    b. Cells remain activated after they are exposed to the antigen in the primary response.
    c. Memory cells were produced during the first exposure to the antigen.
    d. Antibodies produced during the first exposure remain in the blood for decades and can take part in the secondary response.

12. During the secondary immune response, some of the memory B cells that respond migrate to a specialized area in the lymph node, where they undergo rapid mutation. This process is referred to as:
    a. clonal expansion
    b. clonal selection
    c. somatic hypermutation
    d. inflammation

13. What role does somatic hypermutation play in the immune response?
    a. It helps the body to prevent other infections while the immune system responds to a specific antigen.
    b. It helps the immune system become more efficient as antibodies are produced that can bind even more tightly to the antigen during the second exposure.
    c. Somatic hypermutation is a defect in the immune system, which produces an inefficient secondary immune response.

14. The HIV and influenza viruses have been difficult to fight with vaccines. Explain why these viruses are such a problem.
    a. The epitopes on these viruses change frequently as the viruses replicate.
    b. The vaccines would cause serious infections such as HIV infection and the flu.
    c. These viruses do not contain any epitopes on their surface.
    d. These viruses cause infections that have only recently been discovered; an effective vaccine has not yet been found.

15. During an allergic response, the body produces a large number of IgE antibodies. What do these antibodies bind to en route to producing the well-known symptoms of an allergic response?
    a. T_H1 cells
    b. plasma cells and memory B cells
    c. T_H2 cells
    d. mast cells and basophils

16. During the innate response to bacteria that enter a wound, macrophages at the site phagocytize some of the invaders. What are the macrophages doing in this process?
    a. They are engulfing or "eating" the bacteria.
    b. They are producing chemicals that kill the bacteria.
    c. They are producing antibodies that bind to the bacteria.
    d. They are releasing histamine that causes an inflammatory response around the bacteria.

17. The production and maturation of T cells occurs in two different places in the body. Where do T cells become mature?
    a. bone marrow
    b. thyroid
    c. thymus
    d. pancreas

(2) Integrating Your Knowledge

(a) Outline the main steps in the inflammatory response.

(b) What is the bursa, and how does it relate to the immune system?

(c) Outline the steps involved in the light-chain locus changes.

(d) What is clonal expansion?

(e) Compare and contrast the cell-mediated response and the humoral response.

(f) How does a vaccination impart immunity to an individual?

CHAPTER 49—ANSWER KEY
D. Assessing What You've Learned
(1) Testing Your Knowledge

1. b; 2. c; 3. a; 4. d; 5. d; 6. b; 7. a; 8. c; 9. a; 10. d; 11. c; 12. c; 13. b; 14. a; 15. d; 16. a; 17. c

(2) Integrating Your Knowledge

(a) Following are the main steps in the **inflammatory response**:
   1. Bacteria and other pathogens enter a wound.
   2. At wound, clotting reduces bleeding.
   3. Mast cells secrete factors that mediate **vasodilation** and **vascular constriction**, which bring blood, plasma, and cells to the injured area.
   4. **Neutrophils** secrete factors that kill and degrade pathogens.
   5. Neutrophils and **macrophages** remove pathogens by **phagocytosis**.
   6. Macrophages secrete hormones called **cytokinins** that attract immune system cells to the site and activate the cells involved in tissue repair.
   7. Inflammatory response continues until the foreign material is eliminated and the wound is repaired.

(b) Bruce Glick and colleagues investigated the immune system's response to *Salmonella typhimurium*. They found that antibodies are important in neutralizing the antigen and that the bursa is critical for antibody production. The bursa-dependent and thymus-dependent lymphocytes became known as B cells, which produce antibodies, and T cells, which are involved in an array of functions.

(c) The following steps summarize how the light-chain locus changes (see **Figure 49.11**, page 1131):
   1. Start with immature DNA containing a full complement of V, J, and C segments.
   2. One V segment joins with one J segment and the C segment by recombination.
   3. Transcription occurs.

4. RNA processing occurs (removal of introns, poly (A) tail, etc.).

5. Translation results in a protein with a unique amino acid sequence. Depending on which V and J segments join during recombination, 629 other amino acid sequences are possible.

(d) Once a T cell receives signal 1 and 2, it begins to divide and produce a series of daughter cells. This is called clonal expansion—a crucial step in the acquired immune response because it leads to a large population of lymphocytes capable of responding specifically to the antigen that has entered the body.

(e) The immune system has two major ways to eliminate viruses, the **cell-mediated response** and the **humoral response**. In a cell-mediated response, cytotoxic T lymphocytes (CTLs) contain granules that punch holes in cells. CTL makes contact with the virus-infected cell and releases granules, causing the cell to lyse. In the humoral response, antibodies coat free virus particles. The virus envelope cannot fuse with the host-cell membrane. The antibody-coated virus is recognized and phagocytosed by macrophage.

(f) A vaccine contains epitopes from a pathogen, but not the pathogen itself. After vaccination occurs, the body mounts a primary immune response that results in the production of memory cells.

50

An Introduction to Ecology

A. KEY BIOLOGICAL CONCEPTS

50.1   Areas of Ecological Study

Organismal Ecology

- Ecology focuses on how individual organisms interact with their environment. Ecologists explore the morphological, physiological, and behavioral adaptations that allow individuals to live successfully in a particular area.

Population Ecology

- A **population** is a group of individuals of the same species, living in the same area at the same time. Population ecologists focus on how the numbers of individuals in a population change over time.

Community Ecology

- A biological **community** consists of the species that interact with each other within a particular area. Community ecologists ask questions about the nature of the interactions between species and the consequences of those interactions.

Ecosystem Ecology

- An **ecosystem** consists of all of the organisms in a particular region along with nonliving, or **abiotic**, components. Biologists analyze how chemical elements that act as nutrients cycle through ecosystems and how energy flows through the organisms in a region.

How Do Ecology and Conservation Efforts Interact?

- **Conservation biology** is the effort to study, preserve, and restore threatened populations, communities, and ecosystems.

50.2   The Nature of the Environment

Climate

- **Climate** is the prevailing long-term weather conditions in a particular region. **Weather** consists of the specific short-term atmospheric conditions of temperature, moisture, sunlight, and wind.

Why Are the Tropics Warm and the Poles Cold?

- Areas of the world are warm if they receive a large amount of sunlight per unit area; they are cold if they receive a small amount of sunlight per unit area.
- The shape of Earth dictates that regions at or near the equator receive a great deal of sunlight per unit area relative to regions that are closer to the poles.

Why Are the Tropics Wet?

- The **Hadley cell** is responsible for making the Amazon River basin wet and the Sahara dry.
- Air that is heated by the strong sunlight along the equator expands and rises. Warm air can also hold a great deal of moisture. When the air cools as it rises, it can no longer hold moisture and thus, it rains (**Figure 50.3**).

- As old cold air is pushed out of the way, it falls and warms over the desert areas. Thus little rain occurs in these areas.

What Causes Seasonality in Weather?

- **Seasons** are regular, annual fluctuations in temperature, precipitation, or both.
- Seasons are the result of Earth's tilt on its axis. Tilting areas toward the sun causes summer; tilting areas away from the sun causes winter (**Figure 50.4**).

Mountains and Oceans: Regional Effects on Climate

- The most important regional effects on climate are mountains and oceans.
- The presence of mountain ranges tends to produce extremes in precipitation, while the presence of oceans tends to have a moderating influence on temperature (**Figure 50.5**).

50.3   Types of Terrestrial and Aquatic Ecosystems

Terrestrial Biomes

- A **biome** is a type of terrestrial ecosystem that is unique to a given region and characterized by a distinct type of vegetation (**Figure 50.6**).
- Biologists identify climactic regimes in two ways (**Figure 50.7**):
  1. The average annual temperature and precipitation
  2. The variation occurring in temperature and precipitation throughout the year

Tropical Wet Forests

- **Tropical wet forests**, or rain forests, are found in equatorial regions where temperatures and rainfall are high and variation is low. The plants are **aseasonal**, growing all year round (**Figure 50.8**).
- **Productivity** is the total amount of photosynthesis per unit area per year. **Aboveground biomass** is the total mass of living plants excluding roots.
- Equatorial forests are also renowned for their **species diversity**. A single plot measuring 10 m $\propto$ 100 m may contain over 200 tree species, and some researchers contend that the

rain forests house over 30 million arthropod species.

Subtropical Deserts

- Subtropical desert bands exist in areas about 30± of latitude north and south of the equator because of the Hadley cell. Warm air at the equator rises and dumps rain as it cools. The cool air moves poleward and warms as it descends, bathing the desert areas in warm air (**Figure 50.9**).
- This pattern of air movement is responsible for the world's great deserts, including the Sahara, Gobi, Kalahari, and Australian Outback.
- Desert species generally cope with the extreme temperatures and aridity in one of two ways:
  1. Species have evolved morphological and physiological mechanisms that allow them to tolerate the conditions and grow at an extremely low rate throughout the year (e.g., cacti, with reduced surface area to minimize water loss).
  2. Other species have traits that allow them to escape drought (e.g., spadefoot toad).

Temperate Grasslands

- Precipitation is quite low, and the temperature profiles in temperate grasslands are quite seasonal. Temperature variation is important because it dictates a well-defined growing season. These regions usually exist because conditions are too dry to support tree growth but too cold and seasonal for drought-adapted desert species (**Figure 50.10**).
- Although the productivity of temperate grasslands is generally lower than that of forest communities, grassland soils are often highly fertile.

Temperate Forests

- In temperate areas with high precipitation, grasslands give way to forests (e.g., the region around Chicago, Illinois). The abundance of moisture allows trees to dominate the landscape.
- Trees experience a dormant period in the winter; therefore, temperate forests are dominated by deciduous species whose leaves fall

in the autumn and regrow in the spring (**Figure 50.11**).

- Productivity and diversity of temperate forests are moderate.

Boreal Forests

- The **boreal forest**, or **taiga**, stretches across most of Canada, Alaska, Russia, and northern Europe. This climate is characterized by very cold winters, cool, short summers, and extraordinarily high annual variation in temperature. Precipitation is virtually identical to that in the temperate grasslands (**Figure 50.12**).
- The subarctic landscape of boreal forests is dominated by highly cold-tolerant conifers, including spruce, pine, fir, and larch trees. With the exception of larch, these species are evergreen.
- Although aboveground biomass is high, productivity and diversity in boreal forests are low.

Arctic Tundra

- The growing season in the arctic tundra is 6–8 weeks at the most; otherwise, temperature is below freezing. Annual precipitation is also very low.
- The arctic tundra is treeless; however, the soils are saturated year round due to the low evaporation rates.
- Arctic tundra has low species diversity, productivity, and aboveground biomass (**Figure 50.13**).

Aquatic Environments
Lakes, Ponds, and Wetlands

- These biomes are classified as bodies of standing freshwater.
- Lakes and ponds are distinguished by size, with lakes being larger.
- Wetlands are shallow-water habitats where the soil is saturated with water for at least part of the year. They can be distinguished from lakes and ponds in two ways:
  1. They have shallow water.
  2. They have **emergent vegetation** (plants grow above water surface).

- There are three types of wetlands (**Figure 50.17**):
  1. **Marshes** lack trees and have slow, steady water flow.
  2. **Swamps** are like marshes but have trees and shrubs.
  3. **Bogs** develop in depressions where there is no water flow. The water is low in oxygen and very acidic.
- **Streams** are bodies of water that move constantly in one direction (**Figure 50.18**).
- An **estuary** is the environment that forms where rivers meet the ocean (**Figure 50.19**).

Marine Environments

- Water depth itself and distance from shore are used to distinguish the three major types of environments found in the ocean:
  1. The **intertidal zone** (littoral zone) is along the shore. Nutrient levels are relatively high because sunlight can penetrate the waters.
  2. The **neritic zone** extends from the intertidal zone to an ocean depth of about 200 m (the edge of the **continental shelf**).
  3. The **oceanic zone** encompasses the remainder and is the largest of the ocean environments.
- Within all three of the preceding environments, there are three vertical zones (**Figure 50.20**):
  1. **Photic zone**—receives sunlight to support photosynthesis
  2. **Benthic zone**—the area at the bottom
  3. **Aphotic zone**—the area too dark to support photosynthesis

50.4   Biogeography

- **Biogeography** is the study of how organisms are distributed in space.

The Role of History

- **Dispersal** refers to the movement of an individual from the place of birth to the location where it lives and breeds as an adult.
- If an exotic organism is introduced to a new area, spreads rapidly, and eliminates native

species, it is defined as an **invasive species** (**Figure 50.22**).

Biotic Factors

- The distribution of species is often limited by **biotic factors**—interactions with other organisms (**Figure 50.23**).

Abiotic Factors

- The distribution of species is also limited by **abiotic factors**—particularly temperature and moisture (**Figure 50.24**).

B. CROSS-CUTTING THEMES

Looking Back—
Concepts from Earlier Chapters
Temperature—Chapter 3
Temperature is critical because the enzymes that make life possible work at optimal efficiency only in a narrow range of temperatures.

Water Balance—Chapter 42
As outlined in **Chapter 42**, aquatic organisms are surrounded by water and are exquisitely sensitive to changes in ionic composition.

Desert Plant Adaptations—Chapter 10
Desert plant species have adapted in that they have small leaves, a waxy coating on the leaves, and the CAM pathway for photosynthesis, as introduced in **Chapter 10**.

Bacteria and Archaea—Chapter 27
Bacteria and archaea can grow profusely in deep ocean environments, using hydrogen gas, hydrogen sulfide, and other inorganic molecules as a source of energy.

Speciation—Chapter 25
Chapter 25 explored how continental drift has contributed to the current distributions of species; here, the focus is on how interactions with the abiotic and biotic environments affect where a particular species is able to live.

Looking Forward—
Concepts in Later Chapters
Animal Interactions—Chapters 51–53
Chapters 51, 52, and 53 investigate how different species interact with one another and with their physical environment. Here in **Chapter 50**, we have introduced the topics.

C. DIFFICULT TOPICS

The most challenging concept in this chapter, apart from memorizing all the new terms, is the dynamics of the Hadley cell. Remember that the Hadley cell is responsible for maintaining the wet conditions in the Amazon River basin and dry conditions at 30 degrees north and south latitude. The easiest way to learn the dynamics of the Hadley cell is to go over **Figure 50.3** and read the accompanying text. The figure allows to you really understand the movements of cold air and warm air. Drawing the movements without the aid of the figure and explaining the concept to a fellow student are good ways to ensure that you have a grasp of this process.

D. ASSESSING WHAT YOU'VE LEARNED

(1) Testing Your Knowledge

1. The ecology that focuses on how individuals in a group change over time is
   a. organismal ecology
   b. population ecology
   c. community ecology
   d. ecosystem ecology

2. The effort to preserve, study, and restore threatened populations, communities, and ecosystems is called
   a. community ecology
   b. biogeography
   c. conservation biology
   d. climate biology

3. The overall factor dictating whether an area will be generally warm or cold is
   a. the amount and intensity of sunlight
   b. the amount of wind
   c. the amount of precipitation
   d. volcanic activity

4. Seasonality in climate is caused directly by
   a. geomagnetic differences year round
   b. Earth's tilt on its axis
   c. ocean currents
   d. Sun's position relative to other planets

5. The most productive biome is
   a. subtropical desert
   b. temperate forest
   c. boreal forest
   d. tropical wet forest

6. The biome dominated by deciduous trees
   is
   a. subtropical deserts
   b. temperate forests
   c. boreal forests
   d. tropical wet forests

7. The biome with the lowest species
   diversity is
   a. subtropical deserts
   b. temperate forests
   c. boreal forests
   d. tundra

8. The wetlands with the slowest water flow
   and the most acidic water are typically
   a. marshes
   b. bogs
   c. swamps
   d. streams

9. The ocean zone over the continental shelf
   is called
   a. the oceanic zone
   b. the benthic zone
   c. the intertidal zone
   d. the neritic zone

10. If an exotic organism is introduced to a
    new area and becomes successful, it is
    called
    a. an invasive species
    b. a foreign species
    c. a foreign successful species
    d. an alien species

(2) Integrating Your Knowledge

(a) What is the difference between population and community ecology?

(b) What is the difference between climate
   and weather?

(c) Describe the phenomenon of the Hadley
   cell.

(d) How do biologists identify climactic
   regions?

(e) How do desert species cope with extreme
   temperatures?

CHAPTER 50—ANSWER KEY

D. Assessing What You've Learned

(1) Testing Your Knowledge
1. b; 2. c; 3. a; 4. b; 5. d; 6. b; 7. d; 8. b; 9. d

(2) Integrating Your Knowledge

(a) A **population** is a group of individuals of
   the same species that live in the same
   area at the same time. Population ecologists focus on how the numbers of individuals in a population change over time.
   A biological **community** consists of the
   species that interact with each other within a particular area. Community ecologists ask questions about the nature of the
   interactions between species and the consequences of those interactions.

(b) **Climate** refers to the prevailing long-
   term weather conditions in a particular
   region. In contrast, **weather** consists of
   specific short-term atmospheric conditions of temperature, moisture, sunlight,
   and wind.

(c) The **Hadley cell** is responsible for
   making the Amazon River basin wet and
   the Sahara Desert dry. Air that is heated
   by the strong sunlight along the equator
   expands and rises. Warm air can also
   hold a great deal of moisture. When the
   air cools as it rises, it can no longer hold
   moisture, and thus it rains. As old cold air
   is pushed out of the way, it falls and
   warms over the desert areas. As a result,
   these areas receive little rain.

(d) Biologists identify climactic regimes by:
   1. The average annual temperature and
   precipitation
   2. The variation that occurs in temperature and precipitation throughout the
   year.

(e) Desert species generally cope with the
   extreme temperatures and aridity in one
   of two ways:

1. Species have evolved morphological and physiological mechanisms that allow them to tolerate the conditions and grow at an extremely low rate throughout the year (e.g., cacti, with reduced surface area to minimize water loss).
2. Other species have traits that allow them to escape drought (e.g., spadefoot toad).

51

Behavior

A. KEY BIOLOGICAL CONCEPTS

51.1 Types of Behavior—an Overview

- **Behavior** is action—a response to a stimulus.
- **Proximate causes** determine how actions occur.
- **Ultimate causes** determine why actions occur.
- **Learning** is defined as a change in behavior that results in a specific experience in the life of an individual.

Fixed Action Patterns

- These highly stereotypical behavior patterns, called fixed action patterns (FAPs), have three characteristics:
  1. There is almost no variation in how they are performed.
  2. They are species specific.
  3. Once the sequence of actions begins, it typically continues until completion.
- FAPs are examples of **innate behavior**—types of behavior that are inherited and show little variation based on learning (**Figure 51.3**).
- Marla Sokolowski discovered an important behavioral trait controlled by a single gene (**Box 51.1**). She noticed that some fruit-fly larvae tended to move away from food after eating, while others remained in place. By breeding "rovers" and "sitters" that possessed other distinct genetic markers, Sokolowski and colleagues were able to map the gene responsible for the behaviors.

What Is the Adaptive Significance of FAPs?

- Natural selection cannot produce adaptations to unusual or "trick" situations.
- Even though the sign stimuli that release FAPs are extremely simple, they are exceptionally important for the individual's fitness—its ability to survive and produce offspring.

Conditional Strategies

- An example of flexible behavior is female fish that change their sex and become male.
- In some species of coral-reef fish, a group of female fish live inside a boundary dominated by one large male fish. When that male dies, the largest female fish changes sex and becomes the dominant male (**Figure 51.5**).
- **Conditional strategies** are behavioral responses that depend on conditions.

51.2 Learning

Simple Types of Learning: Classical Conditioning and Imprinting

- In **classical conditioning**, individuals are trained by experience to give the same response to more than one stimulus—even a stimulus that has nothing to do with the normal response.
- Upon hatching, ducklings and goslings adopt as their mother the first moving thing they see—this is called **imprinting** (**Figure 51.6**).
- Imprinting occurs only during a **critical** or **sensitive period**.

More Complex Types of Learning: Birdsong

- Depending on the species involved, song-learning behavior falls at various locations along the learning continuum described in **Figure 51.2**.
- For example, song learning is innate in certain species (phoebes), and in other species (such as white-crowned sparrows), it has to be learned during a certain critical period.

Can Animals Think?

- **Cognition** can be defined as the recognition and manipulation of facts about the world.
- New Caledonian crows have the ability to make tools and solve complex problems, which suggests that they can think (**Figure 51.8**)

What Is the Adaptive Significance of Learning?

- Inflexible behavior is adaptive when mistakes would be costly, as when a kangaroo rat jumps to avoid a rattlesnake.
- Innate behavior is advantageous when opportunities to learn are few.
- Learning tends to have an important influence on behavior if individuals have the opportunity to make mistakes without dying and if they have parents or other sources from which to learn.

51.3   How Animals Act: Hormonal Control

- *Anolis carolinensis* live in the woodlands of the southeastern United States. Males emerge in January and establish breeding territories, and females become active a month later. By May, females are laying an egg every 10–14 days and will have produced an amount of eggs equaling twice their body mass (**Figure 51.11**).
- Testosterone and estradiol produce these behaviors in males and females, respectively. If production of these hormones is the proximate cause, what environmental cues trigger these breeding events?

- Researchers brought a large group of inactive lizards into the laboratory and divided them into five groups (**Figure 51.12**):
  1. Single females isolated
  2. Females placed in female-only groups
  3. Females paired with males
  4. Females placed with groups of castrated males
  5. Females placed with groups of uncastrated males
- Females exposed to breeding males began producing eggs much earlier than the females placed in the other treatment groups. Females need to experience springlike light and temperatures and exposure to breeding males.
- To test this idea, researchers repeated the previous experiment, but added a twist. They placed females with males that had intact dewlaps or with males whose dewlaps had been surgically removed. The result? Females grouped with dewlap-less males were slow to produce eggs—just as slow, in fact, as the females in the first experiment that had been grouped with castrated males. These females had not been courted at all. The result suggests that the dewlap is a key visual signal.

51.4   Communication

Modes of Communication

- **Communication** is defined as any process in which a signal from one individual modifies the behavior of a recipient individual.
- A **signal** is any information-containing behavior.
- Communication can be acoustic, visual, olfactory, or tactile.

A Case History: The Honeybee Dance

- Karl von Frisch suspected that successful food finders communicate the location of food to other individuals. He observed bees displaying a "round dance" that contained information about the location of food. Workers got the information about the location of food through the dancer's movements.
- He also observed a "waggle run" that tells the length of distance the workers must fly, as well as the direction (**Figure 51.14**).

Honesty and Deceit in Communication
Deceiving Individuals of Another Species

- The anglerfish has an appendage that looks remarkably like a minnow and dangles near its mouth (**Figure 51.16**).
- Predatory *Photinus* fireflies can mimic the pattern of flashes by females of several other species.
- Individuals increase their fitness by providing inaccurate or misleading information to members of a different species.

Deceiving Individuals of the Same Species

- In some cases, natural selection also favors the evolution of lying to the same species.
- A male bluegill will imitate a female and get close to another male while he is courting an actual female. Although the male bluegill thinks he is courting two females, while he does so, the female mimic secretly fertilizes the real female's eggs but does not have to care for them (**Figure 51.17**).
- Lying works only when it is relatively rare. If deceit becomes extremely common, natural selection will strongly favor individuals that can detect and avoid or punish liars.

51.5   Orientation, Navigation, and Migration

- A movement that results in a change in position is called **orientation**.
- The simplest type of orientation is called **taxis**, which involves positioning the body toward or away from a stimulus.
- **Phototaxis** is orientation toward light; **phonotaxis** is orientation toward sound.

Migration: Why Do Animals Move with a Change of Seasons?

- **Migration** is defined as the long-distance movement of a population associated with a change of seasons.
- Arctic terns nest along the Atlantic coast of North America, fly south along the coast of Africa to wintering grounds off Antarctica, and then fly back along the eastern coast of South America.
- Many of the monarch butterflies that are native to North America spend the winter in the mountains of central Mexico or southwest California.
- Salmon that hatch in rivers along the Pacific Coast of North America and northern Asia migrate to the ocean when they are a few months to several years old. They return to the stream where they hatched to mate and then die.

Navigation: How Do Animals Find Their Way?
Piloting

- **Piloting**, or using familiar landmarks, is used by many species to find their way. Young offspring follow their parents during migrations and appear to memorize the route (**Figure 51.19**).
- Homing pigeons can always find their way back home; but if their eyes are covered, they fail to navigate only the final stage, which suggests that they may use piloting for the final part of a journey.

Compass Navigation

- To determine where north is, animals appear to use the Sun during the day and the stars at night.
- During cloudy conditions, birds appear to use Earth's magnetic field to orient.

51.6   The Evolution of Self-Sacrificing Behavior

- **Altruism** is an act that has cost to the actor, affecting his or her ability to survive and reproduce, and a benefit to the recipient of the altruistic act.
- The **coefficient of relatedness** is a measure of how closely the actor and beneficiary are related. Hamilton's rule states that if the benefits of altruistic behavior are high, if the benefits are dispersed to close relatives, and if the costs are low, the alleles associated with altruistic behavior will be favored by natural selection and will be spread throughout the population.
- **Kin selection** is natural selection that acts through benefits to relatives (**Figure 51.21**).
- Individuals can pass their alleles on to the next generation not only by having their own

offspring but also by helping close relatives to produce more offspring. This is called Hamilton's rule.

- **Reciprocal altruism** is an exchange of fitness benefits that are separated in time.

B. CROSS-CUTTING THEMES

Looking Back—
Concepts from Earlier Chapters
Behavioral Traits—Chapter 13
To use the term introduced in **Chapter 13**, most behavioral traits are polygenic—exhibiting a quantitative variation, with individuals differing by degree and many different phenotypes observed when the population is considered as a whole.

The *dg2* Gene—Chapters 8, 19, 20, and 47
Recall from **Chapter 20** that the entire fruit-fly genome has been sequenced. **Chapter 19** introduced the techniques used by Sokolowski to map the *for* gene. The protein product of *dg2* is involved in the types of signal cascades introduced in **Chapters 8 and 47**—meaning it is involved in the response to an as yet undetermined signal.

Flexible Behavior—Chapter 45
Chapter 45 introduced how learning works at the proximate level and concluded that learning occurs because individual neurons and connections between neurons are modified in response to experience.

Looking Forward—
Concepts in Later Chapters
Population Dynamics—Chapter 52
Behavior is at the heart of reproduction; consequently, the size and stability of natural populations—the subject of **Chapter 52**—depends on many of the reproductive behaviors described in the current chapter.

C. DIFFICULT TOPICS

The study of animal behavior seems to be a more tangible subject in biology because we can directly observe data with our own eyes as it is occurring. However, there are still a few topics that require extra time and thought. The concepts of **altruism** and **kin selection** are relatively straightforward to define, but even professors have a difficult time applying these terms to the observations they record in nature. It will be wise to locate a current ecology text and learn how to apply altruism or kin selection to examples given in the text. An ecology text will give you much more detail and reinforce the knowledge that you have already acquired.

D. ASSESSING WHAT YOU'VE LEARNED

(1) Testing Your Knowledge

1. A change in behavior that results in a specific experience in the life of an individual is termed
   a. behavior
   b. proximate cause
   c. ultimate cause
   d. learning

2. An innate behavior is one that
   a. is not inherited
   b. shows some variation
   c. is inherited and shows little variation
   d. is inherited and shows wide variation

3. Individuals are trained by experience to give the same response to more than one stimulus. This is termed
   a. classical conditioning
   b. imprinting
   c. proximate conditioning
   d. sensitization

4. The recognition and manipulation of facts about the world is termed
   a. imprinting
   b. learning
   c. cognition
   d. memory

5. Inflexible behavior is adaptive when
   a. mistakes would be costly.
   b. there are many safe opportunities to learn.
   c. mistakes can be made without dying.
   d. parents can teach their young.

6. The key signal prompting a hormonal response and breeding in female *Anolis* lizards is
   a. light
   b. male dewlap display
   c. female dewlap display
   d. male hormones

7. Any process in which the signal from one individual modifies the behavior of a recipient individual is called
   a. a connecting signal
   b. communication
   c. imprinting
   d. learning

8. Natural selection favoring the deceit of the same species is best described by which of the following examples?
   a. anglerfish
   b. fireflies
   c. monkeys
   d. bluegills

9. A movement that results in a change in position is called
   a. orientation
   b. taxis
   c. phototaxis
   d. instant taxis

10. An exchange of fitness benefits that are separated in time is termed
    a. kin selection
    b. altruism
    c. reciprocal altruism
    d. coefficient of relatedness

(2) Integrating Your Knowledge

(a) What are the three characteristics of a fixed action pattern (FAP)?

(b) Describe what Karl von Frisch found about honeybee dances.

(c) What is the difference between classical conditioning and imprinting?

(d) What is migration? Provide a few examples.

(e) Define innate behavior and give an example.

CHAPTER 51—ANSWER KEY

D. Assessing What You've Learned

(1) Testing Your Knowledge

1. d; 2. c; 3. a; 4. c; 5. a; 6. b; 7. b; 8. d; 9. a; 10. c

(2) Integrating Your Knowledge

(a) Fixed action patterns are highly stereotypical behavior patterns that have three characteristics:
   1. There is almost no variation in how they are performed.
   2. They are species specific.
   3. Once the sequence of actions begins, it typically continues until completion.

(b) Von Frisch observed honeybees displaying a "round dance" that contained information about the location of food. The dancers' movements give the workers information about where food is located. He also observed a "waggle run" that tells the workers not only the distance but also the direction they must fly to the food.

(c) In **classical conditioning**, individuals are trained by experience to give the same response to more than one stimulus—even a stimulus that has nothing to do with the normal response. Upon hatching, ducklings and goslings adopt as their mother the first moving thing they see—this is called **imprinting**.

   Imprinting occurs only during a **critical** or **sensitive period**.

(d) **Migration** is defined as the long-distance movement of a population associated with a change of seasons. Arctic terns nest along the Atlantic coast of North America, fly south along the coast of Africa to wintering grounds off Antarctica, and then fly back along the eastern coast of South America. Many of the monarch butterflies that are native to North America spend the winter in the mountains of central Mexico or southwest California. Salmon that hatch in rivers along the Pacific Coast of North

America and northern Asia migrate to the ocean when they are a few months to several years old. They return to the stream where they hatched to mate and then die.

(e) FAPs are an example of **innate behavior**—types of behavior that are inherited and show little variation based on learning. Marla Sokolowski discovered an important behavioral trait controlled by a single gene. She noticed that some fruit-fly larvae tended to move away from food after eating, while others remained in place. By breeding "rovers" and "sitters" that possessed other distinct genetic markers, Sokolowski and colleagues were able to map the gene responsible for the behaviors.

52

Population Ecology

A. KEY BIOLOGICAL CONCEPTS

52.1 Demography

- A **population** is a group of individuals from the same species that live in the same area at the same time.
- **Population ecology** is the study of how and why the number of individuals in a population changes over time.
- **Demography** is the study of factors that determine the size and structure of populations through time. Analyzing birth rates, death rates, immigration rates, and emigration rates is fundamental to demography.

Life Tables

- Formal demographic analyses of populations are based on a type of data set called a **life table**, which summarizes the probability that an individual will survive and reproduce in any given year over the course of its lifetime.

Survivorship

- **Survivorship** is the proportion of offspring produced that survive, on average, to a particular age (l_x, where x represents the age class being considered). It is calculated by dividing the number of individuals in that age class by the number of individuals in the first age class.

    $$l_x = N_x / N_0 \qquad \textbf{(Box 52.1, Eq. 52.1)}$$

- Type I survivorship curve—pattern in which survivorship throughout life is high.
- Type II curve—pattern in which mortality is constant throughout life.

- Type III curve—pattern in which there is a high death rate early in life (**Figure 52.2**).

Fecundity

- **Fecundity** is the number of female offspring produced by each female in the population.

    $$R_0 = \sum l_x m_x \qquad \textbf{(Box 52.1, Eq. 52.2)}$$

- Lifetime reproduction is a function of fecundity at each age (m_x) and survivorship to each age class (l_x). If R_0 is greater than 1, the population is increasing. If R_0 is less than 1, the population is declining.

The Role of Life History

- An organism's **life history** consists of how an organism allocates resources to growth, reproduction, and activities related to survival (**Figure 52.4**).

52.2 Population Growth

Exponential Growth

- **Exponential growth** occurs when r does not change over time. It does **not** depend on the number of individuals in the population. Thus it is **density independent** (**Figure 52.5**).
- In reality, it is not possible for growth to continue indefinitely. Eventually, the habitat would not have enough resources for the number of individuals present. When a population stabilizes at the maximum number that can be supported by the resources available, that population has reached the habitat's **carrying capacity**.

Logistic Growth

Discrete Growth

- Biologists use N to symbolize population size. N_0 is the population size at time zero; N_1 is the population size one breeding interval later.
- $N_1/N_0 = \mu$ (**Box 52.2, Eq. 52.3**)
  where μ is the **finite rate of increase**.
- $N_t = N_0\mu^t$ (**Box 52.2, Eq. 52.5**)
  This equation summarizes how populations grow when breeding takes place seasonally.

Continuous Growth

- A population's per capita increase is symbolized r and is defined as the per capita birth rate minus the per capita death rate.
- $\mu = e^r$ (**Box 52.2, Eq. 52.6**)
  where e is the natural logarithm, or about 2.72.
- $N_t = N_0e^{rt}$ (**Box 52.2, Eq. 52.7**)
  This equation summarizes how populations grow when they breed continuously. Because r represents the growth rate at any given time and because r and μ are so closely related, biologists routinely calculate r, even for species that breed seasonally.

What Limits Growth Rates and Population Sizes?

- **Density-independent factors** change birth rates and death rates irrespective of the number of individuals in a population.
- **Density-dependent factors** are usually biotic in nature and change in intensity as a function of population size.
- Researchers studied the bridled gobie, a coral reef fish (**Figure 52.7**). They stocked artificial reefs with varying densities of adult gobies and after 2.5 months, they captured all the gobies and computed the growth rate of individuals, the survival rate, and the immigration rate.
- Adult gobies survive better when population density is **low**. More juvenile gobies immigrate successfully when population density is low. Higher rates of predation and disease might occur in dense populations.

52.3   Population Dynamics

Population Cycles: The Case of the Red Grouse

- **Population cycles** are regular fluctuations in size that some populations exhibit.
- Researchers hypothesized that the roundworm is responsible for the population fluctuations seen in red grouse. Control populations showed a dramatic four-year cycle, whereas populations treated with a drug that kills roundworms maintained high population densities.

Age Structure

- A population's **age structure**—meaning the proportion of individuals that are at each possible age—has a dramatic influence on the population's growth over time.

Age Structure in a Woodland Herb

- The common primrose has a complex population dynamic:
  1. Populations dominated by juveniles experience rapid growth and then are limited by the shading of larger trees.
  2. The long-term trajectory of the primrose population seems dependent on the frequency and severity of windstorms (**Figure 52.10**).
  3. This species seems to live in small, isolated populations of populations—or **metapopulations**.

Age Structure in Human Populations

- In nations where industrial and technological development is advanced and average incomes are relatively high, the age distribution of the population tends to be even (**Figure 52.11a**).
- The predicted population structure in 2050 highlights a major policy concern in developed countries: how to care for an increasingly aged population.
- In contrast, the age distribution is "bottom heavy" in the less-developed countries (**Figure 52.11b**).
- The projected age distribution in 2050 illustrates the major public policy concerns in

these countries: providing education and jobs for an enormous influx of young people who will want to be starting families.

Analyzing Change in the Growth Rate of Human Populations

- The growth rate for humans has increased over time since about 1750, leading to a very steeply rising curve over the past few centuries. The highest values occurred between 1965 and 1970, when population growth averaged 2.04 percent per year (**Figure 52.12**).
- The population growth has slowed over the past few years; it is currently 1.2 percent per year. Humans are ending a period of rapid growth that lasted well over 500 years. How quickly growth rates decline, and the maximum population size ultimately reached, will be decided by changes in fertility rates and the course of the AIDS epidemic (**Figure 52.13**).
- Extrapolating the world's population to the year 2050 is based on three different fertility rates:
  1. High fertility = nearly 11 billion
  2. Medium fertility = about 9 billion
  3. Low fertility = 7.4 billion

52.4   How Can Population Ecology Help Endangered Species?

Preserving Metapopulations

- Over time, each population within the larger metapopulation is likely to be wiped out. Migration from nearby populations can reestablish populations in these empty habitats, thereby stabilizing the overall population.
- Ilkka Hanski and colleagues have shown that metapopulations exist in nature (**Figure 52.14**). They performed a mark-recapture study on the Glanville fritillary butterfly. They found a migration rate of 9 percent, which was high enough to recolonize extinct populations. They also confirmed that some populations had gone extinct while others had been created.
- A small, isolated population—even one within a nature reserve—is unlikely to survive

over the long term. Based on this realization, conservation biologists are attempting to design reserves for threatened species that are sizeable enough to maintain large populations. When this is not possible, an alternative is to establish systems of small reserves that are connected by corridors of habitat, so that migration between patches is possible. Hanski also emphasizes that it is crucial to preserve at least some patches of currently unoccupied habitat as well, to provide future homes for immigrants in the metapopulation.

Using Life-Table Data to Make Population Projections

- Population projections allow biologists to alter values for survivorship and fecundity at particular ages and assess the consequences (**Figure 52.15**).
- Studies performed on this aspect support some general conclusions:
  1. Whooping cranes, sea turtles, spotted owls, and many other endangered species have high juvenile mortality, low adult mortality, and low fecundity.
  2. In humans and species with high survivorship in most age classes, rates of population growth are extremely sensitive to changes in age-specific fecundity.

Population Viability Analysis

- **Population viability analysis (PVA)** is a model that estimates the likelihood that a population will avoid extinction for a given time period. PVAs attempt to combine basic demographic models for the species in question with data on geographic structure and the rate and severity of habitat disturbance (**Figure 52.16b**).
- Researchers constructed PVAs for an endangered marsupial, Leadbeater's possum (**Figure 52.16a**). Their analysis allowed them to assess the impact of migration and to simulate the affects of logging, fires, storms, and other types of disturbances.

B. CROSS-CUTTING THEMES

Looking Back—
Concepts from Earlier Chapters
Animal Interaction—Chapter 51
Chapter 51 explored how organisms respond to environmental stimuli and how they interact with the environment; however, here in Chapter 52, our focus is on population size.

Looking Forward—
Concepts in Later Chapters
Animal Interaction—
Chapters 53 through 55
Chapters 53 through **55** investigate how different species interact with one another and with their physical environment; however, here in Chapter 52, our focus is on population size.

C. DIFFICULT TOPICS

The difficult aspect of this chapter is undoubtedly the mathematical equations and quantitative nature of this chapter. For those of us who have trouble with math and quantitative analysis, this chapter will present challenges.

The amount of time you will spend on working through equations and solving problems will ultimately depend on how much your instructor focuses on these issues.

If you are required to solve equations and word problems associated with population size, mortality, and fecundity, then you will need to invest a significant amount of time in learning these concepts and solving problems.

As with any quantitative subject, you need to learn the concepts and the origins of the equations before you can solve word problems off the top of your head. Study qualitatively, learning all the definitions of terms and the meanings of the equations. Then start plugging in numbers and solving equations. Do as many problems as you can; and then start making up your own problems for your classmates. Ask them to do the same. In problem solving, practice makes perfect!

D. ASSESSING WHAT YOU'VE LEARNED

(1) Testing Your Knowledge

1. The study of how and why the number of individuals in a group of animals changes over time is called
    a. demography
    b. population ecology
    c. population
    d. fecundity

2. The proportion of offspring produced that survive, on average, to a particular age is called
    a. survivorship
    b. life table
    c. demography
    d. fecundity

3. An organism's life history:
    a. is the number of female offspring produced by each female
    b. is the probability an organism will survive and reproduce
    c. consists of how an organism allocates resources to growth, reproduction, and activities related to survival
    d. consists of how an organism interacts with other organisms

4. Factors that are usually biotic in nature and change in intensity as a function of population size are termed
    a. density-dependent factors
    b. density-independent factors
    c. density-neutral factors
    d. exponential growth factors

5. Some populations exhibit regular fluctuations in size called
    a. age structure
    b. population crashes
    c. population tides
    d. population cycles

6. In nations where industrial and technological development is advanced:
    a. Age distribution is top heavy.
    b. Age distribution is bottom heavy.
    c. Age distribution is even.
    d. Age distribution is bimodal.

7. Adult gobies survive better when population density is
   a. low
   b. high
   c. intermediate
   d. dynamic

8. A huge population of animals consisting of smaller isolated populations is termed a
   a. megapopulation
   b. metapopulation
   c. metropolis
   d. cohort

9. The model that estimates the likelihood that a population will avoid extinction for a given time period is termed
   a. mortality
   b. age structure
   c. population viability analysis
   d. population fecundity

10. A small, isolated population
    a. will do extremely well if protected
    b. is unlikely to survive, even within a reserve
    c. will grow, but extremely slowly
    d. will explode given the right conditions

(2) Integrating Your Knowledge

(a) Describe the characteristics of exponential growth.

(b) When did the highest rates of human growth occur in developed countries?

(c) What is a metapopulation?

(d) Define the terms *demography* and *life table*.

(e) What is fecundity, and how does it relate to population growth? Include the equation in your answer.

CHAPTER 52—ANSWER KEY

D. Assessing What You've Learned

(1) Testing Your Knowledge

1. b; 2. a; 3. c; 4. a; 5. d; 6. c; 7. a; 8. b; 9. c; 10. b

(2) Integrating Your Knowledge

(a) **Exponential growth** occurs when r does not change over time. It does **not** depend on the number of individuals in the population. Therefore, it is **density independent**.

(b) The growth rate for humans has increased over time since about 1400, leading to a very steeply rising curve over the past few centuries. The highest values occurred between 1965 and 1970, when popu-lation growth averaged 2.04 percent per year.

(c) A **metapopulation** is made up of many small populations isolated in fragments of habitat. Over time, each population within the larger metapopulation is likely to be wiped out. Migration from nearby populations can reestablish populations in these empty habitats, thereby stabilizing the overall population.

(d) **Demography** is the study of factors that determine the size and structure of populations through time. Formal demographic analyses of populations are based on a type of data set called a **life table**, which summarizes the probability that an individual will survive and reproduce in any given year over the course of its lifetime.

(e) **Fecundity** is the number of female offspring produced by each female in the population.

$$R_0 = \sum l_x m_x$$

Lifetime reproduction is a function of fecundity at each age (m_x) and survivorship to each age class (l_x). If R_0 is greater than 1, the population is increasing. If R_0 is less than 1, the population is declining.

53

Community Ecology

A. KEY BIOLOGICAL CONCEPTS

53.1 Species Interactions

Competition

- **Competition** is a –/– interaction that occurs when individuals use the same resources and when those resources are limited (**Figure 53.2**).

The Consequences of Competition—Theory

- The principle of **competitive exclusion** states that it is not possible for species within the same niche to coexist. However, if the affected niches do not overlap completely, then the weaker species should be able to retreat into an area of non-overlap, a mechanism called **niche differentiation** (**Figure 53.3**).
- The **fundamental niche** is the combination of conditions that the species will occupy in the absence of competitors, and its **realized niche** is that portion of resources used when competition occurs.

The Consequences of Competition—Experimental Studies

- Joseph Connell noticed two species of barnacles with interesting distributions on an intertidal rocky shore of Scotland. This was one of the first experimental examples of how competition can modify a distribution of a species. In this case, the poorer competitor (*Chthamalus*) is restricted to the upper intertidal zone, which is a more severe habitat (**Figure 53.6**).

Consumption

- Consumption occurs when one organism eats another:
  1. **Herbivory** takes place when herbivores consume plant tissues.
  2. **Parasitism** occurs when a parasite consumes small amounts of tissues from another organism, or host.
  3. **Predation** occurs when a predator kills and consumes most or all of another individual.

How Do Prey Defend Themselves?

- Some species defend themselves by running or hiding in response to predators; others sequester or spray toxins or employ weaponry for defense (**Figure 53.7a**).
- **Mimicry** takes place when one species closely resembles another species (**Figure 53.7b**).
- **Mullerian mimicry** occurs when harmful species resemble each other. For example, brightly colored wasps resemble each other; to defend themselves, wasp predators tend to avoid these colors.
- **Batesian mimicry** occurs when a harmless species resembles a harmful one.
- Researchers hypothesized that if blue mussels employ **inducible defenses** against crabs, they should occur in the low-flow area where predation pressure is higher. The researchers found that mussels in the high predation area had thicker shells (**Figure 53.9**).

Are Animal Predators Efficient Enough to Reduce Prey Populations?

- Species interactions have a strong impact on the evolution of predator and prey populations.

- A wolf control program in Alaska during the 1970s decreased predator abundance to 55–80% below pre-control density. Concurrently, the population of moose tripled. This observation suggests that predators reduced this moose population far below the maximum number that the habitat can support.

- In the arctic, tests have shown that hare population is limited by the availability of food as well as by predation, and that food availability and predation intensity interact—the combined effect of food and predation is much larger than their impact in isolation (**Figures 53.10 and 53.11**).

Why Don't Herbivores Eat Everything—Why Is the World Green?

- Biologists routinely consider three possible answers to the question of why the world is green:

1. The **top-down control** hypothesis states that herbivore populations are limited by predation or disease. For example, after a dozen pairs of European rabbits were introduced to Australia in 1859, the population exploded. Similar consequences have occurred when goats and pigs have escaped from captivity and established large, wild populations in areas that lack their natural predators or diseases.

2. In the **poor-nutrition hypothesis**, the quality of the resources available at the base of the food chain limits herbivore density. More specifically, the growth and reproduction of herbivores could be limited by the availability of nitrogen. A recent meta-analysis examined 185 studies of how insect herbivores had responded to plant fertilization. In over half of the cases, herbivores showed a significant increase in growth rate or reproduction when the plants that they feed on were fertilized. Based on this result, the researchers concluded that nitrogen limitation is an im-

portant factor in a large fraction of plant-herbivore interactions.

3. In the **plant defense hypothesis**, plants produce toxins to repel herbivores. Researchers studied interactions between cottonwood trees, beavers, and a leaf beetle called *Chrysomela confluens* in northern Utah. They found that individual cottonwood trees that had resprouted in response to browsing by beavers contained significantly higher concentrations of several defense compounds compared to individuals of similar size and age that had never been browsed (**Figure 53.12**).

Adaptation and Arms Races

- A **coevolutionary arms race** between parasites and hosts begins when a parasitic species develops a trait that allows it to survive and reproduce in a host. In response, natural selection favors host individuals that are able to defend themselves against the parasite.

- The *Plasmodium* are unicellular protists that cause malaria. They have a complex life cycle that includes the mosquito as the main vector.

- Researchers have found a strong association between a certain allele in the major histocompatibility complex (MHC) and protection against malaria. HLA-B53 proteins bind to a particular protein found in sporozoites (**Figure 53.13**).

- Researchers have also shown that *Plasmodium* populations in West Africa now have a variety of alleles for the protein recognized by HLA-B53. This suggests that natural selection has favored the evolution of *Plasmodium* strains with weapons to counter HLA-B53.

Can Parasites Manipulate Their Hosts?

- Snails infected with flukes become attracted to light and are therefore more easily eaten by birds, thus continuing the life cycle of the fluke (**Figure 53.14**).

Mutualism

- **Mutualisms** are +/+ interactions that involve a wide variety of organisms and rewards. For example, bees visit flowers to harvest nectar and pollen, and the flowers benefit by being fertilized.

- Other examples of mutualism are the association between fungi and roots, bacteria and certain species of plants, and rancher ants that protect aphids in exchange for sugar (**Figure 53.15**).

53.2 Community Structure

Clements and Gleason: Two Views of Community Dynamics

- Clements promoted the view that biological communities are stable, integrated, and orderly entities with a highly predictable composition. He argued that communities develop by passing through a series of predictable stages that culminate in a **climax community**.
- Gleason contended that the community found in a particular area is neither stable nor predictable. He claimed that plant and animal communities are associations of species that just *happen to share climatic requirements*.

Historical Data on Community Structure

- Historical data on plant communities showed that groups of species did not change their distributions in close association. Instead, species tended to change their ranges independently of one another.

Experimental Tests

- Researchers constructed 12 identical ponds, sterilized them, and then innoculated them with various species of zooplankton. They found a total of 61 species in all the ponds but only 31 to 39 species in individual ponds (**Figure 53.18**).
- Researchers contended that some species are particularly good at dispersing and are likely to colonize all or most of the available habitats. Other species disperse more slowly and tend to reach only one or a few of the available habitats. As a result, the specific details of community assembly and composition are contingent and unpredictable. As Gleason maintained, communities are a product of chance and history.

Disturbance and Change in Ecological Communities

- **Disturbance** is defined as an event that removes some individuals or biomass from a community. The important feature of disturbance is that it alters the light levels, nutrients, unoccupied space, or some other aspect of resource availability.

What Disturbances Occur, and How Frequent and Severe Are They?

- Each community experiences a characteristic type, frequency, and severity of disturbance, known as a **disturbance regime**. For example, forest fires kill all or most of the trees in a boreal forest every 100 to 300 years, whereas disturbances in temperate and tropical forests are more frequent.

How Do Researchers Determine a Community's Disturbance Regime?

- Ecologists use two strategies to determine a community's natural pattern of disturbance:
  1. They infer **long-term patterns** from data obtained in short-term analysis. Unless sampling is extensive, it is hard to avoid errors caused by extrapolating from particularly disturbance-prone or disturbance-free years or areas.
  2. They reconstruct the **history** of a particular site. Flood frequency, for example, can be estimated. Forest fires are most often studied by using historical techniques.

Why Is It Important to Understand Disturbance Regimes?

- Researchers studied the fire history of giant sequoia groves in California (**Figure 53.19**). They found that in most of the groves, 10 to 50 fires had occurred each century for the past 1500 to 2000 years. They also established that fires are extremely frequent and of low severity in this community.
- Biologists are now able to better manage these forests by allowing, monitoring, and controlling burns in this area.

Succession: The Development of Communities after Disturbance

- Severe disturbances remove all or most of the organisms from an area. The recovery that follows is called **succession**. **Primary succession** occurs when the disturbance removes the soil and its organisms as well as the organisms above the surface. **Secondary succession** occurs when a disturbance removes some or all of the organisms from an area but leaves the soil intact (**Figure 53.20**).

Theoretical Considerations

- Biologists focus on three factors in predicting the outcome of succession in a community:
  1. Particular traits of the species involved
  2. How species interact
  3. Historical and environmental circumstances such as the size of the area involved and short-term weather conditions

A Case History: Glacier Bay, Alaska

- An extraordinarily rapid and extensive glacial recession is occurring at Glacier Bay. In just 200 years, glaciers have retreated about 100 km, thus making this an important site for studying succession (**Figure 53.21**).

53.3   Species Diversity in Ecological Communities

Global Patterns in Species Diversity

- The communities of the tropics contain more species than do temperate or subarctic environments. As latitude increases, the species diversity decreases (**Figure 53.23**).
- **Hypothesis 1**: *High productivity in the tropics promotes high diversity.* Increased biomass supports more herbivores, thus more predators. Data are able to refute this hypothesis, so it is probably not sufficient.
- **Hypothesis 2**: *Habitats with complex physical structures have more niches.* Tropical forests are much more complex because their canopies are uneven in height and their understory is laced with epiphytes and vines. This does not explain why there are more tree species.
- **Hypothesis 3**: *Tropical regions have had more time for speciation to occur.* Ice sheets repeatedly scoured temperate and arctic latitudes over the last 2 million years.
- **Hypothesis 4**: Tropical areas experience fewer disturbance events than do temperate or subarctic regions. Data are able to refute this hypothesis.

The Role of Diversity in Ecological Communities

Does Species Richness Affect Productivity?

- **Primary productivity** is the total amount of photosynthesis per unit area per year. **Net primary productivity** is the amount of primary productivity that ends up in biomass (**Figure 53.24**).
- Productivity is a key component of communities; it represents the amount of food that is available to consumers and detritivores.

Does Species Richness Affect Community Stability?

- **Resistance** is a measure of how much a community is affected by a perturbation such as a drought. **Resilience** is a measure of how quickly a community returns to its prior state following such a disturbance.
- How does species diversity affect resistance and resilience? Two hypotheses have been proposed (**Figure 53.25**):
  1. Diversity is positively correlated with stability.
  2. Diversity leads to high productivity.

B.   CROSS-CUTTING THEMES

Looking Back—
Concepts from Earlier Chapters
Animal Interaction—Chapter 51
Chapter 51 explored how organisms respond to environmental stimuli and how they interact with the environment. Here in Chapter 53, the focus is on large numbers of species within a community.

Population Dynamics—Chapter 52
Chapter 52 explored the dynamics of populations—how and why they grow or decline, and how they change over time and space. Although that chapter considered populations in isolation,

the reality is that individuals of different species constantly interact within a community. As a result, the fate of a community may be tightly linked to the other species that share its habitat.

Growth Rings—Chapter 37
Researchers estimate the frequency and impact of storms by finding wind-killed trees and determining the date of their death. This is done by comparing patterns in the growth rings of dead trees with those of living individuals nearby.

Looking Forward—
Concepts in Earlier Chapters
Ecosystems—Chapter 54
Understanding ecosystems depends on an understanding of the total network of living things, which begins with the foundations laid down in the current chapter on community ecology.

C. DIFFICULT TOPICS

The development of communities after disturbance is probably the most complex event described in this chapter. Because communities respond differently after disturbances, you cannot simply memorize one type of succession. Instead, learning various examples will help you understand how succession proceeds in different communities and after different types of disturbances. Also, by understanding the theoretical considerations (traits of species, species interaction, and environment) you will gain an appreciation for the difficulty involved in determining the exact events of succession in various areas.

D. ASSESSING WHAT YOU'VE LEARNED

(1) Testing Your Knowledge

1. It is not possible for species within the same niche to coexist. This statement describes
   a. competition
   b. niche theory
   c. competitive exclusion
   d. mimicry

2. The combination of conditions that a species will occupy in the absence of competitors is

   a. the fundamental niche
   b. the realized niche
   c. competitive exclusion
   d. mimicry

3. Batesian mimicry can be best described as
   a. when one species closely resembles another
   b. when a harmless species resembles a harmful one
   c. when harmful species resemble each other
   d. when harmless species resemble each other

4. When predation or disease limits herbivore populations, this is
   a. top-down control
   b. the poor-nutrition hypothesis
   c. the plant-defense hypothesis
   d. the disease hypothesis

5. +/+ interactions that involve a wide variety of organisms and rewards are called
   a. predation
   b. competition
   c. parasitisms
   d. mutualisms

6. An event that removes some individuals or biomass from a community is called
   a. competition
   b. predation
   c. disturbance
   d. resistance

7. The recovery that follows a severe disturbance is called
   a. succession
   b. growth
   c. colonization
   d. seeding

8. The primary difference between primary and secondary succession is
   a. the complete removal of species during the disturbance
   b. the complete removal of soil during the disturbance
   c. time
   d. nothing

9. Which of the following statements is *not* representative of a global pattern in species diversity?
   a. High productivity in the tropics promotes high diversity.
   b. Habitats with simple physical structures have more niches.
   c. Tropical regions have had more time for speciation to occur.
   d. Tropical areas experience fewer disturbance events than other areas do.

10. Net primary productivity is
    a. the total amount of photosynthesis per unit area
    b. the total amount of plant mass per unit area
    c. the amount of primary productivity that ends up as biomass
    d. the amount of primary productivity above the ground

(2) Integrating Your Knowledge

(a) What are some key observations about the distribution of community types?

(b) Briefly state Clements' view on community dynamics.

(c) What is the difference between a disturbance and a disturbance regime?

(d) Distinguish between succession, primary succession, and secondary succession.

(e) Compare and contrast resistance and resilience.

CHAPTER 53—ANSWER KEY

D. Assessing What You've Learned

(1) Testing Your Knowledge
1. c; 2. a; 3. b; 4. a; 5. d; 6. c; 7. a; 8. b; 9. b; 10. c

(2) Integrating Your Knowledge

(a) 1. Trees dominate every area where it is not too cold or too dry.

2. Evergreen trees dominate climatic regions at two extremes: tropical forests and subarctic forests.
3. Productivity is positively correlated with temperature and precipitation.
4. Species diversity generally declines from the tropics to the poles.

(b) Clements promoted a view that biological communities are stable, integrated, and orderly entities with a highly predictable composition. He argued that communities develop by passing through a series of predictable stages culminating in a **climax community**.

(c) **Disturbance** is an event that removes some individuals or biomass from a community. The important feature of disturbance is that it alters light levels, nutrients, unoccupied space, or some other aspect of resource availability. Each community experiences a characteristic type, frequency, and severity of disturbance, known as a **disturbance regime**. For example, forest fires kill all or most of the trees in a boreal forest every 100 to 300 years, whereas disturbances in temperate and tropical forests are more frequent.

(d) Severe disturbances remove all or most of the organisms from an area. The recovery that follows is **succession**. **Primary succession** occurs when the disturbance removes the soil and its organisms as well as the organisms above the surface. **Secondary succession** occurs when a disturbance removes some or all of the organisms from an area but leaves the soil intact.

(e) **Resistance** is a measure of how much a community is affected by a perturbation such as a drought. **Resilience** is a measure of how quickly a community returns to its prior state following such a disturbance.

54

Ecosystems

A. KEY BIOLOGICAL CONCEPTS

54.1 Energy Flow and Trophic Structure

- An **ecosystem** consists of the organisms that live in an area along with certain nonbiological components. The relevant abiotic components include the nutrients used by growing organisms and the energy supplied by the Sun.
- **Energy** is defined as the currency of an ecosystem. Every ecosystem consists of four components that are linked by energy flow (**Figure 54.1**):
  1. **Primary producers** use the solar energy or chemical energy contained in reduced inorganic compounds to manufacture their own food.
  2. **Consumers** eat other organisms. **Herbivores** are consumers that eat plants; **carnivores** are consumers that eat animals.
  3. **Decomposers** obtain energy by feeding on the dead remains of other organisms or waste products.
  4. The **abiotic environment** includes the soil, climate, atmosphere, and the particulates and solutes in water.

Global Patterns in Productivity

- **Tropical rain forests** have the highest productivity, whereas desert and arctic regions have the lowest (**Figures 54.2 and 54.3**). Marine productivity is the highest along nutrient-rich coastal areas.

What Limits the Productivity of Terrestrial Ecosystems?

- Overall productivity of terrestrial ecosystems is limited by a combination of temperature and the availability of water and sunlight (**Figure 54.4**).

What Limits the Productivity of Marine Ecosystems?

- Overall productivity of marine ecosystems is limited by the availability of nutrients (**Figure 54.5**).

How Does Energy Flow through an Ecosystem?

- A **grazing food web** is made up of the network of organisms that eat plants, along with the organisms that eat the herbivores.
- Consumers that eat herbivores are called **secondary consumers**.
- The **decomposer food web** is comprised of species that eat the dead remains of organisms (**Figure 54.6**). Plant litter and dead animals, or detritus, are consumed by bacteria, archaea, fungi, protozoa, and millipedes.

Trophic Levels, Food Chains, and Food Webs

- Organisms that obtain their energy from the same type of source occupy the same **trophic level**. Productivity declines from one trophic level to the next (**Figure 54.7**).
- **Food chain** describes the species occupying each trophic level in an ecosystem. Food chains are often embedded in more complex **food webs** because multiple species are present at several trophic levels (**Figure 54.8**).

Why Is Energy Lost at Each Trophic Level?

- Productivity at the second trophic level (primary consumers) must be less than productivity at the first trophic level (primary producers). This produces a pyramid of productivity—and productivity is always highest at the lowest trophic level (**Figure 54.9**).

What Limits the Length of Food Chains?

- **Hypothesis 1: Energy Transfer** Food-chain length is limited by productivity, so there should be more trophic levels in ecosystems with higher productivity or higher energy transfer. To date there is no support for this hypothesis.
- **Hypothesis 2: Stability** Long food chains are easily disrupted by environmental perturbations and thus tend to be eliminated.
- Pimm demonstrated that longer food chains took longer to return to their previous state following a disturbance. He also predicted that food chains should be longer in more stable environments. There have been challenges to this hypothesis, and so far support remains tentative.
- **Hypothesis 3: Environmental Complexity** Food-chain length is a function of an ecosystem's physical structure.
- Researchers predicted that three-dimensional ecosystems should have longer food chains than two-dimensional sites. Data support their predictions, and they concluded that dimensionality does affect food-web structure. The mechanism for the pattern remains to be determined, however (**Figure 54.10**).

Analyzing Energy Flow: A Case History

How Does Energy Flow through the Hubbard Brook System?

- Researchers extensively measured how energy flows through a forest ecosystem (**Figure 54.11**). They found that:
  - 0.8% of the light was captured by photosynthesis.
  - Plants used 55% of energy for growth and 45% for maintenance (lost as respiration).
- In chipmunks, 80.7% of energy intake goes to maintenance, 17.7% is excreted, and only

1.6% goes into the production of new tissue (**secondary production**).

54.2   Biogeochemical Cycles

- The path taken by an element as it moves from abiotic systems through living organisms and back again is referred to as its **biogeochemical cycle** (**Figure 54.13**).
- Three aspects are studied in biogeochemical cycling:
  1. The nature and size of the reservoirs where elements are stored for a period of time
  2. The rate of movement between reservoirs and the factors that influence these rates
  3. How different biogeochemical cycles interact

Biogeochemical Cycles in Ecosystems

- The combination of breakdown products and microscopic decomposers forms the soil organic matter. Soil organic matter is a complex mixture of partially decomposed detritus rich in a family of carbon-containing molecules called **humic acids**. (Soil organic matter is also called **humus**.)
- A key feature of nutrient cycling is that nutrients are reused. Not all of the nutrients are reused, however. Nutrients leave ecosystems when animals leave or when flowing wind or water removes nutrients.

What Factors Control the Rate of Nutrient Cycling in Ecosystems?

- Decomposition of detritus most often limits the rate at which nutrients move through an ecosystem. Decomposition rate is influenced by (**Figure 54.14**):
  1. Abiotic conditions—temperature, precipitation, and so forth
  2. The quality of detritus as a nutrient source for the fungi, bacteria, and archaea that accomplish the decomposition

What Factors Influence the Rate at Which Nutrients Are Exported from Ecosystems?

- **Nutrient loss** is an important characteristic of an ecosystem. Farming, logging, and soil erosion accelerate nutrient loss.

- **Devegetation** has a huge impact on nutrient export (**Figure 54.15**). In the absence of plants to recycle nutrients, they are quickly washed out of the soil and lost to the ecosystem.

Global Biogeochemical Cycles

The Global Water Cycle

- Water typically evaporates from the ocean and then precipitates back into the ocean. The key here is that evaporation exceeds precipitation—a net gain of water into the atmosphere (**Figure 54.16**).
- This water vapor moves over land and precipitates. The cycle is completed by water that moves from land to the oceans via streams and groundwater.

The Global Carbon Cycle

- The **global carbon cycle** involves a relatively rapid movement of carbon between terrestrial ecosystems, the oceans, and the atmosphere (**Figure 54.17**).
- **Photosynthesis** is responsible for taking carbon from the atmosphere and incorporating it into tissue. **Respiration** releases carbon that has been incorporated into living organisms to the atmosphere.
- The geological reservoir of carbon-rich sediments (fossil fuels) is estimated to contain 5000–10000 gigatons of carbon. Burning fossil fuels moves carbon from an inactive geological reservoir to an active pool—the atmosphere (**Figure 54.18**).
- Carbon dioxide is called a **greenhouse gas** because it traps heat that has been radiated from Earth and keeps it from being lost to space, in the same way that the glass of a greenhouse traps heat underneath it.

The Global Nitrogen Cycle

- The key aspect of the **global nitrogen cycle** is that plants are able to use nitrogen only in the form of ammonium or nitrate ions (NH_4^+ or NO_3^-).
- **Nitrogen fixation** (converted from N_2 to NH_4^+ or NO_3^-) results from lightning-driven reactions in the atmosphere or enzyme-catalyzed reactions in bacteria that live in the soil and oceans.
- Three sources of **human-fixed nitrogen** are (**Figure 54.19**):
  1. Industrially produced fertilizers
  2. Release of nitric oxide during the combustion of fossil fuels
  3. Cultivation of crops such as soybeans and peas that harbor nitrogen-fixing bacteria

54.3   Human Impacts on Ecosystems

Global Warming

- In 1998, researchers summarized the studies and data available on global warming and concluded that current evidence suggests a "discernible human influence on climate."
- In 2001 the new report stated that "there is new and stronger evidence that most of the warming observed over the last 50 years is attributable to human activities" (**Figure 54.21**).
- The panel has taken the position that the rise in concentrations of greenhouse gases is at least partially—if not primarily—responsible for global warming.

Productivity Increases

- If grassland habitats are fertilized with nitrogen, productivity increases but species richness declines over time.
- Nitrate ions derived from fertilizers are washing into the Gulf of Mexico. This has stimulated the growth of planktonic organisms, whose subsequent decomposition has used up available oxygen and triggered the formation of anaerobic "dead zones."
- Increased nitrate runoff from fertilizers has also cause toxic algal blooms.
- Massive phosphate runoff in the 1960s and 1970s resulted in explosive growth in cyanobacteria, large die-off, and eventual lake **eutrophication**—the conversion of a lake to a highly productive ecosystem with rapid decomposition, low oxygen levels, and rapid filling with decomposing organic matter.

B. CROSS-CUTTING THEMES

Looking Back—
Concepts from Earlier Chapters
Chemosynthesis—Chapter 2
In some ecosystems, energy flow begins not with photosynthesis but with chemosynthesis. For example, as introduced in **Chapter 2**, bacteria that live near the deep sea vents derive energy from hydrogen sulphide rather than sunlight.

Producers—Chapter 10
Producers form the basis of ecosystems by transforming the energy in sunlight or reduced inorganic compounds into the chemical energy stored in sugars. As **Chapter 10** pointed out, producers are called autotrophs or self-feeders.

Nitrogen Fixation—Chapter 37
Nitrogen fixation results from lightning-driven reactions in the atmosphere or enzyme-catalyzed reactions in bacteria that live in the soil and oceans. **Chapter 37** explains how bacteria fix nitrogen.

Nitrogen-Laced Run-Off—Chapter 27
In some cases, a massive increase in nitrogen availability is beneficial. In other cases, however, it is not. **Chapter 27**, for example, details how excessive applications of nitrogen-containing fertilizers on farmlands have produced nitrogen-laced run-off, which has had disastrous impacts on numerous aquatic ecosystems.

Animal Interaction—Chapter 51
Chapter 51 explored how organisms respond to environmental stimuli and how they interact with the environment. Here in Chapter 54, the focus is on large numbers of species within an ecosystem.

Population Dynamics—Chapter 52
Chapter 52 explored the dynamics of populations—how and why they grow or decline, and how they change over time and space. Although that chapter considered populations in isolation, the reality is that individuals of different species constantly interact within a community, and communities within an ecosystem. As a result, the fate of an ecosystem may be tightly linked to the other species that share its habitat.

C. DIFFICULT TOPICS

The most complex information you will process in this chapter relates to the biogeochemical cycles. It is not difficult to memorize the facts and specifics about each cycle, but you may find it difficult to understand concepts when these types of questions are asked on a test.

It is not sufficient simply to memorize the carbon or nitrogen cycle; you must learn the concepts and trends as well. For example, what happens to the carbon cycle when fossil fuels are burned? Not only do you need to know that atmospheric levels of CO_2 would increase, but you also need to understand how this will affect the rest of the cycle.

D. ASSESSING WHAT YOU'VE LEARNED

(1) Testing Your Knowledge

1. These use solar energy or chemical energy contained in reduced inorganic compounds to manufacture their own food.
   a. consumers
   b. primary producers
   c. secondary producers
   d. decomposers

2. The areas that typically have the highest productivity are
   a. tropical rain forests
   b. grasslands
   c. arctic regions
   d. marine ecosystems

3. The overall productivity in terrestrial ecosystems is limited by
   a. temperature
   b. water
   c. sunlight
   d. a combination of the above

4. Productivity at the second trophic level is always
   a. greater than the productivity at the first trophic level

b. less than the productivity at the first trophic level

c. equal to the productivity at the first trophic level

d. extremely variable compared to the productivity at the first trophic level

5. Which hypothesis regarding what limits food-chain length has the most support?
   a. energy transfer hypothesis
   b. stability hypothesis
   c. environmental complexity hypothesis
   d. All of the above have equal support.

6. The path that an element takes as it moves from abiotic systems through living organisms and back again is referred to as its
   a. biogeochemical cycle
   b. biological cycle
   c. nutrient cycle
   d. geochemical cycle

7. Which of the following is *not* one of the three aspects studied in biogeochemical cycling?
   a. the nature and size of element reservoirs
   b. the rate of movement between reservoirs
   c. how different biogeochemical cycles interact
   d. how new species create their own biogeochemical cycles

8. Which of the following *most often* limits the rate at which nutrients move through an ecosystem?
   a. species composition
   b. decomposition rate
   c. primary production
   d. none of the above

9. Perturbation of which of the following cycles contributes most to global warming?
   a. the global carbon cycle
   b. the global water cycle
   c. the global nitrogen cycle
   d. All of these cycles contribute equally.

10. The most recent evidence on global warming concludes:
    a. nothing about human activity causing global warming

b. possible links to human activity linked to global warming

c. strong evidence of human activity linked to global warming

d. humans are responsible for every catastrophe for the past 1000 years

(2) Integrating Your Knowledge

(a) Define *ecosystem*.

(b) What is the difference between a food chain and a food web?

(c) What are the three aspects studied in biogeochemical cycling?

(d) What is the main factor that controls the rate of nutrient cycling in an ecosystem? What influences this main factor?

(e) How do photosynthesis and respiration contribute to the global carbon cycle?

CHAPTER 54—ANSWER KEY

D. Assessing What You've Learned
(1) Testing Your Knowledge
1. b; 2. a; 3. d; 4. b; 5. c; 6. a; 7. d; 8. b; 9. a; 10. c

(2) Integrating Your Knowledge

(a) An **ecosystem** consists of the organisms that live in an area along with certain nonbiological components. The relevant abiotic components include the nutrients used by growing organisms and the energy supplied by sunlight.

(b) A **food chain** describes the species occupying each trophic level in a particular ecosystem. Food chains are often embedded in more complex **food webs** because multiple species are present at several trophic levels.

(c) Three aspects are studied in biogeochemical cycling:
   1. The nature and size of the reservoirs where elements are stored for a period of time

2. The rate of movement between reservoirs and the factors that influence these rates

3. How different biogeochemical cycles interact

(d) Decomposition of detritus most often limits the rate at which nutrients move through an ecosystem. Decomposition rate is influenced by:

1. Abiotic conditions—temperature, precipitation, and so forth

2. The quality of detritus as a nutrient source for fungi, bacteria, and archaea that accomplish the decomposition

(e) **Photosynthesis** is responsible for taking carbon from the atmosphere and incorporating it into tissue. **Respiration** releases into the atmosphere the carbon that has been incorporated into living organisms.

Biodiversity and Conservation

55

A. KEY BIOLOGICAL CONCEPTS

52.1   What Is Biodiversity?

Biodiversity Can Be Measured and Analyzed at Several Levels

- **Biodiversity** defines all distinctive populations and species living today (**Figure 55.1**).
- There are three levels of biodiversity:
  1. **Genetic diversity**—the total genetic information contained within all individuals of a species
  2. **Species diversity**—the variety of life-forms on Earth
  3. **Ecosystem diversity**—the variety of biotic communities in a region along with abiotic components.

Why Is Biodiversity Important?

Direct Benefits of Biodiversity: Providing Goods and Services

- Research programs, collectively known as **bioprospecting**, focus on assessing bacteria, archaea, plants, and fungi as novel sources of drugs or ingredients in consumer products.
- Agricultural scientists are preserving diverse strains of crop plants in seed banks and continue to use wild relatives of domesticated species in breeding programs aimed at improving crop traits.
- Strategies for cleaning up oil spills, abandoned mines, and contaminated industrial sites are incorporating **bioremediation**—the use of bacteria, archaea, and plants to metabolize pollutants and render them harmless.

Indirect Benefits of Biodiversity: Ecosystem Services

- **Ecosystem services** are processes that increase the quality of the abiotic environment.
- Economists are attempting to quantify the dollar value of ecosystem services as a way of justifying the cost of preserving natural areas and biologists continue to document how the loss of biodiversity is affecting the quality of the abiotic environment.

55.2   How Do Biologists Study Biodiversity?

Quantifying Genetic Diversity

- Researchers are trying to quantify levels of genetic diversity in entire communities or ecosystems rather than in individual species.

Estimating the Total Number of Species Living Today

Taxon-Specific Surveys

- **Key Research**: Terry Erwin and J. C. Scott used insecticidal fog to knock down species from the top of a rain forest tree called *Luehea seemannii*. They identified over 900 different species of beetles among the individuals that fell.
- If each of the 50,000 tropical tree species harbors the same number of arthropod specialists, then the world total of arthropod species would exceed 30 million. Based on these types of studies, biologists estimate there are over 100 million species.

All-Taxon Surveys

- To obtain a more direct estimate of total species numbers, the first effort to find and catalog *all* forms of life present at a single site is now under way. The location is the Great Smoky Mountains National Park in the southeastern United States (**Figure 55.2**). A consortium of biologists and research organizations initiated this all-taxon survey in 1999. When it is complete, in 2015, biologists will have a much better database to use in estimating the extent of biodiversity.

Steps in Understanding Biodiversity

- A thorough understanding of biodiversity involves several levels of research:
  1. Name the species.
  2. Ask what other species are related to it.
  3. Assess the species' geographic range.
  4. Understand the species' ecology—how it interacts with other organisms and the physical environment (**Figure 55.3**).

Biodiversity and Ecosystem Function

- The **redundancy hypothesis** of ecosystem function holds that many species perform similar roles in an ecosystem, such as primary producer or decomposer. If so, then a minimal level of diversity is required for an ecosystem to function properly.
- In contrast, the **rivet hypothesis** describes an ecosystem as being like an airplane wing held together by rivets. An ecosystem's function may be compromised if certain species are lost (**Figure 55.4**).

55.3   Threats to Biodiversity

Humans Have Affected Biodiversity throughout History

- Fossil evidence on islands in the South Pacific suggests that about 2000 bird species were wiped out as people colonized this area between about A.D. 400 and 1600.
- When European settlers arrived on Easter Island in 1722, about 1000 people lived there. The island was treeless. Soil samples analyzed confirmed that the island was once a lush forest (**Figure 55.5**).

Current Threats to Biodiversity

- As a growing human population consumes the Earth's resources to fulfill its needs and desires, biodiversity is being degraded or destroyed at the genetic, species, and ecosystem levels (**Figure 55.6**).
- **Invasive species**—an exotic species that is introduced to a new area, spreads rapidly, and eliminates native species—are implicated in the demise of one-third to one-half of all species currently listed as endangered or threatened in the United States.
- We cause **habitat destruction** by logging and burning forests, damming rivers, dredging and trawling the oceans, plowing prairies, grazing livestock, filling in wetlands, excavating and extracting minerals to build housing developments, golf courses, shopping centers, office complexes, airports, and roads.
- This turns large, contiguous areas of natural habitats into small, isolated fragments. **Habitat fragmentation** is a concern to biologists for these reasons:
  1. Habitats can be too small to support some species.
  2. Fragmentation reduces the ability for some animals to disperse.
  3. Fragmentation creates a large amount of vulnerable edge habitat (**Figure 55.7**).
- Biologists are also beginning to document population declines due to **domino effects**—impacts on other species from the loss of a different species (**Figure 55.6g**).

How Can Biologists Predict Future Extinction Rates?

- **Species-area relationships** measure rates of habitat destruction to projected rates of species loss (**Figure 55.8**).
- Researchers compared satellite images of the Amazon Basin from 1978 and 1988. Their analysis showed that 5000 km^2 was deforested in the Amazon each year during this decade.
- Researchers have used species-area curves to analyze the number of bird species found on

islands in the Bismarck Archipelago near New Guinea (**Figure 55.9**).

- Biologists want to understand how many species are currently threatened so they can estimate how many will go extinct in the near future.
- The **red list** compiled by the International Union for the Conservation of Nature tracks species that are extremely rare or whose populations are in steep decline.

55.4 Conserving Biodiversity: Biology, Sociology, Economics, and Politics

The Role of Governmental and Private Agencies

- One of the most important biological goals in conservation work is to establish a **baseline**, or reference point that provides a standard for evaluating changes in diversity of alleles, species, and ecosystems.

International Agencies

- The Convention on Biological Diversity, or CBD, was the first global agreement on the conservation and sustainable use of biological diversity. CBD has now been formally ratified in over 175 countries.
- Here are the three main goals of the CBD:
  1. Conserve biodiversity.
  2. Promote sustainable use of biodiversity.
  3. Share commercial benefits of biodiversity equitably—especially genetic diversity.

National Agencies

- The U.S. Department of the Interior has several agencies charged with managing biological resources, including the Bureau of Land Management, the Fish & Wildlife Service, and the National Park Service.
- In the ocean realm, the National Marine Fisheries Service, under the U.S. Department of Commerce, is charged with managing and conserving marine organisms.

Nongovernmental Organizations

- NGOs play a vital role in drawing public attention to specific conservation issues and generating grassroots awareness and support.

- Some of the largest and most familiar NGOs involved in biodiversity conservation are Conservation International, The Nature Conservancy, and the World Wildlife Fund.

Conservation Strategies

In Situ Conservation: Protected Areas

- In situ conservation is usually achieved by governments protecting designated areas from development. In theory, it allows wild populations to continue to evolve and to be used by indigenous people who depend on them for food, fuel, or medicine.

Ex Situ Conservation: Zoos, Aquaria, and Botanical Gardens

- These locations can serve to propagate species in captivity for possible reintroduction into the wild.
- They also play a key role in educating the general public about the importance of biodiversity and the need to conserve it. People have to care about biodiversity before they will work to protect it.

Looking to the Future

- The conservation community's common goal was **sustainable development**, or economic progress for local communities based on using certain species or resources carefully enough that they could regenerate and not decline over time.
- This concept has recently been expanded to include **adaptive management**, which recognizes that the design, management, and monitoring of conservation programs are inseparable.

B. CROSS-CUTTING THEMES

Looking Back—
Concepts from Earlier Chapters
Extinction Rates—Chapter 23
Chapter 23 analyzed evidence that extinction rates are now accelerating to between 10 to 100 times the normal, or "background," extinction rates, indicating that a mass extinction event is under way.

Inbreeding—Chapter 13

Genetic factors play a role in the fate of populations. You may recall from **Chapter 13** that small populations tend to become inbred, and that inbreeding is usually detrimental to the fitness of organisms.

Animal Interaction—Chapter 51

Chapter 51 explored how organisms respond to environmental stimuli and how they interact with the environment. Here in Chapter 55, the focus is on large numbers of species, their diversity, and their conservation.

Population Dynamics—Chapter 52

Chapter 52 explored the dynamics of populations and metapopulations—how and why they grow or decline, and how they change over time and space. Though that chapter considered populations in isolation, the reality is that individuals of different species constantly interact within a community, and communities within an ecosystem. Thus the fate of an ecosystem may be tightly linked to other species that share its habitat.

C. DIFFICULT TOPICS

The most challenging aspect of this chapter is the concept of species-area relationships. Curves that are drawn from these relationships indicate how many species may be lost by the destruction of a particular size of habitat. Theoretical and conservation biologists derive these curves from complex data sets so that amateur biologists can calculate the effect of habitat loss on a particular species. The tricky part of these curves is that they are based on a logarithmic scale. If you are not familiar with logarithms, it may be a good idea to review them. If you are familiar with logarithms, this should help you in drawing and analyzing the species-area curves.

D. ASSESSING WHAT YOU'VE LEARNED

(1) Testing Your Knowledge

1. Research programs that focus on assessing bacteria, archaea, plants, and fungi as novel sources of drugs or ingredients in consumer products are called:
    a. bioremediation
    b. bioprospecting
    c. bioprocessing
    d. ecosystem services

2. Why are economists trying to quantify the dollar value of ecosystem services?
    a. This allows them to sell abiotic components.
    b. This allows them to manage the area.
    c. This allows them to justify the cost of preservation.
    d. This allows them to prevent habitat loss.

3. What is genetic diversity?
    a. the total genetic information contained within all individuals of a species
    b. the total phenotypic information contained within all individuals of a species
    c. the variety of life-forms on Earth
    d. the variety of biotic communities in a region along with abiotic components

4. Which of the following is *not* a step in understanding biodiversity?
    a. naming the species
    b. looking at other related species
    c. assessing the species' geographic range
    d. quantifying the species' genome

5. Many species perform similar roles in an ecosystem. This concept is a portion of what hypothesis?
    a. the species hypothesis
    b. the rivet hypothesis
    c. the redundancy hypothesis
    d. the niche overlap hypothesis

6. An exotic species that is introduced to a new area, spreads rapidly, and eliminates native species is called
    a. an exotic species
    b. an immigrant species
    c. an invasive species
    d. a destructive species

7. Which of the following does *not* constitute habitat destruction?
    a. fishing
    b. logging

c. burning forests
d. filling in wetlands

8. Which of the following is *not* a concern regarding habitat fragmentation?
   a. Habitats can change type.
   b. Habitats can be too small to support some species.
   c. It reduces dispersal ability of some animals.
   d. It creates a large amount of vulnerable edge habitat.

9. The impact on other species from the loss of a different species is called
   a. a linked effect
   b. a dependent effect
   c. a domino effect
   d. an edge effect

10. Which of the following is not an ex situ conservation area?
    a. zoo
    b. aquaria
    c. botanical garden
    d. reserve

(2)   Integrating Your Knowledge

(a) Who tracks threatened species, and what is the red list?

(b) What is bioremediation?

(c) What are the three main goals of the CBD?

(d) What is habitat fragmentation, and why are biologists concerned about it?

(e) Contrast sustainable development to adaptive management.

CHAPTER 55—ANSWER KEY

D.   Assessing What You've Learned

(1)   Testing Your Knowledge

1. b; 2. c; 3. a; 4. d; 5. c; 6. c; 7. a; 8. a; 9. c; 10. d

(2)   Integrating Your Knowledge

(a) The **red list** is compiled by the International Union for the Conservation of Nature, which tracks species that are extremely rare or whose populations are in steep decline.

(b) Strategies for cleaning up oil spills, abandoned mines, and contaminated industrial sites are incorporating **bioremediation**—the use of bacteria, archaea, and plants to metabolize pollutants and render them harmless.

(c) The three main goals of the CBD are
   1. Conserve biodiversity.
   2. Promote sustainable use of biodiversity.
   3. Share commercial benefits of biodiversity equitably, especially in regard to genetic diversity.

(d) **Habitat fragmentation** is the transformation of large, contiguous blocks of habitat into isolated parcels surrounded by pasture, cropland, or urban development. **Habitat fragmentation** concerns biologists because:
   1. Habitats can be too small to support some species.
   2. It reduces the ability for some animals to disperse.
   3. It creates a large amount of vulnerable edge habitat.

(e) The conservation community's common goal was **sustainable development**, or economic progress for local communities based on using certain species or resources carefully enough that they could regenerate and not decline over time. This concept has recently been expanded to **adaptive management**, which recognizes that the design, management, and monitoring of conservation programs are inseparable.